CULTURAL
GEOGRAPHY

CULTURAL

THEMES · CONCEPTS · ANALYSES

GEOGRAPHY

WILLIAM NORTON

OXFORD
UNIVERSITY PRESS

OXFORD

UNIVERSITY PRESS

70 Wynford Drive, Don Mills, Ontario M3C 1J9
www.oupcan.com

Oxford   New York

Athens   Auckland   Bangkok   Bogotá   Buenos Aires   Calcutta
Cape Town   Chennai   Dar es Salaam   Delhi   Florence   Hong Kong   Istanbul
Karachi   Kuala Lumpur   Madrid   Melbourne   Mexico City   Mumbai
Nairobi   Paris   São Paulo   Singapore   Taipei   Tokyo   Toronto   Warsaw

and associated companies in Berlin   Ibadan

Oxford is a trade mark of Oxford University Press
in the UK and certain other countries

Published in Canada
by Oxford University Press

Copyright © Oxford University Press Canada 2000

The moral rights of the author have been asserted

Database right Oxford University Press (maker)

First published 2000

Canadian Cataloguing in Publication Data

Norton, William, 1944–
Cultural geography : themes, concepts, analyses
Includes bibliographical references and index.
ISBN 0-19-541307-5

1. Human geography.   I. Title.

GF41.N665 2000   304.2   C99-932954-5

BK
$32.95

Cover image: Normand Cousineau
Cover & text design: Tearney McMurtry

1 2 3 4 — 03 02 01 00

This book is printed on permanent (acid-free) paper ∞
Printed in Canada

Contents

List of Figures, Tables, and Boxes

Figures

List of Tables

List of Boxes

Preface

This textbook aims to offer instructors of courses in cultural geography—understood as one of the subdisciplines of human geography—a comprehensive student resource that is conceptually diverse and that reflects the substantial volume of empirical work in both the landscape school and new cultural geographic traditions. Unlike other subdisciplines of human geography such as economic geography and political geography, cultural geography has received rather limited support from textbook authors. Today the need for such textbook support is especially clear as there is much evidence that the ideas and content of the subdiscipline are, perhaps for the first time, at the center of contemporary human geography. This textbook has been prepared with these circumstances in mind.

Why are there relatively few textbooks on cultural geography as one subdiscipline of human geography? First, much traditional cultural geography has a strong empirical focus and many practitioners have centered their energies in that arena rather than attempting to produce syntheses of empirical content. Second, the challenge of producing a text is a real one, especially because of the need to incorporate both traditional landscape and new cultural geographic themes and because of the related need to integrate North American cultural and European social traditions.

Necessarily then, both the planning and writing of this textbook have been real challenges. The aim is to present cultural geography as a coherent subdiscipline concerned with making sense of people and the places they occupy through analyses and understandings of cultural processes, cultural landscapes, and cultural identities. There is a focus on the geographic expression of culture in landscape and also a focus on the social and spatial constitution of culture. Links with other human geographic subdisciplines, with physical geography, and with other disciplines are regularly stressed.

Notwithstanding the challenges, a number of colleagues and friends ensured that preparing this textbook was a source of sustained pleasure. My colleague, Barry Kaye, provided me with much stimulating reading material. Marjorie Halmarson expertly produced the maps and diagrams. Once again I received unfailing support and encouragement from the always professional and cheerful staff at Oxford University Press, especially Valerie Ahwee, Euan White, and Phyllis Wilson. Most critically, I have as always been supported in my endeavors by Pauline, to whom I owe an unfailing debt of thanks.

Acknowledgements

Reprinted by permission of National Council for Geographic Education: Figure 2.9.

Used by permission of The American Geographical Society: Figure 3.2, Figure 3.11, Figure 6.4, Figure 8.1, Table 8.1.

Reprinted by permission of Blackwell Publishers Ltd: Figure 3.3, Table 6.1.

Reprinted by permission of Economic Geography, Clark University: Figure 3.5.

Reprinted with the permission of Pearson Education Australia: Figure 3.7.

Reprinted by permission of The Canadian Association of Geographers: Figure 3.8.

Reprinted by permission of Academic Press: Figure 3.10.

Reprinted by permission of Prentice–Hall, Inc., Upper Saddle River, NJ: Figure 4.4.

Reprinted by permission of University of Toronto Press: Figure 4.5.

Reprinted by permission of Yale University Press: Figure 4.8, Figure 4.9.

Reprinted by permission of University of Oklahoma Press: Figure 4.12.

Reprinted by permission of The University of California Press: Figure 4.16.

Reprinted by permission of Macmillan Press Ltd and Barnes & Noble Books: Figure 5.2.

Reprinted by permission of Cambridge University Press: Figure 5.3.

Reprinted by permission of Taylor & Francis Books Ltd: Figure 7.3, Figure 7.4, Figure 7.5, Figure 7.6, Figure 7.7, Figure 7.8.

Reprinted by permission of The McGraw-Hill Companies: Figure 8.2.

Introducing Cultural Geography

Cultural geography is emerging as an increasingly central concern within the larger discipline of human geography. One eminent cultural geographer noted: cultural geography is a 'scholarly discourse that has shifted from the comfort of placid marginality toward the overheated vortex of ferment and creativity in today's human geography' (Zelinsky 1996:750). Another distinguished practitioner identified an ambitious goal: 'a theoretically well-grounded, intellectually vigorous, and practically effective social and cultural geography might well assume, in time, a major role in guiding and guarding the evolution of humanity's environments' (Wagner 1990:41). By the time you have finished working with this textbook, I am sure you will agree with Zelinsky's verdict and have a clear opinion concerning whether or not Wagner's goal might be achieved.

This introductory chapter aims to provide the student of cultural geography with the necessary broad context for tackling the substantive chapters that follow. The sequence of material in this chapter is as follows.

- Two brief examples of the type of work conducted by cultural geographers help set the scene.
- There is a working definition of cultural geography and an explanation for the material included in this textbook.
- The larger scholarly context of cultural geography, both as one area of social science and as an interest within human geography, is discussed. This content allows you to appreciate that cultural geography does not function in a scholarly vacuum—it has origins and links to other academic interests.

- Key terms and ideas are introduced and it is noted that multiple interpretations of such terms and ideas are usual.
- A basis is established for the different themes that are evident in analyses of the subject matter of cultural geography and the conceptual bases for these themes are briefly outlined. It is explicitly acknowledged that multiple and ever-changing themes to a changing subject matter are both welcome and necessary.
- The chapter concludes with a summary.

Doing Cultural Geography

Before we attempt to define cultural geography in a relatively formal manner and justify the contents of this textbook, two examples of cultural geographic analyses are briefly summarized to provide you with at least a flavor of what is to come. These two are selected as representative of the diversity of interests that are evident in contemporary cultural geography. The first reflects the traditional approach to the study of cultural geography (what you will come to understand as the landscape school) and is largely concerned with delimiting a cultural group and describing the landscape they have created. The second reflects some more recently developed concerns (what you will come to understand as the new cultural geography) and is largely concerned with matters of cultural identity especially as a group identity can be formed with reference to other groups of people. Notwithstanding this important distinction, it is clear that the two examples share a fundamental concern in that both focus on the way in which cultural groups create landscapes and, in turn, have their cultural identity reinforced by that landscape.

Describing the Visible Landscape

Landscape is a contested term. Difficulties concerning landscape arise both because of uncertainties about the initial meaning of the term, and because of the various challenges to the mainstream Sauerian meaning that was adopted in cultural geography. (The classic Sauerian interpretation of landscape refers to the cultural transformation of the natural world, emphasizing visible and material characteristics and the close links between land and life—this is the landscape that is lived in. This interpretation will be expanded on later in this chapter.)

Initial uncertainties about the meaning of landscape relate primarily to the fact that landscape is the English rendering of a composite German word—*Landschaft. Land* refers to the area used to support a group of people, and *schaft* refers to the moulding of a social unity such that *landschaft* expresses 'the experience and intention of a social group tied by bonds of custom and law to a determined territory' (Cosgrove 1998:66). Much German and French cultural geography has been concerned with locating and describing these groups and their territories, and it is this interpretation that was introduced into North America by Sauer. But a second meaning of the term 'landscape' also became popular in the English language. This is the pictorial, visual meaning of landscape as scenery, referring to something that is beyond rather than a part of ourselves. It is this landscape that we are able to relate to in aesthetic terms. This is the landscape that is looked at. Both meanings of landscape serve to privilege vision.

A classic concern of cultural geography is to describe and explain the visible material landscapes that different groups of people have fashioned from the physical geographic environment that they occupy. One of the most distinctive of such landscapes is evident in that part of the American intermountain West that was settled by members of the Church of Jesus Christ of Latter-day Saints, popularly known as Mormons. What is distinctive about this landscape? What does it

look like? Why is it different from surrounding areas?

The Mormon small town and rural landscape contains some quite distinctive features that are absent elsewhere. These are:

- especially wide streets in town
- barns and granaries inside town
- ward chapels in town
- dominant use of brick
- many I-style homes
- unpainted farm buildings
- roadside irrigation ditches
- open field landscape around towns
- hay derricks
- crude, unpainted fences

This Mormon landscape is distinctive primarily because members of the Mormon cultural group share some distinctive characteristics that affect their activities and it is these activities, related to their cultural identity, that explain the way the landscape has evolved and become what it is today. For the Mormons, the landscape that they have created has been a critical component of their cultural identity.

More generally, cultural geographers try to make sense of the visible and material landscapes that are associated with cultural groups that are relatively easy to label and describe.

Dominant and Other Cultural Identities

A more recent concern of cultural geography is with the tendency for some groups of people, who may or may not be associated with a specific place and who may lack a clear identity, to involve themselves in a struggle to establish a distinctive identity for themselves, often in opposition to some other dominant identity. For example, in many areas of Europe, Gypsies are acknowledged as a distinctive group living on the margins of dominant cultures. Gypsies are not easily defined in conventional cultural terms and are distinctive principally because of their lack of interest in conventional wage labor, a characteristic that has many implications for their lifestyle and for the landscapes that they occupy. Thus, Gypsies are usually highly mobile and are viewed and treated as outsiders when they stay in an area for a period of time. Indeed, they are often perceived by resident majority dominant populations as a threat. This fear of the group becomes in turn a distaste for the landscapes that they occupy. Such landscapes are typically marginal. They may be waste areas inside cities or may be roadside areas. Regardless, they might appropriately be described as landscapes of exclusion. Further, these landscapes are different because they are characterized by disorder and are difficult to understand for those who do not belong to the group, especially when compared to the order of a suburban housing estate. Both Gypsies and their landscapes are in some sense outside of mainstream society and mainstream space and are viewed negatively.

More generally, cultural geographers seek to comprehend how a particular group identity may be created or constructed in opposition to the identities of other groups and how various related landscapes might be seen as reflective of group identity.

What This Book Is About

Question: What is cultural geography?

Answer: Cultural geography is concerned with making sense of people and the places that they occupy, an aim that is achieved through analyses and understandings of cultural processes, cultural landscapes, and cultural identities. There is a long-standing interest in culture as a causal mechanism—especially the geographic expression of culture in landscape—and a more recent concern with cultural politics—especially the social and spatial constitution of culture. There is a concern with both the local and the global, and an explicit acknowledgment that much of what is evident at the local scale is linked to global matters. Cultural geography incorporates both traditional and newer conceptual bases, and is closely related to other areas of geographic interest (especially social, economic, political, and physical geography), to other academic disciplines (especially history, anthropology, psychology, and sociology), and to such inter-

disciplinary concerns as women's studies and ethnic studies.

Explaining the Organization and Content of This Textbook

Following this introductory chapter, this textbook is organized as follows.

- Chapter 2 covers a broad range of ideas under the heading 'Humans and Nature'. Such a heading invites considerable and diverse content, but there is a clear focus on those ideas that have informed the work of cultural geographers and that facilitate the subsequent thematic discussions contained in chapters 3 through 8. You may choose to simply peruse this chapter at first in order to appreciate the basic content, returning to study parts of it in more detail as necessary in your reading of subsequent chapters.
- Chapters 3 through 8 are the six substantive thematic chapters, each focusing on a particular cultural geographic theme and typically including both conceptual material that explains the rationale for the theme and empirical material that provides examples of analyses conducted by cultural geographers.
- Chapter 9 provides an opportunity for some evaluations of the textbook content and for observations on the current and possible future status of the cultural geographic enterprise.

The rationale for this structure and for the content of the chapters is as follows.

- It is essential to reflect teaching material and research interests that are traditionally considered to be cultural geographic in emphasis. Accordingly, this textbook discusses the classic work in cultural geography, mostly in the tradition established and exemplified by the doyen of American cultural geographers, Carl Sauer (1889–1975). Exemplars of such work have typically focused on the evolution of landscapes, the regionalization of landscapes, and human and land relations, often with an underlying assumption that culture is some form of causal mechanism. This landscape or Sauerian tradition continues to be important, especially in North America where it has long been the dominant concern.

- It is similarly essential to reflect recent developments, both conceptual and empirical, that have added to, perhaps even transformed, cultural geography. Some of these comprise what has been called the new cultural geography, informed by a variety of new or previously ignored concepts that, in turn, inspire a different type of empirical concern. The focus on landscape as an object of study, evident in the tradition initiated by Sauer, changes such that the interpretation of landscape becomes more a concern with symbolic and social identity than with the visible material landscapes of the earlier tradition. Further, there is an increasing recognition that landscape is not the sole object of study—there is also a concern with cultural identities. This tradition was initiated primarily by British practitioners and is appropriately seen as a part of the **cultural turn** that has been evident in many areas of social science.

- There is a need to demonstrate, where appropriate, that the older (but continuing) and the newer approaches share some fundamental concerns about peoples and places. This is accomplished in the most compelling of ways—by organizing the textbook around a series of six cultural geographic themes, teaching and research interests, each of which has been evident, to a greater or lesser degree, from the beginnings of modern cultural geography in the 1920s through to the present. The discussion of each theme thus incorporates content that is both relatively traditional and relatively new. Traditional and new concepts, along with traditional and new analyses, coexist in each of the six themes.

- There is a need to acknowledge and incorporate the fact that cultural, social, political, and economic processes are interrelated, such that the traditional insistence in human geography of teaching and researching these subdisciplines separately, while understandable, is unfortunate. Indeed, the cultural turn that is taking place in contemporary social science can be interpreted as extending the boundaries of interest of cultural geography into political, economic, and social

spheres. The phrase 'cultural turn' refers broadly to a wide range of advances in philosophy and social theory, such as those of postmodernism and poststructuralism, that combine to encourage increased appreciation of the importance of culture in studies of human life. Political geography is comfortable incorporating cultural, social, and economic content as needed to explain and understand the ways in which the political world and political life are changing. Similarly, much contemporary economic geography includes cultural, social, and political processes. Finally, social geography is increasingly adopting a cultural perspective. This textbook recognizes both the importance of the cultural in other traditional subdisciplines of geography and the relevance of social, political, and economic processes in cultural geography.

- The **globalization** that is central to discussions of the contemporary political and economic worlds needs to be at the forefront also of cultural analyses of people and place. Both identity and landscape are affected by globalization, a term that refers to the functional integration of internationally dispersed activities, not necessarily in the sense that there is any lessening of the importance of local identity or of place, but rather in that there is an ongoing tension between the global and the local.

THREE TERMINOLOGICAL CHALLENGES

Having made decisions about structure and content, a textbook on cultural geography immediately confronts challenges concerning the meanings attached to three key terms. Most tellingly, there is considerable debate about precisely what the word '**culture**' means—one review stated that culture 'is one of the two or three most complicated words in the English language' (Williams 1976:87). It is fair to say that other compositional subdisciplines of human geography—that is, subdisciplines concerned with some particular subject matter within the larger field such as political and economic geography—do not encounter this definitional problem to anything like the extent that cultural geography does. Expressed simply, unlike the terms 'politic' and 'economy', the term 'culture' is contested. Thus, there is much uncertainty at present concerning an appropriate working definition of culture in geography. This uncertainty about the culture concept is confronted directly later in this chapter and it is concluded that the debate, while important, is not a central one as it is essentially diversionary. To anticipate the conclusion reached, multiple meanings of culture are legitimate, indeed valuable.

Another challenge concerns the evident confusion and uncertainty about the closely related term 'society' and about the differences between 'cultural' and 'social' as geographers use these words. Generally, this debate is rooted in the fact that there have long been uncertainties about these terms in other social sciences, especially anthropology and sociology. More specifically, the former term has long been favored by North American geographers, and the latter term by British and other European geographers. Again, as with the uncertainty about the word 'culture', this terminological issue is addressed later in this chapter. In this instance, it is concluded that the debate is of no substantive intellectual merit for the contemporary cultural geographer, being essentially an accident of disciplinary history, and the matter is debated and left to rest.

The third terminological challenge relates to the meanings of the word 'nature'. Culture is indeed a complex term, but 'nature is perhaps the most complex word in the language' (Williams 1976:184). Given that one of the principal concerns of cultural geography is with the analysis and understanding of relationships between nature and culture, this poses real difficulties. At this stage of the textbook, the easiest meaning of nature to focus on is that contrasted with culture; thus, nature is the material world excluding humans. (See Your Opinion 1.1.)

Your Opinion 1.1

As these opening remarks make clear, cultural geography is a changing and always contested body of knowledge, but is this a good thing? Is this the way cultural geography ought to be? The answer given here is, yes. As a scholarly interest that functions to serve society, cultural geography is an applied academic concern that necessarily responds to changes in society as well as being inspired by an ever-changing, and hopefully always improving, set of concepts. Logically, then, cultural geography is always changing and, further, is always subject to varying interpretations by practitioners with particular interests and various concerns.

Providing a Context

Cultural geography is, of course, but one of many areas of academic and practical interest that are concerned with human activities and human identities, and this section outlines a larger intellectual and disciplinary context to facilitate understanding of the particular concerns of cultural geography. Central to an appreciation of this larger context is an awareness of the **social construction** of knowledge. Ideas, including theories, necessarily develop in a specific social, cultural, and historical context, a context that affects the form of the ideas, such that all our knowledge—including our knowledge of what is considered to be real—is socially constructed. This is a very important idea that is closely associated with some areas of the new cultural geography and that resurfaces on several occasions in this textbook.

Although the scholarly history of cultural geography is a long one, with speculations about cultural and social matters (especially human behavior) being traced back to Greek, Chinese, and Islamic civilizations, it was not until the mid-eighteenth century in western Europe that such issues began to receive widespread attention. Throughout the extended period from classical Greece to the mid-eighteenth century, the European intellectual climate first emphasized humanism, with history as the principal study of humans and, second, emphasized physical science.

The Rise of the Social Sciences

By about 1750, social science, as opposed to the social sciences, was an established realm of intellectual concern, but not as yet focused in universities. Although there were various proposals as to the intent and practice of social science, it is typically asserted that these two basic assumptions prevailed.

- First, the aim of social science was the discovery of a few general principles in order to achieve conceptual and analytical unity. **Empiricism** was to prevail in this endeavor; this is an approach that asserts that all factual knowledge is based on experience, with the human mind being a blank tablet (*tabula rasa*) before encountering the world.
- Second, a central aim of social science was to enhance social progress through revealing truths about ourselves.

Together, these two assumptions reflected the prevailing philosophical ideas of the **Enlightenment** period. The subsequent rise of social science disciplines was concurrent with, indeed one part of, the many changes that occurred with the onset of the **Industrial Revolution** and the social and economic system of **capitalism**. It is not surprising that this period of rapid and dramatic change spurred the establishment of a number of disciplines, each of which functioned to serve the society of which they were a part by identifying and proposing solutions to problems. Thus, it was during the nineteenth century that the various social science disciplines were differentiated, institutionalized in universities, and began to articulate their

Box 1.1: Explaining Human Behavior

*A dominant tradition, beginning with Plato (c.428–c.348 BCE) and Aristotle (384–322 BCE), is that of metaphysical **dualism**.* This theory asserts that substances are either material or mental, and has typically been used to distinguish between a physical, material world, and a human, mental world. Beginning in the sixteenth century with the ascendancy of physical science, a **mechanistic** view of the physical world emerged with outside forces regarded as responsible for physical motion. It proved to be but a short conceptual step to the assumption that human behavior might be similarly explained, an idea that was first advanced by René Descartes (1596–1650) with a distinction made between voluntary and involuntary behavior, a distinction that maintained the dualistic tradition. Voluntary behavior was seen as governed by the mind, while involuntary behavior was seen as mechanical. This **Cartesianism** dualism, the distinction between mind and body, prompted the reference by the philosopher, Gilbert Ryle (1900–76), to the human mind being a 'ghost in the machine'. Thomas Hobbes (1588–1679) departed from this dualistic position to extend the mechanical view to cover all human behavior. Hobbes had no Cartesian inhibitions and regarded the entire universe, including humans, as mechanical. The next logical steps were the derivation of the physical laws that were assumed to determine human behavior and the development of the argument that all of the ideas that humans have could be explained by experience, with both of these steps being taken by other empiricist philosophers, most notably John Locke (1632–1704) and David Hume (1711–76).

A principal dissenting argument was that of G.W.F.

Hegel (1770–1831). For Hegel, each historical period could be summarized in terms of some overarching theme, the *Geist*, or the spirit of the age. In Hegelian thought there was a focus on the meaningful behavior that is followed for voluntary reasons. A number of other German philosophers pursued these ideas, most notably William Dilthey (1833–1911). This **idealist** perspective is not dissimilar to a contemporary subjective concept of culture, as it is suggesting that behavior is related to something that is inside the person, for example, to emotions, feelings, and perceptions. Karl Marx (1818–83) addressed this argument, favoring the more mechanistic approach, making a distinction between consciousness and social being.

Thus, broadly speaking, there are two conceptual poles that have been used as the basis for understanding human behavior, with the dominant nineteenth-century view favoring the mechanistic argument based on physical science procedures rather than the idealist argument.

Thinking about the ideas outlined in this box so far, along with the text comments concerning the assumptions of Enlightenment philosophy and the rise of the social sciences, facilitates understanding of the intellectual climate that was in place during the nineteenth-century rise of social science disciplines. A principal feature of this intellectual climate was the assumption that all human behavior, the key subject matter of the emerging social science disciplines, was explicable in terms of a mechanistic physical science approach; that is, in terms of a cause and effect logic. This assumption of **naturalism** followed from the demonstrable success

particular methods and interests. This specialization was a trend that involved a rapidly increasing world of facts, the rise of universities, the appearance of specialized societies and journals, and the desire of groups of scholars to have their discipline placed on an equal footing with other emerging disciplines. All of these developments may be characterized as part of the rise of **modernism**. All involved some claims about the ability of the new social sciences to generate truths about humans in much the same way that the physical sciences were seen to generate truths about nature and in contrast to earlier approaches that focused on religion or **metaphysics**. Certainly, there were arguments against this social science acceptance of a physical

scientific approach, arguments that were mostly based on the grounds that it was dehumanizing—that is, neglecting what it means to be human—but the dominant trend was clear.

By about 1900, the process of social science discipline creation was effectively complete and since that time, additions and changes have been limited. As R.J. Johnston (1985:5) noted, once created, academic disciplines 'have a defined existence and are invested in by individuals, who wish to protect their capital'. In principle, each of the social sciences can be traced to the recognition of some group of interrelated questions concerning human behavior, a term that is intimately linked with our central concern, culture. Box 1.1 provides an

of physical science and received powerful additional support from Darwinian theory, which effectively confirmed the validity of the scientific study of humans. The principal nineteenth-century advances in social science thus confirmed what might be described as a privileging of science. Given this context, it is not surprising that the new social sciences, in order to explain their general subject matter of human behavior, introduced versions of physical science **determinism** at early stages in their disciplinary histories.

In sociology, Auguste Comte (1798–1857) and Herbert Spencer (1820–1903) argued for a view of society as an integrated whole comparable to a physical system that determined the behavior of all members, while Émile Durkheim (1858–1917) saw society as separate from the qualities of the individual members. In anthropology, the superorganic concept was introduced by Alfred Kroeber (1876–1960) in 1917. Certainly, by the early twentieth century, both sociology and anthropology were seen as natural sciences with the cultural or social as a force constraining the behavior of individuals. Comparable viewpoints emerged in psychology in the form of various behaviorisms, and in geography in the form of environmental determinism, the idea that physical geography was the cause of human behavior. Each of these approaches is a particular version of the idea that the human world can be studied using cause and effect logic, a logic that, expressed simply, presupposes that a single principal variable explains the many and varied concerns of the social scientist. Thus, human behavior and the larger human world are: for the sociologist, explained by reference to society; for the anthropologist, explained by

reference to culture; for the psychologist, explained by reference to a stimulus-response framework, and for the human geographer, explained by reference to physical geography.

Although these approaches to the study of human behavior, derived from physical science, were highly influential during the formative periods of the social sciences, they no longer represent dominant viewpoints. Rather, the favored approaches in social theory and research today are associated with other perspectives. The general but by no means complete rejection of physical science methodology in the social sciences occurred gradually during the twentieth century and is related to the rise of Marxist, humanist, cognitive, critical science, poststructuralist, and postmodernist approaches. Each of these, to varying degrees, rejects the notion that the social sciences bear any methodological resemblances to the physical sciences. Indeed, it is not uncommon today to refer to these academic changes, along with the various cultural, economic, and political changes that began after the Second World War, as representing the decline of the modernist phase—a phase that had roots in the Enlightenment and the Industrial Revolution—and the onset of a postmodernist phase. This can be seen as a transition from modernism to **postmodernism**.

Nevertheless, notwithstanding the current fall from favor of physical science-inspired approaches, an awareness of their great importance during the formative phase of evolution of the various social sciences is important to our concern with the character of cultural geography as one interest within human geography.

overview of the history of attempts to explain human behavior in the Western world.

The Rise of Human Geography

What of geography in the light of these general observations about the larger group of social sciences? The study of geography, as opposed to a more specific human geography, has a long and distinguished academic pedigree, with substantive contributions made by the early Greek, Chinese, and Islamic civilizations, and a consistent history of growth from the fifteenth century onwards in Europe. There were central concerns with mapping and written descriptions of lands and peoples. The seventeenth-century scholar, Bernhardus Varenius (1622–50)—confronted both with the explosion in European geographic knowledge related to overseas activity and with the need to provide a framework for organizing this knowledge—recognized that geography was both a physical and a human science, and that it involved studies of regions and studies of particular systematic interests. In the late eighteenth century, the philosopher, Immanuel Kant (1724–1804), identified geography as essentially a concern with regions.

The nineteenth-century experience paralleled that of the other social sciences, involving as it did an ever-increasing factual base, the founding of geographic societies, and a gradual recognition as a university discipline with the creation of, first, individual chairs and, second, academic departments; in addition, professional associations served to further confirm the practical importance of the discipline.

- The ever-increasing factual base was especially evident in the attempts by two great German geographers, Alexander von Humboldt (1769–1859) and Carl Ritter (1779–1859), to produce comprehensive world geographies, a task that proved to be no longer within the reach of a single scholar.
- Examples of early societies include the Paris Geographical Society (1821) and the American Geographical Society (1851).
- Chairs of geography were created in 1809 at the Sorbonne in Paris and in Berlin in 1820.

- The definitive stage of this process of institutionalization involved the creation of university departments of geography, first in Prussian universities in 1874, followed quickly by other European countries and by the United States beginning in 1903.
- Professional associations were typically formed following the creation of university departments in any given country.

This capsule account of the emergence of geography in the nineteenth century raises a key point for understanding the character of the new university discipline. Put simply, it is not appropriate to consider geography only in the context of the social sciences because geography was, of course, much more than human geography. Thus, human geography is different from the other social sciences because of the explicit and long-standing associations with physical geography and because of the presence of such earlier contributions as those of Varenius and Kant, which had acknowledged the physical science and human science content of geography.

These circumstances certainly served to distinguish geography from the other social sciences in that they combined to create a discipline that was more varied than the other emerging disciplines and that already had much intellectual capital invested. But, more important, they explain a great deal about the late nineteenth-century uncertainty as to the identity of geography. Martin and James (1993:164) noted: 'There was no professionally accepted paradigm to serve as a guide to the study of geography.' Thus, the need to answer the question 'What is geography?' was paramount. Each of the other social sciences encountered a comparable question, but they were not similarly disadvantaged by the quantity of intellectual baggage that accompanied geography and hence were able to respond to their question largely in the context of late nineteenth-century circumstances, specifically the rise of social science questions related to aspects of human behavior. It is in this light that the various definitions of geography that appeared at the end of the nineteenth century are most properly considered. Four rather different definitions appeared, namely:

- geography as physical geography
- geography as the study of regions
- geography as the influence of physical environment on humans
- geography as the study of the human landscape

These definitions bear witness to the contradiction that geography faced, namely accommodating the legacy of the past with the pressing demands of late nineteenth-century science. Nowhere is this more evident than in the matter of the unity of physical and human geography, with most practitioners questioning whether it was possible and appropriate to include radically different concepts and methods in a single discipline. The consensus answer, institutionally speaking, was *yes*.

Thus, geography has a long history such that during the late nineteenth-century period of formal discipline creation, geographers had much to consider when defining their discipline. Furthermore, this history was not one that could be easily accommodated into the emerging structure of disciplines that largely confirmed the long acknowledged division between physical and social sciences. The majority of geographers responded by asserting that geography continued to be a unifying discipline dealing with both physical and social facts. This collective 'decision' by geographers to remain both a physical and social science or, perhaps more correctly, to be neither, is an ongoing source of discussion with debates about the need for and viability of an integrating discipline.

Human Geography in the Twentieth Century

THE REGIONAL APPROACH

It is hardly surprising, in light of the preceding comments, that contemporary human geography exhibits a diversity of subject matter and method; it is perhaps the inevitable result of the character of geography as an institutionalized discipline. The need to secure an intellectual niche for the new university discipline prompted the several interpretations noted earlier. By the 1920s, it was regional geography that assumed the dominant status, a dominance that continued until the mid-1950s. The emphasis was on the delimitation and description of regions, and the underlying philosophy was a form of empiricism, a loose set of ideas that regards the acquisition of knowledge as a gradual correction or verification of facts. There was an implicit acceptance of **environmental determinism**, which is the argument that the physical environment is a principal cause of human activity.

SPATIAL ANALYSIS

By the mid-1950s, some human geographers were expressing concern regarding the apparent failure to keep pace with advances in other social sciences, especially with regard to the absence of an explicitly scientific focus. The response was to borrow heavily from economics in the introduction and application of what became known as spatial analysis. This approach involved both description and explanation, with a particular focus on answering the question of why things are located where they are. In accord with some other social science interests, the approach was essentially derived from physical science, was objective in character, and had a strong theoretical and quantitative content. The underlying philosophical support came from **positivism**, which was earlier accepted by most other social sciences as well as by the physical sciences. It is appropriate to regard human geography as being late in accepting the logic of positivism, indeed, so late that it was rejected by the social science mainstream at the very time it was being introduced into human geography. One consequence of this larger shift in conceptual emphases was that spatial analysis was a major concern for only a brief period, from the late 1950s until about 1970. This approach has never been of major importance in cultural geography, although it has always exerted an impact and continues to do so as will be noted in Chapter 6's discussion of behavior and landscape.

THE LANDSCAPE SCHOOL

Although it is fair to characterize the human geography of the period from about 1900 until about 1970 as dominated, first, by a regional approach and, second, by spatial analysis, such a capsule account from the perspective of the cultural geographer has omitted the single most important tradition. In the early 1920s Sauer formulated what has become known as the landscape or Sauerian school, and it is this approach that is central to the twentieth-century evolution of cultural geography. It has proven to be a long-standing approach that continues to be important today, although it is now further enriched by the addition of a range of newer philosophical movements.

This landscape school developed on the premise that culture, operating on physical landscapes through time, was responsible for the creation of cultural landscapes. It is therefore explicitly opposed to the premise of environmental determinism, although it can be seen as conceptually similar in that it incorporates a specific cause and effect logic. Foundations for this school are evident in the early nineteenth-century work of Humboldt and Ritter (noted earlier), in the pioneering studies of human impacts on environment conducted by the American geographer, George Perkins Marsh (1801–82), and in the work of such geographers as Paul Vidal de la Blache (1845–1918), Friedrich Ratzel (1844–1904), and Otto Schlüter (1872–1952). Much of the material discussed in this textbook, especially relating to landscape evolution, to landscapes as regions, and to cultural ecology, is primarily informed by this research tradition.

MARXISM AND HUMANISM

As noted, the demise of spatial analysis was at least partially a reflection of changes in the other social sciences, especially the move away from objective physical science–inspired analyses toward a range of conceptual concerns that explicitly focused on human beings as subjects, not objects. In short, spatial analysis was criticized for being dehumanizing. Hence, since about 1970 onwards, human geography has proved a willing recipient of, and participant in, a wide range of concerns that in varying ways place humans at the center of analyses. The two principal philosophical approaches that embraced this emphasis are Marxism and humanism. Both of these are discussed more fully later in this textbook, and it is sufficient at this stage to emphasize that, notwithstanding the many differences between them (and indeed within each of them), they both typically reject any suggestion that humans and human behavior can be studied in objective physical science terms.

These approaches proved to be refreshing additions to the conceptual arsenal available to all human geographers, including cultural geographers and their close colleagues in social and historical geography, and they continue to stimulate empirical work. Indeed, cultural geographers who were disenchanted with perceived inadequacies of the landscape school were among those at the forefront of both of these approaches. In this textbook Marxism is considered especially in the chapter on unequal groups and unequal landscapes; humanism is considered especially in the chapter on behavior and landscape and the chapter on symbolic landscapes.

FEMINISM, POSTSTRUCTURALISM, AND POSTMODERNISM

Most recently, since about the early 1980s, at least three additional philosophical movements have added to the conceptual repertoire of the human geographer. It is most significant that feminism, poststructuralism, and postmodernism have had an impact particularly on cultural geography, as suggested by the Zelinsky quote at the beginning of this chapter—a clear indication of the increasing importance of this subdiscipline. Again, as with Marxism and humanism, each of these is explicitly opposed to any physical science analogies. But, more important, they combine to assert the relevance of an understanding of human cultural identity, to emphasize the need to break through the conventional gender blindness of human geographic analyses, and to insist upon a questioning of the bases for the studies and interpretations that human geographers accomplish. Discussions in this textbook that are particularly informed by these approaches include those relating to unequal groups and unequal landscapes and to questions of human identity and symbolism as these relate to landscapes. Accordingly, each of these bodies of thought is discussed in the opening conceptual section of Chapter 7.

Introducing Culture

'Few scholars using the concept of culture have rigorously defined it. Here geographers have been exemplary rather than exceptional. Most users agree that the term resists simple definition'

(Mathewson 1996:97). The most important yet confused concept in cultural geography is that of culture. Indeed, culture is one of the most difficult concepts to interpret in all of social science. In other words, it is not only geographers who have difficulties with the term, but also anthropologists, sociologists, and others. Many of the difficulties that cultural geographers encounter with the term stem from issues that arise in the other social sciences.

There are several good and continuing reasons for the uncertainty about the meaning of this basic term. There was an early association with the biological term 'cultivation', the tending of natural growth, and a logical extension of this usage of the term to refer also to human growth or development. By the mid-eighteenth century, culture began to refer primarily to human organizations, gradually replacing the closely related term 'civilization' over the next 100 years. By 1871, culture, or civilization, was defined by the anthropologist, Edward Tylor (1832–1917), as 'that complex whole which includes knowledge, belief, art, morals, custom, and any other capabilities and habits acquired by man as a member of society' (quoted in Kroeber and Kluckhohn 1952:81). By 1952, two leading anthropologists combined to distinguish definitions proposed by 110 writers on the basis of fifty-two discrete concepts employed in those definitions (Kroeber and Kluckhohn 1952).

Culture and Society

Society and culture are closely related concepts but do relate to different phenomena, as was confirmed in 1958 when two leading figures, the anthropologist, Kroeber, and the sociologist, Talcott Parsons (1902–79), together produced a statement that aimed to disentangle the terms and propose authoritative definitions. It is worth quoting their opening statements:

There seems to have been a good deal of confusion among anthropologists and sociologists about the concepts of *culture* and *society* (or, *social system*). A lack of consensus—between and within disciplines—has made for semantic confusion as to what data are subsumed under these terms; but, more important, the lack has impeded theoretical advance as to their interrelation.

There are still some anthropologists and sociologists who do not even consider the distinction necessary on the ground that all phenomena of human behavior are sociocultural, with both societal and cultural aspects at the same time (Kroeber and Parsons 1958:582).

The confusion identified is explained in terms of the earlier uses of both terms. Thus, in the formative periods of disciplinary evolution, in the late nineteenth and early twentieth centuries, leading anthropologists such as Franz Boas (1858–1942) defined culture as human behavior that was independent of genetic characteristics, while sociologists such as Spencer, Durkheim, and Weber (1864–1920) treated society in much the same way. 'For a considerable period this condensed concept of culture-and-society was maintained with differentiation between anthropology and sociology being carried out not conceptually but operationally' (Kroeber and Parsons 1958:583). The following definitions were proposed and have been generally accepted within the social sciences (Kroeber and Parsons 1958:583):

- culture refers to 'transmitted and created content and patterns of values, ideas, and other symbolic-meaningful systems as factors in the shaping of human behavior and the artifacts produced through behavior'
- society refers to the 'specifically relational system of interaction among individuals and collectivities'

Thus, in anthropology, it is usual to stress the non-genetic character of culture, referring to only those things that people invent, develop, or pass down through generations. In this sense, culture is extrasomatic—literally, beyond the body. Culture is conventionally seen as having three aspects: the values and abstract ideals that members of a human **group** hold; the norms and rules that they follow; and the material goods that they create. In sociology, society is understood as referring to the system of interrelationships that connects individuals who share a common culture. This meaning of society is in accord with earlier usage that distinguished society (an association of free individuals) from the term 'state' (an association based on power relations). Both disciplines incorporated the idea of culture or society as a causal mechanism during formative stages of their development in response to the attractions of physical science modes of explanation, although these views are less important today. There are also many other interpretations in anthropology and sociology that need not concern us at this time. Some of these have carried over into cultural geography and are considered in later chapters as needed.

Cultural Geography, Human Geography, and Geography

Cultural geography, as a compositional subdivision of human geography, is concerned with some particular subject matter within the larger field. Note, however, that the flexibility of the term 'culture' has permitted two broader interpretations.

- First, because culture can be equated with everything produced by humans, as distinct from everything that is a part of nature, there has been a tendency, especially in the United States, to treat cultural geography as synonymous with human geography, such that many courses and textbooks labeled as cultural are more appropriately seen as introductory human geography courses and textbooks.
- Second, for many geographers, cultural geography, when broadly interpreted as the study of humans and land, *is* geography, both human and physical. This interpretation is supported by an alternative view of culture that includes nature. The editors of the *Companion Encyclopedia of Geography* make this point explicitly: 'geography is, and at root always has been (despite excursions into spatial science and other exotic themes), about the interdependence of people and their environment, and about the evolving intercourse between humans and their earthly, and to a lesser extent celestial, habitat' (Douglas, Huggett, and Robinson 1996:ix). Identification of the study of human and land relationships as the core of geography is both a hindrance and an aid. It is a hindrance because it opens the cultural geographic door to such a diverse body of material that the subdiscipline almost disappears—it does in fact become all of geography. But it is an enor-

mous aid because it obliges cultural geographers to acknowledge that cultural geography is not only about humans as individuals and as group members but also about the land they live on and with.

Both of these interpretations are at least partially a consequence of the uncertain status of the term 'culture'. Both are minor points concerning usage that need not concern us further, but they are undoubtedly indicative of the rather casual way key terms can be employed. Neither is a substantive issue because there has always been a majority interpretation of cultural geography as one part of geography and of human geography, and that interpretation is dominant today. Cultural geography, as defined earlier, focuses on peoples and places with particular reference to cultural processes, cultural landscapes, and cultural identities. Human geography, as noted, focuses on peoples and places more broadly with reference to culture, economics, and politics. Thus, human geography includes such subdisciplines as cultural, economic, and political geography. We now consider some of the more substantive issues surrounding the term 'culture' as it is used in cultural geography.

Culture in Cultural Geography

The meanings that cultural geographers attach to the word 'culture' reverberate throughout this textbook, and it is inappropriate to attempt any detailed appraisal in this introductory chapter. Rather, the relevant meanings are discussed as needed for the six substantive theme-based chapters, and the current discussion is essentially a sketching of the big picture with details to follow.

CULTURE AND THE LANDSCAPE SCHOOL

Sauer introduced the dominant view in the 1920s in a series of writings that effectively set the agenda for North American cultural geography until the first serious reservations were voiced in the 1970s.

- Three factors were regarded as basic to an understanding of landscape, namely the physical environment, the character of the people, and time. Sauer (1924:24) thus defined geography as 'the derivation of the cultural area from the natural area'.

- Perhaps the classic statement of intent was: 'Culture is the agent, the natural area is the medium, the cultural landscape the result' (Sauer 1925:46).
- Although culture is thus the 'shaping force', it is affirmed that the physical landscape 'is of course of fundamental importance, for it supplies the materials out of which the cultural landscape is formed' (Sauer 1925:46).

For Sauer, then, culture was a given, such that the principal focus of research was investigating the impact of culture on landscape, an impact that was evident in the creation of distinctive material landscapes that permitted the delimitation of cultural regions. But this concern with the material and visible landscape did not imply an exclusion of cultural identity, and practitioners frequently identified groups according to their way of life, what Vidal earlier labeled *genre de vie*, and based on such characteristics as language, religion, and ethnic identity. Thus, culture was also viewed as a system of shared values and beliefs. This view, or perhaps more correctly these views, of culture remained largely uncontested in geography until the 1970s.

In this interpretation of culture there is an emphasis on culture as cause, on the landscape outcomes, and on the identifying characteristics. This view prompted a cultural geography, the landscape or Sauerian school, that functioned as a leading subdivision of human geography with three principal foci, namely studies of landscape change, of landscapes as regions, and of landscapes as the outcomes of human and land relations—the first three themes included in this textbook. Until relatively recently, most leading cultural geographers traced their intellectual antecedents to the ideas articulated by Sauer. Most important, the first fully fledged textbook in cultural geography was unabashedly Sauerian in focus (Wagner and Mikesell 1962).

QUESTIONING THE MEANING OF CULTURE
Challenges to the Sauerian view of culture, and hence to the landscape school of cultural geography, emanated from several directions, but the most effective involved the claim that Sauer reified culture; that is, treated culture as a reality and not as a concept or human construct. Expressed differently, concern was raised at the use of culture as a cause of landscape rather than focusing on human decision making. This criticism was central to a seminal article by Duncan (1980), in which it was argued that Sauer was reflecting a view popular in anthropology in the 1920s and beyond, namely the **superorganic** or cultural determinist perspective of the anthropologist, Kroeber. This view asserted that any understanding of the cultural world had to be rooted in culture because it was there that decisions were made. Kroeber, along with the likeminded Robert Lowie (1883–1957), was a contemporary of Sauer at the University of California, Berkeley.

It is useful to see Duncan's argument as a major critique of the landscape school conception of culture and as contributing to the rise of the new cultural geography. Certainly, the argument struck a chord in a 1980s intellectual environment that was very different from that of the 1920s. These changes are, of course, but one component of the larger disenchantment with modernism in social science. The Sauerian tradition was criticized also for focusing on the rural and antiquarian and on material landscapes, and a more complex and interpretative view of landscape as cultural construction was proposed. The term '**new cultural geography**' is used in this introductory chapter to refer to this substantial body of ideas that developed beginning in the 1970s.

Significantly, the prompting for a new concept of culture was not by any means the exclusive province of scholars criticizing from outside the landscape school tradition and who had an agenda that was both cultural and social in inspiration. Rather, the landscape school was also criticized from within, as several of the major figures associated with the landscape school asked for and proposed new ideas as early as the 1970s. For example, Mikesell (1978:13) asked practitioners to 'give more serious thought to how they wish to use the concept of culture', while the 'intense preoccupation with the visible material landscape' that 'led to an unfortunate neglect of the less obvious, invisible forces which in some cases form cornerstones in the explanation of spatial patterns of human behavior' was bemoaned by English and Mayfield (1972:6). The important point here is that changes were being initiated by scholars immersed in the landscape school tradition, not only by those looking from outside of the tradition. Furthermore, it is rather misleading to imply that opposition to the Sauerian tradition and prompting for new cultural traditions was evident throughout English language geography. Writing with specific reference to New Zealand, Berg and Kearns (1997:1) noted that 'many cultural geographers here, but especially those with interests in "Maori issues", have always had a political edge to their work' such that 'the simple "old/new" binary constituted in representations of American cultural geography fails to capture the complexity of the situation' in New Zealand.

CULTURE AND THE NEW CULTURAL GEOGRAPHY

There is no doubt, however, concerning the impact on North American cultural geography made by the group of scholars who objected to the Sauerian view principally because of the perceived tendency to reify culture. Uncertainty concerning how to correct this perceived deficiency of the earlier view contributed to the emergence of several understandings of culture following criticism of the Sauerian perspective, a reflection of the diverse theoretical landscape of the times. One particularly distinctive feature was the interest shown in cultural geography by British social geographers; indeed, a 1987 conference on New Directions in Cultural Geography was organized by a Social Geography Study Group that subsequently became a Social and Cultural Geography Study Group. Not surprisingly, then, several of the new conceptions of culture had a strong social flavor. The appearance of several and varied concepts does not, however, disguise a fundamental similarity.

For what soon became known as the new cultural geography, culture was not regarded as a unitary variable that was able to explain. Culture was not approached as a cause. Rather, emphasis was placed on the plurality of cultures, defined as those values that members of human groups share in *particular* places at *particular* times. In brief, culture was seen as a medium or process rather than as an object. In an influential agenda–setting article, Cosgrove and Jackson (1987:99) defined culture as 'the medium through which people transform the mundane phenomenon of the material world into a world of significant symbols to which they give meaning and attach value'. It is useful to appreciate that this view of culture is not substantively different from earlier views; it reflects a difference in emphasis rather than a difference in substance. This change of emphasis is in accord with a larger suspicion of what can be seen as oversimplistic causal arguments and a related preference for acknowledging cultural and spatial differences. This is an important idea that is discussed in Chapter 7 as one aspect of the current move from modernism to postmodernism.

The resultant transformation of cultural geography that developed following the concerns expressed about the Sauerian view of culture encouraged humanistic studies of behavior and landscape and also paved the way for the interests in unequal groups and unequal landscapes and in symbolic landscapes—the final three themes identified in this textbook. This transformation also resulted in much questioning as to the directions to be taken by cultural geography and the possible fragmentation of the subdiscipline. Regardless, there is little doubt that the new cultural geography was institutionalized by the end of the 1980s. This institutionalization is confirmed by the publication of cultural geography textbooks that chose to exclude much of the Sauerian tradition, focusing largely on some aspects of the new cultural geography. Further, in the fifth edition of the influential book, *Geography and Geographers*, there is a chapter titled, 'The Cultural Turn' (R.J. Johnston 1997).

NEW CULTURAL GEOGRAPHY AND CULTURAL STUDIES

The understandings of culture in the new cultural geography are not simply critical reactions to the earlier understandings in the Sauerian tradition, but are also inspired by developments in the area of **cultural studies**. This is a term that could have been introduced earlier during our discussion of twentieth–century human geography, along with feminism, poststructuralism, and postmodernism, but it is perhaps most appropriately seen as an umbrella term incorporating each of those and also aspects of Marxism and humanism. It is an area of scholarly concern that is both rich and diverse, hardly surprising given the preceding statement and the complexity of the culture concept itself. As an interdisciplinary field, cultural studies embraces numerous concepts, mostly involving a rejection of modernism, and is represented in a number of schools of thought, four of which are noted briefly at this time.

- A British tradition developed, focusing attention on working–class culture, on the consequences of the Industrial Revolution, and on new cultural forms, such as film and television. Much work conducted under the auspices of the Birmingham Centre for Contemporary Cultural Studies is central to this tradition.
- An American tradition evolved from a concept of culture articulated within anthropology, a concept that focused attention on the study of the meaningful ordinary lives of people who are from a culture that is different from that of the investigator. Anthropology from this perspective is thus an encounter with otherness.
- There is the Marxist–inspired Frankfurt school of critical theory associated with a number of German philosophers and focusing attention on a wide range of ideas, including a critique of modernity. A basic inspiration is the conviction that social science must not be the instrument of particular ideologies, notably those associated with dominant groups, including the political state; rather, the social sciences must be emancipatory.
- There are a number of traditions, such as postcolonialism, feminism, and studies of racism, that are linked by their explicit concern with exposing the Eurocentrism that has been and often continues to be associated with the production of knowledge. For example, it can be argued that the Orient is really an invention of those who study it from outside, from the Occident, and that America and Europe employ ideas from within their cultures, ideas such as freedom and individualism, to facilitate their conquest and domination of other regions.

The impact of cultural studies on cultural geography has been considerable and is continuing. Indeed, many of the specific areas of interest within cultural studies have been introduced to help create the new cultural geography, for example, studies of working–class culture, of new cultural forms, of otherness, of ideologies of domination and oppression, and of Eurocentric attitudes. Perhaps the single most important cultural geographic idea derived from the cultural studies tradition is a concept of cultures as maps of meaning, as the 'codes with which meaning is constructed, conveyed, and understood' (Jackson 1989:2).

EVALUATING THE CULTURE CONCEPT

Clearly, culture is a contested term today, an important observation that this textbook explicitly acknowledges and willingly incorporates. There are many views of culture, and Box 1.2 both summarizes and expands on the discussion so far by listing some representative definitions.

Understanding what is meant by the word 'culture' is clearly not a simple task and undoubtedly provides both challenge and promise for cultural geography. There are at least three general reasons for this state of affairs:

- First, there has been much debate concerning the realness of culture, and this has contributed to confusion. The modernist approach has been to regard culture as a thing that is capable of explaining other things, an idea that is now rejected by most cultural geographers and other social scientists.
- Second, culture has never been easy to define separately from other related words, especially society, but also economy and politic.
- Third, the dominant current view is to acknowledge that culture has different meanings for different groups in different places and at different times.

Despite these comments, there are three basic statements with which most would agree.

- Culture refers to our human way of life in the sense that it is what makes one group of people different from some other group.

- Culture is the principal means by which humans relate to nature and to each other; we are able to adapt to and change physical environments and are able to cooperate with and sometimes conflict with other groups.
- Culture is extrasomatic, distinguishing humans from other species; it is thus the reason why humans are able to put other species on display in conservatories, aquaria, and zoos.

There continue to be differences of opinion

concerning the claims that the culture concept used by the landscape school involved an almost unquestioned acceptance of the realness of culture, while the concept used by new cultural geographers is more flexible. Indeed, it might be claimed that most concepts of culture, not only the Sauerian version but also those advocated by new cultural geography, involve reification. These are important issues relating to the interpretation of much of the empirical work conducted by cultural geographers and they are

Box 1.2: Some Geographic Definitions of Culture

This box details some representative geographic definitions of culture organized chronologically. The intent is to provide an indication of the variety of definitions available, to indicate changes in definition through time, but also to suggest that, variety and change notwithstanding, there is an essential consistency.

Sauer (1925:30)

Culture is *'the impress of the works of man upon the area'*.

This is one of the classic statements in the seminal 1925 article by Sauer, emphasizing the idea of culture as a cause of landscape.

Sauer (1941a:8)

'Culture is the learned and conventionalized activity of a group that occupies an area'.

This less deterministic definition is closer to the empirical work accomplished by landscape school geographers than is the 1925 statement above.

Gritzner (1966:9)

'A culture is a human society bound together by a common complex of culture traits, each trait being anything which to the culturally-bound group has either material form and applicable function, or an expressed value.'

Clear statement reflecting the consensus view and activities of the landscape school; Gritzner stresses that culture is transmitted spatially and temporally through the unique human ability to symbolize.

Spencer and Thomas (1973:6)

'Culture is the sum total of human learned behavior and

ways of doing things. Culture is invented, carried on, and slowly modified by people living and working in groups as each group occupies a particular region of the earth and develops its own special and distinctive system of culture.'

A textbook definition that expands on the idea that culture is learned behavior.

Zelinsky (1973:70)

Culture is *'a code or template for ideas and acts'*.
This brief definition, taken from a larger statement, is a clear assertion of culture as a causal mechanism.

Wagner (1974:11)

'Learned behavior is pretty much what we mean by culture.'

Wagner (1975:11)

'The fact is that culture has to be seen as carried in specific, located, purposeful, rule-following and rule-making groups of people communicating and interacting with one another.'

Together, the two statements by Wagner reiterate the importance of learning, stress that culture refers to human behavior, and emphasize the importance of communication and interaction between and among group members.

Jackson and Smith (1984:205)

'Culture, in the sense of a system of shared meanings, is dynamic and negotiable, not fixed or immutable. Moreover the emergent qualities of culture often have a spatial character, not merely because proximity can encourage communication and the sharing of individ-

discussed more fully in later chapters. (See Your Opinion 1.2.)

Constitutive Elements of Human Identity—Who We Are

So far in these introductory comments, the word 'culture' has been employed to convey some idea of what cultural geographers study, namely the values, norms, and material goods of humans. However, the degree of uncertainty about that term means that it has not been possible to convey either a clear or a comprehensive picture. One way to address this difficulty is to recognize that a principal concern of cultural geographers is with individual and collective human identity, and also to recognize that identity can be and has been discussed in terms of a number of characteristics, with different characteristics being important in different places and at different times. Each of these characteristics is acknowledged to be socially constructed, that is to say, they are made or acquired rather than inherited. Each of these characteristics

---

ual lifeworlds, but also because, from an interactionist perspective, social groups may actively create a sense of place, investing the material environment with symbolic qualities such that the very fabric of landscape is permeated by, and caught up in, the active social world.'

This definition, one of the first from the new cultural geography, incorporates elements of earlier definitions, notably Wagner (1975), and includes some additional ideas, notably a shift from a concern with behavior to a concern with meaning.

Jackson (1989:ix)

Culture is 'a domain, no less than the political and the economic, in which social relations of dominance and subordination are negotiated and resisted, where meanings are not just imposed, but contested'.

One of the first statements to explicitly recognize that culture needs to be understood in the context of struggles between different groups of people. In both this and the preceding statement, there is explicit reference to the social world.

Shurmer-Smith and Hannam (1994:5–6)

Culture is 'that negotiated intersubjectivity which allows human beings as individuals to reach a tenuous understanding of one another, to experience each other jointly'.

A definition that rejects a cultural geography concerned with mapping traits onto landscapes, emphasizing instead ways of being in the environment.

McDowell (1994:148)

'Culture is a set of ideas, customs and beliefs that shape people's actions and their production of material artifacts, including the landscape and the built environment. Culture is socially defined and socially interpreted. Cultural ideas are expressed in the lives of social groups who articulate, express and challenge these sets of ideas and values, which are themselves temporally and spatially specific.'

After observing that culture is 'a notoriously slippery concept', McDowell (1994:148) suggests that 'a broad consensus' concerning the above definition 'might not be difficult to achieve'.

Jordan, Domosh, and Rowntree (1997:5)

Culture is 'learned collective human behavior, as opposed to instinctive, or inborn, behavior. These learned traits form a way of life held in common by a group of people.'

This statement, taken from a highly influential textbook, maintains the idea of culture as it evolved within the landscape school. Note that this definition is essentially that currently prevailing in anthropology. There seems little doubt that this view remains popular with many contemporary cultural geographers, notwithstanding the undoubted achievements of the new cultural geography.

Summary Comments

Although these selected definitions are but a few of many, they do combine to provide a meaningful overview of the approaches taken by geographers. No one single definition serves the purposes of this textbook, it is appropriate to suggest that the final two reflect many of the current interests of contemporary cultural geography. However hard geographers and others try to achieve a definitive wording, such is unlikely to be attained. As Jackson (1989:180) acknowledged at the end of a book length study, 'the stuff of culture . . . is elusive'.

is also contested in the sense that there are no unequivocal meanings. Given the earlier comments about culture, nature, and society, this is not at all surprising.

Eight such characteristics are now identified, namely place, language, religion, ethnicity, nationality, community, class, and gender. A key question for many social scientists today concerns the question of how it is possible to comprehend identity using group labels, but at the same time avoid essentialism; that is, attributing any essential characteristics to a group. While it is typically necessary to group people, the resultant implication that groups have common traits or behavior patterns may be unwarranted given the often uncertain basis for defining group membership. This question surfaces frequently in the pages of this textbook.

PLACE

Cultural geographers of all persuasions, and increasingly many other social scientists, are sympathetic to the idea that place matters. It is not simply a case of *who* we are, but also *where* we are. Humans in groups occupy areas and thus tend to create material and symbolic landscapes that might be susceptible to delimitation as regions. For most of us, place is an important element in our individual and collective identity.

LANGUAGE

North American cultural geographers, often working in a landscape school tradition, typically distinguished groups of people according to what were viewed as primary cultural characteristics, notably language and religion. Related to classifications of peoples employed by anthropologists,

---

Your Opinion 1.2

What is your reaction to these comments? Does it seem surprising that this textbook is explicitly acknowledging uncertainty about the single most important concept in cultural geography? Do you feel that this might present a serious problem for you as a student as you commence on your studies of cultural geography? Let me suggest quite the opposite to you, namely that it is appropriate to regard the uncertainty as indicative of a healthy concern for seeking meaningful understandings of the world around us. Indeed, as you may well appreciate from your other studies, similar situations prevail throughout the ever-changing arenas of social science. To cite three other notable examples—there are multiple meanings of society, of behavior, and of personality, as any perusal of relevant sociology and psychology textbooks readily confirms. Viewed from this perspective, some uncertainty seems both normal and even desirable. Do you agree?

---

such variables are particularly appropriate to studies of tribal peoples and to areas of European overseas settlement. For many groups, language is the principal expression of a culture and may embody a particular view of the world. The language of a minority group is often dominated by that of a larger group in the same political unit, and it is common to argue that the loss of language means the loss of the related culture. During the nineteenth century, language was one of the principal bases employed for asserting a distinctive national identity.

RELIGION

Religion, the second of the two traditional cultural universals, addresses matters of ultimate significance, such as the question of human meaning, of human presence in the world, and the place of humans in the larger context of god, or gods, and the physical world. Thus religion plays an important role in any consideration of human relations with nature and with the landscapes associated with particular groups of people. Religion is also playing an increasingly important role in the formation of presumed ethnic identities.

ETHNICITY

As is the case with culture, ethnicity is a problematic term. It refers to the perceived distinctiveness of one group relative to others, usually based on a shared history that may be partially mythical and that has political associations. Many groups strive to build ethnic solidarity and enhance ethnic pride by emphasizing origins, history, claims to a homeland, and distinctive material and other cultural traits. In landscape school cultural geography, ethnicity is often employed as a surrogate for language and/or religion. The assumption that ethnic groups exist typically implies that such groups have their own distinctive culture—another instance of reification. There are also close links to the even more problematic term 'race'. This biological term is often treated as a synonym for ethnicity, or the impression is created of ethnic differences as a direct consequence of supposed racial differences.

NATIONALITY

Humans are divided into cultures, however defined, and also into political units known as nation states. However, the two are rarely congruent such that, contrary to the aspirations of nineteenth-century nationalist thought, cultures are rarely if ever neatly packaged into a political unit and political units are rarely if ever comprised of a single culture.

COMMUNITY

Unlike the word 'culture' community has never had a privileged position in the discourse of social science, although, like culture, it has been granted a great many different meanings. A term with traditionally positive connotations, such as loyalty, shared concerns, and personal contact, community is today more frequently employed as an alternative term for ethnic group.

CLASS

In the nineteenth century 'class' was used to refer to people with a common social or economic status. Marx moved beyond classification with the distinction between those who produce surplus (laborers) and those who appropriate it (capitalists) and a realization that each class has a collective consciousness and organization such that class struggle is always occurring. Marx used class con-

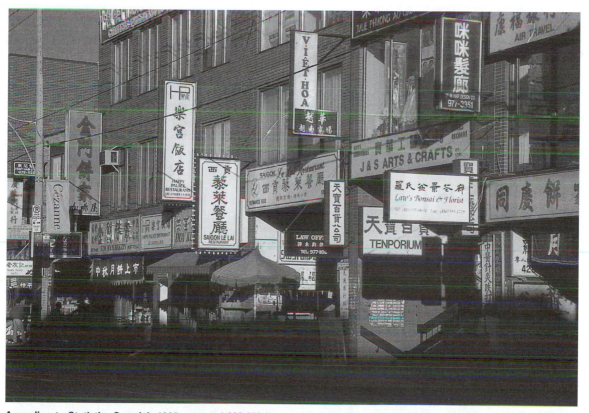

According to Statistics Canada's 1996 census, 1,039,000 immigrants who came to Canada between 1991 and 1996 reported a non-official language as their mother tongue. Chinese was the mother tongue of almost 25 per cent of these immigrants (*Victor Last*).

sciousness in the sense that others used culture in the late nineteenth century. Today, class is a key variable in the area of political economy, some ideas of which have entered cultural geography in the guise of the political ecology that is discussed in Chapter 5. Class is not, however, central to much of the work in contemporary cultural geography.

GENDER

Gender refers to attributes that are culturally ascribed to women and men; it is a cultural construct subject to change. Thus, studies of gender have emphasized the role played by masculine power and domination and assumed feminine subservience in the creation and maintenance of gender roles. It is usual to stress that there is no necessary link between gender and biological sex, such that women may have masculine characteristics and men may have feminine characteristics. But not only is there recognition that gender is not a given—that is, it relates to culture—there are also arguments to the effect that sexuality and sexual preference relate to culture such that it also can be considered a social construction. (See Your Opinion 1.3.)

Introducing Themes and Analyses

Rationale

The six themes that this textbook employs to structure and organize the ideas and work of cultural geographers are not formally institutionalized within cultural geography (see Box 1.3 for a selection of earlier attempts in this direction). Rather, they are simply general themes that can be identified when looking retrospectively at the concepts and practice of cultural geography from about the 1920s onwards. Is this circumstance, a lack of clearly evident and universally agreed upon research themes, peculiar to cultural geography? Of

Your Opinion 1.3

Do the components of human identity noted here capture the spirit of that term as you might apply it to yourself or to others, or do you sense that identity is better captured by reference to some other criteria? Are you inclined to agree with the idea that identity is socially constructed through the experiences of our lives and that it is fluid, changing with our changing lives, or do you feel that there are important aspects of our identity that are effectively presented to us at birth and that are unchanging? Perhaps you favor a view of identity that is capable of encompassing both essentialist as well as socially constructed components?

course not. Probably all of the principal interests in social science are appropriately viewed in this way. To quote one obvious and not unrelated example, authors of personality theory textbooks are confronted with many difficult decisions about precisely which ideas and analyses are related and therefore belong together in a textbook format.

There are at least two general reasons why such circumstances are usual.

- First, themes are not typically identified and then carefully adhered to; rather, researchers conduct analyses that, although often prompted by earlier work, may be quite different in focus compared to that earlier work, or that may be prompted by new ideas imported from other disciplines.
- Second, the nature of much social science work is such that commonalities often only become apparent as a body of work evolves, which is a principal reason why it is often difficult for social science to classify new work; put simply, new work may require a new classification.

Logically, then, the six themes identified and employed in this textbook could not easily have been identified, say, twenty years ago, and are unlikely to be appropriate, or at least sufficient, twenty years hence. Certainly, the association of particular ideas and empirical work with a single theme is necessarily a simplification.

What is being accomplished in this textbook is the identification of six themes, and the assignment of the work conducted by cultural geographers to a particular theme. Other themes could be identified, and in some instances work could be assigned to more than one theme. Analyses are being placed in one of the six categories for heuristic and organizational reasons. There is nothing 'correct' about the classification; it does not reflect some unchanging truth. Rather, it is an attempt to interpret the literature of cultural geography as it has developed and changed since about 1920. The current classification is thus being imposed on a scholarly tradition that has evolved through time. These cautionary statements do not serve to invalidate the themes identified, but they do make it clear that the classification that results must not be treated as sacrosanct.

The six themes represent alternative, often complementary, routes to understanding the issues that are of concern to cultural geographers. Different themes incorporate different conceptual bases and stimulate different empirical concerns. There is not one correct theme, and the six do not compete with one another in some search for mythical universal truths. Each of the themes has made and is likely to continue making contributions to the fun-

Box 1.3: Cultural Geography: Some Organizational Frameworks

Although cultural geographers 'have not been prone to defend their interests or issue programmatic statements' (Mikesell 1978:1), there have been several efforts to clarify both subject matter and approaches. This box notes some of the principal efforts that have been made in an abbreviated format; further details of particular issues are included in this textbook as needed. Students seeking fuller information are advised to refer to the sources noted below, to the Progress Reports on cultural geography and/or social geography published annually in the journal, *Progress in Human Geography* (see Mathewson 1998), and to the assessments of cultural geography included in Dohrs and Sommers (1967), Mikesell (1967), Miller (1971), Salter (1972), Spencer (1978), Norton (1987), and Ellen (1988).

Wagner and Mikesell (1962)

In their pioneering and highly influential book of readings, Wagner and Mikesell (1962:1) proposed the following five themes as constituting 'the core of cultural geography': culture, cultural area, cultural landscape, culture history, and cultural ecology.

Gritzner (1966)

Cultural geography:
- begins with anthropological concept of culture
- considers culture traits and groups in terms of development
- studies culture/nature
- interprets landscapes
- divides the world into culture regions and subregions

Wagner (1972, 1974, 1975)

Renounced the five themes outlined with Mikesell in 1962, and offered a revised focus based on institutions and communication.

Spencer and Thomas (1973)

Rather than specifying themes, four conceptual entities were recognized as follows:
- population
- physical environment
- social organization
- technology

This more process-oriented schema also proposed six interoperative relationships, namely those involving:
- population and environment
- population and social organization
- population and technology
- social organization and technology
- physical environment and social organization
- physical environment and technology

Mikesell (1978)

Identified seven persistent preferences:
- historical orientation
- humans as agents of environmental change
- focus on material culture
- rural bias in North America, non-Western or preindustrial bias elsewhere
- links to anthropology
- substantive research
- fieldwork

Identified three recent developments:
- environmental perception
- cultural ecology
- focus on United States

Norton (1989)

Proposed four themes:
- evolutionary
- ecological
- behavioral
- symbolic

damental goals of cultural geography, namely making sense of the real worlds of both people and place.

The Six Themes

Each of the six thematic accounts, chapters 3–8, provides historical and conceptual details of the theme, discusses links to other disciplines, notes past and current status, and contains examples of analyses. All relate to the Chapter 2 discussion of humans and nature because visible, material, and symbolic landscapes can be interpreted as outcomes of the humans and nature relationship.

As noted above, the landscape school of cultural geography, articulated by Sauer in the 1920s, built on some mid- and late-nineteenth century ideas in European geography and was also related to the anthropology practised by Kroeber and Lowie. Cultural geographers working in this landscape school tradition were involved especially in the first three of the themes employed in this textbook. But each of these three themes has also been enriched by ideas that have evolved from landscape school concepts or that have been added to those concepts. Thus, all three continue as important and changing themes.

The remaining three themes are more closely identified with developments since the 1970s than with the landscape school, but in all three cases there are continued interests in culture and in landscape, and it would be grossly misleading to imply that all aspects of these themes are separate from the landscape school. Since the 1970s, British geography in particular has taken a cultural turn and felt a need to add to and transform the agenda of cultural geography as practised, principally in North America, in the landscape school tradition. The fourth, fifth, and sixth themes reflect these additions and transformations, being typically informed by rather different meanings of the culture concept and, more generally, by a rejection of the larger intellectual environment of the landscape school. Table 1.1 identifies the themes and provides notations concerning conceptual inspirations, principal figures, key concepts, duration, and current status.

LANDSCAPE EVOLUTION

The first theme involves studies of the evolution of landscapes. Landscape evolution has origins in the landscape school, and much work has been accomplished in this area. Indeed, work that is usually labeled historical geography clearly belongs to this theme. The basic idea informing much of this work is the recognition that, in accord with Sauer, cultures transform physical landscapes through time to create human landscapes. Thus, it is usual to identify close links between the characteristics of a culture and the landscapes that are related to cultural occupancy.

REGIONS AND LANDSCAPES

Second, there are studies of the regionalization of landscapes, especially as reflections of cultural occupance over time. Again, such analyses were favored in the landscape school tradition, and there is a large body of related research. But there is also some important and influential work that is clearly outside of the landscape school and that contributes effectively to this theme. There are, of course, close links between the first and second themes in the sense that landscape regionalization can be interpreted as one outcome of evolution; the content difference between the two themes is one of emphasis. Regionalization is accomplished at local, regional, and national scales; note that global cultures and local ethnically defined areas are parts of a single dialectic, not opposites. On a global scale, world systems theory has been developed to explain the evolution of the modern political and economic system.

ECOLOGY AND LANDSCAPE

This third theme reflects and continues the rich geographic tradition of studying human and land relationships. It is especially closely related to Chapter 2 because the ecological theme is concerned explicitly with the relationships between humans and nature. Although there are again important conceptual origins in the landscape school, this theme is now informed by new theory, especially in the form of political ecology. Further, this theme is one of the most obviously practical, both environmentally and socially relevant areas

Table 1.1: Cultural Geography: Six Principal Themes

Theme	Conceptual Inspirations	Principal Figures	Key Concepts	Duration	Current Status
Landscape Evolution	landscape school history *Annales* school world systems	Sauer Clark Darby Kniffen Meinig Wallerstein Carter J.B. Jackson	landscape culture time	1920s to present	continuing and changing
Regions and Landscapes	landscape school regional geography new cultural geography culture worlds	Sauer Hartshorne Jordan Meinig Zelinsky	culture region landscape globalization	1920s to present	continuing and new concerns emerging
Ecology and Landscape	landscape school anthropology Marxism political ecology	Sauer Barrows Butzer	humans/nature culture ecology way of life	1920s to present	continuing and new concerns emerging
Behavior and Landscape	psychology humanism spatial analysis	Kirk Relph Tuan Wagner Wright	behavior perception cognition behaviorism	1940s to present	continuing and new focus emerging?
Unequal Groups, Unequal Landscapes	new cultural geography sociology Marxism feminism postmodernism	Blaut Cosgrove P. Jackson Wallerstein	power authority control patriarchy	1970s to present	major focus today
Landscape, Identity, Symbol	new cultural geography sociology humanism postmodernism	Cosgrove Daniels Duncan P. Jackson Ley Tuan	place sense of place identity landscape as text	1970s to present	major focus today

As is the case with any classification, this identification of six themes is necessarily a simplification of a complex reality. It is employed to facilitate the discussion of contemporary cultural geography. Three of these themes—evolution of landscape, regions and landscapes, and landscape and ecology—may be considered traditional in three respects:

- their principal conceptual inspiration is in the landscape school
- they include a view of geography as an integrating physical and human discipline
- they include the idea that culture is a causal variable

continued...

All three, however, incorporate some new and distinctive content.

The other three themes—behavior and landscape, unequal groups and unequal landscapes, and landscape, identity, symbol—are more recent in origin and reflect:

• generally closer contacts between cultural geography and other academic disciplines
• an increasing concern with social theory and cultural studies
• a view of culture that explicitly rejects the idea that culture is a causal variable

All three are introducing some new and different content to the ideas and practice of cultural geography.

of cultural geographic research. There are, of course, particularly close links with physical geography, and it is ecology that comes closest to fulfilling the ambitions of many geographers to work within a unified physical and human science tradition.

BEHAVIOR AND LANDSCAPE

The first three themes all incorporate considerations of human behavior, especially as that behavior relates to cultural group membership and influences landscape change and activities in landscape. In this fourth theme, however, the concern with behavior is more explicit, and it is typically recognized that behavior is related to individual as well as to group characteristics. There are close links to a number of traditions in psychology, especially to perception concepts and analyses. In geography, an interest labeled behavioral geography was initiated in the 1960s with some close links to spatial analysis, but since about 1970 the interest has developed within a more humanistic tradition. This theme thus questions the value of culture as a central concept, preferring instead to focus on behavior as it relates to landscape.

UNEQUAL GROUPS, UNEQUAL LANDSCAPES

One of the principal interests initiated by the new cultural geography is a concern with power and authority, both in terms of the control exercised by some groups over other groups, and in terms of the uneven quality of life evident in different places. Much of this interest is concerned with ethnicity, class, or gender, and the conceptual inspiration is from such areas as Marxism and postmodernism. But the links with landscape school interests are also evident, with discussions of landscapes and the recognition of regional differences. This theme also flows from Chapter 4, in

that much of the material discussed can be considered from a regional perspective and leads to distinctive landscapes. The single most important distinction between this theme and the landscape school is the embracing of different conceptions of culture.

LANDSCAPE, IDENTITY, AND SYMBOL

Although the landscape school tradition considered both material and symbolic landscapes, there was a clear emphasis on the visible material world. The sixth and final theme extends this landscape school emphasis and focuses more explicitly on human identity and on the symbolic meanings attached to landscape. The principal conceptual inspirations are from the humanistic ideas that began to flourish in the 1970s and from postmodernism, and much of the work conducted is most closely identified with the new cultural geography movement. This symbolic theme focuses on landscapes, both as they express human and nature relations and as they recreate culture.

Concluding Comments

Cultural geography can be fairly described as a contested area in that there are a plethora of approaches and research interests, with a particular distinction to be made between traditional and newer concerns. The Sauerian view of culture and the related landscape school, first introduced in the 1920s and linked to the larger modernist project, continue to dominate much American activity in particular, with ongoing interests in landscape evolution, regionalization, and human and land relationships. The new cultural geography that emerged in the 1970s and that is linked to reservations about the larger modernist project emphasizes alternative meanings of culture, especially concerning the earlier possible reification of culture, and additional areas of research activity, especially concerning the inequality of groups and landscapes and the need to study symbolic as well as material landscapes.

During the period from the 1920s through to the 1970s a Sauerian view of cultural geography dominated in the United States but was essentially disregarded in other English–speaking areas and in Europe possibly because of the unpalatable superorganic interpretation of culture, which never appealed to either anthropologists or geographers outside the United States. A comparable area of investigation in Britain was known as social geography. The key concept of society was similar to that of culture in that it allowed for discussions of group ways of life and of related landscapes.

There is no doubt that the landscape school interests in the evolution and regionalization of landscapes and in human and land relations continue to be major areas of research endeavor. Indeed, even after the new cultural geography had been around for about a decade, an American survey showed that 'mainstream cultural geography seems satisfied with the superorganic' (Rowntree, Foote, and Domosh 1989:212). These interests are undoubtedly more conceptually diverse than was the case early in the twentieth century, but the intellectual antecedents are clear. The new cultural geography interests are less established and necessarily have not generated as much research activity to date, but they are clearly the major growth areas.

But to emphasize the contested nature of cultural geography at the expense of the shared interests is misleading. As one part of the larger social scientific endeavor, cultural geography changes as the needs of human society change and as additional ways of conceiving the subdiscipline change. It can be argued that the differences between older and newer versions of cultural geography are less than has sometimes been asserted, and it may be that many of the purported differences result from the understandable tendency of those introducing new ideas to do so through a rejection of earlier ideas. Accordingly, this textbook:

- acknowledges the presence of different ideas and approaches, but explicitly rejects any suggestion that these are grounds for excluding some of these ideas and approaches
- reflects the tradition of cultural geography as initiated by Carl Sauer in the 1920s, and capitalizes on the reassertion of cultural geography and of the cultural in geography that has been such a feature of geography since the 1970s

For this textbook, the new cultural geography is seen as a welcome addition to, not a replacement for, the more traditional interests.

Further Reading

The following are useful sources for further reading on specific issues.

Francaviglia (1978) on the visible Mormon landscape.

Sibley (1995) on Gypsy identity and landscape.

Philo (1988) on the need to integrate cultural, social, economic, and political processes in human geographic research.

Taylor (1994a), Knox and Agnew (1994), Berry, Conkling, and Ray (1997), and Lee and Wills (1997) are examples of textbooks that include cultural processes in their accounts of political or economic geography.

Berger and Luckman (1966) on the social construction of knowledge.

Sopher (1972) on spatial analysis and cultural geography.

Feder and Park (1997:10) on the meaning of culture in anthropology and Giddens (1991:35) on the meaning of society in sociology. Kuper (1999) for a comprehensive account of the culture concept in anthropology.

Cosgrove (1978) and Jackson (1980) are important early statements of the new cultural geography; McDowell (1994) and Kong (1997) offer overviews; Price and Lewis (1993a) and Mitchell (1995) provide critical accounts.

Jackson (1989), Shurmer–Smith and Hannam (1994), and Crang (1998) are cultural geography textbooks that choose to exclude much of the Sauerian tradition.

Hoggart (1957), Williams (1958), and Thompson (1968) are important works anticipating the British cultural studies tradition.

Geertz (1973) on the anthropological concern with the study of otherness.

Said (1978, 1993) on the biases evident if a specifically European view of the world is adopted.

Blaut (1993a) on cultural geography as a contested area with a particular distinction to be made between traditional and newer concerns.

Humans and Nature

This chapter outlines the material transformation of the world by human groups, the ideologies by which humans understand their activities, and the broad issue of human relationships with nature. As noted in Chapter 1, you might choose simply to peruse this chapter at first, returning to selected parts of it as your study of later chapters deems necessary. The sequence of material in this chapter is as follows.

- There is an overview of changing technologies and changing population numbers as these are two of the principal factors relevant to discussions of relations between humans and nature and discussions of changes regarding the organization of society.
- Reasons for viewing humans and nature as separate—what is often called the dualistic preference—are summarized.
- There is an account of environmental determinism as the principal example of dualistic thinking in the geographic tradition.
- There is a discussion of human use of the earth— a concern that also implies a separation of humans and nature; the approaches of possibilism and the landscape school are noted in this context.
- Alternative ecological traditions that favor attempts to integrate humans and nature—what is often called the holistic preference—are summarized.
- The understanding of nature is expanded by reference to several contemporary ideologies that are proving influential in cultural geography; these are mostly concerned with the question of nature as socially produced.

- A variety of perspectives concerning ways of understanding human behavior in cultural context are noted; these are mostly derived from anthropological and sociological thinking and include sociobiological ideas.
- Some concluding statements reaffirm the value of understanding contemporary concerns in both larger historical and disciplinary contexts.

In addition to aiding our appreciation of humans, nature, and human and nature relationships, a feature of this chapter discussion is that it provides a series of contexts for the detailed identification of themes and related analyses that are contained in chapters 3 through 8. Hence, many of the arguments contained in this chapter resurface in the later chapters. Also included in this chapter discussion are some ideas concerning the ways in which humans strive to control nature and to control others, ideas that refer to some of the typically hidden themes in much of the practice of cultural geography.

Any discussion of humans and nature immediately raises a series of challenging questions.

- When we refer to nature, are we limiting the concept to what geographers typically label the physical environment—geology, landforms, climate, weather, flora, fauna?
- What do we mean by the term 'human nature'?
- What is an appropriate social scale for discussing humans—individuals, groups, the global population?
- What distinguishes humans from other living species, particularly from other primates? Is the

difference biological or is it our capacity for culture?
- Is human behavior affected, perhaps even determined, by nature?
- Why have some cultures often regarded other cultures as being closer to nature than they are themselves?
- Why have most cultures characteristically regarded women as closer to nature than men?
- Are humans *apart from* nature or *a part of* nature?

A preliminary response to the critical final question might note that the phrase 'humans and nature' immediately identifies a separation of the two, a dualism that is a traditional component of Western thought. This emphasis on separation has resulted in what is perhaps an unfortunate tendency, that of privileging one at the expense of the other, and there are numerous examples in this chapter of this tendency to **reductionism**. Add to these issues the multiple interpretations of the word 'nature' (noted in Chapter 1) and it becomes clear that the seemingly simple title to this chapter is in fact but the tip of a conceptual and interpretive iceberg. The broad concern evident here relates to the way in which we divide our world through our language as one means of naming, defining, and imposing order on our world.

To what extent have cultural geographers developed, or indeed felt the need to develop, coherent responses to these issues? For many working in the classic landscape or Sauerian tradition, the terms and the relationship between them are not a matter of substantive debate—their meanings are a part of the taken-for-granted world of the cultural geography being practised. The one debate that took place concerned the causal direction of the human and nature relationship, especially the relative merits of environmental and cultural determinism. However, since the various challenges to the landscape school were initiated in the 1970s, there has been a realization that there is indeed much more that needs to be considered in a discussion of humans and nature. Accordingly, cultural geographers are now turning to the diverse philosophical and social science literature that considers these issues, and are also making their own contributions to the various debates. Consensus

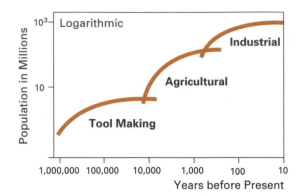

Figure 2.1: Human population growth during the past 1 million years. The relationship between technological change and population growth is summarized in this figure. The use of a logarithmic scale enables each of three technological transformations—tool making, agriculture, and industry—to be placed in context.

responses are not yet evident, although it is becoming increasingly popular to talk about the social construction of nature, as this term was introduced in Chapter 1 (also see Glossary).

Population and Technology

Population and technology are two critical and closely related variables in any consideration of humans and nature as they serve to affect human transformation of nature (figures 2.1 and 2.2). Expressed simply, the more people in a given area, the greater the likelihood of change, and the higher

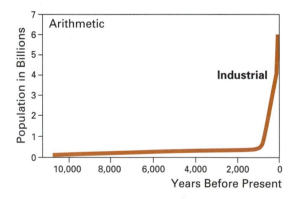

Figure 2.2: Human population growth during the past 10,000 years. The overwhelming impact of the transformations that occurred during the industrial revolution is evident in this overview of population change. Note that the scale employed in this figure is arithmetic, unlike that in Figure 2.1.

the level of technology, the greater the likelihood of change. A useful way to provide an overview of these relationships is to outline changes through time, especially over about the last 10,000 years, and to further inform the account by referring to related changes in political and social systems.

As this discussion makes clear, cultural change over the long period of human occupance of the earth does not take place separately and apart from natural change. Butlin and Roberts (1995:10) noted: 'Nature does not create landscapes, stop its work, and then hand over to human agency to complete the transformation.'

Foraging

Our understanding of human origins is always changing, but the dominant current view is that the human evolutionary line separated from that of the apes some 6 million years ago, with the first appearance of modern humans before 100,000 years ago, and with most of the earth colonized by 10,000 years ago (Figure 2.3 and Table 2.1).

Most current evidence suggests that hominids were restricted to east and southern Africa until *Homo erectus*, which first appeared about 1.8 million years ago, moved into warmer parts of Europe and Asia at the beginning of the Pleistocene about 1.6 million years ago. As different environments were encountered, cultural adaptations relating especially to clothing and shelter were devised. There is evidence of permanent pair bonding and substantial advances in language about 400,000 years ago. At this time, a new species evolved, archaic *Homo sapiens*, similar to modern humans but with some physical differences—Neanderthals are usually considered a subset of this species.

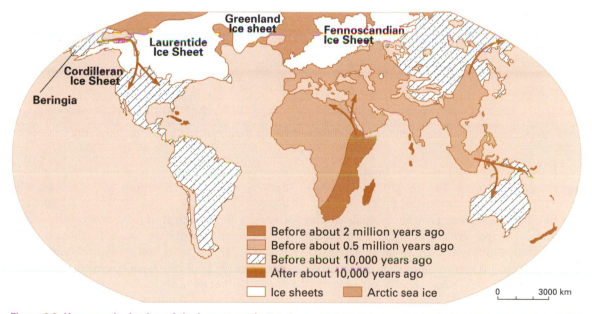

Figure 2.3: Human colonization of the ice age earth. This figure and Table 2.1 suggest a basic chronology of human evolution and human spread across the surface of the ice age earth. Following a probable origin in east Africa about 6 million years ago, the genus *Homo* spread throughout most of the world by the end of the Pleistocene 10,000 years ago. The principal region unaffected by humans was Antarctica. All this expansion was based on a foraging way of life and occurred in a physical environment that changed over time but that was typically different from that of today. The figure locates the principal ice sheets present during the Pleistocene glaciation.

Modern humans settled much of Africa, Europe, and Asia by 40,000 years ago, reached Australia about 40,000 years ago, and reached the Americas perhaps as early as 25,000 years ago, but no later than 15,000 years ago. Movement through northern Canada, Greenland, and some islands occurred during the last 10,000 years.

Source: Adapted from N. Roberts, *The Holocene: An Environmental History* (Cambridge: Blackwell, 1989):56; W. Norton, *Human Geography*, 3rd edn (Toronto: Oxford University Press, 1998):74.

Table 2.1: Chronology of Human Evolution

Time	Event
6 mya	Earliest date for evolution of first hominids, the bipedal primate, *Australopithecus*; in Africa
3.5 mya	First fossil evidence of *Australopithecus afarensis*; in east Africa
3 mya	Probable appearance of *Homo habilis*; in east Africa; evidence of first tools
1.8 mya	First appearance of *Homo erectus*; in east Africa
1.5 mya	Evidence of *Homo erectus* in Europe; evidence of chipped stone instruments in Africa; beginning of Pleistocene
700,000 ya	First evidence of chipped stone instruments in Europe
400,000 ya	*Homo erectus* uses fire; develops strategies for hunting large game; first appearance of archaic *Homo sapiens*
100,000 ya	Appearance of *Homo sapiens sapiens*; in Africa only according to favored replacement hypothesis
80,000 ya	Beginnings of most recent glacial period
40,000 ya	Arrival of humans in Australia
25,000/15,000 ya	Arrival of humans in America
12,000 ya	First permanent settlements
10,000 ya	End of most recent glacial period; beginning of Holocene period

Note: mya refers to million years ago; ya refers to years ago

Source: W. Norton, *Human Geography*, 3rd edn (Toronto: Oxford University Press, 1998):74. Copyright © Oxford University Press Canada 1998.

The appearance of modern humans, *Homo sapiens sapiens*, occurred before 100,000 years ago. There is much debate concerning the specific location of origin, with some favoring multiple origins in Africa, Europe, and Asia from pre-existing archaic groups (this is known as the multiregional hypothesis) and others favoring a single origin from archaics in east Africa only (this is known as the replacement hypothesis). Although this debate, like many other debates about human evolution, is not resolved, the weight of available paleontological, genetic, and archeological evidence favors the replacement hypothesis, with modern humans evolving from archaics in east Africa only, and then spreading across the earth replacing earlier archaic groups because of advantages associated with cultural developments.

This process of human movement over most of the earth took place during a period of harsh climate with the most recent glacial period, one of perhaps eighteen such periods in the past 1.6 million years, beginning some 80,000 years ago and ending with the retreat of ice sheets from much of the northern hemisphere about 10,000 years ago. This date, 10,000 years ago, marks the end of the **Pleistocene** geological epoch and the beginning of the **Holocene**. The glacial period was characterized by some significant changes in average temperature, which suggests that humans were able to adjust to different physical geographic circumstances through some process of cultural adaptation.

Culturally, the late Pleistocene is referred to as the **Upper Paleolithic**, beginning about 40,000 years ago and ending about 10,000 years ago. During this cultural phase, humans depended on natural sources of food. Hunting, fishing, and gathering—described collectively as foraging—prevailed as the means of gaining subsistence. In some parts of early Holocene Europe, **Mesolithic** cultures appeared in response to changes in the available animals and plants associated with ice sheet retreat.

The colonization of much of the earth occurred during the Pleistocene and was accomplished by cultures with subsistence economies. This fact, and evidence from the few remaining foraging groups, suggests that such cultures had an intimate knowledge of their natural environment, including the movement and behavior of game animals and the location and uses for humans of different plants. The relative importance of animals and plants in overall diets varied from place to place, with animals more important in the cooler areas of Europe and Asia and plants more important in warmer African and Asian areas. All individuals were involved, with women and children as gatherers and men as hunters, although it is probable that in many societies women also hunted small game and men gathered food. It is possible that this extended foraging phase, lasting tens of thousands of years, was the period during which it became usual to devalue women relative to men because of the perception that women were closer to nature whereas men were closer to culture.

The sharing of food with other group members was usual, encouraging social cohesion and stability. Groups were small, averaging perhaps about thirty, moved regularly, and required access to a sufficiently large area to support the group. Population densities were low, and a necessarily uncertain estimate of world population about 10,000 years ago is 4 million.

ENVIRONMENTAL IMPACTS

Few people in total, low population densities in any given area, and a nomadic lifestyle resulted in minimal impacts on the physical environment, specially long-term impacts. Fire was used to remove unwanted vegetation, to encourage regrowth, and in animal kills. Although much of the evidence is circumstantial, humans may have caused animal extinctions either through overkill or more probably through human modifications of ecosystems. In Europe, several large mammals became extinct toward the end of the Pleistocene,

The use of terrace cultivation on sloping lands may reduce the amount of run-off and erosion (*Victor Last*).

while the Maori settlement of New Zealand is related to the extinction of about twenty species of flightless birds. (See Your Opinion 2.1.)

Agriculture

Animal and plant domestication was a slow and gradual process, perhaps beginning 12,000 years ago toward the end of the Pleistocene period, and involving increasing numbers of people and expanding areas. By about 2000 years ago, most of the world population was dependent on agriculture. During the cultural phase of the **Neolithic**, agriculture spread throughout many parts of the world, replacing the earlier foraging activities of Upper Paleolithic and Mesolithic groups. What is often called the **agricultural revolution** is perhaps best described as the human use of **artificial selection** to modify animals and plants according to the needs of the human group.

WHY THE TRANSITION FROM FORAGING TO AGRICULTURE?

A possible explanation is as follows. At the end of the Pleistocene, climatic change and associated ice sheet retreat prompted some groups to locate in areas that permitted a more sedentary way of life, coastal and river areas, for example. Increasing population numbers or at least denser populations resulted from this new permanence, perhaps because of increasing fertility and a higher infant survival rate. In order to feed more people, these groups, which were already harvesting wild grass seeds such as wheat, rice, or corn, applied their knowledge of plant propagation to increase food supply by artificially selecting products useful to them.

Your Opinion 2.1

Traditionally, foraging has been understood as providing the bare minimum needed for survival with a popular image of groups eking out a meager livelihood and being obliged to share in order to avoid starvation. Do you feel that this view arose because of cultural relativism and because of observations of the experiences of some contemporary foraging groups that neglected to acknowledge that such groups may not be characteristic in that in many instances they have been gradually pushed into the most difficult of physical environments? Certainly, this understanding has been substantially revised and the favored current view is of an original affluent society with adequate food, considerable group and personal security, and leisure time. Although for some groups the acquisition of food was an ongoing activity, for others, especially those in resource rich areas, there must have been much leisure time.

Previously, it was widely assumed that there was one single origin area for plant domestication and that the technologies diffused from that one hearth. The favored candidate was southwest Asia, although Sauer (1952) argued for a southeast Asian origin. These diffusionist models were inherently **ethnocentric**, and the current consensus, supported by considerable archeological evidence, is that these developments occurred independently in several areas, namely southwest Asia, Africa south of the Sahara, southeast Asia, southern Europe, Central America, coastal South America, and central North America. Each of these areas had a different farming system; animals, for example, were a key part of the agriculture that evolved in southwest Asia, but were less important elsewhere (Figure 2.4).

The transition from foraging to agriculture was related to the use of previously unused **energy** sources and the associated development of new **technology**. The means by which humans became aware of energy sources and acquired the ability to use those sources was through the development of new technology. Thus, the domestication of plants involved humans acquiring control over a natural converter—plants convert solar energy into organic material via photosynthesis. Similarly, animal domestication involved control over another natural converter—animals change one form of chemical energy, usually inedible plants, into another form usable by humans, such as animal protein. In this sense, the domestication of plants and animals involved the use of new energy sources as a result of technological change. By about 2000 years ago, the earth included areas of

Figure 2.4: Global distribution of possible hearths of domestication. Although necessarily simplified, this figure provides an approximate chronological and spatial summary of animal and plant domestication. All dates are years before present.

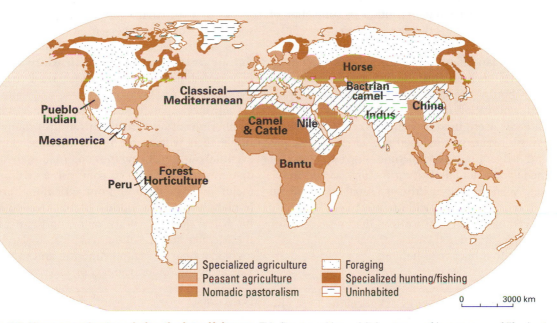

Figure 2.5: Humans and nature during the later Holocene. This figure provides a global summary of human ways of life about 2,000 years ago. At this time there was a broad range of activities: sedentary specialized agriculture in such areas as the Mediterranean; a few areas of specialized hunting/fishing; peasant, often shifting, agriculture in more humid areas; nomadic pastoralism especially in semiarid areas; and other large areas that continued to rely on foraging.

Source: Adapted from N. Roberts, *The Holocene: An Environmental History* (Cambridge: Blackwell, 1989):121.

relatively specialized agriculture, of peasant agriculture, of nomadic pastoralism, and areas that continued to emphasize foraging (Figure 2.5).

ENVIRONMENTAL IMPACTS

The environmental consequences of agriculture were considerable with the greatest impact on those plants and animals domesticated. There was an increasing human dependence on a few species, which initiated conflict with others. Because an agricultural landscape is an artificial environment, it favors some species at the expense of others. Animal and plant extinction resulted from loss of habitat as agriculture expanded. There were also impacts on soil as land was cleared for seeding, especially in areas of permanent agriculture. Notwithstanding these ecological changes, agricultural groups were tied to land and their survival was dependent on maintenance of a productive environment.

At first, agricultural groups grew crops in small areas close to their homes, and in some cases found it necessary to move regularly because of declining yields associated with continuous cultivation of the same land. In other areas, permanent settlement was possible because regular river flooding fertilized land naturally, while in some other areas products such as human and animal dung and household waste were used to fertilize land. Gradually cultivation increased in many areas, and fire and axes were used to remove vegetation as necessary. Terracing was used on some steep slopes and irrigation was sometimes used in areas of inadequate rainfall. During the first several thousand years, prior to the rise of a capitalist social and economic system, agriculture was typically a cause of ecological diversification that replaced, for example, the limited variety of many forest ecosystems with a mosaic of habitats.

CIVILIZATION

The cultural consequences of agriculture were also considerable. Food surpluses were produced in some areas, allowing population growth and encouraging increased labor specialization. Thus, population increased in the major agricultural areas more rapidly than at any previous time, and there were about 100 million people by 500 BCE,

and about 170 million at the onset of the Christian era; growth rates then lessened until the sixteenth century. Unlike foraging, agriculture required a degree of permanence such that in some areas a sedentary village lifestyle became increasingly common. This new permanence, along with food and labor surpluses, initiated a transition from essentially **egalitarian** social groups to situations characterized by social **stratification**. In a foraging society, human activities varied primarily according to age and sex, and those in the same age and sex categories were essentially equal in terms of both power and wealth. But in agricultural societies, these circumstances were dramatically altered. Agriculturalists produced food surpluses, freeing others to be involved in non-food-producing activities. Fixed layers of power and wealth became usual, with a few members of society exercising control over the majority and having access to a disproportionate share of the wealth; in many cases the privileged few used their position to oblige others to work for them as personal servants or as laborers in construction projects.

It is these circumstances—the acquisition of wealth by some members of society, the rise of cities, and the eventual rise of states and empires—that are components of the **civilizations** that first appeared about 6,000 years ago. Any small change—for example, a population increase or technological advance—that disturbed the delicate state of equilibrium of an agricultural group prompted a ripple effect with many further changes, with the eventual outcome being the combination of developments that are seen to comprise a civilization. Major civilizations arose from agricultural groups in many parts of the world, notably Mesopotamia, the Nile Valley, the Indus Valley, northern China, the Mediterranean, southern Africa, lowland and highland areas of Central America, and mountain valleys in the Andes (Figure 2.6).

FEUDALISM

The rise of civilization involved substantial social change, especially the already noted change from egalitarian to stratified societies, which have continued to be the norm through to the present. The earliest stratified societies sometimes incorporated forms of involuntary labor, notably **slavery**. Many

Figure 2.6: Global distribution of early centers of civilization. Although necessarily simplified, this figure provides an approximate chronological and spatial summary of the rise of civilizations. All dates are years before present.

included a ruling group based on noble birth, and in these cases a form of **feudalism** evolved, with several groups and estates distinguished. In the case of Europe, there was a land–owning aristocracy or gentry with considerable power and wealth; the clergy comprised a separate estate with some distinctive privileges and power; and the third estate (the great majority of the population) was comprised of serfs, free peasants, artisans, and merchants.

Unequal power, often involving exploitation of a majority by a minority, has been a key feature of most societies since the beginnings of agriculture. Feudalism transparently qualifies for this description as a proportion of the production of serfs especially was often transferred directly to the aristocracy.

EUROPEAN OVERSEAS EXPANSION

The period between about 6,000 years ago and about the fifteenth century involved the rise and decline of numerous civilizations. Several of these civilizations expanded to new areas affecting other populations, but none expanded globally. At the beginning of the fifteenth century three civilizations had the technological capacity to initiate large-scale movement—China, the Islamic world,

and Europe. Of these, China and Europe were the most densely populated, with each having perhaps 25 per cent of the global population.

It was the European civilization that began a process of overseas expansion. Explanations of this phenomenon are varied, but most rely on such circumstances as a favorable physical geography, a broadly shared culture of Christianity, a fragmented economic and political framework, gradual advances in preceding centuries in agriculture and trade, and the emergence of **mercantilism**. A number of technological advances, notably paper making, printing, the mariner's compass, and gunpowder, were also important. Although some of these advances were evident earlier elsewhere, it was western Europe particularly that used them to help begin a process of global movement and activity that continued until the late nineteenth century. Chapter 7 includes a critical discussion of these ideas.

Once begun, this process of European overseas expansion and related developments rapidly proceeded to dramatically affect the global population, estimated at about 350 million in the mid–fifteenth century, ending the situation that had prevailed previously, namely numerous groups of people liv-

The compass, like this Italian example (c.1580) was every mariner's most valuable navigational tool when out of sight of land (© *National Maritime Museum, London*).

ing essentially separate, local rather than global lives. Since the fifteenth century, there has been unprecedented biological, cultural, and technological movement related to this European activity. The first great European powers to enhance their wealth and status were Spain and Portugal, but subsequently it was France, the Netherlands, and Britain.

Industry

In addition to Europe's expansion throughout much of the world and its diffusion of Christianity and several European languages, it was also the site of a second major revolution in energy sources used and technological achievement—the first having been the agricultural revolution. Between the onset of agriculture and the industrial changes initiated in the eighteenth century, there were many agricultural changes, such as new animal and plant domesticates and new tools and techniques, permitting ever-increasing human control of energy sources. Further, three new energy converters were invented, namely the water mill, windmill, and sailing craft. But these were evolutionary rather

than revolutionary technological advances, and it was not until the Industrial Revolution that another dramatic set of changes occurred.

The Industrial Revolution involved the large-scale use of new energy sources via inanimate converters. The new energy sources were coal in the second half of the eighteenth century, oil in the second half of the nineteenth century, electricity in the second half of the nineteenth century, and nuclear power in the middle of the twentieth century. The critical initial technological change was the adoption of the steam engine in the late eighteenth century. The first developments occurred in England, spreading first to elsewhere in western Europe, second to the United States, and eventually globally.

ENVIRONMENTAL IMPACTS

But the Industrial Revolution was much more than simply a series of changes in energy sources and technology.

- Use of machines involved a factory system, massive increases in output, location of manufacturing in key resource locations, and the rapid growth of towns with localized areas for working-class residences.
- Transport links increased, improved, and diversified, with new canals, roads, and the invention of rail.
- Mechanization contributed to increases in agricultural productivity.
- Enhanced textile industries created a great demand for cotton and wool, both imported from outside of Europe.

CULTURAL IMPACTS

There were also numerous cultural changes.

- The percentage of the total labor force engaged in agriculture declined rapidly.
- There was a dramatic reduction in death rates, followed by reductions in birth rates with the period between these two declines being one of rapid population growth.
- From about 500 million in 1650, the world population increased to 1,600 million by 1900.
- New areas of high population density appeared in industrial and urban areas, there was much

rural to urban migration, and considerable European movement overseas.

- Feudal societies disappeared to be replaced by the social and political system of capitalism, labor was transformed into a commodity to be sold, and there was a related separation of the producer from the means of production.
- This was also the time during which the modern nation state assumed dominance, with several of these developing into world empires through processes of conquest and subjugation of other peoples and places.
- Global population has continued to increase rapidly throughout the twentieth century with a 2000 total of more than 6 billion (Figure 2.7).

More generally, this period, building onto the two Enlightenment assumptions concerning the practicality and desirability of scientific knowledge and the seeking of social truths, witnessed the rise of modernism. The world in which we now live, notwithstanding some notable changes since the end of the Second World War, is very much a prod-

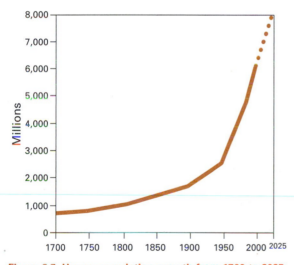

Figure 2.7: Human population growth from 1700 to 2025.
The global population growth since 1700 relates to: (a) a decline in death rates followed by a decline in birth rates in what has become the more developed world, with the interval between these two declines thus being one of high rates of natural increase; and (b) a later decline in death rates followed by a decline in birth rates that is only now occurring in what has become the less developed world, with the still continuing interval between these two declines being a period of high rates of natural increase. As of 2000, a reasonable prediction for the year 2025 is 8 billion people.

uct of nineteenth-century industrialization and associated modernity (Figure 2.8).

A Postmodern World?

According to Marx, societies are stable if there is a balance between economic structures, social relationships, and political systems, and are unstable if tensions or contradictions develop. The transition from feudalism to capitalism involved tensions related to the new manufacturing activities that then caused **class** conflict—most notably the French revolution—such that the capitalist system assumed dominance. But, in turn, the capitalist system involved new tensions, specifically because of the increasing gap created between rich capitalists and poor workers that would lead to a further transformation to **socialism** or communism. However, the experience of capitalist societies has not followed the pattern anticipated by Marx, and it does not appear that industrial capitalism is heading in such a direction. Rather, the current evidence is of a transition to a postmodern society and economy, also variously labeled the postindustrial, information, knowledge, or service society.

The rise of modernism is linked to the scientific revolution, the Enlightenment, and the Industrial Revolution, with their shared emphasis on the practicality and desirability of scientific knowledge. Indeed, the rise of an industrial way of life can be interpreted as a transition from tradition to modernity, or from *Gemeinschaft* to *Gesellschaft*. Modernism, the modern period, is based on the idea that the world is knowable and that there is a need to pursue some presumed knowable truth. Modernism assumed that social reality is logical, subject to laws waiting to be discovered. Explaining the human world would eventually lead to further technological advance, personal liberty, social equality, and the removal of such things as poverty and oppression.

Certainly the contemporary world continues to show the effects of the dramatic changes initiated during the past 200 years, and modernity continues to be the goal for many countries and groups, especially in the less developed world. However, there are reasons to suggest that a new set of processes are in evidence today in the Western world, namely postmodern forces that first began

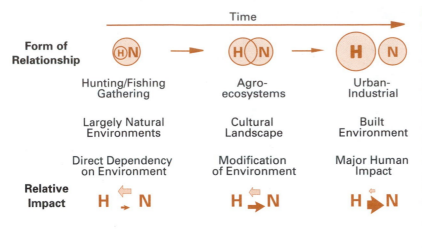

Figure 2.8: Human and nature relationships through time. Different social and economic systems involve different human and nature relationships. Foragers were directly dependent on wild animals and plants, implying an intimate relationship. With domestication of animals and plants, humans began to assume some control over nature and to dramatically and permanently alter nature. The two remained integrated, but the bonds were weakened. It was with the onset of industry that human and nature became separated. H refers to humans and N to nature. Refer also to Table 2.2.

Source: Adapted from N. Roberts, *The Holocene: An Environmental History* (Cambridge: Blackwell, 1989):184.

to appear after the Second World War and that gradually quickened pace through the final decades of the twentieth century. This concept of the postmodern is an especially difficult one, at least partly because it is always difficult to clearly appraise current circumstances.

Essentially, it is suggested that contemporary societies are moving beyond industrial development, the period of modernism, to a new social form emphasizing the use of information or knowledge rather than machines and industrial energy sources. Employment in the service sector is increasing at the expense of the manufacturing sector. But there is much more than this social change associated with the concept of the postmodern. Rather, as the term implies, the postmodern is contrasted with the modern. Postmodernism rejects the grand claims associated with modernism because of a loss of faith in the modern world.

Three postmodern claims are noted.

- The possibility of achieving objective, value-free knowledge is denied.
- It is claimed that knowledge is associated with power, and that existing social institutions are not in any sense natural or right, but rather have been created to ensure the maintenance of power. Both of these ideas surface later in this chapter in the discussion of the culture of nature.
- The modern Western conception of self as unified and given is replaced with a postmodern concept

of self as fluid, continuously reinvented in discourses and experienced anew in life—recall the brief introduction to identity in Chapter 1.

Given these claims, postmodernism rejects the grand attempts at explanation involved in science and Marxism. These postmodernist themes are particularly evident in the academic world, prompting a blurring of traditional—that is, modern—disciplinary boundaries and the creation of new interdisciplinary areas. This circumstance is clearly evident in the cultural geography that this textbook is reflecting.

Separating Humans and Nature

This chapter now proceeds to discuss some of the ways in which humans have thought about themselves, about nature, and about human relations with nature. (See Your Opinion 2.2.)

Humans, as members of cultural groups, have probably always wondered about their place in the world and about their relationship with nature. Questions have been raised in various cultural contexts and in many different ways, reflecting especially the prevailing religious attitudes and advances in scientific knowledge. One characteristic feature of Western thought concerns the presumed separation of humans and nature, a contrasting of the one with the other to emphasize difference. The authoritative account of the changing views of humans and nature in the Western world from the classical period through 1800 is that by Glacken, who began:

In the history of Western thought, men have persistently asked three questions concerning the habitable earth and their relationships to it. Is the earth, which is obviously a fit environment for man and other organic life, a purposefully made creation? Have its climates, its relief, the configuration of its continents influenced the moral and social nature of individuals, and have they had an influence in molding the character and nature of human culture? In his long tenure of the earth, in what manner has man changed it from its hypothetical pristine condition (Glacken 1967: vii)?

Thus, three ideas have predominated in discussions of humans and nature:

- the idea of a designed earth, a view of nature reflecting the doctrine of **teleology**
- the idea of environmental influence
- the idea of humans as agents of environmental change

For the period of the nineteenth century, Glacken (1985) revised this scheme by including the idea of a designed earth in the larger idea of interrelationships in nature, further developing the ideas of environmental influence and of humans as agents of change, and adding a fourth idea about esthetic, subjective, and emotional attitudes to nature.

Why Separate Humans and Nature?

Certainly, the view that humans and nature are separate and distinct has a long tradition in Western thought, and it was principally from these intellectual traditions that cultural geography emerged in the late nineteenth and early twentieth centuries. Accordingly, it is helpful to understand something about this tradition.

GREEK AND CHRISTIAN THOUGHT

In accord with the design idea referred to above, **Stoicism**, a philosophy that flourished in Athens after about 300 BCE and in Rome some 400 years later, held that because the natural world was clearly suited to human life, then nature must have been created for humans. All things had a purpose, with plants created for animals and animals created for humans. Many of these views, and similar ideas from Judaism, were absorbed into early Christian thought and they have often been discussed in terms of the biblical story of the Garden of Eden and the Fall. From this perspective, humans were granted dominion over nature: *Be fruitful and multiply, and fill the earth and subdue it; and have dominion over the fish of the sea and over the birds of the air and over every living thing that moves upon the earth.* But as soon as Adam and Eve ate from the forbidden tree, Adam removed himself from the natural world of the garden, emancipating the human soul from nature. This separation was followed by God again granting humans dominion over all animals after the flood: *Every moving thing that liveth shall be meat for you.* The Fall in the Christian world, and the resulting dualism of humans and nature, can be contrasted with other **creation myths**, such as those of many indigenous cultures that involve seeing humans as a part of the natural world.

THE SCIENTIFIC REVOLUTION

These Stoic, Jewish, and Christian traditions, namely the idea of a designed earth and the right

---

Your Opinion 2.2

This chapter is about humans and nature, but what do these terms mean to you at this stage of your reading? For most people socialized in a Judeo-Christian tradition, it is quite usual to conceive of the two as separate entities with humans in some way superior to nature. But such a view of the world does not prevail in some other traditions, such as those of Hinduism and Buddhism. Is one view right and another wrong? Or is one merely a better interpretation than another? Or are they simply different from one another—different ways of thinking? These are simple but challenging questions that ask you to think about what is often called our taken-for-granted world. You will be asked to reconsider this question at the end of this chapter.

of humans to rule over all other living things, were widely accepted in the European world for many years. They were influential in the birth of modern science in the sixteenth century, in the philosophical movement of the Enlightenment in the seventeenth and eighteenth centuries, and in the rise of positivism in the nineteenth century (Box 2.1). Each of these developments further contributed in some way to the idea that humans and nature were separate. Descartes, Hobbes, Locke, and Hume continued the early expression of the scientific method by the empiricist philosopher, Francis Bacon (1561–1626), and, broadly speaking, a mechanistic world view and a belief in human control of nature were central ideas. For Bacon, a key purpose of the scientific endeavor was to emphasize human dominion over nature, a dominion that some saw as partly lost as a consequence of the Fall. Indeed, natural history, comprising both botany and zoology, was studied not only for reasons of scientific curiosity, but also for explicitly practical and utilitarian reasons.

ENLIGHTENMENT AND NINETEENTH-CENTURY THOUGHT

The empiricist philosophy by which this early science was conducted asserted that all factual knowledge is based on experience, thus denying the view that the human mind is provided with a body of ideas before encountering the world. This empiricism has proven to be the dominant **epistemology** (see also **ontology**) and method of science, including the early social sciences that assumed disciplinary credibility in the nineteenth century. An outgrowth of the development of science, Enlightenment philosophy continued the established empiricist tradition. Nature was seen as explicable in terms of some system of universal

Box 2.1: Science and the Enlightenment

The seventeenth century began with Bacon asserting the need to study nature scientifically and ended with Isaac Newton (1642–1727) vindicating this faith in the power of science with an explanation of the workings of the solar system—the *Mathematical Principles of Natural Philosophy* was published in 1687. Newton demonstrated the power of human reasoning as a vehicle to understand the world, a scientific revolution that facilitated the rise of the Enlightenment, ensuring that science had a philosophical partner.

The term 'Enlightenment' refers to an intellectual movement—the age of reason—that began with Locke in England in the seventeenth century. A contrast is made with the earlier medieval period of presumed darkness, irrationality, and lack of reason. Principal doctrines of Enlightenment thought included the ideas that:

- reason enables humans to think and act correctly
- humans are by nature both rational and good
- all humans are equal and all ways of life should be tolerated
- local prejudices are devalued because they result from particularities rather than from reason

Some consider that the French Revolution established the Enlightenment ideals of liberalism, equality, and popular sovereignty. There is no clear end date, although the late nineteenth century certainly witnessed considerable strain on many Enlightenment ideals. It was, for example, all too clear that, contrary to expectations, much of the success of Europe in the modern age was built on the exploitation of others rather than on any principle of equality.

Enlightenment philosophers looked forward, initiating the modern age, rather than back to Greek or biblical thought. Together, the scientific revolution and the Enlightenment allowed scholars to believe that the world, physical and human, was accessible and comprehensible. Humans could understand all things—there were physical laws and chemical laws, so why not human laws also? It was also anticipated that the new knowledge would permit increasing control over the environment, improvement in the human condition, and prediction of the future, both of which are components of the emerging modernism. The confidence of Enlightenment thinkers in the ability of science and human reasoning did not go unchallenged. Practically, there was powerful opposition from both established religion and from many authoritarian European monarchs. Philosophically, there was opposition to applications of the naturalism of physical science to social science. As noted in Box 2.2, the principal philosophical opposition was from the idealist perspective.

laws, such as the law of gravity, rather than as a number of unrelated parts understood only by reference to God. Continuing Christian tradition, humans were seen as possessing an immortal soul and as dependent on the natural world—this is the concept of **materialism**. Their behavior was explicable by reference to laws similar to those being discovered in physical science; this is the concept of naturalism. These closely related views (some philosophers argue that they are virtually synonymous) further contributed to the rise of positivistic social science in the nineteenth century.

The positivism that came to the fore in nineteenth-century social science is properly associated with the writings of Comte, although the basic ideas were anticipated by the empiricism of Bacon, and there are close links with both materialism and naturalism. Recall that the key assumption of empiricism is that knowledge is dependent on experience. Expressed simply, Comtean positivism claimed that all knowledge, human and natural, is derived from systematic study of the world and the explication of laws about the world. Not all scholars accepted that human observers had access to an objective nature, unaffected by subjective considerations. Friedrich Nietzsche (1844–1900) insisted that it was not possible to discover nature, only to imagine and interpret nature (Box 2.2).

Nevertheless, the principal social science disciplines that developed in the nineteenth century became typically empiricist and positivistic in emphasis—economics, sociology, anthropology, psychology, and, to a lesser degree, geography. This linking of human and natural worlds within one method of science did not lead to any real integration of the two subject matters; although they were to be studied by the same method, they remained separate. Inevitably, there is a close association between the acceptance of positivism as a method for studying humans and the larger nineteenth-century world of European colonialism, capitalism, and industrial growth.

What It Means to Be Human

It has long been believed that humans are fundamentally different from other forms of life. Aristotle observed that only humans possessed a rational or intellectual soul, and Christianity insisted that to be human was to have an immortal soul. But such requirements did not make it easy to distinguish humans from animals and, from Aristotle onwards, considerable thought has been given to the way in which humans are different from animals. Regardless of detail, it has been usual to see the two as different in some fundamental way and to regard humans as superior. In addition to stressing anatomical distinctions, three specific human characteristics were proposed—speech, the ability to reason, and the possession of a conscience and religious instinct. Descartes added to the debate with the proposal that, like animals, humans were machines in that many of their actions were instinctual, but that only humans also possessed an intellect. Ideas such as these served to confirm the separation of humans and animals and, more generally, of culture and nature. Animals were awarded low moral status, justifying such activities as hunting, domestication, the eating of meat, and extermination of undesirable species.

Different Humans?

Related ideas sometimes relegated others, such as different cultural groups, the insane, poor people, children, or women, to the lesser animal status. Thus, the various attempts to explicitly define human had implications not only for the way in which the natural world was regarded and treated but also for human treatment of other humans. Further, it was also usual in both Western and other traditions to regard nature as female. There were two contrasting components to this image. On the one hand, especially prior to the birth of modern science, nature was seen as a nurturing mother providing humans with their various needs. On the other hand, there was the wild aspect of nature, suggesting the necessity for control.

Women are devalued in most societies, a fact that has been interpreted in terms of the association of women with nature—nature is something less than human, women are closer than men to nature, therefore women are inferior humans. In foraging societies, women were seen as closer to nature because it is women who procreate and because of the related domestic social roles that women play. Men, on the other hand, occupied a position closer to culture. An additional distinction that can be made is that men are concerned with the welfare of the social whole, while women are concerned with more particularistic family matters. This is the classic distinction made between the male public domain and the female private domain.

It is difficult to exaggerate the significance of this 'transcoding of feminine qualities to Nature, and of naturalness to women' for the cultural geographic enterprise, both traditional and new (Rose 1993:69). Indeed, it might be argued that all of human geography is structured around the separation of humans and nature, especially as this is reinforced by the distinction between urban and rural ways of life that became increasingly apparent in the nineteenth century.

These various ideas concerning human authority over nature, the perceived variable

human quality of particular people, and the linking of women with nature recur frequently throughout this textbook.

GEOLOGICAL, BIOLOGICAL, AND CULTURAL EVOLUTION

During the nineteenth century, the means by which humans were distinguishable from nature became clearer. Natural scientists began to clarify the **evolution** of the human species, while social scientists began to focus explicitly on the concept of culture as an exclusively human attribute. Although it is correct to note that neither biology nor the culture concept provide absolute distinctions, both aid in clarifying some of the ideas previously employed in this endeavor. More important, a consideration of these two topics indicates that both are relevant

Box 2.2: Two Philosophical Debates

Students of cultural geography may well be forgiven for asking why some philosophical concepts are being presented in this textbook. The answer is straightforward—the questions that cultural geographers ask, the approaches that they employ, the data that they choose to discuss, the results that they consider relevant—all of these are affected by philosophical preferences.

For our purposes, an understanding of twentieth-century cultural geography is enhanced by awareness of two fundamental but unresolved philosophical debates. Issues related to these two debates arise frequently in this chapter and generally throughout this book. Accordingly, this box provides brief summaries of these debates.

First: There is an epistemological debate about naturalism. Is it or is it not appropriate for social scientists to adopt a naturalist perspective? The success of the scientific revolution and of Enlightenment claims encouraged the idea that the methods of science could be applied to humans as well as to physical objects. As a result, for many nineteenth-century social scientists, the search was on for laws of human behavior. This application of mechanistic physical science methods to social science is what is meant by the term 'naturalism'. The history of social science, including cultural geography, is full of examples of competing claims concerning the validity of a naturalistic perspective. These debates are not resolved today, although naturalism is certainly out of favor relative to alternative approaches.

This text has already touched on this issue in Chapter 1 with, for example, references to deterministic approaches, which are naturalistic, and references to versions of **humanism** that are opposed to naturalism.

Naturalism incorporates an empiricist focus, stresses the search for causes, and is typically positivistic. Opponents of naturalism stress that humans cannot be treated as though they were akin to physical objects, and they favor instead approaches that focus on **hermeneutics**, sometimes called *verstehen*. Hermeneutics refers to the study and interpretation of meaning in everyday life. There are many versions of hermeneutics, but all share a commitment to the centrality of the mental quality of humans, asserting that, unlike physical objects, humans have needs and desires that affect their actions, and that these must be considered. From this perspective, human behavior is motivated by something inside the individual, such as feelings or perception. For many social scientists, this is *the* philosophical debate. Traditionally, naturalism has been strong in English-speaking areas, in the disciplines of economics, psychology, and sociology, and in the writings of Durkheim and Parsons. Hermeneutics has been more evident in Germany, in humanistically inclined social sciences, and in the writings of Weber.

Second: There is an ontological debate about realism. **Realism** is the philosophical view asserting that the existence of material objects is independent of sense experiences. In this sense, realism confirms what

in any attempt at defining what it means to be human.

The study of natural science underwent dramatic change in the late eighteenth and nineteenth centuries.

- The biblical account of the earth, involving a belief in **catastrophism**, including events such as the Flood, and a maximum age of about 6,000 years, began to be seriously questioned. In 1774 the French scientist, Count Buffon (1707–88), proposed the perspective of **uniformitarianism**, the idea that the earth was what it was as a result of such observable physical processes as erosion and deposition, not because of a number of unobservable catastrophes. This perspective was further developed in 1788 by James Hutton

(1726–97) and in 1830 by Charles Lyell (1797–1875), by which time it was becoming evident that the age of the earth was much more than the 6,000 years previously believed, but this was not the only change to established thought at this time. It was also becoming clear that not only had the earth evolved slowly into its present form, but animal and plant species had also evolved into their present form.

- The first natural science proposal for biological change was by Jean-Baptiste de Lamarck (1744–1829). Lamarck correctly noted that animals and plants were adapted to their specific environments, but his explanation, known as **Lamarckianism**, as to how such evolution occurred, namely the idea that organisms develop into more complex forms, was incorrect. The correct answer

is essentially the common-sense view of the world, but it is considered to be a philosophical position because an important case against it has been made. The case against realism is the idealist perspective, a term that, in this context, refers to a group of philosophical doctrines claiming that what humans know of the world is fundamentally the product of the human mind (see Glossary). This idealist perspective does not object to the view that material things exist, but it does object to the claim that this material world is completely independent of the human mind.

This debate about realism may seem abstract and unimportant, and it is certainly less easy to translate directly into the practice of cultural geography. Nevertheless, it does have a significant epistemological component concerning what role, if any, mental constructions play in our knowledge of the world. To be realist implies a rejection of the view that social science *must* involve a search for regularities and the testing of hypotheses, and a rejection of the view that social science *must* involve the interpretation of meaning. If this sounds a little like having your cake and eating it, note that in a sympathetic discussion, Yeung (1997) acknowledged a difficulty that a realist researcher faces, namely that realism is a philosophy that lacks a method.

In conclusion? These two philosophical debates are not easy to summarize in a few words because they reflect complex issues. However, the debates—and they are debates, they are not resolved by philosophers—are fundamental considerations to be taken

into account by cultural geographers and others who attempt to develop ideas about culture and nature. Indeed, it is because of these unresolved debates that, in an introduction to a discussion of social theories, it was stressed that the current lack of consensus might be endemic to social science: 'At the very least, whether there can be a unified framework of social theory or even agreement over its basic preoccupations is itself a contested issue' (Giddens and Turner 1987:1).

Examples of naturalistic approaches noted in this chapter include environmental determinism, the superorganic, sociobiology, and evolutionary naturalism. But most approaches currently favored in cultural geography are unsympathetic to naturalism, choosing instead to adopt some version of the hermeneutical perspective, such as one of the several humanistic approaches or postmodernism. Indeed, the challenge from hermeneutics has not left even the physical sciences unscathed, with some postmodernists arguing that both the physical and social sciences involve an interpretive ordering of reality. Such claims are evident in some feminist work. The position of cultural geographers concerning the debate about realism is less clear, partly because the philosophical issues are more difficult. Thus, some versions of realism can be naturalist in emphasis, while other versions can incorporate hermeneutics; indeed, critical realists insist that hermeneutics is the starting point, but not the sole concern, of research. Although both naturalism and realism are opposed to the idealist perspective, they are not equivalent approaches.

to this problem, **natural selection**, was the great achievement of Charles Darwin (1809–82) and Alfred Russel Wallace (1823–1913). Natural selection refers to the process by which nature effectively selects members of a species that are most suited to the environment; hence the most adaptive traits are reproduced and increase while the least adaptive traits decrease. Darwin published his monumental *On the Origin of Species by Means of Natural Selection* in 1859.

Note that **Darwinism** is a non-teleological theory in the sense that the contribution of an event to some result does not explain the occurrence of that event. Thus, species variations occur randomly, and those variations chosen by the environment were not designed with some particular result in mind. For many, this absence of teleological explanation means that the basic logic cannot be directly applied to cultural change, an important consideration in assessing attempts to apply Darwinism to the social sciences. On the other hand, the Lamarckian evolutionary process, involving the idea that the organs and behavior of species can change and that new organs and behavior can be transmitted from individual species members to their offspring, may be more appropriately applied in a cultural context.

With these advances in natural science in place, natural and social scientists began to ask and answer questions about the evolution of one particular species, humans.

- The natural science answer, still incomplete, was based on biological developments in such areas as brain size and the ability to walk upright—a brief account of current understanding of human biological evolution appears later in this chapter.
- The social science answer was based on the always uncertain and contested concept of culture and on the assumption that cultural evolution had occurred. The achievements of those scholars concerned with the evolution of culture, with possible links between biological and cultural evolution, and with some alternative ideas about culture, are also discussed later in this chapter.

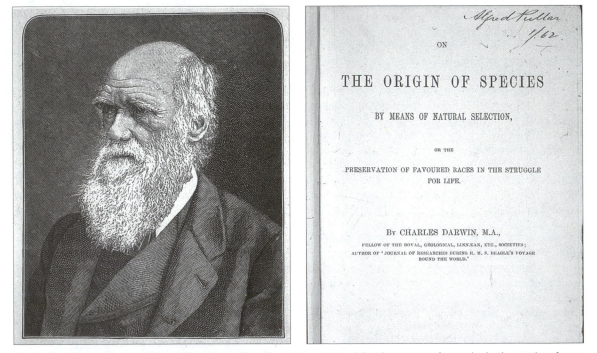

Charles Darwin's classic, *The Origin of Species*, published in 1859, maintained that the process of natural selection tends to favour the survival of those who are best-adapted to their environment (*Thomas Fisher Rare Book Library*).

The traditional way of defining human was to distinguish between humans and other animals, identifying some presumed distinctive characteristics such as having been made in the image of a Christian God, possessing a soul, or being spiritual as well as corporeal. But at the same time, humans have always been intrigued by the possibilities of intermediate species, of creating human life from non–living things, or of creating machines that are akin to humans. The basis for distinguishing humans and other animals lessened with mecha-nistic logic, which provided explanations of the whole material world, humans and nature, and with the advances made by Darwin that showed that human biological evolution was subject to similar laws as was the rest of the natural world. To offer a conclusion on this matter is difficult; it may be that a persuasive definition of what it means to be human is impossible. For a particular interpretation of our contemporary understanding of nature and culture, refer to the ideas proposed in Box 2.3.

Box 2.3: **Understanding Nature and Culture**

According to Gellner (1997), effective knowledge of nature exists, but effective knowledge of humans and their institutions does not. During the long period between the initial domestication of plants and animals and the scientific and industrial revolutions, European culture was based on the production and storage of food, on a rather unchanging technology, and on minimal advances in science. Accordingly, a Malthusian situation prevailed with resources remaining essentially constant and any population increases or food supply decreases prompting stress; this is, of course, the first stage of the demographic transition model. Under these circumstances, lack of sufficient food for the least powerful was an ever-present possibility such that the social hierarchy functioned as a queue to the storehouse. The principal concern for individuals and groups was, therefore, not with the issue of total food production, but rather with their ability to exert power over others, their social status, or, more simply, their place in the queue. Accordingly, the prevailing cultural values were those that enhanced personal and group authority and strength, and not those related to technological advance and innovation.

Under these conditions, humans lacked any real knowledge of either nature or themselves. There was necessarily some understandings of parts of the natural world, but integration of ideas was not encouraged in circumstances where the most powerful assumed ownership of new knowledge and innovation in order to further enhance their status. So how did this situation change? The answer lies in the rise of northwest Europe beginning in the seventeenth century, which involved a transition from coercive or martial to productive or commercial. In Marxist terms, this is a change from a feudal to a bourgeois order. There had been similar transitions at earlier times that proved unsuccessful in changing the larger cultural circumstances. This particular transition was successful because it occurred in conjunction with a second development, namely the rise of science.

Accordingly, since the seventeenth century, success in European cultures has been based not exclusively on power, but also on a sound knowledge of nature, on how to increase production, on progress. The contemporary world thus has a sound grasp of nature and an ability to change nature, but no equivalent grasp of ourselves or how to change ourselves in desirable ways—what some call social engineering. There are five possible reasons for what might be regarded as our continuing failure to achieve an understanding of ourselves comparable to our understanding of nature. These are:

- the inherent complexity of the human world
- the idea that human behavior is affected by human ideas and interpretations (refer to Box 2.2)
- the complex feedback processes in human processes
- the idea that cultural evolution is Lamarckian
- the role played by chance in human affairs

In discussing the prospects for advances in our understanding of ourselves comparable to the advances that have occurred in our understanding of nature, Gellner (1997:16) concluded that 'we simply do not know whether the social sciences will have their 17th century, whether the breakthrough and the subsequent fall-out will occur. What we do know is that if it did occur, it would, once again, completely change the rules of the game, as radically—perhaps more so—than was the case when technologically based natural science made possible Perpetual Growth and the vision of World-as-Progress.'

Environmental Determinism

Although by the nineteenth century thinking about humans and nature clearly incorporated some contradictory elements, the dominant view was to see the two as separate. Viewing them as separate did not, however, prevent scholars from holding the view that both were susceptible to study by the same scientific method as expressed in the philosophy of naturalism and the associated idea of positivism. For the discipline of geography, the separation was most evident in the explicit distinction typically made between physical geography and human geography, but also in the acceptance and application of the idea of environmental determinism, an approach to the study of geography that was closely related to naturalistic epistemology.

Definition

Environmental determinism asserts that the physical environment, nature, determines or at least influences the human world—both culture and cultural landscape (Figure 2.9a). Thus, all of our human actions are to be understood primarily if not exclusively with reference to the physical world of climate, landforms, soils, and vegetation. This is an idea that was often treated as though it were a law of science rather than an assumption, an article of faith. Recall that this theme was one of the three identified by Glacken (1967) as characterizing Western thought about humans and nature from Greek times onwards.

Environmental determinism is one particular interpretation of the larger philosophical position of determinism—the idea that all events are effects of earlier events. In physical science this was the dominant view for at least 200 years after Newton, although it has now been replaced by the indeterminism of contemporary physics. As noted, the social science disciplines found this position very appealing at the time of their foundation as deterministic arguments appeared to give credibility to their specific disciplinary subject matter and to render their enterprise scientific in the nineteenth century sense of science as empiricist and positivistic. (See Your Opinion 2.3.)

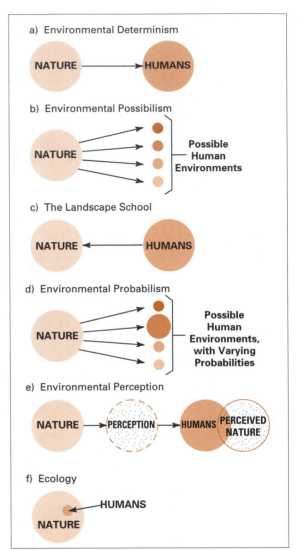

Figure 2.9: Some formulations of the human and nature relationship. This figure summarizes several of the approaches to the study of humans and nature detailed in this chapter. These schematic diagrams necessarily simplify the very real complexity evident in the various debates. It is not appropriate to interpret these approaches as different in any absolute sense; rather, each lies somewhere on a continuum ranging from the dualism most evident in extreme environmental determinist arguments through to the holism evident in some ecological arguments. This figure summarizes various conceptualizations of human and nature relations, whereas Figure 2.8 summarizes changes through time related to changes in social and economic systems. The landscape school is discussed and diagrammed more fully in Chapter 3 and the ecological approach is discussed more fully in Chapter 5.

Source: Adapted from D.N. Jeans, 'Changing Formulations of the Man-Environment Relationship in Anglo-American Geography', *Journal of Geography* 73 (1974):36–40.

Historical Perspective

Environmental determinist thought has a long history with many scholars, often writing from very different philosophical positions, expressing such views. Indeed, some scholars combined thoughts about environmental influence along with thoughts about human impact on the earth, the two being divergent but not contradictory ideas.

GREEK AND CHRISTIAN THOUGHT

Early instances of environmental determinism in classical Greece are included in the writings of Hippocrates (fifth century BCE). 'From early times there have been two types of environmental theory, one based on physiology ... and one on geographical position; both are in the Hippocratic corpus' (Glacken 1967:80).

The physiological theories derive from the idea that human health was related to specific locations, such as proximity to water and altitude, and include discussions of, for example, the best locations for houses. More significant for our purposes are the geographical position theories that derive from comparisons of large areas and concern the character of groups of people rather than the health of individuals. Thus, there are discussions of environments and related human characteristics, with a particular contrast between the relaxed populations of the eastern Mediterranean and the more penurious populations of Europe north of the Mediterranean area. In similar fashion, Aristotle described some eastern Asiatic populations living in warm areas as thoughtful and skillful but lacking in spirit and courage, in contrast to the less intelligent but braver and more warlike northern Europeans. The inhabitants of Greece, living in an intermediate environment, combined the better—interestingly, not the worst—qualities of both these populations.

Your Opinion 2.3

Applied to the human world, determinism implies that events unfold according to laws. Clearly such an idea raises many questions, notably those concerning human free will and moral responsibility for human actions. If the argument that all of our individual behaviors are predetermined and necessitated by prior events is accepted, then humans cannot be held responsible for their behaviors. How do you respond to this argument? One thing is clear: the burgeoning social sciences were attracted to the broad idea of determinism without offering any substantial critique of these larger philosophical implications, and environmental determinism is best approached in this light.

Many other Greek scholars, especially Posidonius (c.135–51), a Stoicist who also discussed the influence of the position of the sun and stars on both physical geography and humans, continued this tradition of thought. In this sense, early environmental determinist thinking does not necessarily imply a humans and nature separation because all things are related. Using more contemporary terminology, much of this early thought was ecological in character in that all aspects of the physical and human worlds were seen as linked; humans grow in an environment and are affected by that environment.

Strabo (c.64 BCE–20 CE), a member of the later Stoic school, adopted a broader cultural perspective in discussing the geography of the known world. Although environmentalist overtones are evident in much of his writing, especially concerning the physical geographic reasons for the rise of Rome, there is also recognition of the human ability to overcome environmental challenges. The writings of Strabo highlight the important general point that many scholars who contributed environmental determinist arguments also recognized other human and nature relationships.

After the demise of the Roman empire in the fifth century, many of the Greek and Roman contributions were unavailable to European scholars and the most studied and most influential book for about the next 1,000 years was the Bible. The design argument proved most influential and there were also accounts of human use of the earth; original environmental determinist ideas were less common, although there were contributions from such scholars as Albertus Magnus (c.1200–80) and St Thomas Aquinas (c.1225–74). Accounts of humans and nature played an important role in Islamic scholarship with ibn Khaldun (1332–1406), one of the greatest philosophers of the period, emphasizing the impact of environment on social organization and economic life.

THE SCIENTIFIC REVOLUTION

In Europe, the onset of the scientific revolution and the beginnings of significant overseas movement combined to stimulate additional examples of the presumed impact of physical environment, especially climate, on humans, both because the basic methodology of cause and effect was reinforced by the advances in science and because there were increasing numbers of applications possible in the expanding known European world. Interestingly, one twentieth-century environmental determinist was fond of quoting Bacon: 'True knowledge involves the study of Causes' (Taylor 1928:3). Certainly, this way of thinking flowered in many European works during this period. The French philosopher, Jean Bodin (c.1529–96), was the most prolific, building on earlier work and arguing that cultural differences were a result of astrological, latitudinal, and local factors. One conclusion was that people in temperate climates were both more talented than those in the colder North and more energetic than those in the warmer South. However, even such an enthusiastic proponent of environmental determinism as Bodin also acknowledged the role of other factors, such as the consequences of cultural contacts related to war and migration.

ENLIGHTENMENT AND NINETEENTH-CENTURY THOUGHT

Montesquieu (1689–1755) restated many of the earlier positions using new information that was becoming available because of European overseas movement. Climate was used to explain cultural differences (people in cold areas were strong and courageous while those in warm areas were more suspicious and cunning) and to explain cultural persistence, as in the presumed unchanging cultures of Asia. *By this time, environmental determinism was much more than simply a way of explaining culture— it also justified cultural relativism.* Thus, Montesquieu, although opposed to slavery, understood that it could develop where a population was so indolent that slavery was the only means available by which

they could be used as labor. Again, as with Bodin, Montesquieu was not an environmental determinist at the expense of recognizing other aspects of the human world; included in his writings are discussions of human impacts on environment.

The idea of natural controls is also evident in the writings of Thomas Malthus (1760–1834), who argued that it was not possible for a genuinely happy society to evolve because of the universal circumstance that population numbers increased more rapidly than did the supply of food, which was limited by nature. In his famous 1798 essay, Malthus referred to the 'struggle for existence'.

One of the most explicit statements was that by the French philosopher, Victor Cousin (1792–1867):

> Yes, gentlemen, give me the map of a country, its configuration, its climate, its waters, its winds, and all its physical geography; give me its natural productions, its flora, its zoology, and I pledge myself to tell you, a priori, what the man of that country will be, and what part that country will play in history, not by accident but of necessity; not in one epoch, but in all epochs; and, moreover, the idea which it is destined to represent (quoted in Febvre 1925:10)!

The second half of the nineteenth century provided many examples of the apparent relevance of the new ideas about biological evolution for human beings. Darwinian theory served to confirm the largely non-scientific arguments of earlier writers, with the result that many versions of environmental determinism—such as that proposed by the English historian H.T. Buckle (1821–62) explaining the onset of the scientific age in Europe in terms of climate, soil, and configuration—began to assume increased prominence at about this time. One proposal for a positivistic social science incorporating environmental determinism was especially important, namely that associated with the evolutionary perspective most fully developed by Spencer. The second half of the nineteenth century was the heyday of evolutionary theories, with major contributions also from the anthropologists, Lewis Henry Morgan (1818–81) and Tylor. But it was Spencer, in his articulation of what is usually labeled **Social Darwinism**, who argued for an analogy between animal organisms and human groups with regard to the struggle to survive in a physical environment. The logic of Social Darwinism is straightforward—cultural groups evolved in accordance with their ability to adjust to physical environments. For example, Spencer saw history as a steady move from warmer to colder areas.

During the nineteenth century, then, environmental determinism continued to be popular, and indeed became a component of other conceptual schemes, such as cultural evolution and Social Darwinism. But the century culminated in the adoption of the approach in a quite traditional straightforward form by the newly institutionalized discipline of geography. This adoption was notwithstanding the fact that several highly influential mid-nineteenth-century geographers had expressed alternative views. The attraction of environmental determinism for geographers was clear:

- there was the long scholarly tradition
- there was ever-increasing evidence indicating cultural differences in different physical environments
- there was the continuing popularity of mechanistic science
- there was the most recent support furnished by Darwinian theory and related ideas applied to cultures

Geography as the Study of Physical Causes

THE IMPACT OF RATZEL

Ratzel accepted many of Spencer's ideas concerning the similarities between human societies and animal organisms, strongly emphasizing that humans were subservient to nature. Indeed, Ratzel is often seen as the originator of modern environmental determinism, proposing the approach as a specifically geographic concern in 1882, in the first volume of *Anthropogeographie*. As restated by Ratzel, environmental determinism proved an immensely popular approach to the study of humans and nature, although various alternatives emphasizing other relationships, even a unity of humans and nature, were available at the time. However, it is clear that Ratzel regarded this concept only as a generalization, not as a scientific law, and in the second volume of *Anthropogeographie* he substan-

tially modified the view. But the early Ratzel appears to have been the more influential, probably because it was the early Ratzel whose views mirrored the needs of geography at the time, namely a subservience of human to physical geography and an apparent scientific emphasis.

Indeed, in addition to the various reasons for the attraction of environmental determinism already noted, it bears mentioning that one compelling reason for other geographers' positive reception of Ratzel was the fact that the approach clearly gave credibility to the geographic enterprise, stressing, as it did, the role of physical geography in human affairs. At a time when the various social sciences were competing to generate explanations, the practitioners of a discipline typically seized upon any approach that highlighted their discipline at the expense of others. A related reason for the use of environmental determinist logic was its implicit incorporation in the dominant regional approach to the study of geography.

The writings of Ratzel played a critical role in what was to become cultural geography, such that it is difficult to exaggerate the importance of environmental determinism during the early years of the newly institutionalized geographic discipline. Thus, the leading British geographer at this time, Halford J. Mackinder (1861–1947), accepted the view that human geography was derivative of physical geography. Similarly, in the United States, the environmental determinist view was articulated by the founder of the Association of American Geographers, William Morris Davis (1850–1934), and, more generally, was an integral component of the new university discipline. But the principal examples of environmental determinist thought are contained in the works of two American geographers, Ellen Semple (1863–1932) and Ellsworth Huntington (1876–1947).

THE WORK OF ELLEN SEMPLE

Semple visited Germany in the 1890s to study with Ratzel and subsequently introduced some of his ideas to American geographers; she rejected the idea, based on Social Darwinism, that the state was an organism, focusing instead on the more direct physical influence ideas. In *Influences of Geographic Environment*, Semple famously began:

Man is a product of the earth's surface. This means not merely that he is a child of the earth, dust of her dust; but that the earth has mothered him, fed him, set him tasks, directed his thoughts, confronted him with difficulties that have strengthened his body and sharpened his wits, given him his problems of navigation and irrigation, and at the same time whispered hints for their solution. She has entered into his bone and tissue, into his mind and soul (Semple 1911:1).

These opening sentences set the tone for a book that includes examples of deterministic explanations throughout. Many of these explanations maintain the long-standing interest in climate as a cause of human character, human activities, and human landscapes, and also reflect prevailing thought about racial differences between groups of people. The following brief quotation is not atypical of the thrust of the 1911 book: 'The northern peoples of Europe are energetic, provident, serious, thoughtful rather than emotional, cautious rather than impulsive. The southerners of the sub-tropical Mediterranean basin are easy-going, improvident except under pressing necessity, gay, emotional, imaginative, all qualities which among the negroes of the equatorial belt degenerate into grave racial faults' (Semple 1911:620).

THE WORK OF ELLSWORTH HUNTINGTON

In similar fashion, Huntington accepted the basic logic of environmental determinism, but employed the idea to focus directly on the supposed links between climate and the growth and decline of civilization. The basic theme was that areas of stimulating climate encouraged the growth of civilization, whereas the monotonous tropical climate was especially inimical to such growth:

... a prolonged study of past and present climatic variations suggests that the location of some of the most stimulating conditions varies from century to century, and that when the great countries of antiquity rose to eminence they enjoyed a climatic stimulus comparable with that existing today where the leading nations now dwell. In other words, wherever civilization has risen to a high level, the climate appears to have possessed the qualities which today are most stimulating (Huntington 1915:4).

But Huntington did not simply produce general arguments; his writings were also full of details of climatic change, of ways of measuring civilization, of maps of civilized areas, and of the implications of climatic controls for humans. As an example of his measures of civilization, Huntington (1945:6) noted, 'intermediate peoples, such as the Chinese, occupy intermediate places, with more cars than the pygmies, less than the Bulgarians'. The concern with the present human condition is well stated as follows: 'The climate of many countries seems to be one of the great reasons why idleness, dishonesty, immorality, stupidity, and weakness of will prevail. If we can conquer climate, the whole world will become stronger and nobler' (Huntington 1915: 294).

APPRAISAL

Certainly, the single most disturbing aspect of environmental determinism as an approach for the newly institutionalized discipline of geography is that it quickly proved to be embarrassingly simplistic. Nevertheless, although environmental determinism may well be an old and essentially discredited idea, it continues to reverberate in many discussions about humans and nature. (See Your Opinion 2.4.)

Generally speaking, applications of the approach assumed the argument to be correct and sought examples rather than approaching geographic facts in a more objective and open-minded manner. Thus, the attraction of environmental determinism was well expressed by Griffith Taylor (1880–1963): 'as young people we were thrilled with the idea that there was a pattern anywhere, so we were enthusiasts for determinism' (Taylor, in Spate 1952:425). Indeed, Taylor made an important contribution in arguing for a modified determinism: 'Man is able to accelerate, slow or stop the progress of a country's development. But he should not, if he is wise, depart from the directions as indicated by the natural environment. He is like the traffic-controller in a large city, who alters the *rate* but not the *direction* of progress; and perhaps the phrase "Stop and Go Determinism" expresses succinctly the writer's geographic philosophy' (Taylor 1951:479).

Although advocates such as Semple, Hunting-ton, and Taylor acknowledged the need for cautious applications and also recognized the role played by humans, it is fair to say that in many geographic studies, culture, technology, and economics were subservient to physical environment. Further, a substantial body of literature, including much earlier geographic literature arguing for the unity of humans and nature, was ignored.

One useful way to think about environmental influence is in the context of human technology. Expressed very simply, it is possible to distinguish between two types of nature, benign and difficult, and between two types of technology, high and low. The outcomes of different pairings of humans and nature are identified in Table 2.2. The table suggests that the variable 'determining' the outcome of a particular relationship between humans and nature is the human variable of technology, not nature. This is because when the human

variable changes from one level of technology to another, the outcome, the nature of the relationship, changes. When nature changes, from one type to another, the outcome is unaffected.

Human Use of Nature

Recall that many scholars, from classical Greece onwards, combined thoughts about environmental influence with thoughts about human impact on the earth, the two being divergent but not contradictory ideas. Indeed, there is a shared con-

ceptual focus between these two ideas in that both imply the presumed separation of humans and nature. Nevertheless, until the eighteenth century, it is not usual to find scholars addressing the question of human impacts in the same detailed philosophical terms as they discussed physical influences. For this reason, and because the larger philosophical context is covered in the earlier discussion of environmental determinism, this discussion of writings about human use of the earth is relatively brief.

Historical Perspective

GREEK AND CHRISTIAN THOUGHT

The idea that humans were able to change the physical environments in which they lived is a logical consequence of human life itself. Philosophical justifications by scholars in classical Greece typically recognized that humans were different from nature in that they were capable of reasoning. As such, they were orderers or organizers of the natural world, a rationale that was an important component of the designed earth argument: God created the earth for humans who were then granted authority and dominion over it, to change or indeed complete it as they saw fit. As Europeans became aware of environments outside of the European area, they viewed such new lands as simply awaiting their finishing touch.

These classical Greek and Judeo-Christian ideas about human purpose and about humans being partners with God, provided ample justification for human impacts associated with such activ-

Your Opinion 2.4

Are you inclined to agree with the following assessment of environmental determinism?

- *interpreted as geography—it neglects much earlier work*
- *interpreted as geography—it defines the discipline in terms of a specific causal relation, neglecting alternative forms of interpretation*
- *interpreted as geography—it excludes subject matter that does not accord with the method*
- *it is an overly simplistic approach, as are all determinisms, in that it ignores the complexity of most real world situations*
- *human landscapes change through time, yet physical landscapes are essentially unchanging*
- *humans make decisions according to all aspects of environment, cultural, social, political, economic, and physical*
- *although the approach is superficially attractive at a global scale, it is typically problematic at a local scale*
- *certainly, the physical environment is important, with much behavior related to diurnal and seasonal cycles and to major physical events such as earthquakes and floods, but all such behavior still needs to be evaluated in a human context*

Table 2.2: **Humans and Nature**

Relationship	Outcome	Example
High technology/ benign nature	Humans not limited by nature—various activities	Temperate areas today
High technology/ difficult nature	Humans not limited by nature—various activities	Hot dry deserts today
Low technology/ benign nature	Intimate human/nature relationship—foraging	Temperate areas 12,000 years ago
Low technology/ difficult nature	Intimate human/nature relationship—foraging	Hot dry deserts 12,000 years ago

The terms 'high', 'low', 'benign', and 'difficult' are relative. Thus, an industrial technology is considered high, and a foraging technology low; an environment without any physical extremes (for example, of temperature or precipitation) is considered benign, and an environment characterized by physical extremes is difficult. This is really simply a way of saying that the only activity possible with a foraging technology is, of course, foraging, whereas an industrial technology allows for a range of agricultural and industrial activities.

ities as irrigation, drainage, mining, and animal and plant domestication. Such changes were seen in a positive light as there was no real evidence of undesirable changes. This broad interpretation of human impacts prevailed until the nineteenth century. Many of the scholars who commented on the role of physical influences, such as Montesquieu, also commented on the positive contributions humans were making in the sphere of environmental change. Another significant contributor to the literature on human impacts was the geographer, Sebastian Münster (1489–1552), whose 1544 book *Cosmographia Universalis* was the authoritative world geography for about the next 100 years.

THE SCIENTIFIC REVOLUTION

Ideas about human impacts were central to the thoughts of the first scientific philosophers, Bacon and Descartes, and to the tradition that they initiated in that there was a new confidence in the ability of scientific knowledge to control physical environments. Combined with prevailing religious understanding concerning human dominion over nature, there was in the seventeenth century a general sense that humans were both improving themselves and improving nature through their impacts on nature.

ENLIGHTENMENT AND NINETEENTH-CENTURY THOUGHT

But it was in the eighteenth century, beginning with the work of the French naturalist, Buffon, that such an interpretation assumed real importance. Buffon acknowledged that humans needed to change environments in order to enhance their civilization and stressed the improvements and orderings of nature that had been achieved in the past. Distinctions between humans and animals—focusing on the intelligence, creativity, and adaptability of humans—were central to such arguments. Buffon wrote extensively about animal and plant domestication, the removal of forests, changes to soils, and the creation of cultural landscapes in general.

The first European suggestions that humans might be abusing rather than using or indeed improving physical environments surfaced during the eleventh to fourteenth centuries, a period of agricultural expansion involving the transformation of forest and marsh, the establishment of settlements, and increased mining activity. But it was the American geographer, Marsh, who made the first real advances in this area with the 1864 publication of *Man and Nature*. Humans were seen essentially as a destructive power and their hostile influences on environments were noted: 'man is everywhere a disturbing agent. Wherever he plants his foot, the harmonies of nature are turned to discords. The proportions and accommodations which ensured the stability of existing arrangements are overthrown' (Marsh 1864:34). Considerable evidence was presented to demonstrate the frequency, magnitude, and deleterious effects of human activity; examples were detailed concerning plants, animals, forests, water bodies, and coastal areas. Thus, Marsh provided a compendium of factual information focused on the theme of human activities and their unfortunate consequences on nature.

Geographic Interpretations

POSSIBILISM

Recall that Ratzel published the first volume of *Anthropogeographie* in 1882 and that the determinist arguments contained therein were accepted and further developed by Semple and others. The second volume was published in 1891 and advocated a quite different approach to human geography. Rather than focusing on the influences of physical environments on humans, emphasis was placed on the role of humans, in their cultural groupings. As we have seen, this approach, like that of environmental determinism, had a long scholarly precedent and it was also favored by the contemporary German geographer, Alfred Kirchoff (1838–1907). It was this approach, advocated in the second volume of *Anthropogeographie*, which was further developed by the eminent French geographer, Vidal.

In this tradition, the physical environment was regarded not as a determinant of human activities but as a factor that set limits on the range of possible human options in an environment, and it is this emphasis on environmental possibilities that prompted the label, possibilism (Figure 2.9b). More generally, possibilism is one component of *la tradition Vidalienne*, a tradition that also includes concerns with the local region, *pays*, with the immediate physical surroundings, *milieu*, and with the culture or way of life of a group of people, *genre de vie*. For Vidal and those who followed him, such as Lucien Gallois (1857–1941), Emmanuelle de Martonne (1873–1955), but especially Jean Brunhes (1869–1930), *genre de vie* was the key to understanding which of the various possibilities offered by the

environment was selected by a human group. Vidal was also a key influence in the development of the *Annales* school of history in France, early members of which included Lucien Febvre (1878–1956) and Marc Bloch (1886–1994). A consistent theme for these historians is the relevance of environment in human history.

Vidal and his followers—often referred to as Vidalians—recognized close relationships between humans and environment, explicitly rejecting the idea of environmental determinism. The approach continued the dualism evident in much previous thought, although the emphasis on human and nature separation is less explicit than is the case with environmental determinism. Further, possibilism arose in direct opposition to the concept that social phenomena are only explicable in terms of other social phenomena, an approach associated with the French sociologist, Durkheim. For Vidal, neither of the two determinisms, environmental or social, was acceptable. For a fascinating account of the practical relevance of these questions, see Box 2.4.

THE LANDSCAPE SCHOOL

Initially, the possibilist approach was closely identified with, indeed essentially restricted to, a group of French geographers. But there were two other closely related concerns initiated during the early part of the twentieth century, both asserting the primacy of culture over environment. In 1906 the German geographer, Schlüter, built on the ideas of Kirchoff and proposed the study of a cultural landscape, *Kulturlandschaft*, as this was created by a cultural group from a previous physical landscape, *Naturlandschaft*. It was this idea of landscape transformation that attracted Sauer and that provided impetus for the highly influential landscape school. Expressed simply, Sauer proposed that geography be concerned with the development of the cultural landscape from the natural landscape. The emphasis was thus on the evolution of the visible material landscape (Figure 2.9c). This approach—the foundation for much of the content of chapters 3, 4, and 5—is more fully discussed and diagrammed in Chapter 3.

It is helpful to interpret the approaches of Schlüter and Sauer as further moves away from an environmental determinist argument in that the explicit incorporation of culture allowed for an ongoing process of environmental change and adaptation rather than an acceptance of certain environmental options. Thus, in a landscape school–inspired textbook, Carter (1968:562) noted that it is 'human will that is decisive, not the physical environment. The human will is channeled in its action by a fabric of social customs, attitudes, and laws that is tough, resistant to change, and persistent through time'.

VARIATIONS ON A THEME

The basic logic of possibilism, as modified by the landscape school, was generally accepted by cultural geographers. Nevertheless, debate on a suitable approach continued during the first half of the twentieth century and a small number of geographers made additional contributions. In addition to the modification already noted, namely the stop-and-go version of determinism proposed by Taylor, Spate argued that neither environmentalism nor 'possibilism' is adequate and coined the term 'probabilism' as an appropriate middle road (Spate 1952:422). If one accepts the basic logic of possibilism, then this variation makes sense and was indeed implicit in the original possibilism, as it is simply recognizing that the various possibilities have varying probabilities of occurrence (Figure 2.9d).

There was also a continued concern with human impacts in the tradition initiated by Marsh. Thus, Sauer was one of the principal instigators of a seminal volume of readings, the motivation for which was well stated by Thomas:

Every human group has had to evaluate the potential of the area it inhabits and to organize its life about its environment in terms of available techniques and the values accepted as desirable. The identification, use and care of resources is in the end a problem of human values and behavior. Cultural differences in techniques and values, and hence in utilization of the physical biological environment and its conversion into a human habitat, have distinguished one human group from another. The effects of man on the earth are geographically varied and are historically cumulative (Thomas 1956:xxxvi).

A related variation on the possibilist approach involved a more explicit acknowledgment that groups of people perceive and therefore behave in environments according to their particular group

Box 2.4: 'An Historic Encounter'

The geographic concern with the relative merits of different ways of conceiving of relationships between humans and nature is not, of course, only an academic issue. A fascinating early example of the practical significance of the differing viewpoints of environmentalism and possibilism concerned debate about settlement prospects in the arid Australian interior after the First World War. Taylor, a proponent of stop-and-go determinism, was teaching geography in Australia at the time and was an important figure in the debate. Taylor referred to 'Nature's Plan', identifying 'optima' and 'limits', and published a textbook that was banned by the Western Australian education authorities because of the pessimistic views expressed. On the other side of the debate were politicians who favored the expansion of settlement and also a Canadian scholar, Vilhjalmur Stefansson (1879–1962). Stefansson, a popular figure in Canada following time spent in the Arctic and

subsequent expressions of the potential of that area for human activity, visited Australia in 1924.

The very different perspectives held by Taylor, the determinist, and Stefansson, the possibilist, prompted very different assessments of the Australian interior, assessments that were widely reported in the press. Stefansson optimistically referred to the prospects of the area and, without any real justification, compared it favorably to the American West. Stefansson's expansionist views were precisely what the Australian politicians and public wanted to hear. Taylor subsequently described Stefansson as a 'dangerous anthropogeographer', and, becoming a relatively isolated figure, left Australia in 1928 (Powell 1980:181).

Who was right? Certainly, Taylor underestimated the ability of humans to utilize difficult environments, but equally certainly Stefansson greatly exaggerated the potential of the area.

characteristics (Figure 2.9e). Differences between real and perceived environments are emphasized in this approach, which became popular during the 1960s (see Chapter 6).

Controlling Nature and Controlling Others

Human use of nature—the very phrase introduces the idea that nature is something to be manipulated according to human needs and wants. Returning to the Christian ideas of human and nature separation and of humans being granted dominion over all other creatures, and recognizing the long tradition of subsequent religious and scholarly thought, it becomes evident that one feature of the human experience, especially in the Western world, has been that of seeking ever greater control over the natural world. Further, this ambition to control has been extended to include human control of other humans, notably those who are perceived, for whatever reason, as perhaps less than human. These are, then, old ideas that have long been acknowledged, but they are also ideas that are subject to various interpretations.

As noted, in the Jewish, Greek, and early Christian worlds, humans were generally enthusiastic to establish hegemony over nature. Hierarchies were established with Aristotle, for example, speculating about human nature and developing a sliding scale with human males at the top, human females second, and then through other living beings. The principal Christian view was to see humans as masters of animals. Wild animals were there to be hunted, and domesticated animals were there to perform labor. The human experiences of hunting and of animal and plant domestication served to confirm human domination of the natural world.

By the end of the eighteenth century, the growth of science encouraged a rational rather than religious justification for human domination of the natural world and it was also evident that humans were not all equal. Slavery, forced labor, feudal social systems, the isolation of the mentally ill and of lepers, were all instances of identifying differences within the human species and of exercising control over those seen to be inferior. If to be human was to have some qualities defined by those in positions of power, then it was inevitable that some humans would be regarded in some way

as less than other humans. Indeed, just as it was claimed that plants could be domesticated or removed, and just as it was claimed that animals needed to be domesticated, hunted, or displayed in zoos, so it was claimed that humans needed to possess culture, European culture, that is—the alternative was to be exterminated, enslaved, or forcibly moved.

An especially important phase in the unfolding of these ideas is that of European overseas expansion, especially in the later nineteenth century during which several European powers effectively carved up a large portion of the globe among themselves. During this period the idea that humans, specifically European males, could also dominate others became paramount. In some respects, as noted in Box 2.1, this European achievement was the culmination of the Enlightenment, involving a triumph of progress and reason. The domination of others was legitimate because Europeans were superior; indeed, the conquerors were replacing the impoverishment of others with the values of justice and reason. The 'White man's burden' was to civilize the savage. This logic was reinforced by the evolutionary biology of Darwinian theory, and especially by Social Darwinism.

It was not unusual for groups that were relatively powerful, such as the English establishment, to view groups that were relatively powerless, such as the Irish, Africans, infants, women, the poor, and the insane, as inferior, and even to describe them in animal terms. Once seen in animal terms, people were liable to be treated in animal terms—perhaps domesticated, perhaps hunted; such issues, notably those of racism and sexism, resurface especially in Chapter 7. Another way to express this general point is to note that it became possible to measure human progress, advances in civilization, in terms of the domination of nature and of others.

Toward Holistic Emphases

Rationale

- Glacken (1967:550) wrote: 'It is wrong to say that the western tradition has emphasized the contrast

between man and nature without adding that it has also emphasized the union of the two. The contrasting viewpoints arise both because man is unique and because he shares life and mortality with the rest of living creation.'
- Tatham (1951:162) wrote: 'The old dichotomy between man and Nature, the view that environment is an antagonist that must be conquered, or to which one must passively submit can only lead to disaster or stagnation.'
- Taylor (in Spate 1952:425) wrote: 'May I remind you that much faulty thinking on the part of historians, philosophers and others is due to the fact that they always will set man against nature as if they were two distinct categories. Man versus Nature instead of man and his environment being regarded as a single complex.'
- Spate (1952:419) wrote: 'The danger is in setting up a false duality and so involving ourselves in insoluble or unnecessary questions of the chicken-and-egg order.'

These four quotations introduce a different approach to the study of humans and nature, an approach that aims to think in terms of unity rather than separation, in terms of **holism** rather than duality. There is a long scholarly tradition of attempting to view humans as a part of nature, although, unlike the theme of separation, this tradition was never dominant. This is not surprising given the prevalence of separation ideas in the major philosophical traditions as well as in Christianity. Many of the scholars associated with a holistic approach argued for a consideration of humans and nature together, but nevertheless found it necessary to employ both terms and thus were implicitly acknowledging a division between the two in accord with prevailing thought. Indeed, the historical perspective presented below is similar in overall emphasis to that contained in the previous section on human use of the earth.

Historical Perspective

GREEK AND CHRISTIAN THOUGHT
The themes of interdependence between all things and of unity of all living things have been a part of human thought at least since the time of classical Greece.

- Both Herodotus (484–c.425 BCE) and Plato observed that all life functioned as a unity and served to maintain a stable universe—this is the tradition of a mother earth, or Gaia, which is discussed more fully in a contemporary interpretation later in this chapter. This holistic theme is evident also in the work of Posidonius, referred to earlier in the discussion of environmental determinism.
- Philo (c.20 BCE–c.50 CE), known especially for achieving a broad synthesis of Greek and Jewish thought, viewed humans as a part of nature in the sense that all were created by God, with the different positions of species in nature evident by the order of their creation with the superior humans created last. Indeed, there are two somewhat different Christian creation stories in Genesis, with that referred to earlier in this chapter having man created first and all other creatures subsequently. In this sense, it can be suggested that, from the onset of the Christian tradition, humanity is both apart from and a part of the community of all living things.
- St Augustine (354–430), the greatest of the Latin church fathers, similarly saw nature as a continuum and ranked living things above those without life, and living things with intelligence—humans—above living things without intelligence—animals. Again, the Jewish philosopher, Maimonides (1135–1204), observed that humans needed to understand their place in nature rather than believing that nature was created for humans. Related to these ideas about humans being one part of a hierarchy, other scholars, such as Henry More (1614–87), noted that nature was unthinkable without humans.

In addition, then, to the many arguments concerning a separation of humans and nature are numerous other arguments, again often based on religion, emphasizing a harmonious relationship or integration of humans and nature. Such ideas continued to be stated until the late eighteenth century, by which time the first of a group of scholars, often with explicitly geographic concerns, provided a new set of arguments about the unity of humans and nature.

GEOGRAPHICAL THOUGHT

Johann Reinhold Forster (1729–98), who traveled round the world in the 1770s, studied the world as a coherent whole, while Anton Friedrich Büsching (1724–93) wrote a geography of Europe that incorporated the idea that all nature, affected by humans or otherwise, was one. Two of the leading scientists of the late eighteenth century, the geologist, Hutton, and the biologist, Lamarck, both conceived of humans as a part of nature.

But it was the nineteenth-century geographers, Humboldt and Ritter, who made the greatest contributions to the holistic theme. Both were major scholars in the first half of the nineteenth century, achieving important positions in science and society. 'Never before or since have geographers enjoyed positions of such prestige, not only among scholars but also among educated people all around the world' (Martin and James 1993:112–13). Their contributions to geography and to the advance of knowledge in general are considerable. Our concern is with their shared commitment to a view of the earth as an organic whole, and to the idea that all things on the earth's surface are related—*zusammenhang*—literally meaning 'hanging together'. Such a commitment, however, did not necessarily exclude the idea that there could be some environmental impacts on humans: writing in 1808, Humboldt acknowledged that climate and configuration could affect agriculture, trade, and communications.

- To quote Humboldt (in Tatham 1951:44): 'The earth and its inhabitants stand in the closest reciprocal relations, and one cannot be truly presented in all its relationships without the other. Hence history and geography must always remain inseparable. Land affects the inhabitants and the inhabitants the land.'
- Similarly, Ritter (in Dickinson 1969:37) stated: 'My aim has not been merely to collect and arrange a larger mass of materials than any predecessor, but to trace all the general laws which underlie all the diversity of nature, to show their connection with

every fact taken singly, and to indicate on a purely historical field the perfect unity and harmony which exist in the apparent diversity and caprice which prevail on the globe, and which seem most marked in the mutual relations of nature and man.' Unlike Humboldt, Ritter expressed a teleological philosophy, believing that the earth was designed to be the home of humans.

Several other nineteenth-century geographers held similar views to those of Humboldt and Ritter in that they clearly favored a holistic emphasis, especially rejecting any suggestion of environmental influences. Thus, the French geographer, Élisée Réclus (1830–1905), emphasized the processes by which cultures used environments to satisfy their needs, and the Russian geographer, Peter Kropotkin (1842–1921), saw nature as a dynamic whole that included humans and that was always changing.

Geographers were not the only scholars reacting to the traditional conceptual separation of humans and nature. In addition to the idea of a human and nature separation, nature has been seen as external in that it is beyond the realm of humans, and as universal in that there is both external nature and human nature. A universal view implies that humans are as natural as are the external components of nature. Although Marx can be read in various ways, for some he introduced a different way of thinking that rejected the external/universal dichotomy and that emphasized the continuity between humans and nature, with humans viewed as part of nature. Some geographers have recognized this contribution of Marxist thought referring to inner actions within nature rather than to interactions between humans and nature.

Ecological Emphases

Perhaps the most developed contribution to the idea that humans and nature are best treated in a holistic fashion, focusing on the unity of the two rather than on their separation, is contained within the broad theme of ecological thought (Figure 2.9f). Indeed, much of the work previously noted, especially that by such geographers as Marsh, Humboldt, and Ritter, is often interpreted today as being within the ecological tradition.

The word 'ecology' is derived from two Greek terms: *oikos*, meaning 'place to live' or 'house', and *logos*, meaning 'study of'. Literally then, ecology is the study of organisms in their homes. The key impetus for the development of ecology was the third chapter of *The Origin of Species*, which centered on the various adaptations and interrelationships of organisms, implicitly including humans, and environment. Haeckel first used the term in 1869, although it was Darwin, in *The Descent of Man* published in 1871, who explicitly included humans. The broad theme of ecology proved of compelling interest in both physical and social science. Thus, along with the specifically physical approaches such as plant ecology and animal ecology, there also appeared a variety of human and cultural ecologies in the various social sciences. Details of and differences between these approaches are included in Chapter 5.

In principle, these various ecological approaches rejected the dominant late nineteenth-century concern with human and nature separation, especially that of human dependence on nature or human independence from nature. They encouraged studies of humans and nature without privileging one at the expense of the other—a reductionist perspective is avoided. Ecology is based on the premise that things cannot be studied except in context such that relations with other things are important. In practice, however, many specific applications of ecology maintained a separation, at least partly because much of the terminology of the approach originated in physical science. Indeed, several ecological approaches—including that incorporated in the landscape school and also what was probably the most influential ecology, namely the cultural ecology initiated by the anthropologist, Steward—incorporated a view of humans as members of cultural groups dominating nature.

Certainly, it is reasonable to suggest that humans are both constrained and enabled by nature, but are also affected by their culture. The interest in ecology continues to flourish and, in the following section on contemporary ideologies of nature, the first concern is with holistic interests that have clear links to the larger idea of ecology.

Contemporary Ideologies of Nature

The Culture of Nature

It is increasingly being argued that nature is a part of culture, meaning that our human experience of nature is, and always has been, mediated through our cultural lenses. This idea has been evident throughout much of the discussion in this chapter and is a central feature of the classic account by Glacken (1967). For example, the image of nature implied by environmental determinism is mediated through religious (specifically Christian) and philosophical (specifically empiricist and naturalistic) beliefs and values.

This general idea of the culture or social construction of nature is evident especially in some humanistic arguments and in postmodern philosophy. According to naturalistic scientific views, it is possible to achieve an understanding of nature that is independent of the individual observer, but a humanistic view stresses the impossibility of any such objective interpretation of nature, arguing that individuals are capable of constructing many different worlds, and the objective nature of science is but one of these. There are also our individual life-worlds, the worlds of our direct experiences. From a humanistic perspective, it is necessary to interpret life-worlds, not to observe objective nature.

There is also a related emphasis on the need for a culture of nature that is sympathetic to concerns expressed by environmentalists and others about negative human impacts on nature. In a thoughtful overview of aspects of the North American landscape, Wilson (1992:13) noted that 'the whole idea of nature as something separate from human experience is a lie. Humans and nature construct one another. Ignoring that fact obscures the one way out of the current environmental crisis—a living within and alongside of nature without dominating it.' What is needed from this perspective is a new culture of nature based on a revised set of power relations. Box 2.5 discusses some related ideas concerning the invention and reinvention of nature.

New Holisms

The assumption that humans are a part of nature is contained within several recent arguments.

- Building on earlier ecological ideas, there is the concept of a global ecosystem. This is holistic in that it refers to all things, organic and inorganic, and to the web of relationships between all parts. For many geographers and others this concept is proving to be a real stimulus for what is at present the relatively new science of global ecology.
- The controversial Gaia hypothesis, initially introduced by Lovelock in 1979, asserts that the environment and all life combine to form a single unified system, and have indeed done so for about the past 3.8 billion years. Further, it is claimed that life itself has a profound impact on the environment, especially on the composition and temperature of the atmosphere, such that life maintains an environment that is favorable to life. This hypothesis opposes Darwinism in that it can be interpreted as having a teleological component, implying some action that is purposefully directed toward a specific goal.

In addition to these two holistic arguments, there are also views that strive to reconcile the need for integrating humans and nature and the need to acknowledge the distinctiveness of humans. According to this logic, any analysis of humans and of human landscapes must be eco-centric, focusing on the larger environmental context including humans. But, at the same time, analyses need to be **anthropocentric**, centering on humans because humans are indeed different from other parts of the ecosystem, including other animals, in that they have culture. It is in this sense that humans might be seen as both a part of, and apart from, the rest of nature.

A conceptual framework that aims to move beyond dualism and to offer a new way of thinking about society–nature–technology–animal relations is that of actor–network theory. Derived from social analyses of sciences, these concepts are an attempt to respond to the increasing evidence that social theories are becoming increasingly anthropocentric at the expense of ignoring the various intimate relations between humans and animals.

Box 2.5: Inventing and Reinventing Nature

There have been two contrasting components to the image of nature as female, namely nature as a nurturing mother providing humans with their needs, and nature as wildness suggesting the need for control. The latter image became dominant during the scientific revolution: 'As Western culture became increasingly mechanised in the 1600s, the female earth and virgin earth spirit were subdued by the machine' (Merchant 1980:2). Such a mechanistic image encourages human use of the earth, the manipulation of nature, and can be contrasted with an alternative social construction of nature, an ecological image that sees nature as a whole and humans as a part of that whole. These two world views are founded on very different ontological and epistemological premises; the mechanistic view is both masculinist and exploitative of nature as female, while the ecological world view acknowledges the need to break from these social constructions that involve human domination of nature and male domination of female.

Arguments such as these have contributed to a substantial and growing set of ideas on what might be called the invention and reinvention of nature. These ideas center on the way in which nature is constructed in terms of prevailing ideas about race, sexuality, gender, nation, family, and class, and the dominations and inequalities that are contained within these. Such arguments require a re-evaluation of much earlier thought, especially concerning the centrality of female representations of nature. Feminists have now initiated discussions of the character and power of the knowledge that passes for science, stressing that such knowledge—including the idea there is a man (not human) nature difference—is a creation of men, and that science was constructed by men at a time when domination of nature by men was seen as desirable. From this perspective, there is a need to reinvent a truly human and nature relationship, a new scientific perspective that stresses human unity and human unity with nature. Haraway (1991:77) expressed the overall critical concern: 'Facts are theory-laden; theories are value-laden; values are history-laden.' These matters are one part of the philosophical debates referred to in Box 2.2.

Evolutionary Naturalism

There is little doubt that much contemporary cultural geography is comfortable with the various critiques of earlier ways of thinking about humans and nature, but there is another interpretation of the scholarly tradition in social science. For Hutcheon (1996), the problem lies not in the physical science seduction of social science, but rather in the failure of social science to fully articulate a position based on the twin scientific pillars of evolution and naturalism. According to this interpretation, the social sciences have largely failed to utilize the scientific approach, with the various attempts in this direction being steps in the right direction, but never being adequately pursued. Naturalism implies the universality of cause and effect, while evolution is seen to apply to all things. Evolutionary naturalism is a way of thinking based on 'the premise that human beings are continuous with all of nature, and that all of nature is continuously evolving' (Hutcheon 1996:vii). This is an important argument that merits consideration in our discussion of humans and nature.

According to this argument, there are two currents of philosophical thought in the Western world since about the sixth century BCE through to the present. The dominant tradition at all times involves a metaphysical dualism, a separation of the physical and the spiritual, or the natural and the supernatural, which implies two categories of experience, namely secular and sacred, and two ways of knowing, namely experiential and mystical. Application of the scientific method, of causal logic, has been essentially restricted to the physical realm. The minority tradition, described as evolutionary naturalism, sees both humans and the universe as natural, and therefore encourages the study of human behavior and human institutions using procedures similar to those applied in physical science. Roots of this minority tradition are evident in Buddhist thought as well as in antiteleological Greek Epicurean philosophy.

It was Descartes who provided a developed logical base for the dominant dualist perspective stressing the separation of science from religion, and it was Hobbes, Locke, and Hume who, although sharing the concern with a scientific approach, countered dualism with the argument that humans could also be studied scientifically. Hutcheon (1996) traced the history of social science efforts to employ the methods of physical science, specifically in the form of evolutionary naturalism, and argued that such efforts have always been countered by approaches that explicitly accept that humans need to be studied differently from the physical world. Principal examples of social scientists who have included elements of the evolutionary naturalist perspective in their work include Spencer, Durkheim, George Herbert Mead (1863–1931), and B.F. Skinner (1904–90). Hutcheon (1996) also identified a series of links between contemporary physical and social science, evident especially in the sociobiological emphasis that is outlined later in this chapter. (See Your Opinion 2.5.)

In a related interest, the cultural geographer, Wagner (1996:1), introduced the admittedly speculative claim that 'human beings are innately programmed to persistently and skillfully cultivate attention, acceptance, respect, esteem, and trust

---

Your Opinion 2.5

Based on your reading so far, on your work in other social science courses, and on your general impressions, are you inclined to agree with Hutcheon that the social sciences have failed because they have not fully pursued a physical science approach? Certainly, the argument in favor of applications of evolutionary naturalism in contemporary social science is a minority view, but it does merit our consideration. It resurfaces later in this textbook, especially in the account of behavior and landscape contained in Chapter 6. Do appreciate that it is an argument that is not in accord with most contemporary approaches to social science, including those associated with the new cultural geography discussed in the previous chapter, primarily because it rejects the idea that the study of humans requires a different set of concepts and methodology from those used in physical science. But it is an argument against the separation of humans and nature precisely because it advocates a common approach to physical and human phenomena.

from their fellows', an idea that forms the basis for a materialist and evolutionary selectionist perspective on human behavior. This idea, more fully discussed in Chapter 6, is in close accord with the basic argument of evolutionary naturalism in that there is an explicit rejection of the separation of humans and nature.

Humans as Members of Cultural Groups

Cultural Evolution

Earlier in this chapter, during the discussion that outlined means by which humans were distinguishable from nature, the importance of advances in geology concerning the age of the earth and in biology concerning the role of natural selection were emphasized. These evolutionary ideas were received and applied with great enthusiasm by many nineteenth-century social scientists.

The thesis that cultures evolved and thus displayed only limited ties to nature—culture changed while nature remained unchanging—was a dominant feature of the early disciplinary experiences of both anthropology and sociology. Although by no means a new idea—indeed it was evident in classical Greece and became especially important during the period of European overseas expansion—it was the period after about 1860 that saw a flowering of theories of cultural evolution. Three great social thinkers of the nineteenth century, Comte, Marx, and Spencer, were all evolutionists, while the father of anthropology, Tylor, and his early follower, Morgan, were similarly evolutionist in their portrayal of culture and culture change. Since this heyday of evolutionary thought, interest has varied, with an early twentieth-century decline in popularity, a mid-century resurfacing of interest, and a relative loss of appeal in recent years. Although there are many and varied evolutionary theories, all of them have three features in common:

- a number of distinct cultural stages are identified
- these stages are ordered in a sequential manner such that the probability of remaining in a stage is higher than is that of returning to an earlier stage
- a probability of moving through the sequence of stages is identified

Most models of cultural evolution incorporate ideas of cultural selectionism (Box 2.6).

Sociobiology

Human beings have a great deal in common both biologically and culturally and it is critical that cultural geographers not lose sight of this essential human unity. It is in this sense that we can appropriately say that humans are a part of nature, but humans are also apart from nature, and it is culture that prompts that apartness. Thus, it is essential to acknowledge cultural diversity and difference, but not to allow this to result in a loss of recognition of human unity. In short, humans are both natural and cultural beings. It is this recognition that, for many, provides justification for arguing that most studies of humans are too anthropocentric and that much can be learned by treating the human cultural experience as but one part of the larger biological continuum.

Sociobiology is a particular expression of the relationship between biological and cultural or social evolution in that it argues for a reduction of human behavior to biological concepts. In *The Descent of Man*, Darwin applied the biological knowledge he derived from the observation of both animal and human behavior, combined with evolutionary theory, in order to study social organization generally. This early linking of biology and human groups was criticized especially from religious quarters, but also by sociologists such as Durkheim, who argued for the autonomy of sociology and hence for the separation of the human and the natural. Versions of this debate have resurfaced in recent years with a number of sociobiological theories being proposed.

Several theories made explicit use of Darwinian concepts to describe a process of natural selection of cultural characteristics to produce, through time, an increasingly complex, increasingly well adapted, and hence improved culture. Changes in human culture may be seen as similar to organic mutations, with parallels between biological and human cultural evolution along with the added idea that cultural evolution is a form of learning process leading to ever-improving human circumstances. This is a particular interpretation of behaviorism as developed in psychology. But the most elaborate interpretation of biological evolution in human terms incorporates the idea that cultural change is often intentional and purposive, as in the Lamarckian evolutionary process, rather than accidental, as in Darwinian theory. The temptation of relating biological and cultural evolution has also proven irresistible for some biologists, with acceptance of the idea that the process of cultural evolution is Lamarckian. A particular development concerns the articulation of theories that assert that human behavior is a product of the coevolution of human biology and culture.

CRITICISMS

Not surprisingly, sociobiology has been especially criticized by feminists concerned with the male construction of science: 'The patriarchal voice of sociobiology is less the effusive sexism that ripples over the whole plane of the text than it is the logic of domination embedded in fashioning the tool of the word' (Haraway 1991:74). Further, sociobiology represents a major challenge to much contemporary social science because of the clear emphasis on a naturalistic and mechanistic perspective—it is certainly out of tune with the prevailing dominant view. Given the generally critical view of sociobiology from the social sciences, and the related internal disagreements, it seems unlikely that the approach is to make a major impact on the contemporary understanding of culture and culture

Box 2.6: Cultural Evolution

Three great social thinkers of the nineteenth century, Comte, Marx, and Spencer, adopted evolutionist perspectives, while the anthropologists, Tylor and Morgan, portrayed culture and cultural change in terms of **unilinear evolution**—a series of stages from savagery to civilization. Together, Tylor and Morgan established the comparative method of inquiry, collecting ethnographic data from different cultures in order to demonstrate the increasing complexity of culture through time. There was an acceptance of the Enlightenment doctrine of progress, and it was implicitly assumed that European civilization represented the apogee of evolution, with primitive peoples as both biologically and culturally inferior. As one component of evolutionist thought, the anthropological typologies employed new data produced by European overseas expansion to argue for unilinear evolution, assuming that all cultures would, by means of independent invention, pass through a series of increasingly superior stages.

In the book, Primitive Culture, *first published in 1871, Tylor (1924) proposed three stages of cultural evolution and their characteristics as follows:*

• savagery—hunting and gathering, limited technology
• barbarism—agriculture, settled villages and towns
• civilization—writing

More generally, Tylor noted:

On the whole it appears that wherever there are

found elaborate arts, abstruse knowledge, complex institutions, these are results of gradual development from an earlier, simpler, and ruder state of life. No stage of civilization comes into existence spontaneously, but grows or is developed out of the stage before it. This is the great principle which every scholar must lay firm hold of if he intends to understand either the world he lives in or the history of the past (Tylor 1916:20).

Although the three stages are distinguished essentially in terms of technology, the greatest emphasis was placed on language, religion, and myth. Further, Tylor employed the concept of cultural survivals, components of a culture that, for reasons of tradition, persisted from one stage to a later stage.

In similar fashion, in the 1877 book, Ancient Society, *Morgan (1974) proposed seven stages of cultural evolution and their characteristics as follows:*

• lower savagery—beginnings of human life, gathering of wild fruit
• middle savagery—eating of fish, origins of speech, and use of fire
• upper savagery—use of bow and arrow
• lower barbarism—use of pottery
• middle barbarism—agriculture, irrigation
• upper barbarism—use of iron tools
• civilization—writing

change regardless of the validity of the arguments. Certainly, the principal cultural geographic initiative in this direction, namely the habitat and prospect–refuge theories discussed in Chapter 6, have not been widely accepted. At the same time, such ideas should not be ignored as they reflect a clear concern with the idea that humans are a conventional animal species.

The Superorganic Concept

Until the late twentieth century, the most influential way of thinking about humans as members of cultural groups was the superorganic interpretation of culture. Recall that this essentially mechanistic and deterministic way of thinking was developed within anthropology as one component of cultural evolution logic, and that there were sim-

ilar developments in sociology, with both Comte and Spencer conceiving of society as an integrated entity, comparable to a physical system and entirely determining the behavior of the people within it. The superorganic concept flows from this idea, seeing societies as wholes created either by growth from within or absorption of groups from without. Similarly, Durkheim argued for the separation of sociology from both biology and psychology, using the term *social organism* to refer to the social as a phenomenon *sui generis*, a process separate from the qualities of individual members of a society and one that cannot be explained by reference to psychology and biology. Both the social organism of Durkheim and the superorganic of Spencer see the social or cultural as a force constraining individual behavior.

The stages are distinguished in technological terms, although Morgan wrote primarily about the evolution of social and political forms and about a transition from communal to private ownership.

A lack of interest in evolutionism in the early twentieth century ended with a revival prompted by Childe in the 1930s and continued by White in the 1940s and Steward in the 1950s. Childe used evolutionist thought in accounts of the origins of agriculture and the subsequent development of an urban way of life. White employed a superorganic concept of culture, specifically a form of technological determinism, to argue that technology determined energy availability and the efficiency of its uses and, further, that the structure and functioning of culture was directly the result of the amount of energy produced and the manner in which that energy was used. Technology, through energy, was therefore the cause of cultural characteristics and change. Steward, a student of both Kroeber and Sauer, was critical of earlier unilinear evolutionist ideas, especially because of their often grand claims and gross generalizations, favoring instead a more modest **multilinear evolution**. The aim of multilinear evolution was to discover the laws that determined cultural development in the sense of identifying parallels of cultural development.

Boas initiated arguments against evolutionism in anthropology with strong support from Kroeber, Lowie, and Wissler. The basic criticism was that the crucial cause of cultural change was **diffusion** and related culture contact, not independent evolution. There were

also objections to the emphasis on progress and related implicit racism, to the lack of objectivity, and to the rigidity of the sequences proposed by Tylor and Morgan. Boasian anthropologists emphasized diffusion and contact as these related to the creation of cultural areas. Human history was seen 'as a sort of "tree of culture," with fantastically complex branching, intertwining and budding off—each branch representing a uniquely different cultural complex, to be understood in terms of its own unique history rather than compared in cultural complexes in other world regions in some grand scheme of "stages of evolution"' (Pelto 1966:24).

Although the evolution of culture is a central component of the landscape school methodology as evidenced by its inclusion as one of the five themes identified by Wagner and Mikesell (1962) (see Box 1.3), it is certainly an interest that has been ignored by most cultural geographers (but see Newson 1976). It appears that the emphasis placed on the landscape outcomes of the cultural occupance of an area largely precluded any detailed concern with cultural change, and, further, that such discussion was seen as more properly belonging within the discipline of anthropology. Indeed, the principal evidence of evolutionist thought in human geography is in association with environmental determinism. Even in anthropology, evolutionist thinking was quickly challenged by a diffusionist paradigm that incorporated the cultural area concept and the idea of cultural borrowing.

There are close conceptual parallels between what was happening in geography, namely the approach of environmental determinism, and what was happening in sociology and anthropology, namely the approaches of social and cultural determinism. All these approaches were versions of larger deterministic logic, all were mechanistic, all were reductionist, and all were privileging one part of the world to the essential exclusion of other parts. The superorganic concept is important to cultural geographers because of the argument (discussed in Chapter 3) that Sauer accepted the approach such that it became a central part of the landscape school.

KROEBER'S CONTRIBUTION

As used by Kroeber, the term 'superorganic' referred to the non-organic human product of human societies, cultural institutions, modes of production, and levels of technology. Although the term is borrowed from Spencer, it was similar to Spencer's only in the sense that it referred to non-biological aspects of human societies. The superorganic concept of Kroeber made a distinction between social processes and biological or organic processes emphasizing that, in contrast to biological evolution or changes in organic structure, human societies did not utilize the principle of heredity to transmit new adaptations to other members of the species. Further, 'the distinction between animal and man which counts is not that of the physical and mental, which is one of relative degree, but that of the organic and social, which is one of kind ... in civilization man has something that no animal has' (Kroeber 1917:169). Although humans may be biological organisms, they are social animals, and it is through language and social and cultural institutions that they learn the values of the civilization in which they live.

All civilization in a sense exists only in the mind. Gunpowder, textile arts, machinery, laws, telephones are not themselves transmitted from man to man or from generation to generation, at least not permanently. It is the perception, the knowledge and understanding of them, their *ideas* in the Platonic sense, that are passed along. Everything social can have existence only through mentality. Of course

civilization is not mental action itself; it is carried by men, without being in them (Kroeber 1917:189).

In addition to being an argument for the centrality of culture in explanations of the human world, the superorganic interpretation of culture was explicitly opposed to explanations focusing on the natural world and employing environmental determinism, but it was also opposed to explanations focusing on individuals, employing psychological logic. According to the argument against psychological explanations, the length of time an individual was involved in culture precluded the possibility of a paramount role for individuals as initiators of human activities.

EVALUATION

It may be possible to exaggerate the specific meaning of this concept. Although many subsequent commentators stressed the explicit cultural determinism evident in the superorganic, others take a rather different view. In a major overview of anthropological theory, Harris (1968:342) concluded that Kroeber was not a cultural determinist. Indeed, Kroeber (1917:205) himself noted that to infer 'that all the degree and quality of accomplishment by the individual is the result of his moulding by the society that encompasses him, is assumption, extreme at that, and at variance with observation.... [N]o culture is wholly intelligible without reference to the non-cultural or so called environmental factors with which it is in relation and which condition it.' This is what Mikesell (1969:231) called a 'cautious philosophy roughly comparable to the geographic concept of possibilism'.

Other Interpretations

The literature of social science includes many other ways of thinking about humans as members of cultural groups, each implying a rather different understanding of culture. Four of these other ways of thinking are noted.

• Following Boas and Kroeber, some anthropologists offered explicit refutations of Social Darwinism through their comparative analyses of cultures. The principle of cultural relativism was proposed, ideas of cultural evolution were dismissed, and there was a search for cultural universals. Culture was treated as a cause of human behavior.

- A functionalist approach developed in the 1920s, rejecting historical reconstruction of cultures and focusing instead on contemporary cultures with emphasis on the functions of institutions. There were two rather different versions. One view saw the function of social institutions as that of maintaining social structures, with culture functioning not for the individual, but for the larger society. An alternative view saw culture as a response to biological needs, with social institutions functioning to serve the needs of humans; for example, kinship is the response to the reproductive need. It is the first version that has been the more enduring.
- Several views have emerged since the 1940s regarding culture as a system of ideas; **structuralism** is the principal such approach. Based on advances in linguistic theory that saw a sound and the object it represented as entirely arbitrary, culture was viewed as a system of shared symbols. Structuralism is a concern with relationships and not with those things that create and maintain the relationships. Thus, the key to understanding cultures was to see all things in context in relation to the whole culture. For example, a structuralist interpretation of gift giving in primitive societies focuses on the act of giving rather than on what is given. A structuralist approach aims to remove the difficulties of cross-cultural comparisons by showing that elements of all cultures are the product of a common single mental process. The idea has proved attractive to many anthropologists. Despite the impact made by this approach, it appears that many of the arguments are highly impressionistic.
- A symbolic approach, as developed especially by Geertz (1973), sees culture as a system of shared symbols and meanings. There is a clear rejection of the Enlightenment conception of humans in this approach. For Geertz, culture is not a given that shapes the lives of individual members of the group; rather, it is the people in the group who continually shape their culture. They do this through an ongoing process of manipulating conventional symbols to create new meanings. This

symbolic approach sees humans as suspended in webs of significance, cultures that they themselves have spun. Understanding humans in their cultural groups is therefore a search for meaning rather than a search for regularities. It is an interpretation of culture that is explicitly opposed to the superorganic. The specific methodology proposed is known as thick description, meaning that it is necessary for the researcher to create an account that is as faithful as possible to the specific details of the cultural situation being discussed. The emphasis is always on the search for meaning.

Concluding Comments

This chapter first summarized the relevance of population and technology as factors related to human and environmental change and then introduced a number of philosophical concepts and associated scholars in the attempt to clarify our understanding of the meanings of the terms 'humans' and 'nature', and the character of relationships between the two. Necessarily, the philosophical concepts have been discussed in simplistic and selective terms. As noted in Box 2.2, the principal reason for incorporating these concepts in the discussion is that they provide relevant larger contexts for an understanding of many of the issues that concern cultural geographers. Unfortunately, their introduction can also cause confusion, and it is with this thought in mind that the Glossary provides basic explanations and brief discussions of key philosophical and other terms. A complete understanding of these issues is rendered difficult by the various interpretations of key terms, and by the several unresolved issues involving the status of reality and the merits of competing claims.

This chapter also details several variations of the presumed relationship between humans and nature as these are relevant to an understanding of the human world. Versions of each of these variations continue to be evident in contemporary cultural geography.

- Views that argue for cause and effect relationships remain attractive for many cultural geographers. The idea of physical geography as cause, while no longer a widely accepted argument, remains important in some specific contexts. Of course, cultural geographers cannot avoid consideration of the physical environment as physical processes necessarily have an impact on the surface of the earth and on human activities. The view that culture is a principal determinant of human activities also continues to be an important component of some cultural geography.

- However, as has been stressed in both this and the previous chapter, much contemporary cultural geography downplays cause and effect approaches in general, whether physical or cultural, favoring instead approaches that advocate some holistic perspective. The most notable such approaches are those that are ecological in emphasis.

From a European perspective, the emergence of the modern world beginning in the sixteenth century involved and encouraged the application of the new methods of physical science to the social sciences, including the derivation of universalist impersonal processes, objective forces, and dominant ideas, as means of explaining the human world. Culture was one such process, moving from simple to increasingly complex forms in an essentially deterministic manner.

But these ideas were not without contemporary critics. Idealist thought was an important if perhaps less influential tradition, rejecting the claim that it was possible to reflect the reality of the physical and human worlds; they were not there to be discovered and explained, but rather to be interpreted. During the twentieth century a series of developments, notably postmodernism, have served to further question the claims that there is a real and knowable world and that humans can objectively study their own world. Further, both physical and social science have been accused of being nothing more than a politically motivated practice designed to provide a privileged few (usually White males) with power while simultaneously denying power to others.

One transition in the cultural geographic approach to the study of humans is by now very clear. There has been a substantial redirection of scholarly work concerned with human behavior and cultural change from the search for causes, whether the causes are physical or cultural, to the search for meaning—this is the first philosophical debate in Box 2.2. Although both endeavors have always been present, the search for causes achieved paramount status during the period from the seventeenth century until about the 1970s when the search for meaning became and continues to be increasingly central. Significantly, the search for causes was especially influential in the social sciences during the late nineteenth-century formative period, and this emphasis was therefore able to exercise a significant impact on cultural geography at a key stage.

A related transition concerns the general rejection of arguments of nature as cause in favor of arguments that focus on culture and society, not as in the superorganic concept, but rather as providing the context for our understanding of the world. Stressing the presumed dependence of humans on nature involves a privileging of the natural world, while stressing the presumed dependence of humans on culture involves a privileging of the cultural world. Recently, there has been a movement toward a rather different

Your Opinion 2.6

In light of these discussions, have you revised your thoughts about humans and nature? Certainly, it is difficult not to be impressed by the quite remarkable quantity of work addressing this issue. It may be helpful to think of nature in two ways. There is wilderness, nature without humans, and there is nature transformed by humans into a cultural landscape. In both cases, nature has a meaning for those who perceive it, and therefore in both cases nature, wilderness or transformed, is a place. Finally, if we are inclined to be uncomfortable with the traditional separation of humans and nature because of the association of such thinking with the application of the scientific method to humans, then it is worth asking what we are to make of the situation where humans are in nature, which surely prompts concerns about sustaining a sense of autonomy for human actions.

privileging of the human, as in the accounts of nature as socially constructed. It is useful to recognize that all of these emphases can be interpreted as examples of reductionism, in the sense that they seek understanding in terms that are restricted to what is viewed, a priori, as most important.

There are clearly many unanswered questions in the above discussion, such as what nature is, and what it means to be human. Our discussion of nature separate from humans sought to explain how this separation came about and what it implies. Perhaps the easiest way to think about this is to recognize that it simply results from defining nature as everything that is not human. If we are uncomfortable with this approach and prefer to conceive of humans as a part of nature, then an alternative definition is needed, such as viewing nature as everything that there is in the world of experience. According to such an alternative definition, to be natural is to be a part of the world. (See Your Opinion 2.6.)

Further Reading

The following are useful sources for further reading on specific issues.

Atkins, Simmons, and Roberts (1998) on the history of relations between landscape, culture, and environment.

Sahlins (1974) on the argument that foragers are the original affluent society.

Feder and Park (1997:384–426) summarize and evaluate eight proposed explanations for the origin of agriculture, concluding that five of the proposals are complementary, not competing, and together provide a plausible explanation.

Feder and Park (1997:438–43) identified and evaluated seven proposals for the rise of civilization, concluding that no one explanation fits all cases and that myriad factors were involved (see also Scarre and Fagan 1997:1–20).

Teich, Porter, and Gustafsson (1997) and Coates (1998) on the history of humans and nature.

Graham (1997) on philosophical debates in human geography.

Bhaskar (1989) and Sayer (1992a) on the merits and demerits of the realist position.

Kuznar (1997) on the merits of naturalism in anthropology.

Ortner (1972), Ortner and Whitehead (1981), Fitzsimmons (1989), and Johnson (1996) on the links made between women and nature.

Flew (1978) on the difficulty of defining what it means to be human.

Peet (1985) claimed that, because geography followed Social Darwinism rather than Marxism, the discipline was prevented from achieving a meaningful approach to humans and nature.

Olwig (1996a:86) argued that the concept of nature was 'tainted by its use in the context of a nature/culture dichotomy in which nature was seen to have some form of determinant relation to culture—a relation that had to be, and largely was, erased from academic geographic discourse'.

For more on 'stop and go determinism', see Taylor (1937:459).

Rostlund (1956:23) claimed that environmental determinism was 'not disproved, only disapproved'.

Berdoulay (1978) noted that neither environmental nor social determinism was acceptable to Vidal.

Olwig (1980), Bowen (1981), and Breitbart (1981) on nineteenth-century geographers sympathetic to an integrated humans and nature perspective.

Anderson (1995, 1997) on the human interest in displaying animals in zoos.

Whatmore and Thorne (1998) on geographies of wildlife.

Tuan (1997:6) on the acknowledgment by Humboldt that physical geography could affect human geography.

Young (1974:8) noted: 'Human ecology may be defined (1) from a bio-ecological standpoint as the study of man as the ecological dominant in plant and animal communities and systems; (2) from a bio-ecological standpoint as simply another animal affecting and being affected by his environment; and (3) as a human being, somehow different from animal life in general, interacting with physical and modified environments in a distinctive and creative way.'

von Maltzahn (1994) on the social construction of nature from a humanistic perspective.

Evernden (1992) on the need to rethink our culture of nature by refocusing on the wildness that is inherent in nature.

Eder (1996) on the need to understand why humans continue to abuse the environment, proposing that the real culprit is our continuing taken-for-granted idea that nature is there to be dominated.

Oelschlaeger (1991) on the cultural idea of wilderness from a historical perspective, beginning with the experience of preagricultural societies to show that there has long been an idea of wilderness as other.

Gerber (1997) and Proctor (1998) on the social construction of nature.

Murdoch (1997) on actor network theory.

Wolch and Emel (1995) on the claim that social theories are becoming increasingly anthropocentric.

Lovelock (1982) on the Gaia concept.

Wright (1983), Rossi and O'Higgins (1980), Sanderson (1990), and Trigger (1998) on evolutionist arguments; Childe (1936), White (1949), and Steward (1955) for examples of evolutionist writing.

Langton (1979) on the parallels between biological and human cultural evolution.

Cavalli-Sforza and Feldman (1981) on the interpretation of biological evolution in human terms.

Durham (1976:89) claimed that 'both biological and cultural attributes of human beings result to a large degree from the selective retention of traits that enhance the inclusive fitness of individuals in their environments'.

Leaf (1979) and Heyer (1982) on the superorganic.

Rossi and O'Higgins (1980) on the various ways of thinking about humans as members of cultural groups.

Landscape Evolution

This chapter is the first of six to identify and discuss a relatively distinct theme in cultural geographic analysis. Historical and conceptual details of the theme are provided, links to other disciplines are identified, the past and current status of the approach is noted, and examples of related analyses are outlined. This first theme, cultural landscape evolution, has origins in the landscape school, and a huge amount of work has been accomplished in this area since the 1920s. North American geographers have undertaken most such work. The central concern of the theme, the idea that informs most of this work, is the recognition that cultures transform physical landscapes through time to create human landscapes. It is usual to identify close links between the characteristics of the culture and the landscapes resulting from cultural occupancy. Further, much of the substantial body of work that is usually labeled historical geography also belongs to this theme.

The sequence of material in this chapter is as follows.

- There is an extended account of the landscape school that considers the intellectual origins of the school, the possible stress on culture as a causal variable, and the impact of the school on subsequent work in cultural geography. This section also serves as valuable background material for the discussions contained in chapters 4 and 5. It is helpful to appreciate that the ideas covered in this section inform these subsequent analyses in much the same way that the section on conceptual underpinnings, specifically those of Marxism, feminism, and the cultural turn, at the beginning of Chapter 7, informs much of the content of chapters 7 and 8.
- The landscape consequences of cultural change as these relate to processes of spatial diffusion are discussed.
- The landscape consequences of cultural change as these relate to contact between different groups are discussed. This section includes accounts of the movement of Europeans overseas and of the subsequent encounters between Aboriginal and European cultures. There is also reference to our changing understandings of the character of these encounters.
- The related historical geography tradition is addressed with reference to a number of approaches and to selected examples of empirical work; the emphasis here is on the shaping or making of landscape. Included in this section are discussions of frontiers, of the use of narrative, cross-sections, and sequent occupance to analyze change through time, of the evolutionary regional landscape studies that are most closely associated with the Sauer tradition, of an essentially British local history approach to the study of landscape change, and of the tradition of reading landscape as exemplified in much of the work published in the magazine, *Landscape*.
- The chapter concludes with an assessment of the various evolutionary approaches and concepts in the context of contemporary cultural geography, and with a set of ideas that provide a link to the cultural regional content of Chapter 4.

Although the common theme in this chapter is that of landscape evolution, the material included is quite diverse. For example, some studies have identified a distinctive cultural group and then described the related landscape change, while other studies have identified a region and then sought an understanding of landscape evolution in terms of cultural occupance through time. Spatial and related social scales range from the local area and corresponding community to the world and corresponding global culture. Throughout this chapter there is a close link with the Chapter 4 discussions of cultural regions in the sense that such regions can be considered as dynamic outcomes of evolutionary processes.

The Landscape School

Sauer wrote the article that is usually viewed as the seminal statement of the landscape school, 'The Morphology of Landscape', in 1925. The substantial impact of this one piece on most cultural geography prior to about 1970 and much since that date is undeniable, and yet there are a number of intriguing question marks surrounding the article.

- Was it an attempt to delimit a subfield of human geography, which is what was effectively achieved, or an attempt to set a course for all of human geography?
- Was it influenced more by earlier European work such as that of Ratzel in volume 2 of his *Anthropogeographie*, of Schlüter, and of Vidal, or more by the anthropological arguments about the meaning of culture as elaborated especially by Kroeber?
- Did Sauer himself, in his considerable body of empirical work, adhere closely to the methodological principles outlined in the 'The Morphology of Landscape'?

    Discussion of these questions as part of this larger account of the landscape school enriches our understanding of the identity of the cultural geography that built on the ideas of Sauer.

Origins

A CONFUSED SCHOLARLY TERRAIN
At the time when Sauer was expressing his ideas about landscapes, the larger discipline of geogra-

phy had an uncertain identity. Following the first establishment of geography departments in Prussian universities in 1874, there were various claims and counterclaims about the identity, methodology, and subject matter of the newly institutionalized discipline. Such debate took place against a complex intellectual backdrop. Geography, both physical and human, was a long established area of both intellectual and more general concern, with traditional interests in exploration, mapping, and written descriptions of places and peoples. Scholars writing about geography before 1874 contributed to many important debates, such as those about humans and nature, about populations and resources, about human differences, and about changes in ways of life. Further, the late nineteenth century was a period during which two seemingly contradictory trends were evident—there was an increasing separation of physical and human sciences as the social sciences became institutionalized, and yet there was much sympathy in both areas of endeavor for cause and effect analyses. Geography occupied a distinctive position in the organization of knowledge at this time in that it was both a physical and social science, and most geographers who achieved leadership roles found that it was incumbent on them to define the discipline of geography.

    Principal statements about the identity of geography included that by Ferdinand von Richthofen (1833–1905), favoring a geography centered on the regional concept, and also the two ideas promulgated by Ratzel, namely geography as environmental determinism in volume 1 of *Anthropogeographie* (1882) and geography as the study of human landscapes in volume 2 of *Anthropogeographie* (1891). Vidal followed the later Ratzel in advocating the study of human landscapes. For most American and British geographers at this time the central concern was the study of earth and life relationships. In addition, two tasks were seen as important—determining the influence of physical environment on human geography and on human history, and delimiting and analyzing regions. Certainly, both the physical cause and the regional approaches to geographic study were widely accepted in the English–speaking world. The formative physical cause work was accomplished especially by Semple and Huntington, while the formative regional statements were contributed by Mackinder and others in Britain and by Davis in the United States. It was into this somewhat confused and quite uncertain scholarly landscape that Sauer ventured in the 1920s.

ENTER CARL SAUER

From 1915 to 1923 Sauer taught at the University of Michigan, after which he moved to the University of California at Berkeley, continuing there until 1957 when he commenced an active twenty-three-year period of retirement. During his early years at Berkeley, Sauer made his contribution to the debate about the identity of geography. What Sauer accomplished during the period from 1924 to 1931 was the articulation of both a subject matter, namely landscapes as created by humans in their capacity as members of cultural groups, and an approach to that subject matter, an approach that emphasized evolution, regional delimitation, and relations between cultures and the physical world. The interest in physical and cultural landscapes was explicitly anticipated in earlier writings; the evolutionary concern was not dissimilar to the earlier ideas of Vidal and Schlüter; the regional concern was in accord with the dominant view of the discipline; and the human and physical relations theme had antecedents in the great nineteenth-century work of Humboldt and Ritter. *Sauer's methodological achievements were to treat geography as one of the social sciences, to synthesize and expand upon earlier geographic works, to express this synthesis in clear and forceful language, and to influence the activities of those many younger geographers who were attracted to work with him.*

The considerable literature about Sauer and the landscape school includes a variety of interpretations about precisely what was being said. Varied interpretations are perhaps inevitable when a scholar writes often and at length, contributes to knowledge over an extended period, and proves to be highly influential. Most important, two evaluations of Sauer's philosophical stance arrived at quite different conclusions.

- Entrikin (1984:387) stated: 'Culture history became his model of social science, and its underlying methodological and philosophical basis was

drawn from natural science rather than from historiography'. This view is supported by Leighly, Sauer's closest associate at Berkeley, who identified geomorphology and plant ecology as principal influences and noted that, on one occasion, Sauer described himself as an earth scientist.

- A rather different view, to the effect that a historicist perspective was the principal inspiration, was argued by Speth (1987) and has been accepted in broad terms by humanistic geographers who have traced their roots back to the landscape school.

This disagreement is of course but one particular version of the larger philosophical debate that is addressed in Box 2.2. Notwithstanding these two different interpretations, five basic ideas promulgated by Sauer can be identified.

Key Ideas

- *Sauer was adamant in his rejection of environmental determinism at a time when both Semple and Huntington were major figures in American geography and when much regional geography was sympathetic to the approach.* For Sauer, acceptance of environmental determinism involved needless generalization resulting in a loss of objectivity because advocating a particular viewpoint typically produced results that were inferior in value to results that were achieved without some initial assumptions having been made. Thus, facts were the first priority, to be followed, not preceded, by interpretation, with interpretation proceeding without either stressing or ignoring physical geography. Rejection of environmental determinism was based on perceived flaws in the argument and not on any downplaying of the importance of physical environment. Possibilism, as discussed by Vidal and Febvre, was consistently referred to in sympathetic terms: 'Within the wide limits of the physical equipment of area lie many possible choices for man, as Vidal never grew weary of pointing out' (Sauer 1925:46).
- *Sauer was also critical of the regional approach to geography that centered on the question of where.* 'The geographer is hardly likely to prosper by giving his main attention to all kinds of distributional studies' (Sauer 1927:185). Concern was expressed that a

focus on the distribution of any and all phenomena would result in geographers studying some subject matter in which they lacked competence. But regions were important as areas that exhibited a characteristic expression as a result of cultural occupance.

- *Rather than either of the two preferred approaches to geography, namely environmental determinism and regional geography, Sauer favored the study of the facts of landscape as the subject matter of the discipline:*

> It then becomes the task of geography to grasp the content, individuality, and relation of areas, in which man comes in for his due attention as part of the area, but only in so far as he is areally significant by his presence and works. This is a unitary and attainable objective. The landscape is constituted by a definite body of observational facts that may be studied as to their association and origin.... A phrase that has been much used in German literature, unknown to me as to origin, characterizes the purpose perfectly: "the development of the cultural out of the natural landscape" (Sauer 1927:186–7).

Although a regional emphasis was implicit in this landscape methodology, the term 'landscape' was thought to be a clearer alternative concept than either of the then favored terms 'area' or 'region'.

- *A focus on landscape was seen as constituting both physical geography, the physical landscape, and the impacts of humans on physical geography, the cultural landscape.* Sauer consistently recognized the unity of physical and cultural components of landscape.
- *Sauer was insistent on the importance of acknowledging that cultural impacts on landscapes prompted landscape change, and also that occupying cultures underwent change.* 'An additional method is therefore of necessity introduced, the specifically historical method, by which available historical data are used, often directly in the field, in the reconstruction of former conditions of settlement, land utilization and communication' (Sauer 1931:623). Neither cultural landscape nor culture was fixed, and reference was made to a succession of landscapes with a succession of cultures. This evolutionary emphasis was similar to the concept of sequent occupance, relating changing landscapes to changing cultural occupances that was proposed by Whittlesey (1929). Sauer outlined the evolution of both natural and cultural landscapes, the latter being summarized as follows: humans, through their culture, transform the natural landscape (Figure 3.1). Cultural landscape, not culture, is the object of analysis. Recall from Chapter 1 that, although Sauer (1925:46) referred to culture as 'the agent' and the 'shaping force', the physical landscape was referred to as 'fundamental'. For a discussion of the possible use of a superorganic concept of culture in the landscape school, see Box 3.1.

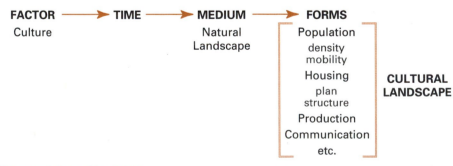

Figure 3.1: The morphology of landscape. This is the classic diagram included in the seminal 1925 article by Sauer, 'The Morphology of Landscape'. It is, in fact, the second of two very similar diagrams, the first outlining the established research area of the morphology of the natural landscape, which Sauer (1925:41) described as 'the proper introduction to the full chorologic enquiry which is our goal'. Thus: 'The natural landscape is being subjected to transformation at the hands of man, the last and for us the most important morphologic factor. By his cultures he makes use of the natural forms, in many cases alters them, in some destroys them' (Sauer 1925:45).

Source: C.O. Sauer, 'The Morphology of Landscape', *University of California Publications in Geography* 2 (1925):46.

Box 3.1: A Superorganic Concept of Culture?

In a retrospective comment, Sauer (1974:191) reflected: 'The morphology of landscape was an early attempt to say what the common enterprise was in the European tradition.' Certainly, the landscape school, as expressed in Sauer's methodological writings, was essentially an American modification and elaboration of established European thought. Emphasis was placed on physical landscape, culture, and time.

In light of some later commentaries on the work of the landscape school, it is important to consider whether or not Sauer was proposing an approach that incorporated a cultural determinist or superorganic perspective. There are three obvious reasons why this might not be the case.

- Sauer was highly critical of environmental determinism because it was seen as 'based on a belief that a single natural law can explain the social order' (Sauer 1927:173), and it would thus be illogical to replace an environmentally centered law with a culturally centered law that shared a similar failing.
- The importance of physical landscape was consistently acknowledged in all of the methodological writings and evident also in many of the empirical studies.
- The European sources that were continually acknowledged in a positive light were largely concerned with questions of human and nature interactions from a possibilistic or similar perspective. Indeed, there was a debate between Vidal and the sociologist, Durkheim, on precisely this issue, with Durkheim objecting to the inclusion of physical landscapes in attempts to explain human behavior.

On this basis, it seems reasonable to conclude that Sauer was not any version of determinist, environmental or cultural. But there is an alternative argument that has support from at least two areas.

- Sauer did refer to culture using terms such as cause. In this context, Figure 3.1, the classic landscape school diagram, has culture acting on physical landscapes to produce cultural landscapes.
- There is no doubt that Sauer was intellectually close to both Kroeber and Lowie, who were in the anthropology department in the same university as Sauer. Leighly (1976:339–40) emphasized these close ties: 'Basic to Sauer's course in world regional geography was the concept of "cultures", now used in the ethnological sense, which he learned from the anthropologists at Berkeley.' Sauer (1974:192) also acknowledged this association and influence:

'Anthropologists were our tutors in understanding cultural diversity and change. Robert Lowie in particular introduced us to the work of such geographers as Edward Hahn and Ratzel as founders of an anthropogeography that I had not known.' The significance of the intellectual association between Sauer and both Kroeber and Lowie is that the anthropologists were sympathetic towards a superorganic concept of culture.

Was Sauer an explicit superorganicist, proposing explanations of cultural landscape exclusively in terms of culture as cause? A balanced answer seems to be no, given the considerable weight that Sauer accords to physical landscapes. The aim of cultural geography was to 'explain the facts of the culture area, by whatever causes have contributed thereto' (Sauer 1931:623). The close links between the landscape school and *la tradition Vidalienne* suggest a shared concern with broadminded interpretations of the human and land relationship that were incompatible with either the superorganic of Kroeber or indeed the similar social organism concept proposed by Durkheim. For both Kroeber and Durkheim, the major concern, respectively, was with cultural and social causes of human behavior, but in cultural geography both culture and society were neglected, relatively speaking, in favor of visible material landscapes as objects of study. Arguments about possibly excessive influences from Kroeber also ignore the fact that Sauer was similarly intellectually close to Bolton, a Latin American historian at Berkeley. More generally, a scholarly association does not necessarily result in indiscriminate acceptance of ideas.

On the other hand, there is little doubt that several later landscape school geographers either explicitly or implicitly accepted the superorganic concept. Duncan (1980) details compelling examples. Thus, Zelinsky (1973:70) claimed: 'The power wielded over the minds of its participants by a cultural system is difficult to exaggerate.' In similar fashion, Carter (1968:562) stated: 'It is human will that is decisive, not the physical environment. The human will is channeled in its action by a fabric of social customs, attitudes, and laws that is tough, resistant to change, and persistent through time.' More generally, many cultural geographers referred to Kroeber and other anthropologists with superorganicist inclinations in their discussions of the concept of culture. In conclusion, there is no unequivocal final word on debates of this type.

These methodological statements combined to encourage three related themes that have been central to most subsequent landscape school research.

- Cultural landscapes change through time, as indeed do the cultures occupying the landscape; hence there is an evolutionary component to cultural geography.
- Cultural activities result in the creation of a relatively distinct cultural landscape; hence there is a regional or cultural area component to cultural geography.
- Landscape results from cultural occupance of a physical landscape; hence there is an ecological component to cultural geography.

Each of these three themes is evident in what is perhaps the most mature methodological statement (Sauer 1941a), and accordingly they are used as the basis for this chapter and the following two chapters of this book. (See Your Opinion 3.1.)

Impact

The impact of these early writings by Sauer was both immediate and long lasting. As early as 1934, a major overview of geography as a social science quoted the 1925 article at length and included discussion of the cultural landscape in regional analysis: 'wherever man enters the scene he immediately alters the natural landscape, not in a haphazard way but according to the culture system which he brings with him, his house groupings, tools, and ways of satisfying needs' (Bowman 1934:149–50).

DEFINING GEOGRAPHY AS THE STUDY OF LANDSCAPE

The landscape school was one of five attempts made to define geography between about 1900 and 1960—the other four were environmental deter-

---

Your Opinion 3.1

On the basis of your understanding of the discipline of geography at the time Sauer was writing, do you feel that Sauer's contribution was innovative and merited adoption by others? Although largely ignored by the older generation of American geographers, many responded favorably to this American proposal for a cultural geography—a term used explicitly in the titles of the 1927 and 1931 articles—involving the study of landscape. It does not seem difficult to understand the attraction to American geography of a specifically American proposal, especially given the uncertain status of the discipline at the time.

---

minism associated with Ratzel, possibilism associated with Vidal, regional geography associated with Hartshorne, and spatial analysis associated with Schaefer. It is in this sense that cultural geography has often served as a synonym for human geography—Sauer was attempting to determine an agenda for the larger discipline, rather than simply generating a subdiscipline. Although landscape school cultural geography never really dominated the larger human geographic scene, it has proven to be an especially enduring set of ideas and practices, initiating a compositional subdiscipline of human geography that has made major contributions from the 1920s onwards to the present.

The prevailing consensus is that the legacy of Sauer is a concern with cultural landscapes, although this is an association that is critically discussed on several grounds. Thus, along with the concept of landscape, Sauer and his followers also focused on links with anthropology, on regional studies in Latin America, and on ecological analyses. Further, many geographers identified with Sauer centered their studies on topics other than landscape, while some other cultural geographers not directly associated with Sauer stressed the landscape concept. Geographers directly associated with Sauer who focused on landscape include Kniffen, Parsons, Salter, and Zelinsky, while Denevan and Peet focused their energies elsewhere; geographers not directly associated with Sauer who are known for their landscape studies include Hart, Jakle, Lewis, and Meinig.

REVISIONS OF THE APPROACH

Before the first indications of a major revision of the agenda of cultural geography in the late 1970s, the landscape school concepts introduced by Sauer were subject to four revisions both by Sauer and by associated scholars.

- The emphasis on cultural landscapes evolving from physical landscapes was lessened at least partly because of the specialized knowledge required to analyze physical landscapes.
- The links with a regional focus, as implied in the concept of sequent occupance, weakened and there was an increasing emphasis placed on visible landscape.
- The concern with landscape evolution prompted a close integration of historical and cultural analyses. The increasing importance of time for Sauer and the landscape school may have been related to the influence of some work in anthropology. Rather than cross-sections through time, anthropologists saw time as continuous, a view to which Sauer had fully subscribed by 1941.
- Relatively little emphasis was placed on culture as cause—an implicit rejection of the superorganic and its unproven assumptions.

Such transformations were relatively minor and did not radically change the character of cultural geography as a concern with the evolution and description of visible material landscapes—landscapes as repositories of **material culture** (see also **non-material culture**). Meaningful criticisms of the historical landscape tradition first appeared in the 1960s and centered on the perceived emphasis placed on visible, material, landscapes, and the related ignoring of values and beliefs. Such criticisms appeared notwithstanding Sauer's explicit proposal to the effect that: 'Human geography, then, unlike psychology and history, is a science that has nothing to do with individuals but only with human institutions, or cultures' (Sauer 1941a:7).

HISTORICAL GEOGRAPHY

One explicit challenge laid down by Sauer was that of studying the evolution of cultural landscapes. This was a genuine challenge to geographers because the orthodox view of historical geography was that of the study of past times, an acceptable version in that it was not in direct contradiction to the dominant regional approach to geography. What Sauer was asking for was, within the American tradition, a less orthodox view because the study of change was a challenge to the hegemony of regional geography. The response, however, was considerable, effectively resulting in the rise of a relatively distinct historical geography, largely separate from the tradition of studying the geography of past times. The historical geography of changing landscapes was accepted into mainstream North American geography as evidenced by the inclusion of a historical geography chapter in the seminal 1954 volume, *American Geography: Inventory and Prospect*. Written, at the suggestion of Sauer, by one of his students, the contents of this chapter spell out a historical geography concerned with both changing landscapes and past times (Clark 1954).

Cultural Diffusion

Carter (1978:56) claimed: 'In the broadest sense diffusion is the master process in human culture change'. There are three principal approaches in geographic diffusion research.

- The traditional approach, focusing on the spread of a specific cultural trait, is identified with the landscape school.
- The spatial analytic approach, seeking to uncover empirical regularities in the diffusion process, evolved from the traditional approach in the 1960s, but is best seen as a part of the theoretical and quantitative movement in human geography at that time.
- The political ecology approach, concerned with links between diffusion, culture, and power, began as a reaction against the perceived dehumanization of the spatial analytic approach.

All three approaches are of importance in contemporary cultural geography, and indeed the history of diffusion research is itself a reflection of the larger twentieth-century history of human geography. In most studies, the concern is with the diffusion of an **innovation**.

The Spread of Culture Traits

This traditional cultural geographic approach has close ties with early twentieth-century anthropology, especially with the Boasian school, and is, of course, an explicit component of the landscape school of cultural geography. Accordingly, the concern is with diffusion that is largely controlled by culture but that does not exclude physical environment. For the landscape school, culture history is the core of human geography, and diffusion is the basis for understanding cultural origins, cultural landscape evolution, and the creation of a cultural region.

Most analyses in this tradition are concerned with a particular culture trait—there were, for example, analyses of agricultural fairs, covered bridges, and house types—and the concern is empiricist and inductive, with origins, routes, and present distributions identified. Other than Sauer, the principal early researcher was Kniffen, the geographer most responsible for applying landscape school ideas in the United States context from about 1930.

THE EXAMPLE OF COVERED BRIDGES

Justification for such studies often emphasized that an understanding of origins and diffusion of a particular trait aids understanding of spatial variations in cultural landscapes. For example, in an analysis of covered bridges in the United States, Kniffen (1951b) stressed the European antecedents, with covered wooden bridges especially evident in the Rhine Valley region of Switzerland. The original reason for covering is uncertain, although it does provide shelter, a location for market stalls, defensive advantages, and significantly extends the longevity of a wooden bridge. Once covering became common, its use probably continued for reasons of tradition.

In North America, covered bridges became popular after about 1810 with the New England to New Jersey area proposed as the **cultural hearth**. Subsequent diffusion of this landscape feature is indicated in Figure 3.2. The diffusion process is explained in terms of:

- physical geographic considerations, specifically the availability of suitable streams and of the necessary timber for building
- cultural considerations with covered bridges seen as just one component of the larger New England culture

The Henry Bridge in Bennington County, Vermont, was built about 1840 and spans the Waloomsac River. Covering bridges protected the trusses from exposure to the elements and thus prolonged the structures' life expectancy (*Vermont Department of Tourism & Marketing*).

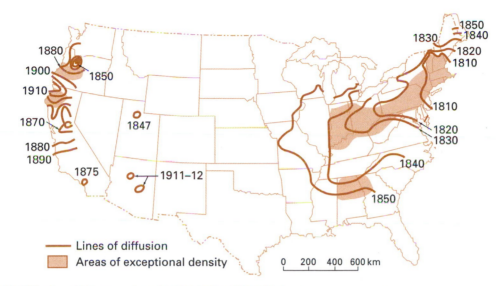

Figure 3.2: Diffusion of the covered road bridge in the United States. This is an example of the type of diffusion map produced by researchers working in the landscape school tradition. Areas of high density are marked and there are isolines indicating the date of acceptance of the landscape feature. In explaining this map, Kniffen (1951b:123) stated: 'The covered wooden bridge tended to spread freely among those possessing kindred patterns of thought and action.' More specifically, the feature spread unevenly west with migrants, reaching southeastern Ohio by 1820, southern Indiana by 1840, and attaining a maximum expansion in the eastern region by 1850. After this date, the number of covered bridges within the eastern region increased substantially and includes many examples in the Canadian province of New Brunswick. The principal area in the western United States extends from southwestern Washington to central California, a largely humid and wooded area west of the Sierra Nevada Mountains. The isolated instances of covered bridges in Utah and Arizona are seen as special cases and not as parts of larger regional patterns.

Source: Adapted from F. Kniffen, 'The American Covered Bridge', *Geographical Review* 41 (1951):119.

- technological considerations with new iron structures appearing after about 1860

This analysis of covered bridges suggested several generalizations about cultural differences in North America: it was, for example, suggested that within Canada, Quebec was more receptive than Ontario to cultural innovations from the United States.

LINKING TRAIT STUDIES TO CULTURE

This tradition of diffusion research typically focused on the description and mapping of a single landscape feature, with emphasis on particular cultural identities and descriptions of related impacts on landscape. However, although the specific concern might be a single trait, analyses often viewed such traits as representative of the larger culture, a tendency that is most apparent in studies of agricultural diffusion.

For example, an analysis of the diffusion of cigar tobacco production in the northern United States demonstrated links between the diffusion of a cultural trait and particular cultural identities (Raitz 1973a). The significance of cultural identity was twofold: first, a relation was uncovered between culture, notably religion, and the interest in cultivating a crop that is labor intensive; second, it was established that local group social networks facilitated communication of the necessary knowledge to encourage cigar tobacco cultivation. The diffusion pattern is shown in Figure 3.3. The introduction of cultivation in Pennsylvania was associated with Germans, especially with Old Order Amish; in the Connecticut Valley with Polish Catholics; in Ohio with German Lutherans, Baptists, and Old Order Brethren; and in Wisconsin and Minnesota with Norwegian Lutherans. In all major production areas today, there continues to be an association between cigar tobacco production and a specific cultural group.

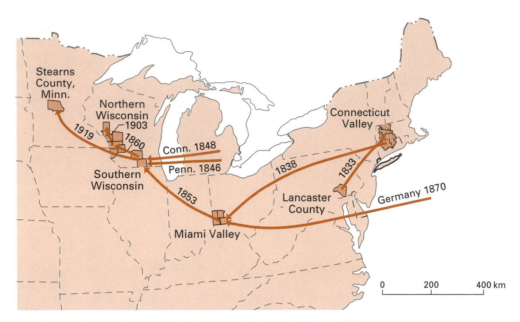

Figure 3.3: Diffusion of cigar tobacco production areas in the northern United States. Beginning in southeastern Pennsylvania, cultivation diffused to the Connecticut Valley of Massachusetts and Connecticut in 1833; a single migrant then diffused cultivation from the Connecticut Valley to southwestern Ohio in 1838, with later German immigrants reinforcing the tradition. Diffusion to southern Wisconsin occurred independently from three sources between 1846 and 1853, namely Pennsylvania, Connecticut, and Ohio (there was also an earlier 1838 diffusion, not shown on the map, from a tobacco area in New York). Finally, there were later diffusions west and north from southern Wisconsin toward northern Wisconsin in 1860 and in 1903 and also to central Minnesota in 1919.

Source: Adapted from K.B. Raitz, 'Ethnicity and the Diffusion and Distribution of Cigar Tobacco Production in Wisconsin and Ohio', *Tijdschrifte voor Economische en Sociale Geografie* 64 (1973):296. Copyright © Royal Dutch Geographical Society.

An ambitious study by Sauer (1952) expanded earlier local studies of plant diffusion in Mexico into a thesis about agricultural origins and diffusion in a global context. Central to this argument was the claim that cultural changes such as the development of agriculture occurred in areas that lacked environmental stress. It was with this argument in mind that Sauer proposed two areas of agricultural origin—southeast Asia and the lands around the Caribbean, both areas offering food from water and land. Vegetative planting cultures developed in and diffused outwards from these hearths. Although most subsequent work on the origin of agriculture does not provide compelling support for these ideas, they remain as a stimulating contribution to an important debate. (See Your Opinion 3.2.)

MIGRATION AND DIFFUSION

Cultural geographers have also been concerned with the movement of a cultural group and the cultural diffusion and related landscape change associated with that movement. The distinction between migration (the movement of people) and diffusion (the movement of a culture trait that may or may not involve the movement of people) is rarely employed in these geographic analyses, a reflection of the fact that the central interest is the effects rather than the process of diffusion. Migration and diffusion links were summarized by Salter (1971a:3–4): 'The cultural geographer views man's mobility with a tripartite perspective: the catalyst for movement, the effect of movement on trait or people in motion, and the consequences of such movement.'

A study of Amish populations addressed aspects of all three of these concerns (Crowley 1978). The Protestant Anabaptist religious movement began in 1525 and included the establishment of the Mennonite religion, from which the Amish group, led by Jakob Amman, split in the 1690s. Amish populations migrated throughout much of western and central Europe by the early nineteenth century, largely as a consequence of religious persecution. About 500 Amish moved to North America between 1717 and 1750 with a further 1500 moving between 1817 and 1861. The first group settled in southeastern Pennsylvania, while members of the second group settled in Ohio, Illinois, Iowa, and Ontario. From these original North American locations, Amish settlements have diffused throughout a much larger area of North America and also into Latin America, often moving long distances, often settling on the frontier, sometimes experiencing failures, and further subdividing into different groups. Despite this diffusion process, most settlements continue to be in the original hearth areas; thus, of ninety-eight Old Order Amish colonies in the United States in 1976, there were twenty-nine in Pennsylvania, eleven in Indiana, ten in Missouri, nine in Ohio, and nine in Wisconsin. This type of diffusion process is especially difficult to explain in general terms because many of the movement decisions reflected particular circumstances.

A Spatial Analytic Emphasis

Although a spatial analytic approach to the analysis of diffusion, as one component of the larger theoretical and quantitative movement in geography, was not popular until the 1960s, the seeds were sown much earlier in the pioneering work of the Swedish geographer, Hägerstrand.

THE WORK OF HÄGERSTRAND

Hägerstrand studied the diffusion of new agricultural practices, innovations, by mapping their spatial patterns through time and, further, by seeking to understand the process of diffusion that was responsible for the changing patterns. It was in seeking to understand process that original concepts and procedures of analysis were introduced.

Your Opinion 3.2

Sauer's contribution to the question of the origins of agriculture can be summarized as 'necessity seldom mothers invention'. What is your reaction to this way of thinking? Are you inclined to agree that an environment offering a sedentary lifestyle and food surpluses affords the necessary leisure time for creativity and cultural change? Similarly, do you feel that human groups need to cope with physical environments such that a difficult environment, one requiring mobility and a full-time search for food, limits the opportunity for cultural change?

Moving beyond the primarily descriptive analyses being accomplished in the landscape school tradition in North America, a series of Monte Carlo **simulation** models was developed—the term 'Monte Carlo' referring to the fact that these models incorporate an element of chance. The basic model, derived from empirical work, assumed that the probability of an innovation being adopted in a particular location is related to distance from the location of first adoption. Hägerstrand thus described a probability surface, comprised of decreasing probabilities with increasing distance, and then simulated the spatial diffusion process.

Although the mechanics of conducting a Hägerstrand-type simulation might appear complicated on the basis of the preceding comments, the logic is straightforward.

- The process of diffusion of an innovation, such as a new agricultural practice, and therefore the changing cultural landscape, is explicitly interpreted as affected by both the prevailing pattern of communication in a region and by chance.
- A probability surface is constructed on the basis of assumptions about the process of communication flow. The simplest such surface assumes that information about an innovation is passed from one person to another. It is a probability surface precisely because chance is being incorporated into the process.
- **Surrogate** data, such as information on telephone calls, are used to calculate specific probabilities. The probability surface is, therefore, a distance decay surface.
- Once the probabilities are mapped, usually in a square grid format, the simulation process is conducted and a surface of adoption dates created.
- The result of the simulation is compared to the known real world pattern of diffusion, and a close

correspondence between the two is interpreted as an indication that the probability model, necessarily a simplification of reality, is a reasonable approximation of the diffusion process.

Related to this work was the recognition of a number of empirical regularities of the diffusion process. Three examples—the neighborhood effect, the hierarchical effect, and the S-shaped curve—are noted in Figure 3.4. This research emphasis was linked to a tradition in rural sociology. There were two principal contributions from this tradition.

- It was typically assumed that adoption of an innovation is not immediate upon receipt of the relevant information. Rather, there is an adoption process that proceeds through stages, such as those of awareness, interest, evaluation, trial, and, finally, adoption.
- It was recognized that individuals varied in terms of innovativeness, that is, in their willingness to adopt an innovation. The standard assumption is that individual innovativeness follows a normal statistical distribution, with a few innovators who often were opinion leaders accepting the innovation quickly, a few laggards resisting the innovation for some length of time, and the vast majority of the population placed between these two extremes.

The Amish lifestyle stresses humility, family and community, and separation from the world. The oldest group of Old Order Amish live in Lancaster County, Pennsylvania (*Pennsylvania Dutch Convention & Visitors Bureau*).

SPACE NOT CULTURE?

Studies that employed the methods pioneered by Hägerstrand were not typically cultural geographic in focus, favoring such topics as urban settlement and disease rather than specific landscape features that were the hallmark of the traditional landscape school approach. Probably the most substantive example of a cultural geographic theme being analyzed in this manner was a study of the diffusion of settlement in the Polynesian region (Levison, Ward, and Webb 1973), but this work, involving use of a comprehensive simulation model to describe and explain a regional history of cultural movement, failed to encourage cultural geographers to follow suit.

The spatial analytic approach initiated by Hägerstrand clearly lacks explicit cultural content despite the fact that he was himself a cultural geog-

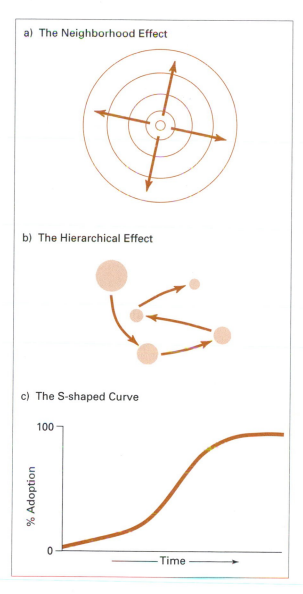

a) The Neighborhood Effect

b) The Hierarchical Effect

c) The S-shaped Curve

Figure 3.4: Three empirical regularities of the innovation diffusion process. A principal concern of the spatial analytic approach to diffusion research was the identification of regularities in the process of diffusion. In Figure 3.4(a), the neighborhood effect refers to a situation where acceptance of an innovation is distance biased, with adoption first achieved by a group of individuals living in close proximity to one another. Subsequent expansion occurs such that the probability of new adoptions is higher among those who live nearer the early adopters than among those who live farther away. This regularity reflects a particular pattern of communications, most characteristic in rural or preindustrial situations. It is useful to think of this regularity as similar to the result of throwing a pebble in a pond.

In Figure 3.4(b), the hierarchical effect refers to a different interpretation of distance, with initial adoption occurring in the largest urban center and subsequent diffusion occurring both spatially and vertically down the size ladder of urban centers. Knowledge and adoption of the innovation leaps from center to center rather than spreading in a wave fashion as is the case with the neighborhood effect.

There are two key considerations affecting which of the two distance effects is more likely to occur. First, the character of the innovation itself—a new variety of corn seed will diffuse throughout an agricultural landscape whereas a new style of fashion will diffuse through the urban system. Second, the larger cultural landscape in which the process is occurring—a face-to-face communication system favors a neighborhood effect while a technologically more advanced system favors a hierarchical effect.

Figures 3.4(a) and 3.4(b) describe alternatives for the spread of an innovation in geographic space. Figure 3.4(c), the S-shaped curve, describes the expected growth through time of innovation adoption. The curve summarizes a process that begins slowly, then picks up pace, only to slow again in the final stage. There is a close relationship between the normal distribution of innovativeness in a population and this S-shaped curve—a cumulated normal distribution produces an S-shaped curve. Note that all three regularities are not restricted to studies of diffusion processes; indeed, they are evident in various other areas of geographic analysis.

rapher in the Swedish tradition and was primarily interested in landscape change. It proved more attractive to North American quantitative geographers than to North American cultural geographers. Specifically, studies that employed the procedures introduced by Hägerstrand largely ignored the cultural content that was a significant component of the pioneering landscape school-inspired work, focusing instead on the mechanics of the diffusion process. Thus, the spatial analytic approach to diffusion, although explicitly derived from traditional interests, was primarily concerned with process and model construction, and less concerned with culture and cultural landscape. This preference was in accord with prevailing interests and preferred approaches in the human geography of the 1960s. Interest in spatial analytic diffusion research declined as one component of the decline of interest in spatial analysis generally after about 1970. Criticisms of this approach to diffusion studies became prominent with several weaknesses identified—the most notable of which concerned the exclusion of culture, indeed of any recognition of human distinctiveness. Consequently, and as one part of the changes taking place in cultural geography in the early 1970s, a rather different approach to diffusion that reintroduced an explicit cultural emphasis came to the fore. (See Your Opinion 3.3.)

Diffusion, Culture, and Power

RATIONALE

One response to criticisms of the spatial analytic approach was to recognize that a process of innovation diffusion does not only result in the presence of that innovation among a cultural group, but that it also affects the use of resources in space and over time. Thus, some innovations are time saving, causing substantial shifts in the daily time budgets of household members, while other innovations are time demanding, such as a village school in an agrarian society. Analyses that emphasize these aspects are less diffusion process oriented and more cultural change oriented. This is not a return to the traditional approach, but it does represent a renewed concern with culture.

Although cultural geographers and others had long acknowledged that all members of a population are not equally receptive to an innovation—recognizing, for example, the fundamental distinction between more innovative urban dwellers and more conservative rural dwellers—the spatial analytic tradition often assumed uniform behavior for reasons of simplicity. This third approach to diffusion explicitly acknowledges that some groups, such as the poor, less educated, aged, and unemployed, may be disadvantaged in having a limited access to innovations. Accordingly, studies in this tradition stress the social, economic, and political conditions over which most individuals have little control. This may require an understanding of the political state and of institutions, arguing that different diffusion processes operate in different contexts and have different causes. In some instances, those who are in a position to affect the innovative behavior of others may preempt valuable innovations.

DIFFUSION IN THE LESS DEVELOPED WORLD

Two analyses of agricultural diffusion in Kenya illustrate the difference between the spatial analytic and the more politically based approaches.

- Garst (1974) described the diffusion of six agricultural products in one district of Kenya, indicating the way in which both spatial diffusion and the characteristic S-shaped adoption curves were affected by a wide range of physical, cultural, and infrastructural factors, including the availability of information, the location of processing plants, and various attempts by authorities to restrict some practices, such as tea growing, to particular areas (Figure 3.5).

- Freeman (1985) discussed agricultural change throughout Kenya, focusing on the diffusion of coffee, pyrethrum, and processed dairy products, and concluded that pre-emption by early adopters was a critical part of the process, explaining both spatial spread and temporal growth (Figure 3.6). Pre-emption—meaning that those who are relatively privileged and hence able to adopt early institute policies and practices that effectively prohibit others from adopting—may often be a critical factor permitting the rise of landed élites and rich peasants within a larger landscape of poverty. In many parts of the less developed world especially, those who adopt new innova-

---

Your Opinion 3.3

On the basis of this brief account, does the spatial analytic approach to diffusion appear to be dehumanizing—to be excluding humans and their cultural identities—focusing instead on questions of process? Certainly, most cultural geographers voted with their feet and either continued to conduct more traditional analyses or sought some alternative approach. In the spatial analytic approach, diffusion is viewed, first, as a narrow process of change, namely the transformation of a landscape that is empty of a particular trait to one that includes that trait, and, second, as a process that relies on patterns of communication to achieve change. Certainly, the earlier interest in cultural change and cultural landscape change—shown by Sauer, Kniffen, the early work of Hägerstrand, and others—is lacking in most applications of the spatial approach.

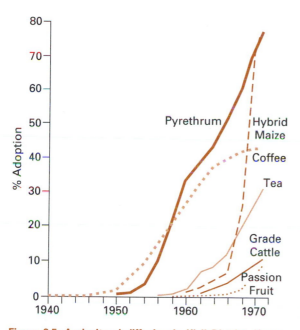

Figure 3.5: Agricultural diffusion in Kisii District, Kenya, 1940–71. The Kisii district of western Kenya, occupied primarily by the Gusii cultural group, is a densely settled agricultural area that has experienced significant landscape change since about 1940 as a result of the introduction of new agricultural practices. For six of these practices, Garst (1974) discussed the diffusion process, identifying the origin location, mapping the spatial spread, and plotting the temporal growth. The figure shows the adoption curves for coffee, pyrethrum, tea, passion fruit, grade cattle, and hybrid maize.

Innovation diffusion among the Gusii follows a classic pattern. It is generally characterized by an initial rapid outward spread of adoption at low intensity with relatively little contrast between adopting and non-adopting areas. Later, new diffusion nodes appear at scattered locations while other nodes develop into peaks of higher adoption above the general low adoption level. As the peaks of higher adoption approach 100 percent the adoption surface becomes highly irregular. Finally, the peaks spread out forming plateau-like surfaces of saturation acceptance that soon coalesce, producing saturated regions (Garst 1974:311).

Source: Adapted from R.D. Garst, 'Innovation Diffusion Among the Gusii of Kenya', *Economic Geography* 50 (1974):303.

tions at an early stage may be entrenched élites who are able to transform initial profits related to early adoption into permanent profits. The impacts on the cultural landscape include a strengthening of the élites within a larger pattern of rural poverty, and a dramatic reduction in the number of adopters of innovations.

These and related themes have been addressed especially in the context of the diffusion of agri-

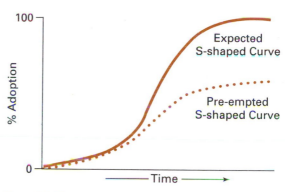

Figure 3.6: The pre-empted S-shaped curve. In a discussion of the importance of being first, Freeman (1985:17) contended that many diffusion analyses overlook 'the frequent cases of preemption of valuable innovations by early adopters'. With reference to the diffusion of coffee, tea, and processed dairy products in Kenya, pre-empted curves were explained by the ability of entrenched élites who were able to adopt early and then to restrict later adoption by others. Restriction is typically accomplished by means of production quotas, limiting the number of processing plants, and the manipulation of credit and farm extension, as well as social overhead capital.

Source: Adapted from D.B. Freeman, 'The Importance of Being First: Preemption by Early Adopters of Farming Innovations in Kenya', *Annals of the Association of American Geographers* 75 (1985):18.

cultural innovations associated with the green revolution. (See Your Opinion 3.4.)

Cultural Contact and Transfer

Much cultural and related landscape change has occurred as a consequence of population movement and related contact with others. Indeed, the expansion of European groups overseas between about 1450 and 1900 has directly or indirectly affected the lives and landscapes of peoples throughout much of the world. Accordingly, this discussion includes a general account of this European overseas movement and a consideration of related Aboriginal–European contacts. Consideration of the links between European overseas movement and global regional concerns, as presented in world systems theory, is delayed until the following chapter.

One feature of the material included in this section is the emphasis on the United States and Canada, along with some references to central and southern parts of the Americas, South Africa, Australia, and New Zealand. In this respect, the discussion is a reasonable reflection of the interests of cultural geographers.

Europe Overseas

'How strange it is to find Englishmen, Germans, Frenchmen, Italians and Spaniards comfortably ensconced in places with names like Wollongong, (Australia), Rotorua (New Zealand), and Saskatoon (Canada), where obviously other peoples should dominate, as they must have at one time' (Crosby 1978:10). Indeed, before about 1450, Europe, Asia, and Africa were linked only by a few land trade routes, while the Americas and Oceania were separate regions; only the Islamic civilization had spread significantly by sea, with expansion east of the Arabian hearth to the southeast Asian island region. However, after about 1450, five European countries—Spain, Portugal, France, the Netherlands, and Britain—embarked on overseas movement to areas outside of Europe previously unknown to them.

WHY EUROPE?

European overseas movement was motivated by more than curiosity as to what lay beyond the horizon; the colonization was not only scholarly, it was also political and economic—in short, it was about power and wealth. Acquisition of land and resources was probably the most compelling motive.

Two fundamental issues, addressed primarily by historians, are why exploration took place in the second half of the fifteenth century, and why this overseas movement started in Europe. After all, many of the advances in late medieval European culture such as the compass and astrolabe were diffused from China, often via the Islamic world. Basic answers refer to the advances in navigation and increased understanding of trade winds and

ocean currents already made by Portugal in the first half of the fifteenth century, to the presence of government or merchant company support, to the desire to expand trading activities and seek wealth beyond the confines of Europe, and to the missionary zeal of Christianity, which was especially apparent in the context of the ongoing conflict with Islam. It can be argued that Europe was able to move overseas because the region is distinctive both in terms of physical geography and culture. The critical facts of physical geography include those of climate, mild summers and cold winters, and adequate rainfall for agricultural development. The cultural factors include those of political fragmentation, systematized investigations in science, technological advances, a competitive market, and the idea of individual property rights. An alternative argument might stress the proximity of Europe to North America, compared to the distance that Chinese or Islamic explorers needed to travel. This important debate is placed into a larger context in Chapter 7.

Necessarily, Europeans carried their conflicts, languages, religions, customs, and economic systems overseas. At first, the Europe that was transferred elsewhere was one characterized by feudal economic and social systems that privileged a few at the expense of many. As suggested in Chapter 2, many segments of the European population, including women, were seen as inferior beings, while the attitude to nature was essentially one of exploiting at will. Most of those who moved overseas, at least prior to the late eighteenth century, were male. Most of the colonial administrators were privileged people, while many of the colonists were religious dissidents persecuted in their home areas. The prevailing attitude to both lands and peoples encountered was in accord with the European **norms** and **values** of the time, and was

---

Your Opinion 3.4

You have probably noticed that the section on diffusion has not been able to provide a substantive account of regional landscape evolution related to diffusion. Why not? Because cultural geographers concerned with diffusion have tended to focus on specific questions and specific approaches rather than on larger regional issues, but also because there has been much interest in cultural contact and transfer as initiators of landscape change. Notwithstanding the claim by Carter that opened this section, it appears that there is much more that cultural geographers might accomplish in this area.

therefore not very different from what was happening in Europe—a not unimportant point. Land was to be used as desired, and weaker peoples were to be used as labor, even enslaved perhaps, and moved as desired. If the indigenous population was judged unsuitable as a labor supply, slaves might be imported. In brief, Europeans, or more correctly those Europeans who moved overseas, saw themselves as superior to those they met. The transition from feudalism to capitalism did little to change these general characteristics.

ECOLOGICAL IMPERIALISM

Of course, it was not only Europeans who moved overseas, it was also European plants, animals, and diseases—a process of demographic takeovers that can be described as ecological imperialism. Europeans were accompanied by 'a grunting, lowing, neighing, crowing, chirping, snarling, buzzing, self-replicating and world–altering avalanche' (Crosby 1986:194). Indeed, many of the most aggressive plants in the overseas temperate humid regions today are of European descent, while horses, cattle, sheep, goats, and pigs were all introduced to overseas areas by Europeans. But it was the introduction of disease, notably smallpox, which was to prove devastating to Aboriginal ways of life. The effects of European diseases were overwhelming, especially in the Americas, which had been isolated from the larger world for an extended time. Aboriginal population estimates for the Americas prior to European contact are uncertain, although Denevan (1992) estimated an 89 per cent reduction between 1492 and 1650 (from 53.9 million to 5.6 million).

The effects of disease varied spatially. Demographic collapse was not usual in either Africa or Asia, which were not as isolated from Europe as were the Americas. Nor did it occur in those more isolated areas, such as New Zealand, that were settled by Europeans after the introduction of vaccines. Even in the Americas where some groups such as the Yahi of California and the Beothuk of Newfoundland disappeared, there were others such as some Maya groups in Guatemala that survived relatively unscathed. Not surprisingly, the debates about population numbers in the Americas prior to 1492 and about the role played by disease in subsequent declines are far from resolved with much of the debate linked to the divergent views of history that a choice of numbers implies.

Aboriginal-European Contacts: Changing Understandings

The long history of human movement has included numerous instances of different cultural groups coming into contact with one another. Contacts between foraging groups or shifting agriculturists were often an incidental consequence of the movement that was an integral part of their lifestyle. Most such contacts were between groups with broadly similar cultures, including comparable levels of technology. In many such cases one result of contact was a gradual agreement, through experience, on territorial limits. Another result was the transfer of ideas and goods, a process of **transculturation**. In other words, a series of linked cultural regions came into being.

The Maoris, who constitute about 9 per cent of New Zealand's population, live mostly in North Island. Many continue to participate in tribal activities in order to maintain their cultural ties (*New Zealand and Australia Travel Information Service*).

However, the context for European movement and the resultant contact with Aboriginal groups was quite different. As noted, the principal motivations for movement included the acquisition of power and wealth, while the expansion was one of cultures that typically viewed themselves as superior to those they contacted. Further, European newcomers were representatives of Europe and not simply fishermen or traders. The commercial system of early capitalism was both a cause and a component of contact. Aboriginal–European contacts were, then, of a different order from previous contacts between cultural groups—they involved very different cultures that had very different motivations for their behavior at the time of contact. The dissimilar cultural views of Europeans and Maoris are diagrammed in Figure 3.7.

An overall context to facilitate understanding of the specific details of cultural contact situations was provided by Meinig (1986:205–13). The proposed usual sequence was contact, followed by Aboriginal population loss and changes to social and ecological patterns, followed by the appearance of some form of dominance–dependence relationship, followed by some population recovery and further changes to social and ecological patterns, and followed finally by some cultural stabilization. At any given time, different areas contacted by Europeans were in different stages in this process of cultural change.

DIFFERENCE AND INFERIORITY

But there is another aspect to this discussion. The accounts written by Europeans were not a fair reflection of Aboriginal peoples and Aboriginals were not the unchanging peoples they were once thought to be. The obvious but typically neglected point is that the European view was necessarily fla-

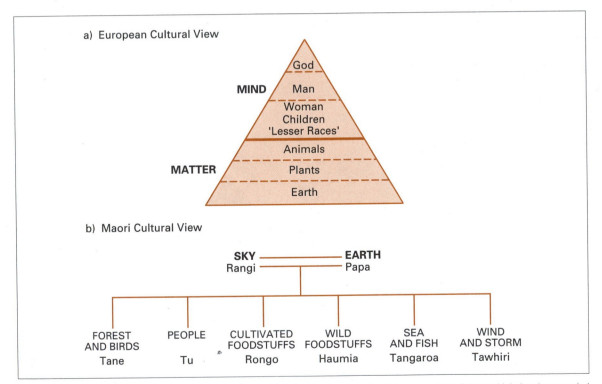

Figure 3.7: European and Maori cultural views. Figure 3.7(a) depicts the generalized European view of the world during the extended period of overseas expansion. There is a clear separation of humans and nature. As an indigenous group, Maoris are included in the 'lesser race' category. Unlike the European view, the Maori view is premised on a close relationship between humans and nature, with nature providing humans with both material and spiritual sustenance.

Source: Adapted from E. Pawson, 'Two New Zealands: Maori and European', in *Inventing Places: Studies in Cultural Geography*, edited by K. Anderson and F. Gale (1992): 18. Copyright 1992.

vored by ideas and assumptions that were a part of their own background and experience. It is because their cultural baggage flavored their perceptions that Europeans frequently lacked respect for Aboriginal cultures and values. Eliades (1987:33) stated: 'From the beginning of white–Indian contact, the Europeans, with only a few exceptions, refused to see the richness and diversity of Indian cultures. Instead of perceiving and accepting the native cultures as different, the Europeans saw them as inferior.' But not all observers accept this generalization.

- Sauer (1971:303) arrived at a rather different conclusion: 'Throughout the eastern woodlands, the observers were impressed by the numbers, size, and good order of the settlements and by the appearance and civility of the people. The Indians were not seen as untutored savages, but as people living in a society of appreciated values.'
- Similarly, with reference to prairie Canada, Friesen (1987:18) observed: 'when contact was made, Europeans were quite prepared to recognize both the legal existence and the military power of the native peoples, to accept native possession of the land, and to negotiate with them for privileges of use and occupancy by the customary tools of diplomacy, including, of course, war'.

But the prevailing consensus in the scholarly literature continues to emphasize the negative assessments made by the European newcomers. *Simply expressed, difference becomes inferiority.* A logical extension of this argument is that Aboriginal voices have been largely unheard because of the power relations inherent in the process of contact and in subsequent events.

The conceptual background for much current research in this vein is derived from contemporary social theories, including the perspectives of poststructuralism and postmodernism. Although these approaches are more fully addressed in Chapter 7 of this book, a brief context is provided at this time. As stated, many Europeans overseas commented on what they observed, with these commentaries informed by their ideological baggage. How, then, are contemporary readers to understand these commentaries? It is helpful to approach this diffi-

cult issue by thinking in terms of a coordinating grid defined by a perspective that was culturally specific—usually White, male, middle class, and with a specific national identity. Perceptions and descriptions of Aboriginals are to be understood by reference to this grid, to what Geertz (1973:3) described as a 'web of significance'. In other words, the accounts provided by Europeans are not to be treated as objective, factual, and truthful because such qualities cannot be achieved; rather, the accounts are mere representations, one of many possible understandings of reality. Work in the poststructuralist and postmodernist traditions often claims, then, that knowledge is necessarily uncertain and that there is no one truth waiting to be discovered by the contemporary scholar. This idea is central to much current work on cultural contact scenarios. It is now usual to emphasize how the voices of the **subaltern** were usually omitted or misrepresented in official documents.

There is a discussion of changes in the scholarly understanding of one component of culture contact, that of cultural change, in Box 3.2. As noted, the larger issue of particular ways of interpreting empirical evidence is approached in greater detail later in this book (see especially the accounts of feminism and postmodernism in Chapter 7 and see also Box 8.1).

Aboriginal-European Contact: Cultural Change

The contact between incoming European populations and Aboriginal populations resulted in both groups experiencing cultural change, and in many cases it is not possible to clearly separate the roles played by the two groups in the details of the changes that followed contact. Certainly, the cultural change that took place involved both groups and did not simply involve the less dominant group almost unwittingly changing to accord with the more dominant group. Thus, Europeans often adopted selected Aboriginal cultural traits that were judged to be to their advantage in the new environmental setting. In early French Canada, for example, the birch-bark canoe was a critical borrowing as it was made from local materials, was able to carry heavy loads, and yet was light enough to be carried when needed. More generally, much

European exploration relied extensively on the knowledge provided by Aboriginal groups; in Australia, for example, inland movement by Europeans was dependent on Aboriginal knowledge of environments and of other groups encountered.

RESISTING CHANGE, ACCEPTING CHANGE

'The Indian bands of southern Saskatchewan . . . had by no means been converted into the competitive agrarian individualists sought by their white guardians. Much had been said to them of the virtues of agriculture. Farming, and not the supposedly demoralizing pursuits of hunting and fishing, was identified as work' (Raby 1973:36). Although contact typically prompted some cultural change by the Aboriginal group, the specifics of such change were often the choice of the receiving group rather than an imposition by the incoming group. Indeed, as this quote suggests, there was often resistance to Europeans' efforts to impose radical cultural change.

Not only did Aboriginal populations resist some of the European attempts to impose change,

they also often accepted only those aspects of European culture that were thought to be to their advantage and that involved the least disruption to their existing cultural context. For example, in the Queen Charlotte Islands of British Columbia between 1774 and the 1860s, the Haida experienced substantial population losses because of disease, a dramatically changed settlement pattern, and detailed changes to their seasonal cycle of activities. The settlement pattern changes involved both abandonment and fusion of villages as a result of population loss and the desire to function better within the fur trade. Impacts on the seasonal cycle involved changes in detail that did not affect the principle of seasonality (Figure 3.8). This was because the cycle served important cultural functions, thus resulting in a strong resistance to substantive change. This is an example of change occurring only if such change was acceptable to the Aboriginals. Some other possible changes, such as the introduction of a sedentary way of life that was forcefully advocated by missionaries, were rejected because they were not in accord with Haida desires.

Box 3.2: Introducing Cultural Change?

Not surprisingly, scholarly understanding of Aboriginal-European contacts has changed a great deal since the first accounts appeared—these were produced by Europeans and necessarily were from a European perspective. The characteristic view, shared by historians and anthropologists, was that civilization encountered savagery. From this point of view, Aboriginals had no history because their cultures were unchanged for long periods prior to contact—in other words, their history began with contact. Static cultures and their related unchanging cultural regions were often assumed to be simple responses to physical environment. This view was gradually replaced by the idea that a more complex civilization encountered a less complex civilization, a distinction that was explicit in a 1930 study of the Canadian fur trade (Innis 1930).

Of course, the suggestion that there was Aboriginal cultural change prior to contact meant that an understanding of Aboriginals before contact could not be based on the observations of Europeans, for those Europeans were describing a dynamic, not a static, culture. Further, a traditional **ethnography** necessarily described cultures after, rather than before, contact, and was there-

fore describing a culture affected by contact. Clearly, archeological data are needed for descriptions of Aboriginal groups prior to contact. Certainly, there is today much archeological evidence to suggest that many of the apparent results of contacts were not really dramatic revisions of earlier culture, but rather logical extensions of already occurring change—that is, change that was occurring before contact. For Huron groups in Ontario, for example, there was 'a significant revival of intentional trade in the late prehistoric period and it was along the networks that supplied traditional prestige goods that European materials first seem to have reached the interior of eastern North America' (Trigger 1985:162). Of course, European movement and diffusion of their goods generated changes, but many of these changes need to be understood in the wider context of already evolving Aboriginal cultures. Further, it is being increasingly recognized that it is not appropriate to assess the consequences of contact for Aboriginals in terms of such values of European culture as progress and change. Certainly, the argument of the traditional cultural enrichment hypothesis needs to be approached with caution as it is both simplistic and Eurocentric.

Similar conclusions are evident concerning the impacts of fur trading on the Canadian prairie interior where continuity of traditional ways was a feature of the early fur trade period. The European trade was conveniently fitted between established annual migrations that followed changes in animal location and plant resources. It is increasingly evident that Aboriginals were often very careful in their interactions with Europeans, adapting to the new challenges posed and often regarding the newcomers with disdain.

IMPOSED CHANGE

In some areas there were active attempts to alter Aboriginal lifestyles. In Guatemala, for example, the military contact between 1524 and 1541 involved a Spanish takeover of both land and people, followed by conscious efforts to impose a Spanish way of life on the conquered groups. This forced cultural change involved both a process of obligatory resettlement that invariably severed ties with ancestral lands, and heavy demands on native labor. Relations between Spanish and Aboriginals can be summarized as those of oppressor and oppressed. An institutionalized exploitation was characteristic; the hacienda system, for example, involved new tools, crops, and animals. Certainly, the colonial experience must have been a traumatic experience for many Aboriginal groups, especially when it was reinforced through an extended period of time. However, attempts to impose change were not always successful. Spanish control of Aztec populations in Mexico did not prevent a successful retention of many tradi-

The Haida village of Masset, as photographed in 1879 by O.C. Hastings. Masset is a small fishing community near the northeastern corner of Graham Island, the most northerly of the Queen Charlotte Islands (*Royal British Columbia Museum* PN *10980*).

HAIDA SEASONAL CYCLE 1774

Nov.   Dec.   Jan.   Feb.   Mar.   Apr.   May   June   July   Aug.   Sept.   Oct.

In Permanent Villages
Sea Lion Hunting—Some Men
Canoe Carving—Some Men
Trading with Other Groups at the Nass R.
Preparation for Fishing
Berry Gathering—Women
Fishing at Summer Camps
Seal Hunting—Some Men
Salmon Fishing—Some Men

HAIDA SEASONAL CYCLE 1915

Nov.   Dec.   Jan.   Feb.   Mar.   Apr.   May   June   July   Aug.   Sept.   Oct.

Crafts and Mission-Oriented Activities
Berry Gathering—Women
Working in Canneries in Alaska
Farming
Salmon Fishing

Figure 3.8: Haida seasonal cycle, 1774 and 1915. In addition to substantial depopulation and related abandonment of villages that began following the first European contact in 1774, the Haida of the Queen Charlotte Islands experienced some detailed changes to the activities of their seasonal cycle. The cycle provided a framework for social and economic life, with the regular movements supporting the basic rhythm of life and confirming the human continuity with nature.

   Although the seasonal cycle depicted for the precontact situation in 1774 and that depicted for the postcontact situation in 1915 are different in detail, the basic logic of the cycle, namely spatial movement on a seasonal basis, is unchanged. Fur trade contacts after 1774 until the 1830s did little to change the basic seasonal pattern with the fur trade incorporated into the established framework of life. By the 1830s, a decline in the availability of sea otter for fur meant reduced Haida-European contact. Renewed contact was initiated in the 1850s with the discovery of gold, copper, and coal, but again these mining activities had little impact on the seasonal cycle. Even the gardening introduced by European missionaries toward the end of the nineteenth century was incorporated into the cycle.

Source: J.R. Henderson, 'Spatial Reorganization: A Geographical Dimension in Acculturation', *Canadian Geographer* 12 (1978):11, 16.

tional cultural values, and it was because of this that Aztec resistance to Spanish rule was less violent than in the cases of the Maya and Inca groups.

DEPENDENCE

One controversial issue concerns the view that the overexploitation of both fur-bearing animals and game by North American Aboriginals was caused not by a developing dependence on European goods but rather by the new Aboriginal belief that animal spirits were responsible for the ravages of disease. This new belief prompted a rejection of traditional beliefs accompanied by an embracing of the new Christianity and a deliberate extermination of animals (Martin 1978). This issue is part of two larger debates.

• There is a debate about the characteristic Aboriginal human and nature relationship prior to European in-movement, with the conventional wisdom being that Aboriginals were typically in harmony with nature, although there is evidence to suggest that some groups abused their environments and that Aboriginal peoples in general are not best described as ecological custodians.

• There is also a debate about the extent to which Aboriginal groups became dependent on the newcomers. Certainly, European culture often failed to satisfy Native aspirations. Referring to Australian Aboriginals, Reynolds (1982:129) stated that 'young blacks who went willingly towards the Europeans fully expected to be able to participate in their obvious material abundance. Reciprocity and sharing were so fundamental in their own society that they probably expected to meet similar behavior when they crossed the racial frontier.'

Cultural change, **acculturation** but not necessarily **assimilation**, was an inevitable consequence of population loss, of the attitudes of Europeans, and of the difference in levels of technology; also, some version of a dominancedependence relationship often arose. Although specific details vary from place to place, it was not unusual for Aboriginals to become economically and/ or politically dependent on the newcomers. Sometimes, as in much of Spanish America, this was effectively a forced dependence, while on other occasions, as in many North American fur trade areas, it was an unintentional effect of contact. Sometimes it occurred quickly, sometimes more slowly. Regardless, some form of dependence is a not uncommon consequence of contact that, in many instances, has carried through in some form to the present.

In the case of the North American fur trade, there is a debate between those who emphasize that Aboriginal groups retained the ability to make choices about economic lifestyles and those who see forms of economic and political dependence. Certainly, in general terms, the ever-increasing exploitation of a limited resource base meant that those Aboriginal groups that became reliant on the trade often suffered once the trade declined. European cultural traits, such as guns and steel traps, facilitated the exploitation that eventually resulted in the termination of the trade. It is possible that involvement in the trade may have prompted the rejection of traditional alternative ways of making

a living. Further, before the fur trade, Aboriginals exchanged commodities that they produced with each other, whereas once the trade began, only furs and European goods were exchanged. In the case of northern Manitoba, it is clear that the fur traders became wealthy while the Aboriginals received only a marginal return for their labor (Figure 3.9). (See Your Opinion 3.5.)

Shaping Landscapes

Recall that several German geographers, including Schlüter and Ratzel, advocated analyses of landscape evolution at the turn of the nineteenth century such that a concern with landscape evolution is a central feature of the German cultural geography tradition. Similar ideas were evident in France with the writings of Vidal and his followers.

Your Opinion 3.5

Understanding the details and consequences of culture contact is far from easy. Facts are slippery and different ideologies prompt different evaluations of the available evidence. Do you agree that, from the outset, Aboriginals and Europeans entered into an uneven and misunderstood competition for land and resources? Europeans' desire to expand frontiers and to settle, in what were for them new areas, necessarily conflicted with the established understandings of Aboriginals and in many areas some forced movement occurred. In other areas, treaties were signed that, from the European perspective, gave them ownership of land, and that, from the Aboriginal perspective, merely allowed Europeans to share available resources. In some cases, then, treaty signing is an excellent example of the different understandings that the two groups had of land, of resources, and of their relationships with each other. The current disputes over Aboriginal land claims in many parts of the world reflect these quite different understandings.

- Schlüter was the principal advocate, defining geography as the study of the visible landscape as it changed through time: 'Schlüter was the first to raise the landscape forming activity of man to a methodological principle' (Waibel, quoted in Dickinson 1969:132).
- Ratzel stressed the importance of the evolutionary perspective also: 'Ratzel did not exaggerate the potency of the physical environment. . . . What saves him from such naiveté is the recognition of the time factor. . . . No one could emphasize more than Ratzel the force of past history' (Lowie 1937:120).
- Vidal introduced the term 'personality' into the geographic literature: 'geographic personality is something that grows through time' to create a distinct regional landscape (Dunbar 1974:28).

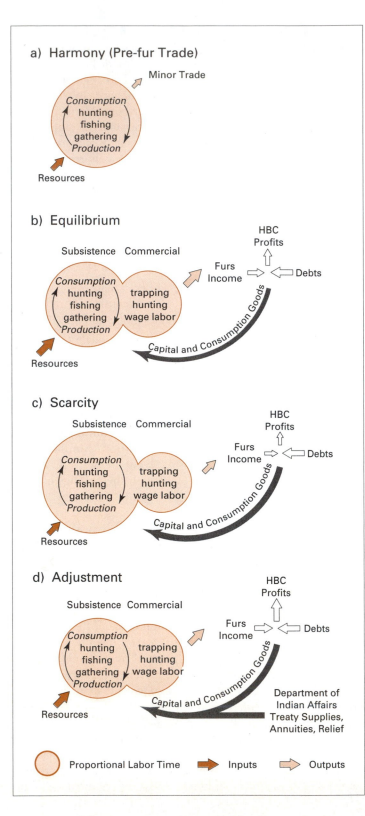

Figure 3.9: Changes in the Aboriginal economy of northern Manitoba. The inter-relationship between the two sectors of the Aboriginal economy, subsistence and commercial, changed over time. Before the fur trade a harmonious economy prevailed with production and consumption in balance, as shown in Figure 3.9(a). After about 1670, with the establishment of the Hudson Bay Company, a commercial economic sector developed and a delicate balance was in place (Figure 3.9[b]). Subsequent scarcity of subsistence resources meant that less time was available for the commercial sector, prompting the import of food products (Figure 3.9.[c]); for the company, it was cheaper to import food than to experience a drop in fur production. In the final phase, the Canadian government assumed responsibility for the import of food, thus both subsidizing the Aboriginal economy and supporting the production of fur (Figure 3.9.[d]). This sequence of stages supports the idea that the 'concept of paternalism, and not partnership, seems to capture the historic relationship' between Aboriginals and traders (Tough 1996:9).

Source: F. Tough, *As Their Natural Resources Fail: Native Peoples and the Economic History of Northern Manitoba, 1870–1930* (Vancouver: University of British Columbia Press, 1996):16. © University of British Columbia Press 1996. All rights reserved by the Publisher.

Indeed, the French *Annales* school of historical research, initiated by Febvre and Bloch in 1929, has close intellectual ties with *la tradition Vidalienne*.

In the United States, the landscape school built on both these German and French ideas to provide an English language conceptual basis for evolutionary analyses. Landscape concepts failed to attract much interest in British geography. The early work of the leading British historical geographer, Darby, for example, lacked both visible landscape and evolutionary content, with a focus on past landscapes rather than a focus on changing landscapes.

Natural landscapes are modified by humans to create cultural landscapes. With this basic thought in mind, cultural geography aims to understand contemporary landscapes as they are the outcome of long-term processes relating to the changing relations between humans and land. It is generally acknowledged that particular cultural systems evolve in particular places. Certainly, the literature of cultural geography includes many studies of areas where the primary concern is the changing landscape with, in some cases, one goal being to demonstrate that an understanding of past circumstances is a prerequisite to an understanding of the present. The following statement from Sauer, recorded by one of his students in 1936, emphasized this focus: 'We are what we are and do what we do and live as we live largely because of tradition and experience, not because of political and economic theory' (Sauer 1985:1).

A number of other valuable concepts, including those of core, domain, and sphere, of first effective settlement, of duplication, deviation, and fusion, of stages of regional evolution, of culturally habituated predisposition, and of preadaptation, have supported studies of changing landscapes, but these are more closely identified with the evolution of cultural regions. Accordingly, discussion of these concepts is delayed until Chapter 4.

Two Approaches to Historical Geography

Historical geographers have identified and employed a number of approaches to facilitate the study of change through time, often recognizing, however, that the differences between approaches are rarely substantive. Box 3.3 lists the relevant approaches as noted by three reviews of work in historical geography. Two of the more popular of these are briefly discussed, namely, narratives and cross-sections.

NARRATIVES

The use of narrative in both history and historical geography is a classic way to describe and perhaps explain change through time. Indeed, for some historians, all history is necessarily narrative in that a narrative both describes and explains change and in that an understanding of change is achieved through the writing of a story. Other historians disagree, seeing the narrative as failing to explain why things happen, and arguing that following a story to a conclusion is not the same thing as arguing that the conclusion necessarily followed from what came before. Historical and cultural geographers have chosen not to enter this debate, preferring instead to integrate a narrative approach with the cross-sectional approach.

CROSS-SECTIONS

The concern in the cross-sectional approach is to select particular moments in time, describe the geography of that time, and then to include a connecting narrative to another moment in time, and so forth. It is important to stress that studies such as these do not necessarily convey details of change through time, but rather focus on amounts of change between selected times—they are concerned with changing geographies rather than geographical change.

A study of Scandinavian historical geography employed a genetic and evolutionary approach. Five stages, cross-sections, were identified and a series of connecting narratives used to link successive stages (Figure 3.10). 'The object is to pause at certain stepping stones in the stream and to look at the surrounding stones. They will be scenes of the same place, but they will present patterns and distributions particular to different times. The changes between the scenes are explained by the processes—physical and human— that have characterized the intervening years' (Mead 1981:1).

Sequent Occupance

Following Sauer, most cultural geographers who have employed the cross–sectional approach have done so in a rather specific format, namely the approach known as **sequent occupance**. Studies in this vein recognize that a landscape remains relatively stable for a period of time, but then undergoes change that is both rapid and substantial leading into another period of stability. Thus, there is a succession of cultures and a related succession of cultural landscapes. Each period of stability reflects a particular cultural occupance, but is transformed into a subsequent period by some evolutionary or diffusionist process. One stage may be transformed into another because of the in-movement of a different cultural group, or because of some substantial change in, for example, technology or economic system. In this sense any landscape is a **palimpsest** in that it comprises the consequences of a series of different occupations. Whittlesey (1929:162) stated: 'The view of geography as a succession of stages of human occupance

---

Box 3.3: Changing Geographies: Some Historical Geographic Approaches

Historical geographers have embraced a wide range of approaches to their studies and there are many methodological writings that attempt to classify these approaches. These approaches refer to the principal historical geographic concerns, namely the study of the past and the study of change. Smith (1965:120) noted four approaches, two of which (those indicated with an *) refer to the study of change:

- operation of geographic factor in history
- reconstruction of past geographies
- study of geographic change through time*
- evolution of the cultural landscape*

Newcomb (1969) noted twelve approaches, six of which (those indicated with an *) refer to the study of change:

- historical regional geography
- areal differentiation of remnants of the historic past
- genre de vie
- theoretical model
- pragmatic preservation of landscape legacies
- past perceptual lenses
- temporal cross-section*
- vertical theme*
- cross-section-vertical blend*
- retrogressive*
- dynamic culture history*
- humans as agents of landscape change*

Prince (1971) noted three themes, namely the study of real worlds, perceived worlds, and theoretical worlds. With regard to real worlds, studies of past geographies, of geographical change, and of processes of change were identified and a total of fourteen approaches were listed, ten of which (those indicated with an *) refer to the study of change:

- past geographies
    géohistoire (reconstructing past landscapes in order to aid understanding of everyday life)
    urlandschaften (reconstructing the prehuman landscape)
    static cross-sections
    sources and reconstructions
    narratives of changes*
- geographical change
    sequent occupance*
    evolutionary succession*
    episodic change*
    frontier hypothesis*
    morphogenesis of cultural landscapes*
    agency of humans*
    rates of change*
- processes of change
    dynamics of change*
    inadequacy of inductivism* (an appreciation that facts do not speak for themselves)

Although historical geographers have found it possible to classify their preferred approaches in a variety of ways, it is clear that many of the approaches refer to change through time (including one that Prince chooses to include under the heading of studies of the past). This preference is notwithstanding the opposition that this approach faced from the then dominant regional approach to geography before about the mid-1950s. Certainly, the popularity of studies of change owes a great deal to the advocacy of Sauer and to his student, Clark.

establishes the genetics of each stage of human occupance in terms of its predecessor.'

A classic sequent occupance analysis in the Sauerian tradition is Broek's (1932) study of the Santa Clara Valley. This aimed to understand landscape change as a result of a succession of differ-

ent cultural occupances within a relatively short (less than 200 years) period.

- the first cultural occupance was Aboriginal
- the second was a Spanish occupation including missions and cattle ranches
- the third was an early American economy of wheat and cattle
- the fourth and final stage was dominated by horticultural activity

Looking back at this work in 1965, Broek (1965:29) noted that it would be appropriate to add a fifth stage:

- one dominated by urbanization

More generally, many world histories that consider human activity during the past 13,000 years or so stress the transitions from foraging to agriculture and from agriculture to industry as relatively brief periods of substantial change separated by a lengthy period of relative stability. Indeed, this was the approach taken in the Chapter 2 discussion of population and technology.

The differences between a historical geographic cross-sectional analysis and a cultural geographic sequent occupant analysis are worth noting. In the cross-sectional analysis, there is no suggestion that each cross-section is a stage separated from the previous and the following stages by some significant change of occupance. Rather, the cross-sections selected are intended to be representative of some larger period; relatively speaking, they are arbitrarily chosen. Thus, in a cross-sectional analysis the cross-sections are a particular moment in time and are separated by much longer periods of change. In a sequent occupance analysis, the stages described are lengthy and separated by briefer periods of rapid and significant change. Expressed another way, a cross-sectional approach emphasizes patterns at particular moments in time, albeit with process-oriented linking narratives, whereas sequent occupance places a higher priority on the processes operating to establish and maintain a particular stage of occupance. This is one difference between historical and cultural geographic traditions.

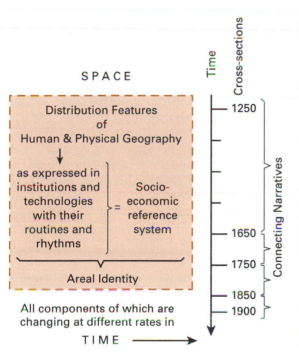

Figure 3.10: Cross-sectional historical geographies of Scandinavia. There are five cross-sectional analyses describing geographic patterns for the years 1250, 1650, 1750, 1850, and 1900. The time between cross-sections decreases because the pace of change increased through time with, for example, more change during the final 100 years studied than during the preceding 500 years. Each of the cross-sections is discussed in a systematic manner, with accounts of political, population, agricultural, resource, and communication geographies. 'At any given cross section, a number of people with particular sources of energy at their disposal will be viewing the options open to them and shifting the directions of Scandinavian development responsively. Simultaneously, they will bequeath a legacy of evidence for the geographical understanding of their times which will reflect their degree of literacy and numeracy' (Mead 1981:3). Connecting narratives that describe processes of change explain the transition from one cross-section to the next, and there is an additional narrative for the period since 1900. Each cross-section and each linking phase is discussed in a separate chapter. This approach is based on that developed by Darby (1973) in a study of the historical geography of England.

Source: W.R. Mead, *An Historical Geography of Scandinavia* (London: Academic Press, 1981):2.

Frontier Experiences

In relation to, and often integrated with, studies of ecological imperialism and analyses of the details and consequences of cultural contact as these are discussed in the preceding section, cultural geographers have paid much attention to European frontiers overseas, especially to matters of definition and to the cultural changes that occurred as European groups embarked on new lives in what were for them new places.

THE FRONTIER THESIS

Writing in the late nineteenth century, the American historian, Frederick Jackson Turner (1861–1932), introduced the **frontier thesis**: the idea that a **frontier** could be defined in terms of population density, with a figure of two per square mile sometimes quoted, and further proposed the view that frontiers were the meeting point between savagery and civilization—note the use of terms popular in anthropology at the time. Within history, these ideas have been further developed in major studies of the Great Plains and world frontiers.

Meinig (1993:258–64) proposed a highly suggestive alternative to Turnerian logic, namely a six-stage transition from a North American traditional system, through a modern phase, to a world system. The six stages are titled Indian society, imperial frontier, mercantile frontier, speculative frontier, shakeout and selective growth, and toward consolidation. Each of these stages is characterized by some particular cultural, economic, and political circumstances.

One substantive issue raised in the cultural geographic discussion of frontiers is the way in which the frontier concept, although usually ill defined, has been seen as a key concept to aid understanding of the larger European experience overseas. Thus, for Turner (1961:38), the frontier was where Europeans became Americans: 'American development has exhibited not merely advance along a straight line, but a return to primitive conditions on a continually advancing frontier line. . . . American social development has been continually beginning over again on the frontier.' Certainly, Europeans borrowed numerous cultural traits from indigenous groups. But neither the Turnerian assertions nor the examples of cultural borrowing

really address the issue of the European frontier experience.

EUROPE SIMPLIFIED?

'It is a striking insight that Europeans established overseas drastically simplified versions of European society' (Harris 1977:469). If this is correct, the question is: why?

- One possible answer concerns the movement of a fragment of larger European society and the subsequent (often lengthy) period of isolation in a new and different environment. Cultural change in Australia might be appropriately interpreted in terms of this fragment concept.
- An alternative explanation centers on the particular conditions, the confrontation with a new land, that Europeans experienced. In early French Canada, South Africa, and New Zealand, for example, Europeans with the usual strong sense of family and a desire for private ownership of land encountered areas of inexpensive land and limited markets for agricultural products, with the result that a homogeneous society based on the nuclear family dominated—a process of **simplification** occurred. The initial transplanted population might have been culturally differentiated, but the resultant settler society was culturally uniform. (See Your Opinion 3.6.)

TRANSFERRING CULTURAL BAGGAGE

Necessarily rejecting the Turner thesis because of the explicit environmental determinism, Sauer (1963:49) noted that the European frontier experience overseas was affected by 'the physical character of the country, by the civilization that was brought in, and by the moment of history that was involved'. This observation raises the larger questions of:

- What cultural baggage did an incoming population bring to an area?
- What happened to that baggage?
- To what extent did the shaping of landscape reflect cultural characteristics?

The simplification thesis suggests a loss of baggage, perhaps implying that settlers in the same area but from different backgrounds did not shape markedly different landscapes. *A loss of cultural traits and therefore of identity is indeed characteristic of many cultural groups after movement to a different area.* The essential explanation for such developments relates to the fact that the motivations behind most settler activity were economic, and appropriate behavior given the market economy tended to be more critical than retention of some cultural traits. Thus, Swedes and Norwegians in the Upper Midwest grew wheat while Germans in Texas adopted cotton and tobacco cultivation. Evidence from eighteenth-century southeastern Pennsylvania showed that differences in land use were not significantly attributable to differences in settler source area, while evidence from Kansas suggested an even more rapid loss of cultural identity than was the case in eastern North American areas settled earlier and under different circumstances.

But there is also a considerable body of literature demonstrating the effects of different groups in areas of European overseas settlement. Studies of European frontiers overseas have often examined the role played by the ethnic background of settlers. For many cultural groups, the most important institution that functioned to minimize trait loss was the church; indeed, in cultures primarily united by a religion, the rate of trait loss was often very low. Several homeland areas, notably the Mormon region in the intermountain West, continue to be relatively distinctive primarily because of the role played by the church. In the case of Swedish settlement in Minnesota, group

membership was able to influence economic decisions, but not sufficiently to prompt farmers to behave in an inappropriate economic manner; the community that evolved from cultural background rather than cultural identity itself was seen as influencing behavior.

IRISH SETTLEMENT IN EASTERN CANADA

An illuminating example of the question of retention or loss of cultural baggage is that of the Irish in nineteenth-century eastern Canada. Taking into account material folk culture and settlement morphology, the movement of Irish groups to Canada typically involved a loss of cultural traits, but the rate of loss varied substantially between three eastern Canadian locations. Retention of cultural baggage was greatest in the Avalon Peninsula of Newfoundland, and loss was greatest in the Peterborough area of Ontario, while the Miramichi River area of New Brunswick represented a middle ground. The different experiences in these areas reflected two principal considerations, namely physical environment and proximity to other incoming groups. Thus, the Avalon Peninsula encouraged retention of traits because the environment was similar to that of the source area and because there were few other newcomers with whom to interact. The environment of the Peterborough area, on the other hand, was unlike that of the source area and also included other cultural groups. The loss of traits in the Peterborough area did not, therefore, necessarily

Your Opinion 3.6

Although the simplification of European cultures on some overseas frontiers is well documented, especially prior to the large-scale movements of the nineteenth century, the significance of these changes is debatable. Is it possible that the frontier was of only minor importance in shaping North American society and civilization? This could be the case given that there were relatively few people involved on the North American frontier, and that the spread of commercial values and systems quickly overwhelmed the frontier experience in most areas such that an extended subsistence phase was not usual. Meaningful frontiers (that is, isolated fringe areas) only remained for any extended period in a few relatively undesirable locations, such as Appalachia or much of South Africa, which experienced population movement over great distances with limited subsequent immigration to assist in the initiation of a commercial way of life. On the other hand, in the Shenandoah Valley in eighteenth-century Virginia, frontier populations included many individualistic opportunists who moved quickly into the new culture of commercial agriculture, shedding unnecessary cultural traits and adopting necessary new traits. In such cases, the frontier experience was important to subsequent change.

imply a simplification of culture as there was trait borrowing from other groups to compensate for the lost traits.

GERMAN SETTLEMENT IN TEXAS

In the case of German settlers in Texas, imported traits were an important factor related to agricultural landscapes, despite the fact that some of these traits were lost soon after arrival, while some new traits were adopted from other groups. Certainly, there are numerous examples of particular agricultural practices associated with ethnic groups—recall the cigar tobacco production referred to in the discussion of diffusion—and with distinctive cultural landscapes. Trait loss may indeed have been usual in response to economic circumstances, but it was certainly possible for some of the new traits acquired to subsequently become associated with a particular group. (See Your Opinion 3.7.)

Evolutionary Regional Studies

Landscapes, cultural regions, or geographic personalities emerge in response to cultural occupance through time, becoming 'as it were, a medal struck in the likeness of a people' (Vidal, quoted in Broek and Webb 1978:32). With this evocative phrase, Vidal captured both the concept of *genre de vie* and the Sauerian idea of humans occupying physical landscapes to create cultural landscapes through time.

Writing in 1930 with reference to the American West Sauer (1963:45) offered a clear statement of the historical geographic or evolutionary approach to landscape study: 'The three major questions in historical geography are: (1) What was the physical character of the country, especially as to vegetation, before the intrusion of man? (2) Where and how were the nuclei of settlement established, and what was the character of this frontier economy? (3) What successions of settlement and land utilization have taken place?' Further: 'In order to evaluate the sites that were occupied, it becomes necessary to know them as to their condition at the time of occupation. Only thus do we get the necessary datum line to measure the amount and character of transformations induced by culture' (Sauer 1963:46). With these arguments in mind, there was a discussion of the physical landscape, the Aboriginal landscape, and the several European frontiers that moved west resulting in a series of cultural successions on the landscape. Frontiers were regarded as secondary cultural hearths.

In a discussion of Mexico, Sauer (1941b) employed the term 'personality' to refer to the relation between human life and land. Because the roots of Mexican life lie in a distant past, an understanding of contemporary Mexico was sought in that past and in the transformations that have occurred during that past. Thus, a pre–Spanish and a Spanish past were identified as the principal bases for the present cultural geography. A critical distinction was made between the two cultural regions of North and South that were evident before, during, and after the Spanish occupation (Figure 3.11). The North was described as less advanced and the South as more advanced. Sauer (1941b: 364) concluded: 'The old line between the civilized south and the Chichimeca has been blurred somewhat, but it still stands. In that antithesis, which at times means conflict and at others a complementing of qualities, lie the strength and weakness, the tension and harmony that make the personality of Mexico.' A work such as this, full of generalizations that were supported with factual detail, is characteristic of much early landscape school work—ambitious,

---

Your Opinion 3.7

What conclusions can be reached concerning this question of trait loss or retention by immigrant ethnic groups in a new environment? Do you think that trait retention by a cultural group would be most likely if:

- *the physical environments of the source and destination areas were similar*
- *a dominant goal was that of seeking a location for the group*
- *there were only limited contacts with others*
- *the group was socially cohesive with some set of shared values*
- *the impact of external institutions was limited*
- *the number of immigrants was substantial*
- *settlement was contiguous*
- *there was either very little or a great deal of prior economic experience*

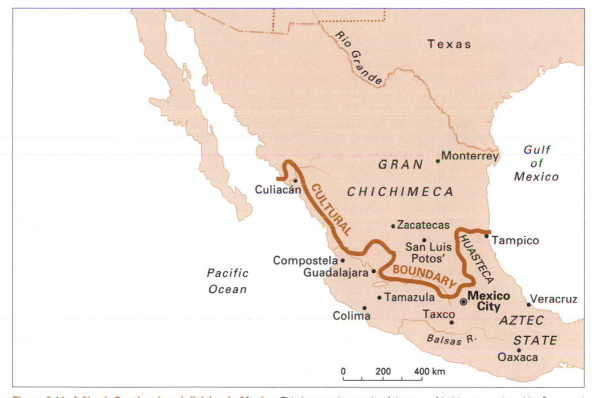

Figure 3.11: A North-South cultural division in Mexico. This is a good example of the type of bold map produced by Sauer and other cultural geographers in the early landscape school tradition. The boundary drawn is explained in cultural, not environmental, terms, although there is also a climatic distinction in that the North is mostly arid or semiarid but includes some alluvial valleys and good upland soils, while the South is generally better watered upland including some rainforest. The Spanish soon distinguished between the Aboriginals of the more cultured South, the Aztec region, and those of the less cultured North, the Chichimeca region. Indeed the southern region is one of the great cultural hearths of the world (see Figure 2.6). Further, the Spanish recognized the South as a region of peace and the North as a region of conflict. At the time of Spanish in-movement, the South was largely agricultural while the North was a mix of agricultural and foraging activities.

Source: C.O. Sauer, 'The Personality of Mexico', *Geographical Review* 31 (1941):355.

thoughtful, and provocative. Box 3.4 notes a few of the many other examples of work in this 'making landscape' tradition.

Reading the Landscape

Observing and reading a landscape is far from easy: 'It has been said that sight is a faculty but that seeing is an art that must be cultivated and kept finely honed through persistent field observation' (McIlwraith 1997:377) (Your Opinion 3.8). Nevertheless, there are two important, seemingly quite different but not really quite so different, traditions concerned with seeing and reading landscapes. The two are:

- An explicitly historical approach, essentially a form of narrative as described above, that is concerned with detailed field investigations of local landscapes. It is primarily the product of British researchers being closely associated with the historian, W.G. Hoskins.
- A more broadly conceived endeavor with strong historical content certainly, but also with some additional defining characteristics, such as a concern with landscape symbolism and with the vernacular in landscape. This is the *Landscape* magazine tradition and is closely associated with J.B. Jackson, a private scholar who described himself as a lay geographer.

THE LOCAL AND REGIONAL NARRATIVE HISTORY TRADITION

One of the classic works concerned with the making of landscape (unrelated to the landscape school tradition) appeared in 1955: *The Making of the English Landscape* (Hoskins 1955) was a trailblazing book that, although initially received without any real acclaim, led in the 1970s to a series of local county landscape histories and, indeed, to a television series. Meinig (1979b:209) quoted Hoskins as saying: 'I once wrote a book with the simple title of *The Making of the English Landscape*, but I ought to have called it *The Morphogenesis of the Cultural Environment* to make the fullest impact.' Certainly, the work is now considered to be a classic. This was pioneer work in the field of local history. The concern was to provide a detailed analysis of the landscape as it was made through time by humans through presentation of a series of chronologically arranged chapters, each dealing with a particular period of landscape-forming activity.

Despite the multitude of books about English landscape and scenery, and the flood of topographical books in general, there is not one book which deals with the historical evolution of landscape as we know it. . . . What I have done is to take the landscape of England as it appears today, and to explain as far as I am able how it came to assume its present form, how the details came to be inserted, and when. At all points I have tried to relate my explanation to the things that can be seen today by any curious and intelligent traveller going around his native land. There is no part of England, however unpromising it may appear at first sight, that is not full of questions for those who have a sense of the past (Hoskins 1955:13–14).

The originality of this work as a piece of historical scholarship lay in the use of the landscape itself as a source of information so much more valuable, given the aims of the study, than documents and books on local history. Hoskins wanted to read the landscape, not simply to read about it. There was also a recurring concern with localities in this work—a concern with the local rather than the larger regional or national scene. Further, Hoskins (1955:231) was greatly concerned about what he perceived to be some of the more unfortunate, even painful, recent developments in landscape: 'especially since the year 1914, every single change in the English landscape has either uglified it, or destroyed its meaning, or both'.

THE LANDSCAPE MAGAZINE TRADITION

Undoubtedly, the most substantive contribution to the literature concerned with the making of land-

Box 3.4: Making the Regional Cultural Landscape

There are several book-length works in historical geography that are closely related to the evolutionary landscape school approach and four examples are noted in this box.

- Davidson (1974) studied a small group of islands—the Bay Islands off Honduras—focusing on a description of the physical landscape and on a reconstruction of past landscapes, explicitly identifying culture as a cause of change. A sequent occupance focus was incorporated, with eight cultural stages noted, each affecting the material landscape. The principal cause of change over the long term was an ongoing conflict between English and Spanish.
- Williams (1974) discussed the making of the South Australian landscape, stressing the details and consequences of land clearing, such as swamp draining and woodland removal, and also the consequences of changing technologies and incoming cultural groups.
- An historical geography of the Netherlands by Lambert (1985) described the physical landscape, followed by an analysis of impacts on that landscape, with the resulting material landscape being the key concern. Notwithstanding the focus, this Netherlands study displayed relatively little concern with culture as a cause of landscape change; rather, economics was the central concern.
- A study of the Cistercian monastic order in medieval England is in a rather different vein, being described as an attempt to 'recreate the changing landscape of a culture rather than a cultural landscape' (Donkin 1997:248).

Several other types of study concerned with regional landscape creation are considered in Chapter 4.

scape, other than that associated with the landscape school, is the tradition initiated and sustained by Jackson that began with the publication of the magazine *Landscape* in 1950, a magazine that he financed, published, edited, and contributed much material to for the next seventeen years. Initially subtitled *Human Geography of the Southwest*, the magazine soon outgrew, first, the regional restriction and, second, the link with a single discipline to develop into an acclaimed outlet for articles adding to our understanding of landscape. There is little doubt that the initial idea behind and the early contents of this magazine struck a chord with many academics and others interested in landscape topics. The contents were both distinctive and thought provoking. Jackson clearly identified a void and, through *Landscape*, successfully filled that void. What were the characteristics of the tradition initiated by Jackson that combined to capture the imagination of readers?

- There was a recognition that landscapes are evolving, such that there is no such thing as a static or finished landscape, a suggestion that was, of course, in accord with the ideas of cultural geographers who were steeped in the landscape school tradition.
- Jackson also stressed that to change a landscape in a desirable direction necessarily implied changing the culture of the occupants, an idea that is always a controversial matter—recall the reference to social engineering in Box 2.3. For Jackson, as with Hoskins, the topic was an important one because of the sometimes undesirable directions being followed in contemporary landscape-forming activity.
- There was an emphasis on the vernacular, the everyday commonplace features in landscape (especially on houses) and it is in this tradition that Lewis (1975:1) employed the phrase, 'Common houses, cultural spoor.'

- Landscapes were seen as related to prevailing ideologies, and were interpreted in relation to the values of their occupants. They were, therefore, composed of material things, but also had symbolic character.
- Recognition of the symbolic quality of landscape introduces the need to read the landscape in order both to learn about that landscape and about the creators and authors of that landscape and about those who authored it. For Lewis (1979), a landscape was seen as an 'unwitting autobiography'. It is also an autobiography that is forever being revised.
- Jackson was also concerned with the ideas that landscape was both a reflection of the unity of humans and nature and a means by which it was possible to understand those who made it.

The tradition initiated by Jackson is original, but it would be misleading to give the impression that it is quite different from the work of more traditional landscape school geographers for there are clear links with the ideas of Sauer. Not surprisingly, Sauer wrote for *Landscape* magazine on several occasions. Recall also that Bowman (1934:149–50), in an early review of human geography as a social science, explicitly identified the landscape school concern as follows: 'Wherever man enters the scene he immediately alters the natural landscape, not in a haphazard way but according to the culture system which he brings with him, his house groupings, tools, and ways of satisfying needs.' Further, the original landscape school diagram prepared by Sauer referred to landscape features including housing plan and housing structure (Figure 3.1). Sauer was by no means unconcerned with the vernacular landscape.

The *Landscape* magazine tradition has proven stimulating to many cultural geographers both because of the familiar, essentially landscape

---

Your Opinion 3.8

Because landscapes are indeed shaped through time, might it be both possible and appropriate to observe a landscape and to seek to read the detailed history of the shaping of that landscape—to regard the landscape as a text? Rather than emphasizing the cultural causes and the landscape consequences in the now classic Sauerian tradition, why not investigate the contemporary landscape in terms of how each of the features came into being, asking questions such as: When was the feature added? Under what circumstances was it added? Why was it added? Who was responsible?

school content and because of the original ideas put forward by Jackson. Even a brief perusal of the cultural geography literature in general makes clear the impact of Jackson and his magazine. In addition to the explicit emphasis on change, there are the important concerns with symbolism and with the concept of landscape reading that clearly anticipate some of the work on symbolic landscapes that is discussed later in this book. Box 3.5 suggests that much of the work accomplished in this tradition might be best thought of as the geography of the everyday landscape.

MAKING THE AMERICAN LANDSCAPE

In introducing *The Making of the American Landscape*, Conzen (1990a:vii) noted: 'This book has its roots in the fertile bicontinental traditions of landscape study nurtured by William G. Hoskins and John Brinckerhoff Jackson, and it was written in the belief that nothing quite like it exists in the American literature, and that there is a place for it.' The decisions taken regarding organization of the book reflected the complex processes that are understood to have contributed to the building of the American landscape. Thus, there is an overview of the physical environment, there is an account of the approximately 15,000–year Aboriginal presence, and there are discussions of the principal colonizing cultures, namely the Spanish, French, and British. Relatively speaking, the Spanish and French cultural occupations were spatially restricted, leading to the formation of cultural regions, the distinctiveness of which is becoming less apparent through time. The British cultural

Box 3.5: Everyday Landscapes

The concern in much of the writing associated with the *Landscape* magazine tradition is a concern with the everyday landscape, with the majority of people in a landscape, and not with those few who are wealthy and exercise power. These everyday landscapes come into being and continually change because they are a reflection of the changing ideas, values, and needs of ordinary people.

- For geographers such as Aschmann, landscapes were studied 'as places where we could live and work and celebrate together' (Jackson 1997b:vii).
- Clay (for example, 1980, 1987, 1994) has produced original interpretive essays on the landscapes and character of urban areas that are very much in this tradition. 'Ready and waiting to be found out there in our visible landscape, there exists an observable and universal order that speaks eloquently to us, that will tell us far more of ourselves and our future than we can see at any one moment' (Clay 1987:1). From this perspective, we only have to see what is right in front of our eyes. This interpretation sees an orderly, indeed grammatical, landscape.
- Hough (1990) focused on the way in which traditional vernacular landscapes reflected the character of an area and the identity of people, thus emphasizing differences between places, while designed landscapes tended to negate the differences between places as a result of the fact that local populations were not responsible for their creation.

There are many other books and articles that are indicative of this *Landscape* magazine interest in the everyday world. A few examples of titles are:

- *House Form and Culture* (Rapoport 1969)
- *The Accessible Landscape* (Jackson 1974)
- *The Look of the Land* (Hart 1975)
- *Ordinary Landscapes* (Meinig 1979a; Groth and Bressi 1997)
- *Common Landscapes* (Stilgoe 1982)
- *Learning from Looking* (Lewis 1983)
- *Discovering the Vernacular Landscape* (Jackson 1984)
- *Wood, Brick, and Stone* (Noble 1984)
- *Right Before Your Eyes* (Clay 1987)
- *Real Places* (Clay 1994)
- *The Gas Station in America* (Jakle 1994)
- *People and Buildings* (Marshall 1995)
- *Landscape in Sight* (Jackson 1997a)

Certainly, capsule phrases such as these reflect this tradition of reading the landscapes that are created by the majority of people to serve the needs of their daily lives rather than a tradition of focusing on the behavior and landscape consequences of élite populations. An interesting corollary of this focus is that unusual landscape features were necessarily of lesser concern, and *Landscape* magazine occasionally published pieces that were critical of some conservationist groups related to their emphasis on the unusual at the expense of the everyday.

occupation involved a series of traditions as a consequence of the various population groups involved and of the variety of physical environments encountered. The most influential British tradition, spreading west over much of the larger American area, was that from New England and southeastern Pennsylvania, while the British plantation tradition expanded west throughout much of the American South.

These sections on the principal colonizing cultures are followed by accounts of the national settlement policy that was initiated after political independence, and of the implementation and effects of that policy on three principal physical environments, namely forest, grassland, and desert. Two landscape preferences are then identified and discussed. The first involved shedding cultural baggage, typically to create a landscape of single-family farms and homes, while the second involved

retention of cultural roots. Industrial, urban, and transport processes are then evaluated as these have involved changes in landscape. The book concludes with sections on the role of central authority, the importance of power and wealth, and the vernacular landscape. The multilayered organization of *The Making of the American Landscape* is an accurate reflection of the complexity of the subject matter, an acknowledgment of the fact that landscapes are shaped through time by a multitude of processes that are not always easily or readily distinguishable.

MAKING THE ONTARIO LANDSCAPE

A second valuable addition to the body of literature on the reading of landscape is the study by McIlwraith (1997) on two centuries of landscape change in Ontario. The stated concern is regional, not local, and the approach taken is very much in

The new City of Toronto, created on 1 January 1998, amalgamated seven municipalities: the regional government of Metropolitan Toronto and six local area municipalities (Toronto, North York, Scarborough, Etobicoke, East York, and York). The concentration of head offices in Toronto makes the city the economic heart of Canada (*Victor Last*).

the *Landscape* tradition, while the achievements are also not dissimilar to those of Hoskins. The aim is to read the landscape as a reflection of the way in which ordinary people lived their lives, hence there is an interest in both material and symbolic landscape. This aim is achieved through two sets of discussions. Thus, a context is established through accounts of attitudes to land, of historical matters, of the land survey system, of the process of determining place names, and of the technologies and resources employed in the making of the landscape. These contextual issues are followed by a consideration of particular landscape features and of the clustering of features. Reflecting the association established by Jackson between people and their homes, there are discussions of house architecture as it reflected the consistency of vernacular behavior, and also accounts of churches, schools, barns, fences, industrial buildings, and gravestones. Clusters of features that are considered include farms, roadsides, transport systems, and town streetscapes.

MAKING THE IRISH LANDSCAPE

There is a tradition of historical cultural geography in Ireland that stands somewhat apart from the various approaches identified so far, but that is explicitly concerned both with understanding how the Irish landscape was shaped through time and with understanding how the landscape reflects the culture of those who did the shaping. Indeed, one reason it is different is because it is so eclectic in character as it is based on an integration of ideas from geography, anthropology, and history. In this respect it stands apart especially from the English historical geographic tradition within which, until recently, very few attempted to integrate geography, anthropology, and history—the principal exception being Fleure (1951). The key figure in the Irish tradition was E. Estyn Evans (1905–89), whose influential university career extended from 1928 to 1968.

For Evans, an understanding of Ireland and the Irish landscape involved a wealth of interests and approaches. There were concerns shared with *la tradition Vidalienne* and with the related *Annales* historical school, especially regarding *genre de vie* and *milieu*. Such interests are not surprising in light of the plurality of cultural and regional identities evident throughout Ireland. There was also a concern with the evolution of the material cultural landscape in response to cultural occupance, in close accord with the landscape school ideas of Sauer. As noted, the willingness to incorporate these ideas was in marked distinction to their general neglect in the more dominant English historical geography tradition. In accord with both Vidal and Sauer, Evans employed the term 'personality' to refer to the idea that through a long history of human occupance, a distinctive character was impressed on the Irish landscape.

In addition to the approaches and interests prompted by these two principal influences, there were a number of other aspects of this tradition that effectively anticipated or paralleled work elsewhere. Thus, there was a focus on fieldwork and related observation of both people and landscape. Such approaches were advocated by Sauer and were also practised in the British local history tradition initiated by Hoskins. The concern with fieldwork also involved interests in common people and their way of life, and therefore in the related vernacular landscape. Further, the explicit linking of culture and landscape included recognition of the symbolic content of landscape, although this was expressly limited to the rural scene. The interests in vernacular landscapes and in landscape as symbol are, of course, central to the *Landscape* magazine tradition.

Although this eclectic approach to historical cultural geography is in accord with the related ideas of both Vidal and Sauer, the additional interests result in it being a distinctive tradition. It has been practised by Evans and others throughout much of the twentieth century with minimal formal interaction with other scholars and with generally little recognition outside of Ireland.

Concluding Comments

This chapter concludes with comments on the various evolutionary approaches to the study of landscape in the context of contemporary cultural geography and with a set of ideas that provide links to the cultural regional content of Chapter 4.

The evolution of the cultural landscape. For many cultural geographers, this one phrase immediately identifies the central concern of the landscape school tradition as initiated by Sauer in the 1920s. As is evident from the discussions in this chapter, many contemporary cultural geographers continue to work in both the general landscape school tradition and with reference to the more specific evolutionary concern.

- A historical orientation was the first of the seven 'persistent preferences' of cultural geography noted by Mikesell (1978:4).
- Williams (1983:3) noted that in 1936 Sauer described historical geography as 'the apple of my eye'.
- Donkin (1997:248) wrote: 'a fascination with "origins" is both primitive and widely shared—the product, perhaps, of "the essential time bond of culture rather than its looser place bond", some rather uncomfortable words of John Leighly, which I've often pondered'.

It bears repeating that this emphasis on time and change was in marked contrast to the regional geographic approach (as favored by Hartshorne and others) that dominated North American geography between about 1900 and the mid-1950s.

Writing about the physical landscape, Sauer (1925:36) noted that it was to be understood in terms of both space and time as it was continually changing in response to both geomorphologic and climatic processes, and also noted that the presence of humans and consequent introduction of cultural processes resulted in the making of a landscape that was no longer only physical but that also reflected cultural occupance. Physical geographers studied the physical processes generating change, while cultural geographers studied the cultural processes generating change. Although Sauer himself apparently showed little interest in such methodological statements after they were written, their impact on others seems undeniable. Indeed, notwithstanding the various earlier European traditions that Sauer acknowledged, it is not an exaggeration to claim that he was responsible for prompting the English language interest in both the evolutionary study of cultural landscapes and the historical geographic tradition of studying change through time. Mikesell (1978:3) asserted that 'Sauer acted as a catalyst for cultural geography, as well as an initiator of specific trends.'

Cultural geographers continue to be concerned with studies of changing landscapes and with the somewhat loosely defined historical geographic tradition. This is evidenced by a considerable body of literature focusing on the evolution of cultural regions, especially in the United States, and on the evolution of groups of similar cultural regions, known as cultural realms, that are evident at a global scale. These contributions are discussed in the following chapter.

But the general Sauerian influence is also accompanied today by a series of other concerns that are associated with developments in geography since the late 1970s. For example, the evolutionary study of cultural regions is enhanced by a consideration of the way in which humans relate to each other and thus identify themselves as members of groups. Similarly, the evolutionary study of cultural realms has been enhanced both by the arguments of world systems analysis and by the cultural studies tradition referred to in the first chapter, especially by postcolonial theory. These contributions are discussed in subsequent chapters.

It is notable that much of the discussion in this and the following chapter centers on North America. There are two reasons for this state of affairs. First, it is North America that has experienced the most dramatic cultural and cultural landscape transformation in recent centuries as a result of the burst of European expansion after about 1450. This expansion affected primarily temperate areas, notably North America, Australia, New Zealand, and South Africa. Second, the cultural geographic tradition is an American one and has accordingly been most readily embraced by American geographers. Although most other regions have a significant tradition of historical geographic research, there is less of an evolutionary cultural geographic and a related regional cultural geographic tradition.

Further Reading

The following are useful sources for further reading on specific issues.

Bowen (1996), Entrikin (1984), Hooson (1981), Macpherson (1987), Mathewson (1996), Mikesell (1968, 1969), Solot (1986), and Martin (1987a) on the methodological writings of Sauer.

Leighly (1976) and Macpherson (1987) on the specific Berkeley context for the growth of geography and related disciplines.

Sauer (1924, 1925, 1927, 1931) outlined the basic arguments of the landscape school.

Andrews (1984) and Berdoulay (1978) on the debate between Vidal and Durkheim.

Parsons (1979:13) and Platt (1962:39) on the links between Sauer and the anthropologists, Kroeber and Lowie.

Price and Lewis (1993a, 1993b) and Duncan (1993, 1998) on the superorganic in landscape school cultural geography.

Aschmann (1987:137) on the links between Sauer's methodological and empirical writings.

Rowntree (1996:130–1) on the diverse character of landscape school cultural geography.

Brookfield (1964) and English and Mayfield (1972) on the possible exclusion of human values and beliefs in landscape school cultural geography.

Unstead (1922) and Smith (1965:128) on historical geography as the study of the past.

Donkin (1997) and Williams (1983, 1987) on the approach to historical geography initiated by Sauer.

Hornbeck, Earle, and Rodrigue (1996), Conzen, Rumney, and Wynn (1993), and Butlin (1993) on the varied methodologies practised in historical geography. These works aid considerably in placing the landscape school emphasis in a larger historical perspective, making it very clear that Sauer initiated a 'sweeping and majestic brand of culture history' (Earle 1995:457).

Hugill and Dickson (1988) on diffusion as a multidisciplinary concern and Entrikin (1988) on diffusion and the naturalism debate.

Analyses of culture trait diffusion include Kniffen (1949, 1951a) on agricultural fairs; Kniffen (1936, 1965), Kniffen and Glassie (1966), Walker and Detro (1990),

and Jordan (1983, 1985) on house types and log construction; Stanislawski (1946) on the grid–pattern town; Leighly (1978) on place names; Seig (1963), Brunn (1963), and Carter (1977, 1980, 1988) on agricultural practices; Arkell (1991) and Carney (1994a, 1996) on musical styles.

Sauer (1968, 1969, 1970) on the origins and diffusion of agriculture.

Leighly (1954), a colleague of Sauer, discussed both the traditional cultural and the spatial analytic approaches to diffusion.

Hägerstrand (1951, 1952, 1967) on the spatial analytic approach to diffusion.

Gould (1969), Abler, Adams, and Gould (1971), and Haggett (1979) include comprehensive accounts of the simulation procedure developed by Hägerstrand.

Rogers (1962) on the rural sociological approach to diffusion.

Morrill (1965) simulated the spatial distribution of towns in Sweden, while Pyle (1969) analyzed the diffusion of cholera during three pandemics in nineteenth–century North America.

Blaut (1977) argued (1) that the traditional approach to diffusion provided the needed foundations for further theory development, precisely because of the breadth of the approach, and (2) that the spatial approach was too narrow because of the tendency to view innovations in isolation and to ignore the consequences of diffusion.

Carlstein (1982) on diffusion and use of resources.

Yapa (1977, 1996), Blaikie (1978, 1985), and Browett (1980) on agricultural diffusion in the less developed world.

Hauptman and Knapp (1977) contrasted Dutch movement and activity in the two areas of Formosa (Taiwan) and New Netherland in North America.

Fox (1991) on the rise of the modern European world; Jones (1987) on the European 'miracle' argument; and Blaut (1993b) for a critique of that argument.

Sauer (1935), Veblen (1977), Lovell (1985, 1992), and Cook and Lovell (1992) on the impact of disease on American populations numbers.

Trigger (1982, 1985) and Ray (1996) on cultural change in North American Aboriginal groups prior to European contact.

Vibert (1997:3–23) and Radding (1997) on how the voices of the subaltern were usually omitted or misrepresented in official documents.

Ray (1974), Judd and Ray (1980), and Peterson and Anfison (1985) on Aboriginals in the fur trade.

Cumberland (1949) on New Zealand; Clark (1959) on Prince Edward Island; and Darby (1973) on England are all examples of the cross-sectional approach.

Mikesell (1976) and Conzen (1993:33–7) on sequent occupance.

Webb (1931, 1964) on frontiers in the Great Plains and in a world context.

Mikesell (1960), Wishart, Warren, and Stoddard (1969), Eigenheer (1973–4), Hudson (1977), Christopher (1982), and Guelke (1987) on the spatial aspects of the frontier, and on the use of the one term to refer to both new land settlement and to cultural interaction.

Mitchell (1977) on the Shenandoah Valley frontier.

Clark (1959), Jordan (1966), Lemon (1972), Mannion (1974), Rice (1977), McQuillan (1978, 1990, 1993), Ostergren (1988), and Conzen (1990a) on the loss or retention of cultural traits, especially agricultural practices, by immigrant ethnic groups in North America.

Norton (1988) and Conzen (1990b) on general considerations relating to cultural trait loss or retention.

Early examples of evolutionary regional studies include those by Kniffen (1932), who focused on physical and later cultural landscapes of the Colorado delta area and recognized three stages of occupation and related landscape, and by Meigs (1935), who focused on the Dominican mission frontier of lower California, also identifying physical and cultural landscapes and the processes of change. Studies of Mexican and American landscapes are included in two valuable collections of the writings of Sauer, namely *Land and Life* (Leighly 1963), and *Selected Essays, 1963–1975* (Sauer 1981).

Other British work comparable to the tradition initiated by Hoskins includes the historical writing of Beresford (1957), the historical geographic writings of Darby (1940, 1956) on the drainage of the fens and the clearing of the woodland, and numerous historical geographic analyses of farms, fields, fences, and other features of the cultural landscape (see, for example, Baker and Butlin 1973; Slater and Jarvis 1982). A popular paperback book in this broad tradition is *The Penguin Guide to the Landscape of England and Wales* (Coones and Patten 1986).

Lowenthal (1968) on the perceptions of the American landscape that emphasized the overwhelming character, the great size, the wild and unfinished feel, and the formless and confused visual impression, especially in comparison to the landscape of England (Lowenthal and Prince 1964, 1965).

Mills (1997) on American physical and cultural landscapes and the way these have been interpreted in popular culture.

Widdis (1993) on the cultural landscape of Saskatchewan; Ennals and Holdsworth (1981) on the vernacular architecture of the maritime province area of Canada; and Francaviglia (1991) on mining landscapes in the United States.

Buchanan, Jones, and McCourt (1971), Glasscock (1991), Graham and Proudfoot (1993) on E. Estyn Evans; the classic work is Evans (1973) and for a collection of essays, see Evans (1996).

Regions and Landscape

This second thematically based chapter builds on material already presented in chapters 2 and 3. The discussion of humans and nature provides an important context because regions can be interpreted as one of the outcomes of the human and nature relationship, while the discussion of landscape evolution anticipated the fact that cultural occupance of an area and related landscape change often involve the creation of a **cultural region**. Indeed, the difference between much of the material in the previous chapter and that in the current chapter is one of emphasis.

The basic idea in this chapter is that groups of people who share some cultural characteristics in common and whose members live in close proximity to one another are likely to produce a landscape that reflects the shared characteristics—that is to say, a cultural region evolves. Such regions are evident at a variety of scales—from the local to the global—in response to the many different bases that humans employ to identify with others. Notwithstanding the presence of cultural regions at a variety of scales, it is important to stress at the outset that the relatively small, local, ethnically defined region and the much larger global **cultural realm** are parts of a single dialectic. Regardless of scale, the existence of a cultural region demonstrates the critical role in landscape creation played by individuals in their capacity as members of groups, and also the critical role that may be played by those members of a group who are able, for whatever reason, to affect the behavior of others to ensure a consistency in behavior among group members.

The sequence of material in this chapter is as follows.

- There is an opening account of the regional theme that includes an overview of this approach to geography and an evaluation of the approach as used by cultural geographers. The several types of region—namely formal, functional, and vernacular—are noted, and some of the problems and limitations of regionalization are identified. Also included are references to criticisms of what is sometimes described, rather inaccurately, as the traditional approach to region identification.
- A number of concepts that have been devised to facilitate the understanding of region creation at the subnational scale are outlined. These concepts build upon some of the discussions of shaping landscapes in the previous chapter.
- Principal examples of regional cultural geography at the subnational scale are discussed. This account includes discussions of vernacular regions and homelands, stressing the importance of a socially cohesive group in region formation. Shared religious beliefs and values most usually link such groups, which are often described as ethnic groups. Principal examples discussed are those of the Mormon and Hispano regions in the American West. There is also an account of ethnicity and related material landscape features.
- There is an account of the regional concept as it has been applied at the global scale, including a discussion of the way in which the modern world has evolved. Central to this discussion are several major analyses of civilizations and also the world systems theory, one aim of which is explaining the evolution of the modern political, economic, and cultural world system. This interest is clearly rather different from the more traditional aim of regional cultural geography, namely describing

and explaining particular visible and material landscapes.

- The contemporary outcome of global evolutionary processes is discussed. The pioneering 1951 global regionalization developed by Russell and Kniffen is presented, as are two other more recent attempts at regionalization.
- There is a brief set of concluding comments that reintroduce the question of how we proceed to delimit cultural groups and, accordingly, how we are able to delimit the regions that those groups occupy.

As noted, much of the discussion in this chapter flows from the preceding chapter on landscape evolution, but the chapter content also provides background material for discussions in subsequent chapters. Thus, the preadaptation concept resurfaces in the ecological focus of Chapter 5 and the homeland concept resurfaces in the Chapter 6 account of human behavior. Further, much of the content of this chapter informs ideas about identity contained in the Chapter 7 account of unequal groups and unequal landscapes, and also ideas about place contained in Chapter 8.

What Is a Cultural Region?

Geography has always been concerned with descriptions of parts of the surface of the earth. A regional approach was formally identified by Varenius in the mid–seventeenth century, explicitly seen as the geographic method by Kant in the late eighteenth century, and then proposed as the basis for the newly institutionalized discipline by Richthofen and others in the late nineteenth century. One consequence of all this activity was that regional geography was the dominant **paradigm** in the discipline during most of the first half of the twentieth century. Indeed, it is not unusual to regard the landscape school of cultural geography as an alternative view of the discipline—recall the Chapter 3 comments concerning competing views of the discipline. Notwithstanding the possibly competing concerns of regional geography and of the landscape school, cultural geographers fully appreciated the need to structure their studies of cultural landscapes in a regional framework. *Delim-*

iting regions was not the goal of cultural geography, but rather was a logical outcome of their primary interest in landscape.

Culture Areas in Anthropology

Interestingly, in stressing the value of a regional landscape focus, cultural geographers turned not to the leading regional geographers, such as Richthofen, Hettner, and Hartshorne, but rather to Ratzel and to related anthropological applications of the culture area concept. Culture area analysis was proposed by Ratzel, but first explicitly used in anthropology to delimit North American ethnic environments. For anthropologists, the concept was used initially as a means to establish some semblance of order on seemingly diverse cultural phenomena. In this sense, culture areas were one way to classify data. There were environmental determinist overtones in some of this work, with close links identified between natural and cultural areas.

Thus, a culture area was defined as: 'A part of the world where inhabitants tend to share most of the elements of culture, such as related languages, similar ecological conditions, economic systems, social systems and ideological systems. The separate groups within the system may or may not all be members of the same breeding population' (Foley 1976:104). In most of the anthropological studies the culture area concept was a tool to assist in the recognition of cultural wholes; cultural cores appeared in the most favorable parts of an area, followed by a diffusion of the culture throughout the area.

Regions in Geography

As discussed in the previous chapter, the landscape school incorporated evolutionary, regional, and ecological concerns. Of these three, the regional concern is the most confusing methodologically as it has been common to conceive of Sauer and Hartshorne, the foremost advocate of geography as the study of regions, as being pitted against each other. The differences between the two were not, however, based on substantive differences about the role of regions, but rather about the legitimacy of incorporating time in geographic studies—the rift between the two was complete in 1941 at the time of Sauer's major address on historical geography—

and also about the Sauerian position that studies of landscape should focus on visible material features. Although Sauer did not accept the regional approach advocated by Hartshorne, the need for regions in geographic study was not questioned. The difference of opinion on the matter of regions was essentially one of terminology, emphasis, and degree, and not one of substance. Indeed, there is some confusion regarding the use of four terms to refer to those parts of the surface of the earth that can be regarded as internally homogeneous and as different from the surrounding parts. (See Your Opinion 4.1.)

- *Landscape.* Sauer favored this term to characterize the geographic association of visible facts, but also used the terms 'area' and 'region'.
- *Area.* Anthropologists favored this term.
- *Region.* Following Hartshorne, and as used by many geographers outside of the landscape tradition, this term was seen as the basic concept of the discipline of geography until the 1950s.
- *Place.* This fourth term was also used by Hartshorne, but is now associated primarily with some of the more recent approaches to cultural geography.

Terminological confusion aside, Sauer was explicit concerning the definition of a geographic culture area:

The geographic culture area is taken to consist only of the expression of man's tenure of the land, the culture assemblage which records the full measure of man's utilization of the surface—or, one may agree with Schlüter, the visible, areally extensive and expressive features of man's presence. These the geographer maps as to distribution, groups as to genetic association, traces as to origin and synthe-

---

Your Opinion 4.1

Terminological confusion (always a source of irritation to students) is often the unavoidable consequence of different scholars favoring particular terms to express their ideas in order to distance themselves from others with similar but different ideas, although it can also be the consequence of rather casual writing. This is something that cannot be avoided in most social science. The terms 'landscape' and 'region' are used in the current discussions, with landscape as a term referring essentially to the visible scene, and region as a term referring to a part of the surface of the earth that has a similar landscape. Do you agree that this is a reasonable use of terms given the understanding of the landscape school that you gained from your reading of Chapter 3?

---

sizes into a comparative system of culture areas (Sauer 1931:623).

Further, Sauer employed this cultural area concept, usually favoring the term 'region' in his lecture courses. Box 4.1 distinguishes between three types of region, namely formal, functional, and vernacular.

Delimiting Regions

Having identified what is meant by the term 'region', the question arises as to how such regions are delimited. Alternatively phrased, what is the regional method? The basis for the answer that follows is the idea that delimiting formal regions (the most usual type) is essentially a process of classification, with classification as a procedure designed to impose some order on complex realities.

CLASSIFICATION

From a conventional scientific perspective, there are two approaches that are used to achieve a classification of formal regions.

- One approach is to classify all possible locations individually according to the criteria employed and then, where appropriate, to draw boundaries around groups of locations that are deemed to be similar, thus delimiting regions. This is described as grouping or classification from below and is a procedure that has often been informally applied in attempts to regionalize a national area, or, more simply, to identify a single region.
- An alternative approach is to treat all the possible locations as a single set, and then proceed to draw boundaries between smaller groups of locations by means of some predetermined criteria. This is described as logical division, or classification from above. Classifying in this manner

requires knowing a great deal about the locations so that sensible distinguishing criteria can be used. Most of the traditional world regional clas-sifications by cultural geographers and others are of this second type.

Interestingly, some efforts at regionalization made by cultural geographers rely on a procedure that is different from both of those associated with the exercise of classification.

- Rather than begin with the finished product (that is, the region or set of regions), an evolutionary approach has been used. Specifically, following Sauer, the aim is to identify a cultural hearth and then trace the diffusion of the culture outwards over some larger area such that a region evolves.

FOUR DIFFICULTIES

There are four general issues that need to be addressed in any attempt to demarcate a cultural region or a set of cultural regions.

- Regions are continually evolving and thus there is a concern with time and the dynamic aspect of region identification. This interest follows closely from the contents of the previous chapter, espe-cially from the idea that regions are ever-chang-ing outcomes of cultural occupance and transformation of landscape, but also relates to the Sauerian interest in identifying cultural hearths and tracing diffusion. This chapter includes two sections that focus on the evolution of cultural regions; one section refers to the sub-national scale, and the other refers to the global scale.
- Region identification also needs to consider ques-tions of spatial scale and cultural scale. Linking these two scales stresses the fact that cultural regions are, logically, usually regions occupied by distinguishable cultural groups, and are not just evident because of different visible landscapes. Indeed, most attempts at regionalizing the world are classifications of people rather than classifi-cations of visible landscape. The significance of

Box 4.1: Types of Region

From our cultural geographic perspective, the terms 'landscape' and 'region' are not necessarily inter-changeable, but they are closely related. For the current purposes, a cultural region may be regarded as an area that possesses a similar visible material landscape and/or a similar cultural identity. With this in mind, it is usual to distinguish three types of region.

First, a **formal region** is characterized by uniformity of a given trait or traits. They can be of any size. The cul-tural area concept as used by anthropologists and the regional concept as used more broadly by geographers are of this type. There are numerous criteria that might be employed to identify such regions. Characteristics of language, religion, and ethnicity may be employed because they are reflected in the landscape. It is usual, for example, to recognize a Mormon cultural region in the intermountain West. Alternatively, the concern may be directly with the landscape, as in the recognition of a corn belt or prairie landscape. Cultural geographers typ-ically identify formal regions employing multiple traits. Lewis (1991:606) noted that formal regions are created through human relatedness, that is, 'the notion that important commonalities unite certain groups of indi-viduals to varying degrees, while separating them from

those in other, similarly defined communities'.

Second, a **functional region** is one ranging in scale from a single home to the entire world, that in some way operates as a unit. Unlike formal regions, func-tional regions result from human connectedness rather than human relatedness. Clearly, a part of the surface of the earth that is integrated in a functional sense, such as a political unit or a religious settlement, is quite likely to also exhibit some landscape similarities as land-scape may be affected by the functional factors unify-ing the region. Accordingly, functional regions often overlap with formal regions in cultural analysis.

There is also a *third* type of region, the **vernacular region**, which is a locally perceived regional identity and name. Although these regions may exhibit some distinct visible material landscape, their distinguishing charac-teristics relate more to a perceived sense of identity than to a visible sense of identity. There is considerable over-lap between formal and vernacular regions, with many cultural regions being both formal and vernacular, and there may also be overlap with a functional region. In this chapter, the discussion of homelands reflects the interest in this type of cultural region. The term is also applied to small, relatively local tourist areas.

different scales, spatial and cultural, is that the tasks of regionalizing the world, regionalizing a country, and regionalizing a subnational area necessarily require the use of different criteria and involve different levels of expectation concerning the internal homogeneity of the various cultures and landscapes delimited.

- A difficulty that is apparent in most regionalization exercises concerns the precise location of boundaries. Most functional regions have relatively precise boundaries, but both formal and vernacular regions are usually much more susceptible to particular interpretations made by the researcher. The reality is, of course, that most cultural regions do not have sharp boundaries separating them from other regions. Rather, the boundaries are zones of transition with a gradual merging into neighboring regions. Although cultural geographers are well aware of this situation, there is often an understandable temptation to map regions with boundaries that imply a line of separation.
- The final difficulty is the most problematic. On what basis are regions delimited? What criterion is—or what criteria are—to be used? How is this choice to be justified? Of course, there are no simple responses to questions such as these. The criteria are typically a reflection of the specific aims of the regionalization—for example, if a set of religious regions is the desired outcome, then some religious criteria are employed. But statements such as these do not really address this issue. The key point is that multiple criteria might be legitimately employed in the act of demarcating cultural regions, and any interpretation of the resultant regionalization must proceed with the

criteria used in mind. There is always a danger of extending a regionalization beyond the purposes intended. What this point tells us is that there is no such thing as a correct system of regions.

These four difficulties are indeed real and often result in a regionalization being the personal product of an individual researcher, such that in many cases the boundaries of regions are located in a relatively arbitrary manner. Hart (1982:21–2), a leading advocate of regional geography, stated: 'Regions are subjective artistic devices, and they must be shaped to fit the hand of the individual user. There can be no standard definition of a region, and there can be no universal rules for recognizing, delimiting, and describing regions.' The extent to which this state of affairs is a problem is, of course, debatable as it is dependent on philosophical persuasion. The more humanistically inclined geographers are comfortable with individual researchers playing key roles, while more scientifically inclined geographers are much less comfortable. (See Your Opinion 4.2.)

THE DECLINE OF TRADITIONAL REGIONAL GEOGRAPHY?

Two quotes provide us with two very different opinions about the need for a regional geography.

- Hart (1982:1) argued for the study of some type of traditional regional geography, asserting that such work was 'the highest form of the geographer's art'.
- Thrift (1994:200) referred to the regional approach needing 'exhumation rather than resuscitation', and bemoaned the fact that it appeared to have

Your Opinion 4.2

Given these difficulties, is the idea that there are regions one that cultural geographers need to pursue? Certainly, the idea of regionalizing has proved a compelling one for many cultural geographers, not only because regions are an integral part of much geographic research generally but also because regionalizing is clearly a means by which some order is, at least apparently, imposed on diversity. Further, and more critically, the favored Sauerian procedure of identifying hearths and diffusion routes does succeed in minimizing the difficulties identified. But, on balance, are we misleading others and ourselves if we begin to conceive of regions as genuine entities with unique characteristics? Are you inclined to agree with the problems implied by the old joke to the effect that the number of boundaries of a region is the square of the number of geographers delimiting that region?

become 'an acceptable form of professional nostalgia, conjuring up memories of a golden age, now (thankfully) defunct'.

It is commonplace in much recent human geography to make two related claims: first, to say that traditional regional geography has declined and, second, to say that a revised regional geography, one that is informed by recent advances in social theory, is at the forefront of much geographic research. The basic logic behind the first claim is clear—the regional approach, as advocated by Hartshorne, effectively dominated English language geography from about the 1920s until about the mid-1950s, at which time it was challenged by the scientifically inspired spatial analysis. Since then it has not succeeded in re-establishing itself as the geographic paradigm.

But although it is correct to say that the regional approach declined in popularity after the mid-1950s, it is also true that regional geography has remained central to much geographic research, and that it continues to be advocated—as the above quote from Hart makes clear. More important, for our purposes, the discussion of regions has continued to be a popular and worthwhile approach for the study of cultural landscapes. Indeed, as this chapter demonstrates, some of the very best cultural geography of recent years is explicitly concerned with identifying and understanding regions at a variety of scales.

Principal criticisms of traditional regional geography are that the approach is of limited value because it is primarily concerned with the observation and recording of facts and because there is an implicit assumption in most studies to the effect that human geographic regions are organized according to differences in physical geography. Using the terms introduced in Chapter 1 and defined in the Glossary, these criticisms relate to the empiricism and naturalism evident in regional geography.

But these criticisms apply less clearly to the regional approach as it is used by cultural geographers who have always been concerned with a level of generalization, notably in the identification of cultural hearths and of diffusion routes, and who have always rejected simplistic physical and human associations. Indeed, the more socially informed new regional geography is partially built on the ideas of cultural geography, specifically on the idea that regions are related to spatial variations in culture and not to physical geography. The principal difference between traditional and new approaches, so far as cultural geography is concerned, relates to the traditional interest in material landscape and the newer interest in culture as socially constructed.

To conclude this brief discussion, although traditional regional geography is out of favor, relatively speaking, having been replaced by a more socially informed regional geography, there seems no doubt that regional cultural geography continues to thrive because, as part of the landscape school, it did not incorporate some of the principal limitations of the larger traditional concern. Certainly, some of the criticisms of the traditional regional approach neglect to consider the substantive achievements of cultural regional geography, a fact that is evidenced by the remaining contents of this chapter.

The Evolution of Cultural Regions: Concepts

The processes by which distinctive areas of cultural landscape, cultural regions, come into being are complex and have been studied in diverse ways. For the United States, the previous chapter noted various circumstances—such as a substantial number of immigrants, contiguous settlement, a set of shared values, and lack of contact with other groups at the time of settlement—that might encourage retention of cultural traits and hence that might contribute to region formation.

This section details a number of useful concepts that have been developed by cultural geographers and employed in the process of region identification and delimitation. Examples of cultural regions that are explicitly related to concepts are considered in this section, while the succeeding section reviews a number of other major regional studies.

Cultural Hearths

In accord with the landscape school tradition and accompanying the interest in diffusion, there has been considerable emphasis placed on the concept of the cultural hearth as an original source area that possesses some distinctive cultural attributes and from which these attributes were diffused. Sauer stressed the importance of locating hearths from which settlement expanded as a basis for region identification, and the concept has been used in numerous cultural geographic analyses. Most such work has been accomplished for American and, to a lesser extent, Canadian landscapes. This preference is related to the importance of the landscape school in North America rather than elsewhere, and to the fact that the European settlement of North America involved distinct cultural groups settling separate areas, thus contributing to the creation of cultural hearths. Examples of hearth identification are discussed later in this section along with examples of some of the more elaborate conceptual formulations.

Core, Domain, and Sphere

A stimulating conceptual contribution to the literature on landscape evolution and related region formation is the core, domain, and sphere model as developed and used by Meinig. The model formulation was introduced in an exemplary analysis of the emergence of the Mormon cultural region. Following accounts of the cultural group, of the movements of the group, and of the evolution of the region, this generic model was outlined as a basis for both delimiting the region and identifying variations within the larger region. There are three components to the model (Figure 4.1).

- There is a cultural *core*. This is not quite the same concept as the hearth area noted above. Rather, the core is the center of cultural control and the zone of most intense activity, but is not necessarily the original hearth area—the core may or may not be the hearth. In the Mormon case, the core was founded following movement of the cultural group away from the area of cultural origin, and hence was not the hearth.
- Surrounding the core is a *domain*, the area over which the culture diffused and became dominant. Although the Mormon domain lacks the intensity of occupance and complexity of development that are evident in the core, it is, because the core

Figure 4.1: Core, domain, and sphere. This is the classic application of the core, domain, and sphere model that Meinig developed in order to accommodate a common problem, namely mapping cultural regions that contained internal variations. The innovative solution was to employ some generic concepts that were able to reflect such variations. The concepts introduced were those of core, domain, and sphere. Because cultures often spread away from a core, the transition from core to domain to sphere is likely to be suggestive of spatial and temporal patterns. The success of the model in the Mormon context seems undeniable, but it is also clear that the Mormon case is far from typical, especially with regard to the isolation of the region and the distinctiveness of the group.

Source: Adapted from D.W. Meinig, 'The Mormon Culture Region: Strategies and Patterns in the Geography of the American West, 1847–1964', *Annals of the Association of American Geographers* 55 (1965):214.

is not the hearth, the area that most obviously displays the culture in the visible landscape.
- Beyond the domain is the *sphere*, an area that only partially belongs to the culture region in question because other cultural influences from adjacent cultural regions are also present.

Certainly, the Mormon region for which this model is constructed is an unusual one because of the initial isolation of the group in the area occupied and the especially distinctive character of the Mormon cultural group. This combination of isolation and a distinctive culture are probably prerequisites for the evolution of a core (or hearth), domain, and sphere regional identity that is as readily recognizable as it is in the Mormon case. Other applications of this model are therefore more limited, precisely because most groups are in close contact with other groups in adjacent areas, and because most groups do not actively seek to emphasize their differences from the wider society as did the Mormon group. The principal value of the model, aside from the specific Mormon application, is the explicit focus on evolution and the acknowledgment, in the reference to a sphere, that the boundaries of cultural regions are zones of transition and not lines.

As one part of a study concerned with urban street patterns in eighteenth-century Pennsylvania,

the Pennsylvanian cultural hearth, domain, and sphere was mapped (Figure 4.2). A hearth rather than a core was identified, as this was the area of cultural genesis. Pennsylvania was probably more typical of the North American situation than was the Mormon region in that several cultural groups, rather than just one group, were present during the period of expansion. Accordingly, the cultural region developed quite differently. This argument was also extended to the larger region of the northeastern United States. Thus, the core, domain, and sphere model was tentatively revised to include, for each of several cultural groups, a hearth, a domain, and a sphere (Figure 4.3).

First Effective Settlement

Whenever an empty territory undergoes settlement, or an earlier population is dislodged by invaders, the specific characteristics of the first group able to effect a viable, self perpetuating society are of crucial sig-

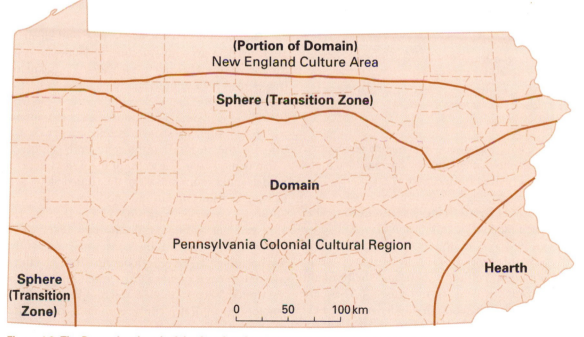

Figure 4.2: The Pennsylvania colonial cultural region. This application of the core, domain, and sphere model is in a quite different historical, regional, and cultural context to the original Mormon application, although the model is used by Pillsbury for essentially the same reason as it was developed by Meinig, namely to accommodate the fact that the cultural region being analyzed displayed some internal variations. The most obvious differences between the two cases relate to the fact that this Pennsylvania colonial cultural region is not isolated from other cultural influences, and to the fact that the core is, in this instance, also the genesis area, the hearth. This figure derives from a study of urban street patterns in Pennsylvania, hence the region is mapped only as it is located within Pennsylvania.

Source: Adapted from R. Pillsbury, 'The Urban Street Pattern as a Cultural Indicator', *Annals of the Association of American Geographers* 60 (1970):445.

Figure 4.3: Colonial cultural regions in eastern North America.
This figure is a suggestive diagram proposing the presence of three colonial cultural hearths in the eastern United States, namely those of New England, Pennsylvania, and Chesapeake Bay. In addition, a fourth cultural hearth was identified in French Canada. The three American hearths were previously proposed by Kniffen (1965) in a study of the diffusion of folk housing. One way to interpret this suggestion that eighteenth-century North America comprised four regional cultural hearths is by reference to the simplification thesis discussed in Chapter 3. According to this thesis, the large number of local cultural regions, or *pays*, evident in Europe were simplified after being transferred to an overseas area, with one result being a smaller number of hearths. One implication of this proposal for four hearths is that some cultural groups, such as Swedes in Delaware and Dutch groups on the Hudson, became assimilated shortly after arrival. This figure also suggests that the three American hearths merged as they diffused west to contribute to the formation of a national American culture.

Source: Adapted from R. Pillsbury, 'The Urban Street Pattern as a Cultural Indicator', *Annals of the Association of American Geographers* 60 (1970):446.

nificance for the later social and cultural geography of the area, no matter how tiny the initial band of settlers may have been (Zelinsky 1973:13).

This concept of **first effective settlement** is similar to the initial occupance proposal suggested by Kniffen (1965:551) in the context of a diffusion analysis of folk housing. Zelinsky used the concept to map modern cultural regions and the resultant regionalization is depicted in Figure 4.4. There are five regions designated as first order, namely: New England, the Midland, the South, the Middle West, and the West.

The significant variations in regional size are related to differences in physical geography and to the different settlement experiences. For each of these five, the date of first effective settlement and of related cultural formation, as well as the major

sources of culture for region formation, are indicated by Zelinsky (1973:119). Each of the first order regions are subdivided into secondary regions, while three areas (Texas, Oklahoma, and peninsular Florida) are considered difficult to classify given the criterion employed. All but one of the first order regions have more than one major source of culture, the exception being New England which has England as the one source area. Because of this, only New England, among the first order regions, has a date of first effective settlement noted; for the other four it was deemed appropriate to indicate first effective settlement dates and sources of culture for secondary and, in three instances, tertiary regions. Further, in most of the cases, multiple sources are identified. The concept of first effective settlement enables this regionalization, but it cannot be applied on a simple one date–one cultural source–one region basis. Certainly, the regionalization is a valuable application of the concept but, not surprisingly, the complexity of the regional scene is more than can be accommodated by a single concept.

The basic logic of first effective settlement is often used in accounts of cultural landscapes. In the case of the Musconetcong Valley landscape of New Jersey, for example, many elements reflect a continuity from the eighteenth-century pioneer period to the present; thus, houses and barns along with many auxiliary structures reflect types established in the area during the eighteenth century.

Again, as with the core, domain, and sphere model, applications outside of the United States are lacking. Certainly, in Europe and in many areas of European overseas expansion where the experience was not as numerically overwhelming as it was in North America, the concept is necessarily less meaningful. This is both because the first effective settlement is of much greater antiquity than in the United States examples, and also because contemporary cultural identity is often a composite product of a series of contributions rather than a relatively clear reflection of only a single dominant contribution.

Duplication, Deviation, and Fusion

The Turner frontier thesis, as introduced in Chapter 3, explained the formation of American cultural

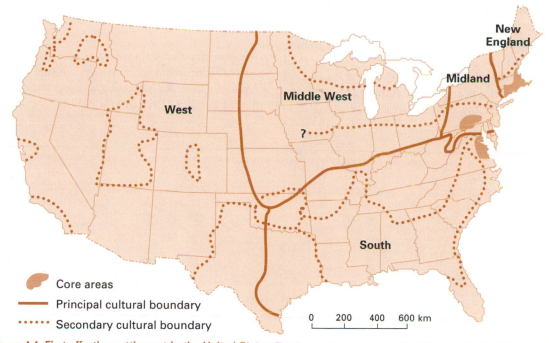

Figure 4.4: First effective settlement in the United States. This is the classic regionalization of the modern United States proposed on the basis of the concept of first effective settlement. Zelinsky preferred this concept rather than the core, domain, and sphere model as a basis for regionalization because of the lack of isolation experienced by most of the core areas identified. This is the same point as made by Pillsbury (see Figure 4.3).

Source: Adapted from W. Zelinsky, *The Cultural Geography of the United States* (Englewood Cliffs, NJ: Prentice Hall, 1973):118. Copyright © 1973.

regions in terms of the westward moving frontier, but there are several problems with this frontier argument, including:

- the assumption of an initial subsistence phase of economic activity, which was not typically the case as many settlers moved quickly into commercial activity
- the assumption of an initial regional isolation, which was not typically the case as most areas were closely linked to other areas from the outset of settlement
- the assumption that American culture was derived from the frontier experience west of the Appalachians

It is possible that the cores of the three colonial cultural regions—those located in southern New England, southeastern Pennsylvania, and the Chesapeake tidewater—acted as dynamic culture regions, first contributing to the creation of intermediate regions, and later contributing to the creation of regions west of the Appalachian regions. Each of these regions included internal differences such that hearths and domains can be distinguished. Three mechanisms might have been responsible for the creation of regions west of the colonial hearths, namely:

- *duplication* of the traits evident in the hearth
- *deviation* from the traits evident in the hearth because of different local circumstances
- *fusion* of traits from two or more hearths with the resulting formation of a different cultural form

Of these three, 'it is the fusion process, aided by the appearance of symbols of national unity, that seems to have been most important in the cultural formation of the early trans–Appalachian west' (Mitchell 1978:67). This proposal, a development of both the core, domain, and sphere model and the first effective settlement concept, allows for a variety of formative processes (Figure 4.5).

Stages of Regional Evolution

In a penetrating analysis of the American West, Meinig (1972) identified a set of six dynamic regions. For each of these regions, four general phases of development were identified for each of four categories of regional features, namely, population, circulation, political area, and regional culture. The sequence of stages for culture was as follows (Figure 4.6).

- *Transplant*, the first stage, was a selected and experimental adjustment to the new environment.
- *Regional culture*, the second stage, saw the formation of a cohesive regional society.

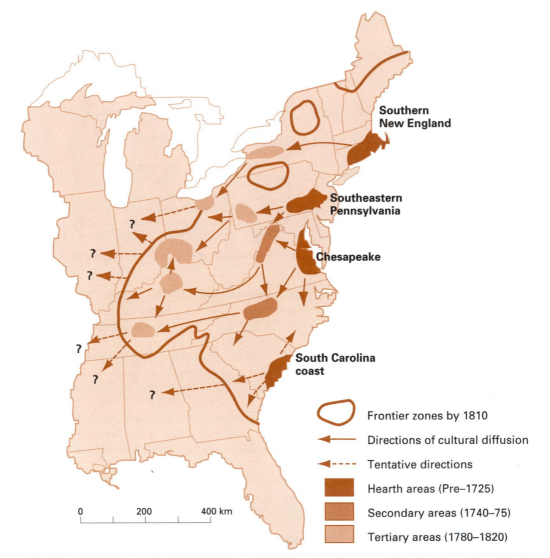

Figure 4.5: Cultural diffusion, eastern United States, *c.*1810. This map outlines some of the directions, and possible directions, of cultural diffusion from three, possibly four, colonial cultural hearths between about 1725 and 1820. Diffusion of traits involving duplication, deviation, and fusion contributed to the formation of other cultural regions.

Source: R.D. Mitchell, 'The Formation of Early American Cultural Regions', in *European Settlement and Development in North America*, edited by J.R. Gibson (Toronto: University of Toronto Press, 1978):75.

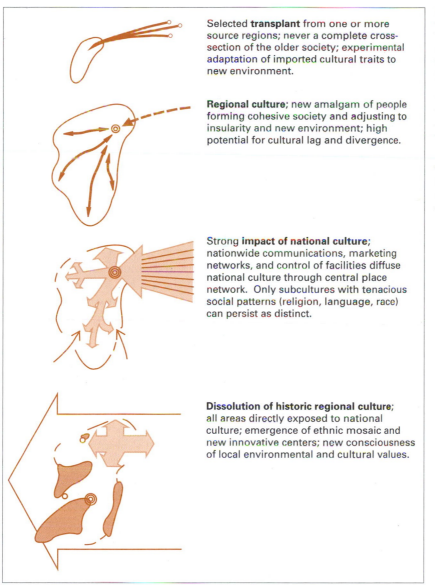

Selected **transplant** from one or more source regions; never a complete cross-section of the older society; experimental adaptation of imported cultural traits to new environment.

Regional culture; new amalgam of people forming cohesive society and adjusting to insularity and new environment; high potential for cultural lag and divergence.

Strong **impact of national culture**; nationwide communications, marketing networks, and control of facilities diffuse national culture through central place network. Only subcultures with tenacious social patterns (religion, language, race) can persist as distinct.

Dissolution of historic regional culture; all areas directly exposed to national culture; emergence of ethnic mosaic and new innovative centers; new consciousness of local environmental and cultural values.

Figure 4.6: Four stages of cultural region development. This diagram shows one of four related schema, with the other three referring to different categories of regional features, namely circulation, population, and political areas. Each of the six western regions (shown in Figure 4.7) is regarded as passing through these four cultural stages.

Source: D.W. Meinig, 'American Wests: Preface to a Geographical Introduction', *Annals of the Association of American Geographers* 62 (1972):163.

- *Impact of national culture*, the third stage, witnessed the consequences of improved communication and marketing.
- *Dissolution of historic regional culture*, the fourth stage, involved a new awareness of regional values.

The distinction between the first two and the later two stages is particularly important as the beginning of the third stage marks the end of insularity and local cultural identity and the onset of a rapid increase in the power of those forces encour-aging national cultural uniformity. Much of the early regional distinctiveness has been eroded because the twentieth-century experience has typically been one of cultural integration related to the impact of a national American culture.

Six Regions in the American West

This approach to regionalization is developmental, synthetic, and generic. Applied to the American West, it facilitated the mapping of six major nuclei (the first stage) and six recognizably distinct regions (the second stage), along with a number of secondary nuclei, some of which were inside and some outside the six major regions (Figure 4.7). Settlers in the American West came mostly from areas in the East and brought with them certain cultural traits, such that behavior was more imitative than innovative. It was this behavior, combined with environmental differences, that contributed to the formation of the six regions. The six major regions identified as present by the end of the second stage, defined as that of regional culture, are as follows:

- Hispano New Mexico was the first region to form, following late sixteenth-century Spanish settle-

Figure 4.7: Major nuclei and regions in the American West.
Six regions were proposed for the late nineteenth century. Hispano New Mexico has a major nucleus in the upper Rio Grande Valley centered in Santa Fe. The Mormon region has a major nucleus at the base of the Wasatch Mountains centered in Salt Lake City. The Oregon country has a major nucleus in the Willamette Valley centered in Portland. In all three of these cases, the nucleus is a prime agricultural area.

Northern California has a major nucleus centered in San Francisco; southern California has a major nucleus centered in Los Angeles; and Colorado has a major nucleus centered in Denver. In addition to the six major regions and their major nuclei, several minor nuclei are located, each of which showed some degree of local autonomy.

Source: D.W. Meinig, 'American Wests: Preface to a Geographical Introduction', *Annals of the Association of American Geographers* 62 (1972):169.

ment in the upper Rio Grande Valley. This was an isolated region of European culture for over 200 years until in-movement by Anglos initiated change. By the late nineteenth century, it was a complex cultural area with relatively limited links to the larger national system.

- The Mormon region formed beginning in 1847, with the initial in-movement of a distinctive religious group seeking isolation from larger American influences. A cultural region was created during about the next fifty years. Both the Hispano and Mormon regions are discussed in greater detail in the following section.

- The region of the Oregon country was formed beginning in the 1840s, primarily by settlers who were a part of the larger westward movement, with no one cultural group dominating the in-movement. The character of the region diverged from that of the migrant source areas in the East because of various differences in location and environment.

- Northern California had limited settlement until the discovery of gold in 1848, an event that precipitated rapid and dramatic population growth accompanied by related cultural and economic change. The incoming population was especially heterogeneous.

- The region of southern California was unaffected by the gold rush and remained as a dual Mexican and Anglo-American culture until a real estate boom in the 1880s initiated a process of rapid growth.

- Regional development began in eastern Colorado in the late 1850s following the discovery of gold and the area developed as a cattle and irrigated farming area.

None of the final four regions is as readily identifiable as a specifically cultural region as are the Hispano or Mormon areas; their identification as regions within the West is based on economic and other criteria rather than primarily on specifically cultural criteria.

The example of Texas can be analyzed similarly. Texas can be interpreted as a distinct cultural region that has evolved through four stages, each of which has left a mark on the present scene. The four stages are:

- *implantation*, reflecting both Spanish and Mexican influences
- *assertion*, including the periods of republic and early statehood
- *expansion*, following the Civil War
- *elaboration* involving more recent developments

Box 4.2 reviews some of the more ambitious uses of the core, domain, and sphere model combined with the synthetic logic of the stages of evolution approach.

Culturally Habituated Predisposition

A pioneering analysis of agricultural, not specifically cultural, region evolution interpreted change in three regions—the American corn belt, the Philippine coconut landscape, and the Malayan rubber landscape—as a result of cultural rather than environmental or economic, processes. Although each case was different in detail, a cultural (specifically psychological) process was evident in all cases. The American corn belt, for example, was seen as the 'landscape expression of . . . the totality of the beliefs

Box 4.2: The Shaping of North America

The fundamental logic behind both the core, domain, and sphere model and the stages of regional evolution approach have been further pursued. Thus, Meinig argued the need to interpret larger American cultural and economic development from a series of initial units:

. . . the most important task in the historical geographic study of colonial America is to define as clearly as possible this sequence of territorial formation from points to nuclei to regions on the North American seaboard and to describe the changing geography of each in terms of spatial systems, cultural landscape, and social geography (Meinig 1978:1191).

Our understanding of North American cultural region evolution is enhanced significantly by the three published volumes, as of 2000, of the proposed four-volume set, *The Shaping of America: A Geographical Perspective on 500 Years of History* (Meinig 1986, 1993, 1998). There are elaborations of the earlier work on core, domain, and sphere, and on the related stages of regional evolution framework. Thus, a core, domain, and sphere model for American national cultural, political, and economic expansion was proposed and applied for the end of the colonial period. The core is on the eastern seaboard, especially New York and Philadelphia; the domain is the area settled by European Americans and is divided into a northern free domain and a southern slave domain; and the sphere is the area claimed but unsettled by European Americans (Figure 4.8). Beyond the sphere were areas of foreign territory in 1800, namely British North America, Louisiana, and Florida.

For the late 1850s, these ideas were further developed at a transcontinental scale. The New York City to Philadelphia axis, with extensions north to Boston and south to Baltimore, continued as the core area, with two important extensions west along the Hudson–Erie canal and through central Pennsylvania (Figure 4.9). The domain is the area of contiguous European American settlement, a much more extensive area than was the case in 1800. At this national scale, it is clear that the domain comprised several parts, each exhibiting some variation of the larger national cultural characteristics. Thus, the domain in the late 1850s included some backwoods areas adjacent to the core that were clearly subordinate to the core, and other major areas (notably the South and West) that were a part of the nation, but that contained their own variants of the national culture and their own regional landscapes.

The remarks accompanying Figure 4.9 include an important distinction between the use of the term 'domain' in the Mormon study referred to earlier, and the use of the same term in these more ambitious analyses. More generally, this application of core, domain, and sphere is rather different from the earlier Mormon application. In the Mormon case, the concern was with a cultural region. In this more ambitious application, the concern is with a developing national culture and those areas that, although linked to the national culture, have some distinctive characteristics of their own. This is simply a way of saying that the spatial and social scales are changed and, therefore, the criteria for identification and the expectations for regional homogeneity are also changed. Later in this chapter the spatial and social scales change again, and in the discussion of regions at a global scale, the criteria and expectations for regional homogeneity adjust accordingly.

of the farmers over a region regarding the most suitable use of land in an area' (Spencer and Horvath 1963:81). Thus, formation of the corn belt followed a decline in sheep numbers, resulting from the general lack of interest in sheep rather than from any crucial environmental or economic factors. The relevant cultural process was that of innovation diffusion—once a process of change began, the mechanics of local communication networks ensured that such change became widespread; this

was particularly effective in an area experiencing settlement where the change was that from an essentially natural landscape.

In a related fashion the coconut landscape of the Philippines resulted from a particular farming mentality, with coconut planting occupying a prominent place in the minds of most farmers in the southern Philippines. This psychological mind-set was a strong force in the evolution of the coconut landscape. Expressed rather differently, it

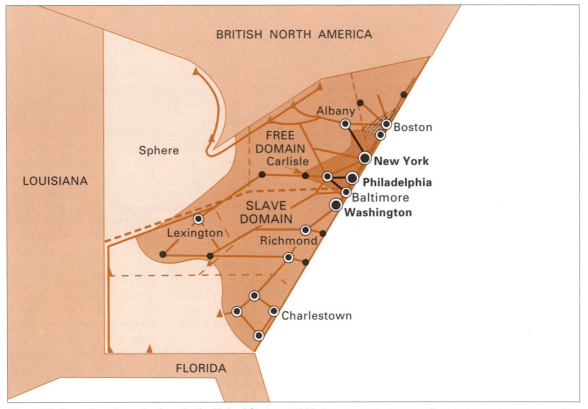

Figure 4.8: Core, domain, and sphere in the United States, *c.*1800. This schematic diagram reflects an innovative extension of the core, domain, and sphere model to the emerging political unit of the United States. The concern here is with nationalism and the new nation rather than with culture and landscape. Philadelphia and New York comprised the true core, the area around which the new United States was beginning to form. Even in 1800, however, this was only an incipient core. Aspects of a new national lifestyle began in the core with little concern for what was happening elsewhere. Boston was not strictly a part of the core because it reflected a regional New England interest, while Baltimore looked south rather than north. The domain was only loosely linked to the core. Within the domain, there was a major distinction between those states with large numbers of slaves and the other states. Beyond the domain was the sphere, an area available for future settlement. The outer boundaries of the sphere were determined by the presence of colonial powers.

Note that the terms 'core', 'domain', and 'sphere' do not have quite the same meanings as they did in the original Mormon application of the model. A revised definition of the term 'domain' is noted in the comments accompanying Figure 4.9. Meinig (1986:404) stated: 'The tentative character of all these regional patterns was evidence that the United States was no more than a nation abuilding and it was being shaped not after the design of a simple model but uniquely, experimentally, within the framework of a federation.'

Source: D.W. Meinig, *The Shaping of America: A Geographical Perspective on 500 Years of History, Volume 1: Atlantic America, 1492–1800* (New Haven: Yale University Press, 1986):402. Copyright © 1986.

The American corn belt spans from Nebraska to Ohio. Development of new and more tolerant strains has extended corn production on either side of the belt (*Doyle Yoder*).

was a culturally habituated predisposition toward a particular crop providing a stable return that initiated the formation of an agricultural region. The origin of this region was similar to, although different in detail from, that of the corn belt, particularly because the coconut had long been known and all farmers were cognizant with the crop and with appropriate techniques. The specific stimulus for the emergence of a commercial coconut landscape was a change in demand and farmers were able to respond accordingly

A third example of cultural causation was that of the Malayan rubber landscape. This landscape was formed as a result of psychological change among the local population when Malays, who initially viewed rubber as an alien system of agriculture, changed their attitude toward this activity and adjusted their behavior accordingly.

Preadaptation

Persistent dissatisfaction and disagreement over the nature of American frontier settlement necessitates replacement of the principal theories by a cultural theory. The American frontier of the eighteenth and early nineteenth centuries is best understood—rather than as a time, place, or process—as the preemption of a vast domain by one preadapted, syncretic American culture. The new culture was the Upland South of Turner and Kniffen; the processes were diffusion and migration; and the mechanism was cultural preadaptation (Newton 1974:143).

The concept of **preadaptation** was introduced in the specific context of attempting to explain the diffusion of culture from the American Upland South, an area stretching from southeastern Pennsylvania to eastern Georgia, to cover much of the

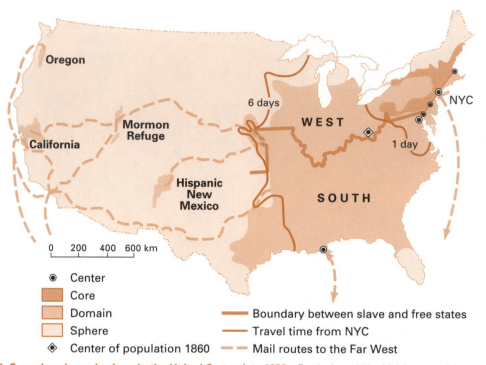

Figure 4.9: Core, domain, and sphere in the United States, late 1850s. For the late 1850s, Meinig mapped the core, domain, and sphere in a more conventional manner than was the case for 1800. The primary concern continued to be with the idea of nation building. The core is the heart of the nation and the source of most of the innovations that diffused throughout the country. The core is also multicentered, with New York City as the financial and commercial capital, Philadelphia as a prestigious city, Boston as a center that sought to shape moral character, and Washington as the political capital.

The domain is the remainder of the contiguously settled area, but is far from culturally uniform. Within the domain, for example, the South exhibited many characteristics that were different from those found in the core. Indeed, Meinig (1993:423) stated: 'Domains are realms in which distinct regional variants arising from different physical environments, resources, local economies, and mixes of people are likely to show through the veneer of national culture.' Expressed rather differently, the domains in this and the previous figure can be conceived of as incipient domains in comparison to the Mormon domain shown in Figure 4.1. The West was more closely linked to the core with principal differences relating more to the recency of settlement and to low population densities. By the late 1850s, the sphere extended continuously to the Pacific.

Source: D.W. Meinig, *The Shaping of America: A Geographical Perspective on 500 Years of History, Volume 2: Continental America, 1800–1867* (New Haven: Yale University Press, 1993):424. Copyright © 1993.

United States as far west as the plains. Preadaptation involves a culture that already possesses the necessary cultural traits to allow successful occupation of a new environment prior to movement to that environment; groups with these characteristics have a competitive advantage. Eleven traits that permitted the culture of the Upland South to expand can be identified (Box 4.3).

Settlers in the Upland South were seen as the initiators of a set of cultural preadaptations that provided a basis for subsequent expansion of an Upland South culture beyond the Appalachians

and into the plains. Further development of this essentially ecological concept included the argument that seventeenth-century lower Delaware Valley Finns were preadapted to life on the American frontier because of their previous European experience. As such, this group can be seen as the single most important contributor to the culture of the American backwoods frontier. Although this identification of a specific, highly localized source area and of a particular ethnic group is debatable, the basic argument is clear—preadapted settlers were necessarily effective settlers and hence, if they were the first arrivals in an area, they established the first effective settlement.

Preadaptation can be seen as but one aspect of general evolution and, as such, can be interpreted as a version of naturalism. Although this idea could have been introduced in the previous chapter during the discussion of evolution, it is included here because the principal issue addressed using the concept is that of regional cultural formation.

The 'Authority of Tradition'

This discussion of conceptual contributions to the understanding of the formation of cultural regions concludes with a useful general idea, the basic logic of which is evident in most regional cultural geographic writing. In an account of the formation and shaping of the American corn belt as it related to cultural hearths in the East, specifically the Upland South, Hudson (1994:3) identified the idea of the 'authority of tradition'. This self–explanatory term is a useful way to summarize a key interest of cultural geographers and has been implicit in much of the preceding discussion.

The authority of tradition is clearly evident in landscapes and any folk components in a landscape can be analyzed as the key to understanding the distinctiveness of a region, although neither reading the landscape nor telling about the landscape comes easily as cultural landscapes reveal little meaning unless each component is observed and understood both for what it is and for how it fits into the larger context. (See Your Opinion 4.3.)

Box 4.3: Cultural Preadaptation

There were eleven preadaptive traits that, according to Newton (1974:152), made it possible for the Upland South culture to expand west to the edge of the Great Plains quickly and successfully. The eleven were:

- dispersed settlement, allowing a relatively small number of people to lay claim to an extensive area
- kin-based dispersed hamlets, providing a readily replicated form that easily adjusted to varying physical and economic circumstances
- dispersed services, such as mills, stores, and churches, offering numerous foci for settlement
- a combined stockman-farmer-hunter economy, encouraging diverse economic activity
- log construction, permitting exploitation of forests
- universal construction techniques, allowing people from diverse ethnic backgrounds to share building tasks
- a productive and adaptable food-and-feed complex, including cattle, hogs, corn, and vegetables and lacking restrictive tree or shrub cultivation
- extreme adaptability concerning the choice of commercial crop to generate income
- an evangelical Protestantism combined with antifederalism, allowing settlements to control their own affairs
- an open class system, allowing White settlers to advance socially

- a courthouse-town system, providing a focus to civil order and emphasizing skills of elite over others

For Newton, it was these eleven cultural traits characteristic of the Upland South culture that allowed the culture to advance westward—each trait was preadaptive. The culture itself was formed east of the Appalachians in the early to mid-seventeenth century, and moved west to the Great Plains where the semiarid and treeless environments required a different set of preadaptive traits.

Jordan and Kaups (1989) addressed essentially the same problem, but chose to identify the preadapted culture by using the term 'backwoods'. Arguing that the culture was formed in the mid- to late-seventeenth century in eastern Pennsylvania, they also emphasized the role the culture played in the settlement of the western United States following a brief pause at the eastern edge of the Great Plains (Figure 4.10). Some additional traits were added to the list of eleven proposed by Newton as follows:

- considerable mixing with Aboriginals
- an almost compulsive mobility, with individuals and families moving on as many as five occasions during their lifetime
- expansion that often leaped ahead of the frontier to create islands of settlement
- no interest in conservation

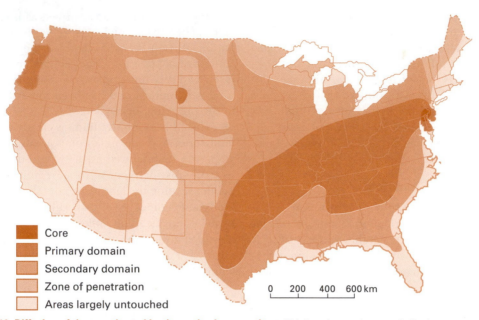

Core

Primary domain

Secondary domain

Zone of penetration

Areas largely untouched

0   200   400   600 km

Figure 4.10: Diffusion of the preadapted backwoods pioneer culture. This is an interesting map, indicating a proposed diffusion of the preadapted backwoods culture, and also employing the concepts of core, domain, and sphere in a modified format. There is a core area where the culture first formed among Finnish settlers between about 1640 and 1680—the date and specific ethnic source continue to be debated. Note that the proposed core area, the lower Delaware Valley south of Philadelphia, is much more localized than that suggested by Newton (1974). It is also smaller than and north of the cultural core proposed by Mitchell (1978), and different in detail from that described by Meinig (1993). Necessarily, then, the ethnic composition proposed by Jordan and Kaups is also different from that proposed by the several other authors, who typically identify a mix of backgrounds, usually including English, Scotch-Irish, and German. The primary domain is the area of most intense cultural transplant, beginning about 1700 and ending about 1850. The secondary domain was settled between about 1725 and 1875 and required considerable cultural adjustment. The zone of penetration is comparable to the sphere, showing even less evidence of the original culture.

Source: T.G. Jordan and M. Kaups, *The American Backwoods Frontier: An Ethnic and Ecological Interpretation* (Baltimore: The Johns Hopkins University Press, 1989):8–9. © 1989 The Johns Hopkins University Press. Reprinted by permission.

Cultural Regions

The preceding section focused on conceptual formulations relating to cultural region formation in areas of European overseas expansion and with their application, usually in an American context. This section focuses on some of the principal regional analyses accomplished by cultural geographers and again, in accord with the interests of practitioners, the emphasis is on American regions.

The difficulties of delimiting regions in areas where the first effective settlement might be 1,000 years or more in the past are substantial, and the problem of determining causal processes often appears insurmountable. Thus, the relative paucity of both conceptual and empirical analyses of European, African, and Asian areas is not surprising, and relates to the greater complexity

evident in researching those areas. But it also appears that, for the United States especially, there has been a notable preoccupation with understanding both national and regional character and landscape, and cultural geographers have responded to that preoccupation. Nevertheless, some of these problems have been tackled in long settled areas and there is a venerable regional tradition in some European countries, notably in France, with the distinctive *pays* landscapes, and also in Germany.

Vernacular Regions

Cultural geographic interest in contemporary vernacular regions focuses on the idea of such regions as the product of the spatial perception of average people. Thus, vernacular regions are perceived to have a regional identity, usually both by those inside and outside the region, and also usually have a particular meaning attached to them. Given this emphasis on self-defi-nition, many attempts at delimitation are based on data collected from indi-viduals because a funda-mental assumption in such studies is that if the con-cern is to understand re-gions as they exist in the minds of the people, then one must ask those people. Thus:

- Jordan (1978) gathered data from 3,860 students in an analysis of regions in Texas.
- Raitz and Ulack (1981a) received 847 responses from college students in a study of Appalachia.
- Lamme and Oldakowski (1982) collected responses from 356 people attending the Florida state fair in a study of regions in Florida.
- The most substantial data collection was that by Hale (1971, 1984) involv-ing 6,800 responses from weekly newspaper edi-tors, county agents, and postmasters.

Although the results of these studies are sug-gestive, the validity of such data-gathering exercises is debatable as the sam-ples may not be representative of larger popula-tions. A different procedure for collecting data involves calculating the ratio between a specific regional term, such as 'Dixie', and a term such as 'National' or 'American' as these appear in tele-phone directories, with the assumption that the greater the relative incidence of the regional term,

the more likely that the location belongs to the region (Figure 4.11).

In addition to region delimitation, the exer-cise of identifying vernac-ular regions may produce some useful general-izations about regional identity. Assessments of regional identity may indicate a tendency to reflect values in regional names—for example, there appears to be a Texan pre-occupation with adjectives such as 'Big' and 'Golden' and there is also a specific Appalachian inclination to avoid the term 'Appa-lachia' because of the neg-ative image that the term suggests. Comparisons of vernacular regionaliza-tions with other attempts at delimiting formal cul-tural regions often show a close similarity between the two, although the for-mal region of the Corn Belt does not correspond with a vernacular region. More generally, vernacular regions may have differing levels of intensity—in the United States there appear to be strong regional feel-ings in the Southeast and weaker sentiments in the Northeast. Certainly, the recognition that a given area has a particular iden-tity, such that it is named, implies that the identity is culturally important and that the area has mean-ing to the occupants.

Cultural Regions as Homelands

The term **'homeland'**, although not well estab-lished in the cultural geographic tradition, is increasingly being employed to refer to a partic-

---

Your Opinion 4.3

Although each of the eight ideas about the for-mation of cultural regions is different in detail from the others, it is clear that there are many areas of overlap and that together they offer real insights. This body of work is indicative of the considerable effort made by North American cul-tural geographers to understand the formation of cultural regions—or, more correctly, cultural regions in the United States. But given the insights offered into American region formation, are you disappointed at the apparent relative lack of interest in other parts of the world? Might one or more of these ideas merit more substantial application elsewhere or are quite different con-cepts needed for other areas and circumstances? One of the attractions of analyzing cultural regions in the United States is the apparent rela-tive simplicity related to the recency of first effec-tive settlement and related region formation. Although the same argument applies to other temperate areas of European overseas expansion, these areas have not received similar treatment. Australia, for example, has been well researched in terms of historical and regional geographic topics, but there has been little interest in cultural regions with many analyses treating Australia itself as a cultural region. The reasoning for this may be that Australia offers fewer opportunities than North America for cultural region delimi-tation because of the relative homogeneity of the first effective settlement with most settlers coming from a single source area, Britain.

Figure 4.11: North American vernacular regions. Zelinsky presented this map of fourteen subnational vernacular regions as a summary statement. Note that two areas are given dual region membership. The correspondence between this map and the map of cultural regions as distinguished using the concept of first effective settlement (Figure 4.4) is unsurprising, although six of the vernacular regions are locational and not cultural in character (Atlantic, Gulf, North, Northeast, Pacific, and East). A study of regions in Kansas combined the same procedure (that is, comparing the incidence of regional and national terms) with questionnaires (Shortridge 1980).

Source: W. Zelinsky, 'North America's Vernacular Regions', *Annals of the Association of American Geographers* 70 (1980):14.

ular type of cultural region. In the introduction to a Special Issue of the *Journal of Cultural Geography*, Nostrand and Estaville (1993:1) defined homelands as 'places that people identify with and have strong feelings about', and identified four necessary ingredients, namely a people, a place, a sense of place, and control of place. A rather different approach to homeland identification might focus on such characteristics of a group as indigenity (the length of time a group had been present in an area) and exclusivity (the extent to which the group is a majority of the population). Homelands may or may not be associated with a specific ethnic identity, although the various examples in the Special Issue noted above are all ethnic in focus and include two urban-based examples, namely Jewish and Cuban-American groups.

To clarify the meaning of the term 'homeland', the examples of French Canada and French Louisiana are used, after which two of the most easily recognized and well-researched cultural regions in North America, namely the Mormon

intermountain region and the Hispano Southwest, are discussed in greater detail.

FRENCH CANADA

'No sooner permanently settled in the St. Lawrence Valley at the beginning of the 17th century than the French spread beyond the great river in quest of continental adventure' (Louder, Morissonneau, and Waddell 1983:44). Unlike most of the American regional examples discussed in this chapter, this French Canadian hearth was quickly deserted by those who first settled there, in this case by ambitious *coureurs du bois* who, encouraged principally by the fur trade, spread throughout much of North America—from the St Lawrence to Louisiana and from Acadia to the Rocky Mountains. This expansion was doomed to failure as occupation was extensive with few permanent locations established. Following the French defeat in North America, formalized by the 1763 Treaty of Paris, a gradual process of retreat to the St Lawrence occurred. Quebec was confirmed as the French homeland, an insular reserve surrounded by Anglos, in North

America. But the idea of expansion remained and by about 1870, three different ideologies prevailed among French leaders in Quebec.

- A *continental* ideology resurfaced, prompted by substantial French Canadian movement into New England to work in textile mills.
- A *Canadian* ideology was supported by some who saw the opening of the Canadian west as an opportunity to expand the French area.
- The third, *northern*, ideology was the one that came to fruition. This was really a retreat into Quebec, but into *all* of Quebec, including the extensive northern area. This commitment to Quebec ensured the survival of French such that Quebec today might be considered both part of and apart from the rest of the North American continent.

FRENCH LOUISIANA

To facilitate understanding of the French Louisiana homeland, seven stages can be identified. The first three stages were French and the final four English.

- *Transplantation* occurred between 1700 and 1760, creating a cultural hearth along the Mississippi between Baton Rouge and New Orleans; this Lafourche hearth was characterized by the French language, Catholicism, various French food and clothing preferences, and included a large slave population and some German settlers.
- Between 1760 and 1800 hearth *expansion* occurred and was reinforced by the in-movement of French settlers deported from Acadia by the English in 1755 (Acadia, on the shores of the Bay of Fundy, was the smaller of the two French colonies in Canada; the larger was on the St Lawrence) who formed a secondary French hearth, the Teche hearth, west of the initial location. These Acadians adjusted to their new environment: they abandoned wheat, barley, and other crops that had been grown in Acadia, switching to corn, rice, and other crops suited to the subtropical environment. Thus, by 1800, there were two rather different hearths, the original Lafourche French Creole hearth, and the secondary Teche Acadian or Cajun hearth.

- Large numbers of Anglos moved into the area between 1800 and 1840, especially following the Louisiana Purchase of 1803, initiating a period of *competition*, especially in the Lafourche hearth, which lost the earlier isolation and purity. The French managed to retain some significant measure of political control until 1845.
- Between about 1840 and 1880, a process of *accommodation* and French cultural loss occurred mostly because of increasing numbers of Anglos and the expansion of the rail network.
- These changes continued in the *attraction* period (1880 to 1920) such that, although the French were able to cling to a sense of place, their cultural intensity was significantly diminished.
- The period between 1920 and 1960 was one of *assimilation*, continuing the dissolution of the original hearth and involving further Anglo intrusions into the secondary hearth. Many of these changes were linked to economic developments, such as the rise of the oil industry.
- Finally, since 1960 there has been evidence of a resurgence of French identity, especially Cajun identity; thus, a process of *revitalization* is occurring. This final phase involved the beautification of the Cajun identity and the Cajunization of French Louisiana.

The Mormon Homeland

An understanding of the Mormon cultural region needs to consider both the religious background and the physical environment, what Francaviglia (1978:3) described as the 'bold, semi-arid backdrop which frames all cultural elements'.

PEOPLE AND REGION

Mormons are members of the Church of Jesus Christ of Latter-day Saints, a Protestant group formed in Fayette, New York, in 1830. In 1847, following a series of movements in Ohio, Illinois, and Missouri, members of the church began a movement to an area well beyond the frontier, the Great Salt Lake area of the intermountain West. The Mormon church is characterized by a strong sense of community with a related emphasis on cooperative endeavors, and also by a hierarchical administrative structure with members typically responding positively to the requests of church

The Mormon rural and small town landscape is a distinctive one characterized by a number of specifically Mormon features and these three photographs provide some sense of that distinctiveness. The exceptionally wide main street of the agricultural small town remains evident in the contemporary landscape; this photograph [top] is of Manassa in southern Colorado. There are temples present in some of the larger settlements and they provide a powerful statement as to the importance of the Mormon church in the local community; this example [centre] is in Cardston, southern Alberta. Less evident in the present landscape (as it is no longer used for the original purpose of stacking hay) is the Mormon hay derrick; this example [bottom] is from Cove Fort, central Utah (*William Norton*).

leaders. The combination of these two characteristics, which are discussed in greater detail in Chapter 6, has contributed to a relative uniformity of behavior in landscape after 1847 that has facilitated the formation of a formal cultural region and a homeland for the Mormon people.

On first seeing the Salt Lake Valley in 1847, Young declared that this was the place, a powerful symbolic assertion that set the scene for the subsequent expansion of settlement, region formation, and, indeed, homeland creation. The physical landscape of the region is semiarid and comprises extensive mountain areas with numerous valleys, meaning that agricultural settlement had to take place in a few selected locations. The first areas settled—valleys close to the initial Salt Lake City location—were relatively benign, but expansion of settlement, diffusion from the core, involved, in addition to some attractive valley locations in central Utah and southern Idaho, other isolated intermountain locations and areas of greater aridity in southern Utah and northern Arizona. Political difficulties also prompted movements into northern Mexico and western Canada.

Despite the very real physical challenges, colonization was both rapid and successful with almost 400 communities pioneered by 1877. This remarkable colonization effort was largely under the personal direction of the church president, Brigham Young, variously called the 'American Moses', the Colonizer', or the 'Lion of the Lord'. The extent of the cultural region is mapped in figures 4.1 and 4.7.

LANDSCAPE

Speaking in 1874, church president George Smith observed:

The first thing, in locating a town, was to build a dam and make a water ditch; the next thing to build a schoolhouse, and these schoolhouses generally answered the purpose of meeting houses. You may pass through all the settlements from north to south, and you will find the history of them to be just about the same (quoted in Church News 1979:2).

Given this relatively uniform behavior, it is not surprising that a cultural region, characterized by a distinctive visible landscape, was formed.

A key feature was the building of towns in preference to dispersed agricultural settlement where all the local population, including farmers, resided. It is probable that in advocating town life, Joseph Smith, the first president of the church, was both reflecting his own eastern American background and anticipating the difficulties of maintaining a community in areas of dispersed farmsteads elsewhere in America. Reasons for favoring nucleated settlement included the original plan of the 'City of Zion' as detailed by Smith in 1833, the solidarity of the group, and the physical environment. Mormon towns were laid out according to a regular grid pattern and oriented to the cardinal compass points. There were square blocks, wide streets, lots as large as 2.5 acres (1 ha) that included backyard gardens, use of brick or stone construction, and central areas for church and educational buildings. Farm dwellings were inside town, giving a cluttered rural appearance. Such towns both enhanced and reflected the community focus.

Within towns, the most distinctive house type was the central-hall house, a well-established house plan in the eastern United States but one that was being modified and even replaced by the time settlement reached the plains. The plan was transferred to the West by Mormon pioneers and became a distinctive landscape feature. The two principal versions of this type are the 'I' house, one room deep and either one-and-a-half or two stories high, and the 'four over four' plan, which is two rooms deep. Both are symmetrical, usually with a chimney at each end.

Arable agriculture dominated the rural landscape and was supported by a network of irrigation ditches. Irrigation was necessary because the area being settled was typically semiarid, unable to support arable agriculture without irrigation, and also because the desired Mormon way of life required village settlement and agriculture. Other distinctive features included the planting of lombardy poplars, unpainted fences and barns, and hay derricks. The combination of these various features clearly distinguished the Mormon landscape—the cultural region.

Interestingly, the uniformity of the Mormon landscape is notwithstanding the diverse origin areas of Mormons, many of whom were converted to Mormonism in Europe before traveling to the Salt Lake area. The willingness of these immigrants to jettison ethnic cultural baggage relates to the overriding importance of their religious identity and to the role played by cooperative effort in the creation of their homeland, the 'Great Basin Kingdom'. More so than many other immigrants, Mormons really were making a fresh start both in terms of place and their identity. Mormon identity was clearly more important than the authority of any earlier tradition.

Further, early experiences in the intermountain West had emphasized the need for Mormon political autonomy and, on arrival in the intermountain West, church leaders initiated a process of seeking statehood. A constitution was written, ambitious boundaries proclaimed, a symbolic name (Deseret) proposed, and a provisional government established. This tentative state was dissolved in 1851, following the creation of the smaller territory of Utah with Brigham Young as the first governor. Further unsuccessful attempts to achieve self-government were made in 1856, 1862, 1867, 1872, and 1882. The boundaries of the hoped-for state of Deseret included all of present Utah, much of Nevada and Arizona, about one-third of California, and parts of five other states.

The Hispano Homeland

Is there a Hispano homeland region in the American Southwest? The titles of two books by cultural geographers provide an answer: *The Spanish-American Homeland* (Carlson 1990) and *The Hispano Homeland* (Nostrand 1992).

PEOPLE AND REGION

Hispanos are those people of Spanish–American ancestry who settled in the upper Rio Grande Valley of New Mexico. The physical environment of the Hispano region comprises arable floodplains and an extensive semiarid highland with a sparse vegetation cover. The core of this Spanish–American or Hispano region is the Río Arriba including Santa Fe, and the Río Abaja including Albuquerque. Prior to the late sixteenth century, about 45,000 sedentary and peaceful Pueblos were settled in perhaps eighty villages in this core area, while the surrounding plateaus and mountains were occupied by foraging groups, Apaches to the west and east and Navajos to the north. Spanish colonization from Mexico began in 1598 and this outpost of the Spanish empire remained largely isolated from other influences for over 200 years.

Relations between Pueblos and Spanish were those of Spanish domination and passive resistance by Pueblos punctuated by revolts. An uprising in 1680 was followed by a reconquest in 1693 and a subsequent increase in the extent of the Spanish-dominated area because of the ever-increasing Hispano population. Expansion was in all directions and was especially marked between 1790 and 1900. By the late 1840s, the larger region contained about 70,000 Hispanos and less than 10,000 Pueblos. Together these peoples were effectively an island in a sea of nomadic Indians, an island that was linked only tenuously to European–American civilization to the east. In 1848 the region became a part of the United States and was immediately converted, politically speaking, from the far north of Mexico to the southwest of the United States. Initial intrusion of Anglos, a term that included French trappers and traders, followed Mexican independence in 1821 and resultant changes in frontier policies, but was most significant beginning in the 1860s. These movements added another dimension, the third of three peoples, to the cultural mix of the region.

The details of regional morphology reflect both Hispano expansion and contact with other groups. The Hispano region was in its heyday in 1900, extending over parts of five states, with Santa Fe at the center, and a population that was two-thirds Hispano (Figure 4.12).

LANDSCAPE

Within the larger Spanish Southwest the Hispano region of the upper Rio Grande Valley is distinctive for two principal reasons:

- the Spanish settlers came earlier and more directly from Spain
- the area was isolated for an extensive period following the Spanish colonization

These two points are not dissimilar to the experience of the Mormon region, an area that

The impress of the Catholic religion is very evident in the Hispanic landscape and the buildings shown in these photographs are representative of that landscape. El Santuario de Chimayo [top] in northern New Mexico is an adobe shrine constructed in 1816 and remains as an important symbolic location today. It is described in some tourist literature as the 'Lourdes of America' and has been designated a National Historic Landmark. The adobe church [centre] is in Abiqui, also in northern New Mexico, and is located at one side of the old plaza. The *morada* [bottom], a much less ornate construction than either the shrine or the church, is one of several in the southern Colorado–northern New Mexico region of Hispanic settlement; the example in the photograph is on the outskirts of the small settlement of San Pablo in southern Colorado (*William Norton*).

Figure 4.12: Hispano homeland, 1900. This map shows the Hispano homeland during its heyday. The demographic center is Santa Fe in the lower Rio Grande Valley, the largest Hispano center and the political and religious capital. In 1900 90 per cent of all Hispanos lived within 150 mi. (241 km) of Santa Fe. 'Thus, like an island on the land, much of the Homeland in 1900 stood tall and flat, so uniformly high were its Hispano proportions. Its edges were like steep cliffs, so few Hispanos lived off the island' (Nostrand 1992:232). Since 1900, regional identity has waned somewhat, especially because of significant intrusions of Anglo populations. It remains, however, as a distinctive region today.

Source: R.L. Nostrand, *The Hispano Homeland* (Norman, OK: University of Oklahoma Press, 1992):231.

received a distinctive group of settlers and that had a domain, although not a core, that was not really shared with others. Mormon settlers, a singularly uniform group, certainly stamped their identity on the landscape of their part of the intermountain West. Hispanos similarly created a quite distinct region both through an adjustment to their natural environment and through a process of impressing their cultural identity on the environment. The principal distinction in this respect between Mormons and Hispanos is that, while much Mormon behavior was community focused and related to the wishes of church leaders, the behavior of Hispanos was more individual and family focused, although various institutions, especially the Catholic Church, also played a role.

During the period of formative colonization, prior to substantial Anglo in movement, a network of missions, presidios, towns, farm villages, and ranches began to dominate the landscape. For the Spanish and later Mexican governments, this was an important frontier area where land policies were implemented to attract settlers and to ensure orderly settlement. The most distinctive landscape feature was the long lot, a system of land subdivision that involved long, narrow, ribbonlike lots often fronting on a river. This system provided for equal access to irrigation water and permitted settlers to live close to each other in village settlements, a desirable state of affairs for both defensive and social reasons. Beyond the long lots were large pastoral areas and transhumance (seasonal movement of animals) was practised.

The landscape of this region is distinctive, especially because of the long lots, irrigation systems, adobe buildings and outdoor ovens, village settlement with plazas, and profusion of Spanish and religious place names. Many of these cultural traits were evident earlier in Spain and New Spain. The early Spanish colonists built adobe buildings, a reflection of the shortage of timber for building purposes in the favored floodplain areas. Houses were frequently added to over time to accommodate additional family members, resulting in L-shaped or even U-shaped forms, often with an outdoor adobe oven. There were also homes built of stone, while log buildings were usual in the upland areas. Early villages comprised either houses on small lots around a plaza or rows of houses on lots along a riverfront. The latter form was especially popular after about 1850 when defense was less of a consideration. The impress of religion is also striking with Catholic churches and religious meeting houses being characteristic features of the village landscape. The religious meeting houses (*moradas*) of Penitente chapters are of special symbolic importance. Spanish (often religious) place names are usual throughout the landscape and provide an especially compelling aspect of the regional identity.

In a discussion of regional morphology, as evident in landscape, Nostrand (1992:226–30) employed the terms 'stronghold', 'inland', and 'outland', and explained their changing form in terms of a series of changing geographical processes.

- The stronghold, defined as the area that is more than 90 per cent Hispano, existed by 1790, peaked in 1900, but had largely disappeared by 1980.
- The inland, 50–90 per cent Hispano, was a series of separate islands in 1850 that enlarged notably by 1900 and coalesced to replace the stronghold by 1980.
- The outland, 10–50 per cent Hispano, was present only between 1870 and 1930.

There is, of course, a similarity here with the core, domain, and sphere model as it was developed for the Mormon region. The intent is the same, namely to acknowledge variations within the region rather than simply presenting the region as internally homogeneous and sharply differentiated from surrounding areas. (See Your Opinion 4.4.)

Ethnicity and Cultural Islands

In addition to the specific concerns with identification, description, and analysis of cultural regions, including homelands, cultural geographers are also interested in material landscape features, especially as these relate to ethnic group identity. Such an interest was reflected in both the Mormon and Hispano discussions and in the references to vernacular landscapes in Chapter 3. This section on cultural regions concludes with reference to the large body of regional cultural geographic literature, which focuses on variables that are essentially **ethnic** in character, as this term refers especially to place of origin, sense of tradition, religion, and language. (For a fuller discussion of ethnicity, see the section 'Understanding Ethnicity' in Chapter 7.)

A concern with those material culture elements that are visible in landscape has been a hallmark of Sauerian cultural geography. The pioneering exponent of this type of study was

---

Your Opinion 4.4

Is it helpful and appropriate for cultural geographers to extend their accounts of cultural regions into the rather more emotive issue of homeland identity? The idea of a homeland can, of course, assume many forms, but one interpretation might be that if a group is sufficiently different from others and occupies a particular territory, then that group both merits labeling as a nation and can argue for some form of separate government. As noted, this was precisely the approach taken by Mormons during the second half of the nineteenth century. It is not difficult to see how relatively politically innocent cultural geographic analyses can become quite controversial.

Kniffen, who employed this focus largely to demonstrate certain generalizations involving questions of regionalization and diffusion. Thus, Kniffen wrote:

> The material forms constituting the landscape are the geographer's basic lore. The cultural geographer deals primarily with the occupance pattern, the marks of man's living on the land. He finds his data, his evidence, in buildings, fields, towns, communication systems and concomitant features (Kniffen 1974:254).

Expressions of ethnic identity in landscape result primarily from the tendency of members of ethnic groups to settle in close proximity to one another, from the importation of distinctive culture traits, and from the presence of particular ethnic attitudes and perceptions. Cultural inertia played a role in the initial creation of landscape such that the presence of an ethnic landscape is often some measure of group conservatism. Certainly, most ethnic groups faced two opposing forces when settling in a new environment—there was, on the one hand, a powerful sentiment that favored the maintenance of ethnic traits and identity while, on the other hand, it appeared clear to settlers that many advantages accrued to those willing to break with their traditions. Distinctive landscapes are most likely to emerge and be maintained where the group abides by certain common rules of organization. This claim is supported by a consideration of the rural landscapes of religious and other ethnic groups, many of which display a preferred architectural style that may be reflected in house, farm building, and church construction, and some of which reflect a particular lifestyle in their choice of clothing and food preferences.

The idea of a cultural island, effectively a small cultural region or local homeland, might be employed to refer to those areas occupied by ethnic groups and exhibiting some ethnically based landscape features. Amish and Mennonite populations are two closely related groups that are among the most frequently studied as many of their landscapes reflect aspects of a particular religious world view, notably the commitment to an agricultural way of life and the preference for a closed and introverted society. Dispersed settlement landscapes were favored by the initial Amish and Mennonite groups that settled in North America, while later groups that settled in southern Manitoba came from Russia and favored settlement in street villages with homes that had large rooms and a centrally located kitchen.

Some other well researched and rather more widespread examples are those of Ukrainian landscapes, evident in a broad belt across the Canadian Prairies from Manitoba to Alberta; of Finnish landscapes in the upper Great Lakes, evidenced by the many Finnish place names (see Box 8.9) and by the sauna that is a typical addition to farms; and of German landscapes in the Texas hill country, where resistance to acculturation is evident in church affiliation and reflected in a landscape of distinctive ecclesiastical architecture, cemeteries, and graveyards. There are many cultural islands outside of North America, and cultural geographers have identified and described many of these, including Japanese areas on the Brazilian frontier and German towns in southern Brazil.

In some areas, the identity of a cultural island is dependent on the ability of the group to impose their form of land survey rather than having their landscape surveyed in common with other areas. This is because land surveys often have lasting impressions on the boundaries of property, on road networks, and on other aspects of cultural and eco-

The first Ukrainian immigrants arrived in Canada in the 1890s. These early settlers built small, modest churches out of logs. As parishes became larger and more prosperous, the churches, such as St George's Ukrainian Greek Orthodox Church, became more elaborate (*Victor Last*).

nomic occupance. Long lot systems in the Hispano area, along the lower St Lawrence, and in Louisiana were a marked contrast to prevailing land survey systems and are vivid indicators of a specific cultural impress. Land subdivision imposed upon an area may have the effect of minimizing potential impacts of local cultures.

Two Criticisms

Some cultural geographers reject the idea that an appropriate way to investigate the landscapes of North America is in terms of a variety of different ethnic groups and related identities. Thus, Hart criticized these types of regional cultural studies based on ethnic identity for their lack of sophistication:

Cultural geographers, who should have been able to make a special contribution to our understanding and appreciation of the importance of noneconomic values, have let us down rather badly, at least in the United States, because their approach to values has been so unimaginative. They have focused on those groups that 'have worn name tags', at the expense of groups that have not been easily identifiable by such obvious distinctions as country of birth, language, and religious affiliation, and they have largely ignored the importance of region of birth, social class, and other more subtle distinctions (Hart 1982:26).

The Chapter 8 discussion of identity that is informed by postmodernism is the most substantive response made to this type of criticism.

A rather different criticism of these cultural geographic analyses favors interpreting the term 'ethnic' to refer to a national identity rather than to a number of more localized identities: 'there is really no serious challenge to a pervasive, if largely

subconscious, code governing the proper ways in which to arrange human affairs over American space' (Zelinsky 1997:158). Such an interpretation may involve either a real or an imagined community and is in accord with a view of ethnic groups that sees them as named human populations with shared ancestry myths, histories, and cultures such that they have a sense of togetherness and also an association with a place. Conceiving of 'ethnic' in this fashion suggests that the United States cultural landscape is best understood in terms of a single dominant culture, namely the Anglo–American ethnic group. From this essentially superorganic perspective, regional variations such as those discussed earlier in this section are acknowledged but should not be exaggerated. (See Your Opinion 4.5.)

Pursuing aspects of both of these criticisms it does appear that many regions might be identified without explicit recourse to ethnicity or to a homeland concept and that some of these might even aspire to separate political status. Indeed, some such regions might not be the sort that cultural geographers have typically considered because they are distinctive for a reason other than ethnicity. Box 4.4 outlines the example of the Riverina region in southeastern Australia noting that, although the region lacks cultural distinctiveness as this is usually understood, there were mid–nineteenth century attempts to achieve some sort of separate status for the region.

Shaping the Modern World

Cultural geographic interest in the evolution of regions has focused largely on the subnational scale, and there has been relatively little concern with the question of global divisions. This seems

---

Your Opinion 4.5

Are you in sympathy with either of these two criticisms? What is your understanding of the role of ethnic identity in the North American and other landscapes? Does it seem appropriate for cultural geographers to focus on these (often local) variations in landscape and identity that link to ethnicity or should there be, as Hart suggests, more concern with other aspects of our identity, or, as Zelinsky suggests, a greater concern with national identity? Whatever your reaction to this section of the chapter, and especially to these two critical ideas, it is evident that cultural geographers have been and continue to be attracted to analyses that center on ethnic identity as an important factor related to landscape creation.

understandable given the intellectual antecedents of the landscape school in the work of such Euro–pean geographers as Schlüter and Vidal and in the culture area concept in anthropology. Indeed, there has also been little interest in world regions in the larger discipline of geography, except as basic teaching aids.

The traditional geographic concern with world regions was essentially with regions as the outcomes of human and land relations, usually

Box 4.4: The Riverina Region

Australia entered the European overseas world in the late eighteenth century, a relatively late involvement that is most easily explained in terms of the distance between Australia and Europe. Movement within Australia was also delayed, again because of the distances involved or, perhaps more correctly, because of the goods being produced, the costs of transport, and the availability of markets.

The early British settlement of southeastern Australia involved three locations, Sydney, Melbourne, and Adelaide (Figure 4.13). In all cases the immediate hinterland was not well suited to agricultural production, largely for topographic reasons, and inland expansion followed exploration in the 1820s. A search for new and better pastures for cattle and sheep was the ongoing stimulus. European penetration and settlement was little affected by Aboriginal occupation, although some friction occurred. From their perspective, these settlers, known as squatters, moved into a new and empty land and were able to lay claim to vast areas. Squatters moving from Sydney reached the eastern Murray Valley by the mid-1830s. Pastoral occupation continued west and was reinforced in the early 1840s by movement north from Melbourne such that the region was fully claimed by squatters by about 1850. The pastoral landscape that emerged included both cattle and sheep until sheep assumed prominence by the 1860s.

The landscape associated with this squatter settlement was not one that demonstrated substantial economic progress. Squatting involved dispersed and low-density settlement and, for many, the insecurity of tenure mitigated against substantial capital investment in landscape. Neither poor nor wealthy had any great incentive to change the landscape, although regulations passed in the 1840s for New South Wales provided some security of tenure, thus encouraging investment, especially in fencing. Squatters had little need for towns, so urban centers were few and small in size.

However, as immigration to Australia continued and population numbers increased, there was substantial objection to the fact that so much land was in the hands of so few, and a series of land acts gradually opened the squatter lands to closer settlement by wheat farmers. The transition from a landscape dominated by squatters to one dominated by selection was not an easy one. By about 1860, squatters were a powerful force throughout much of southeastern Australia and proved very resistant to selector intrusions into their lands. Successful settlement by selectors required revised land laws and was associated with a relatively rapidly developing economic infrastructure of towns and communication lines. The expansion of the selection landscape was encouraged by a variety of factors, including demand for wheat and the need to settle surplus populations on the land. These two factors combined to ensure the success of selection in the long run.

It is in this context that the separation arguments that appeared in the area of New South Wales known as the Riverina can be understood. Arguments favoring the transfer of the Riverina from New South Wales to Victoria appeared in the 1850. Wheat farmers favored this as a means of destroying the security of tenure that Riverina squatters had gained. For precisely this reason, annexation was not favored by the squatters, who began to concentrate in the Riverina to seek relief from unfavorable (from the squatter perspective) Victorian land laws. Separation arguments appeared in the 1860s with squatters at the forefront and favoring independence because of the perceived threat of land acts passed in New South Wales. This second development involved a much larger area than the Riverina proper, essentially all of western New South Wales. Both of these movements proved unsuccessful. The group that favored annexation to New South Wales was never sufficiently powerful, while the squatter claims for independence were short-lived as the land acts were not as detrimental to their interests as had been anticipated. Perhaps the most compelling reason for both failures was that, by the 1860s, Britain was uninterested in additional colony creation in eastern Australia. North Queensland suffered a similar fate in the 1880s.

Figure 4.13: The Riverina region in southeastern Australia.
The Riverina, although located in New South Wales, is mostly closer to Melbourne than to Sydney. The precise boundaries are difficult to locate: 'Generally speaking, in the nineteenth century, every man defined his own Riverina to suit his own purpose' (Buxton 1967:3). Twentieth-century delimitations tend to focus on the area enclosed by the Murray and Murrumbidgee rivers eastward from their junction to a line joining Albury and Wagga Wagga.

with an emphasis on physical geography as cause. A principal exception to this focus was the identification of human regions as outcomes of human energy as well as of physical geography, an approach that allowed for the recognition of regions of hunger, debilitation, increment, effort, industrialization, lasting effort, and wandering (Fleure 1919).

The pioneering cultural geographic contribution to the evolution and identification of major world regions is the textbook *Culture Worlds* (Russell and Kniffen 1951), while a more recent effort in this direction is that of Lewis and Wigen (1997). Other than these important studies of world regions, cultural geographers have paid little attention to this topic. However, there is a substantial literature in comparative history, anthropology, and sociology on the larger topic of the rise of civilizations, and also a concern with the evolution of the modern

world as a system of related areas. Both of these concerns are reflected in this section. (See Your Opinion 4.6.)

Civilizations as Global Regions

Civilizations, or 'giant cultures', have been identified by scholars, often called civilizationists, such as Toynbee, Spengler, Sorokin, Quigley, and Kroeber. Each of these writers proposed different processes, stages, and civilizations, often in significant disagreement with others, but most concurred in seeing cyclical patterns in history. For most writers in this tradition, the distinction between a culture and a civilization was one of scale and level of achievement, with civilizations being larger and more complex. Indeed, civilizations usually involve an integration of multiple languages and religions. Civilizationists have made a major contribution to the study of world history and world geography because of their explicit concern with civilizations globally rather than with the European civilization as it formed in association with other areas. As such, they can be described as scholars of world history.

Probably the best known example in this genre is the twelve–volume work of Toynbee, *A Study of History* (1934–61). Although there was no explicit definition of civilizations given, they were implicitly regarded as large cultural and social groups that evolved from earlier primitive societies. The process by which a primitive society was transformed into a civilization was one of challenge and response, a form of adaptation. An organismic analogy was employed, with civilizations passing through a life cycle of four stages, namely genesis, growth, breakdown, and disintegration.

Your Opinion 4.6

Does the paucity of geographic research on world regions surprise you? It might be seen as one component of the larger weakness of geography in American education, although such an observation does not really explain the similar lack of scholarly concern with global regions in European and some other countries. Perhaps the reason that cultural geographers have rarely ventured into this arena is that world regionalizations are simply so difficult to accomplish, resulting as they do in broad generalizations. Difficulties aside, there does appear to be considerable merit in thinking at the world scale, although it must not be forgotten that the regions delimited, like any other sets of regions, are necessarily mere approximations of a complex reality.

- For *genesis* to occur, five types of challenge or stimulus were proposed, namely difficult physical environment, availability of new territory, military defeat, regular aggression from outside, and discrimination by others. Such challenges needed to be at an appropriate level, that is, sufficient to encourage a creative response without being too devastating.
- *Growth* followed genesis if there were continued responses to subsequent challenges. In both the genesis and growth stages, a creative minority of the population took the lead in initiating responses to challenges.
- Continued growth eventually led to a situation where growth was no longer possible, and the third stage, *breakdown*, began. This occurred when there was a loss of unity and direction related to the demise of the creative minority. New institutions were needed but not created, such that responding to a new challenge consumed large amounts of energy, making it difficult to respond to subsequent challenges.
- Finally, *disintegration* occurred, often involving major social schisms.

In accord with these arguments, Toynbee identified twenty-three civilizations in world history, subsequently increasing the total to twenty-six, ten of which were defined as living. The ten living civilizations were Western Christendom, Orthodox Christendom, the Russian offshoot of Orthodox Christendom, Islamic culture, Hindu culture, Chinese culture, the Japanese offshoot of Chinese culture, Polynesian, Nomad, and Inuit, with the last three defined as living but arrested. As the regional names suggest, the basic criterion employed was that of religion. The exclusion of Africa from this classification results from the fact that it was regarded as primitive, not culturally committed, and hence did not qualify for inclusion.

The typical reaction to this corpus of work has been predominantly negative, at least until relatively recently. This is not surprising given the difficulties of working at a global scale, especially given the generalizations required. Also subject to criticism has been the naturalistic idea that civilizations can be treated as organisms—perhaps they may be more appropriately seen as adaptive systems or ecosystems that can be understood as responses to ecological opportunities.

Notwithstanding these problems, it is now increasingly recognized that there is some value in viewing civilizations at a grand global scale as this may involve a useful integration of economics, politics, society, and culture, and may also offer insights into both past and present that more local scale analyses are not able to achieve. The idea of some dogmatic sequence of stages is not popular, but the idea that civilizations are meaningful units for analysis remains. Civilizations may be more akin to functional than to formal regions in that they are linked by human connectedness rather than by human relatedness or uniformity—recall the important distinction noted in Box 4.1.

The Evolving World System

World system ideas can be seen as a more recently developed alternative way to approach long-term historical change. Whereas civilizationists often stress the isolation and distinctiveness of regional civilizations, the **world system** concept focuses on integration and interaction. A world system is a large social system that possesses three principal distinguishing characteristics:

- it is autonomous, meaning it can survive independently of other systems
- it has a complex division of labor, both economically and spatially
- it includes multiple cultures and societies

Historically, world systems covered only some part of the entire world, but the current capitalist world system can be seen as encompassing the entire world. Even when a world system covered only a part of the world, it could qualify as a world system in the sense that it was a form of self-contained world.

The current capitalist world system that began about 1450 with the onset of European overseas expansion has *broadened*, that is, expanded spatially, to cover the entire world. This broadening has involved a division into three unequal but interdependent regions—a dominant core, a semiperiphery, and a subordinate periphery. In order to maintain a position of privilege, the core needs to maintain the underdevelopment of the periphery. The current capitalist world system has also *deepened*, that is evolved into a more complex system; this deepening involves a set of processes that are more fully outlined in Chapter 7. Further, both the broadening and deepening of the system can be understood in terms of the Marxist concept of ceaseless capital accumulation as this is introduced in Chapter 7. The world system approach has focused primarily on the modern world system, but it is possible to extend the ideas to a much longer time scale comparable to the 5,000-year time scale employed by civilizationists.

CIVILIZATIONS AS WORLD SYSTEMS

Expressed rather simply, the greatest single difference between the civilizationist and world system approaches to the study of long-term historical change and of the shaping of the modern world concerns the civilizationist interest in culture and the world system interest in politics, economy, and connections between regions. This distinction, which was never really that clear, is increasingly less clear as evidenced by the discussion above and also by some attempts to integrate the two approaches.

One extension of civilizationist ideas sees civilizations as multiple cultures, polycultures that are best seen as sociopolitical units comprising multiple states linked either in alliance or conflict. The suggestion that conflict might serve as an integrating mechanism is not a well-established theoretical construct. Extending this general idea, world history can be interpreted in terms of a process of civilization integration that has resulted in a single dominant central civilization.

> The single global civilization is the lineal descendant of, or rather I should say the current manifestation of, a civilization that emerged about 1500 BC in the Near East when Egyptian and Mesopotamian civilizations collided and fused. This new fusional entity has since then expanded over the entire planet and absorbed, on unequal terms, all other previously independent civilizations (Wilkinson 1987:46).

Table 4.1 lists these various civilizations. Thus, central civilization is the forerunner of the current world system. Central civilization formed about 1500 BCE, grew spatially and demographically, and is now global in scale. As such, it can be regarded as the key entity in any consideration of world systems and world economies. The central civilization is the same entity as the modern world system as defined by Wallerstein.

'The Greatest Topic in Historical Geography'

Cultural and historical geographers have chosen not to address the challenging question of a geography of the formation of global regions as one of their principal research topics. Certainly, there is no equivalent to either the civilizationist or the world system approaches. This failing is notwithstanding the plea from Meinig (1976:35) that this is 'the greatest topic in historical geography'. Given the presumed cultural importance of central civilization or of the capitalist world system for any understanding of global cultural issues, there is good reason to suggest that cultural geographic interest in this topic should increase significantly.

There are three aspects to the cultural geographic interest in civilizationist and world system concepts, each of which is noted briefly at this time and more fully discussed later, with the first topic as the focus of the next section and the second and third topics being reintroduced and further discussed in Chapter 7 (see Box 7.7).

- Civilizationist concepts link directly with cultural geographic attempts to delimit regions at the world scale.
- According to the basic logic of both central civilization concepts and of world system theory, the significance of the fact that Europe has played the critical role in the last 500 years of world history is that Europeans have not only effectively taken over the world, but they have also reconstructed the world in their image. This reconstruction has been political, economic, and cultural, and is variously called westernization or modernization. The significance of this situation is difficult to exaggerate; it was referred to in the Chapter 3 account of cultural contact and resurfaces especially in Chapter 7.
- The growth of a capitalist world economy can be seen as a major global transformation in cultural

Table 4.1: **Incorporation of Civilizations into Central Civilization**

Civilization	Duration	Terminus
Mesopotamian	Before 3000 BCE–1500 BDE	Coupled with Egyptian to form Central
Egyptian	Before 3100 BCE–1500 BCE	Coupled with Mesopotamian to form Central
Aegean	2700 BCE–560 BCE	Engulfed by Central
Indic	2300 BCE–1000	Engulfed by Central
Irish	450–1050	Engulfed by Central
Mexican	Before 1100 BCE–1520	Engulfed by Central
Peruvian	Before 200 BCE–1530	Engulfed by Central
Chibchan	?–1530	Engulfed by Central
Indonesian	Before 700–1590	Engulfed by Central
West African	350–1550	Engulfed by Central
Mississippian	700–1590	Destroyed (pestilence?)
Far Eastern	Before 1500 BCE–1850	Engulfed by Central
Japanese	650–1850	Engulfed by Central
Central	1500 BCE–present	?

Note: All dates are approximate.

Source: D. Wilkinson, 'Central Civilization', *Comparative Civilizations Review* 17 (1987):31–59.

as well as in economic terms; indeed, it is often asserted that the world has changed from a mosaic of groups with different cultures to a more culturally unified world. This change is related to the increasing importance during the past 500 years of the nation state and to the roles played by the three hegemons, namely the seventeenth-century Dutch, the nineteenth-century British, and the twentieth-century Americans. Thus, the transition is from a world differentiated in cultural terms to a world differentiated in political terms.

Global Regions

Lewis and Wigen (1997:14) argued, 'we would insist that world regions—more or less boundable areas united by broad social and cultural features—do exist and that their recognition and delineation are essential for geographical understanding'. This section outlines three proposed regionalizations.

- The first is the classic cultural geographic regionalization derived from the Sauerian tradition that recognized a number of *Culture Worlds* on the grounds that the 'logical approach to regional geography is one that is based on cultural outlines' (Russell and Kniffen 1951:viii). This work had a considerable impact on human geography generally as it replaced environmental determinism with a form of cultural determinism.
- The second is the work of a political scientist that identified cultural regions as the appropriate basis for understanding the contemporary world on the grounds that conflicts are no longer based on politics and ideologies but rather on differences in culture, hence the modern world is seeing *The Clash of Civilizations* (Huntington 1996).
- The third is an original and innovative geographic contribution that critiqued earlier work, stressed the heuristic value of global regions, and explicitly identified *The Myth of Continents* (Lewis and Wigen 1997).

In addition, cultural geographers have traditionally emphasized the roles played by language and religion as indicators of various cultural differences from place to place, and Box 4.5 provides a brief overview of these two variables in the global context.

Culture Worlds

As noted earlier, the landscape and cultural region concepts were not devised, methodologically speaking, with a global scale in mind, but there have been attempts to divide the world into cultural regions as a means of facilitating general world comprehension. The pioneering textbook in world cultural regional geography grouped people according to culture, related culture to area in order to derive culture worlds as of the mid–twentieth century, and showed that such culture worlds had formed over a long time period. The problems of regional classification at the global scale were recognized and various subregions identified. The seven culture worlds, shown in Figure 4.14, were summarized as follows.

- The European world was described as aggressive in the sense that Europeans typically assumed that most of their everyday cultural matters were universal, whereas they may of course have been foreign to those outside of Europe. This critical point was raised as the second of three topics at the conclusion of the previous section. Although contemporary discussions in cultural geography

Box 4.5: Language and Religion: Classification and Distribution

There are many human languages, each subject to changes in rules, content, and dialect. Interestingly, there is much debate concerning the number of languages spoken today (with the usual estimate around 6,000), although one recent major report identified about 10,000 (see Carvel 1997). To aid understanding of the origin, diffusion, and current distribution of these languages, cultural geographers employ the concept of a **language family**—a group of closely related languages that show evidence of having a common origin—although it is now widely accepted that the various language families are themselves related. Specific classifications vary in detail, but most recognize at least fourteen principal families. Within language families there are usually several subfamilies. Different languages are by definition mutually unintelligible, but each individual language may comprise several dialects—that is, mutually intelligible varieties of that language.

The numerically largest and spatially most widespread family is the Indo-European, which includes most European languages and also those of the Indian subcontinent. The dominance of this family in the world context is related to the process of European colonial expansion. Principal subfamilies within Indo-European include: Romance, which evolved from Latin and includes Portuguese, Spanish, French, and Italian; Germanic, which includes English, Norwegian, Swedish, and German; Slavic, which includes Polish, Ukrainian, and Russian; and Indic, which includes Hindi and Bengali. There are smaller Celtic and Baltic subfamilies and also some individual languages, such as Greek and Albanian, which do not clearly belong to a subfamily. The common bond between all of these languages is their origin in eastern Europe perhaps some 8,000 years ago in the form of proto Indo-European, itself but one branch of a much larger linguistic tree.

Classifying religions has proven rather more straightforward than is the case for language, with the two major types recognized by cultural geographers as ethnic and universalizing. An **ethnic religion** is closely identified with a particular group and does not actively seek to convert others. Numerically, the principal ethnic religion is Hinduism, the oldest of the major religions originating in the Indus Valley about 4,000 years ago. Other ethnic religions include Judaism, Confucianism, Taoism, and Shinto.

A **universalizing religion** actively seeks converts because of the claim that the religion is proper for all people—the three principal examples are Buddhism, Christianity, and Islam—each of which can be further subdivided. Buddhism, an offshoot of Hinduism developed about 500 BCE; Christianity developed from Judaism at the beginning of the Christian Era, and Islam developed from both Judaism and Christianity about 600 CE. Historically, Christianity has been especially concerned with seeking converts and, because it was associated with Europe during the period of European overseas expansion from the fifteenth century onwards, it has spread with Europe over much of the globe along with some of the Indo-European languages, notably Spanish, Portuguese, French, and English.

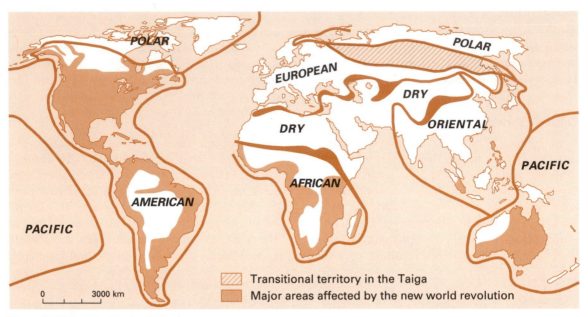

Figure 4.14: World cultural regions I. 'Through the device of recognizing seven culture worlds, the geography of the inhabited parts of the earth is presented in an orderly manner. Each culture world is a reasonably unified subdivision of the earth's surface occupied by peoples who are strikingly alien to inhabitants of other culture worlds' (Russell and Kniffen 1951:viii). The global regions mapped and depicted in this figure are the European world, Oriental world, Dry world, African world, Polar world, American world, and Pacific world.

A transitional area is located between the European and Polar worlds and large areas outside of Europe are seen as affected by a New World Revolution, notably the American, African, and Pacific worlds. Each of the regions is discussed in considerable physical, historical, and cultural detail. This pioneering use of the cultural region concept at the world scale is a major development within the Sauerian tradition. A principal question about this contribution concerns the use of the label 'Dry world' as the use of a climatic term does not accord with the idea of a cultural region. There are many other schemes for world regionalization employed by textbook authors for essentially heuristic reasons, but most are similar to, although different in detail from, this pioneering effort.

Source: R.J. Russell and F.B. Kniffen, *Culture Worlds* (New York: MacMillan, 1951).

are more theoretically and socially informed than they were in the past, it is clear that this point was not ignored in earlier cultural geographic discus-sions. Key European cultural traits were identified as those of field agriculture, industrialization, urbanization, and labor specialization.

- Two separate cultural hearths were noted for the Oriental world, namely northern China and the Indus Valley. Although the Indus Valley hearth showed some similarities to the European world, it was argued that it was closer culturally to northern China.
- Between the European and Oriental worlds, and also extending through northern Africa, is an area of semiarid and arid climate that discouraged interaction between these two worlds and that was the location for the formation of the culture world labeled the Dry world. In this area, nomadic peoples developed ways of life quite dif-ferent from those in the neighboring more humid

areas. The Dry world also served to separate the African world from the European.
- In the African world, people have solved prob-lems in a different manner than elsewhere and have therefore developed different cultures.
- The isolated Polar world, characterized by a severe climate, extends across northern Eurasia and northern North America.
- Both the American and Pacific culture worlds devel-oped largely in isolation from other areas until the beginnings of European overseas movement. Although both of these areas displayed significant internal cultural differences, sufficient similarities were noted to allow recognition as culture worlds.

This regionalization, and many subsequent regionalizations included in introductory human and regional geography textbooks, were essentially attempts to classify the world into a fairly small number of discrete areas that demonstrated a degree of internal homogeneity and that were different from other areas. As such, they served as heuristic devices to facilitate larger textbook accounts of world regional geography. Such regionalizations are easily criticized as attempts at delimiting meaningful cultural regions, principally because the larger the area to be divided into regions, the more superficial or more numerous the regions become. As previously noted, world regional classifications typically result from a process of logical division of a large area. Such classifications require that a great deal be known about the locations prior to the act of regionalizing so that sensible distinguishing criteria can be used.

The Clash of Civilizations

The second world regionalization discussed originated from concerns that were quite different from those of Russell and Kniffen. In an attempt to understand the principal schisms in the world following the termination of the cold war in the early 1990s, it has been proposed that these would be located along the peripheries of the major world civilizations or cultures. The central argument put forward was that cultures and cultural identities, as in the larger civilizationist tradition discussed in the previous section, which can be more broadly conceived as civilizations, are shaping relations in the contemporary world in the sense that groups sharing cultural values are increasingly cooperating with one another. Further, the contemporary world, since about 1990, is seen as multicivilizational—this is an important claim that contrasts with the ideas of central civilization and of world systems, both of which see a single large area, the West, as dominant globally.

The most intriguing aspect of this argument for cultural geographers is the claim that, unlike

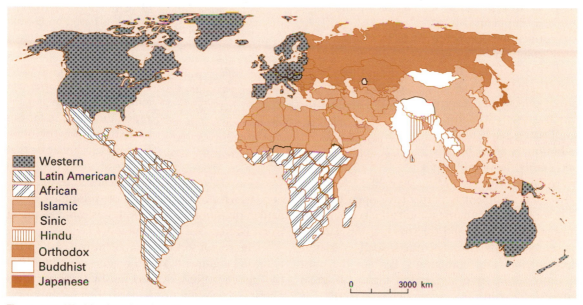

Figure 4.15: World cultural regions II. Although Huntington insisted on the importance of regional civilizations as the basis for understanding contemporary world conflict, the task of identifying those regions was not straightforward. Huntington (1993:25) named but did not map 'seven or eight major civilizations.... Western, Confucian, Japanese, Islamic, Hindu, Slavic-Orthodox, Latin American, and possibly African civilization'. Africa was listed as possible because of the divisions into Islamic and non-Islamic and into Saharan and sub-Saharan.

This figure shows the nine civilizations depicted on the map included in Huntington (1996), but even this second regionalization contains some uncertainties and apparent confusions. The map and the discussion of the map are not in complete accord. Thus, in referring to the map, Huntington (1996:21) noted that there are 'seven or eight major civilizations', although nine are included on the map.

Following the statement, 'The major civilizations are thus as follows', Huntington (1996:45) listed and briefly discussed only seven civilizations—Sinic, Japanese, Hindu, Islamic, Western, Latin American, and African (possibly); neither Orthodox nor Buddhist were identified as major civilizations, although both were included on the map. It was noted separately from the list that there is a Theravada Buddhist civilization in Sri Lanka, Burma, Thailand, Laos, and Cambodia.

Source: Adapted from S.P. Huntington, *The Clash of Civilizations and the Remaking of World Order* (New York: Simon and Schuster, 1996):26–7.

most nineteenth- and twentieth-century conflicts based on economic and ideological antagonisms, post–cold war conflicts are related to differences in culture. Hence, the critical need to map cultures and thus identify those areas (called fault lines) susceptible to conflict. This regionalization contains no great surprises given the earlier work of civilizationists, as discussed in the preceding section, and given the standard regionalizations proposed by cultural geographers, as exemplified by Russell and Kniffen. The key criterion used to identify civilizations is that of religion, as several of the regional names indicate (see Figure 4.15).

This argument has generated substantial discussion in political science as it represents a marked departure from the conventional concern with states as the key units in a conflict situation. There has been some broad agreement concerning the idea that future conflicts are likely to occur in boundary areas, but not necessarily the boundaries between world civilizations. The argument is particularly open to criticism for being an attempt to build the West around conservative values and for stressing the dangers presented to the West by others, such as Islamic fundamentals and immigrants. Perhaps the most credible alternative suggestion (again of great interest to cultural geographers) concerns the prospect of future conflicts related to discord within political units that are prompted by massive variations in quality of life, which are in turn often related to ethnic and cultural differences, or, more directly, to resurgent ethnicities.

Although cultural geographers have not traditionally concerned themselves with conflict issues, these two suggestions about likely future conflict emphasize the close links between political, economic, and cultural themes. As noted in Chapter 1,

good cultural geography, like good political and economic geography, needs, at times, to be broadly concerned with all other aspects of larger human geography.

The Myth of Continents

Attempts to regionalize the world using cultural criteria are problematic. It is not an easy task to delimit large areas that can be meaningfully argued to be in some sense internally homogeneous. Of course, the conventional approach, both within and outside of geography, is to use continental divisions. Most of us tend to think of Europe, Asia, Africa, North America, and Latin America as regions, at least in some informal sense. The principal alternative approach is based on literate civilizations, as discussed earlier and as used by Huntington. It is not difficult to see that the attempts at world regionalization discussed so far can be criticized for the often flawed logic. It may be that these essentially traditional approaches to regionalization, which are often based on presumed continental divisions, are at best misleading and at worst wrong. (See Your Opinion 4.7.)

One response to this problem is based on the following argument: 'World regions are multicountry agglomerations, defined not by their supposed physical separation from one another (as are continents), but rather (in theory) on the basis of important historical and cultural bonds' (Lewis and Wigen 1997:13). This does not mean civilizations as discussed by civilizationists, which typically assume literacy; rather, global regions, large sociospatial groupings, are identified on the basis of a shared history and culture. The resultant regionalization (Figure 4.16) is based on processes and on traits, ignores political and economic boundaries, and employs cultural boundaries. (See Your Opinion 4.8.)

Your Opinion 4.7

One continental division that may be particularly unjustifiable is that between Europe and Asia as Europe is not a distinctive and separate land mass, being essentially a western peninsula of Asia. Is it reasonable, for example, to regard Europe as a continent and India as a subcontinent? The use of continents, or presumed continents, as a basis for regionalizing is certainly flawed, reflecting, as it does, a particularly Eurocentric view of the world—precisely the point made by Russell and Kniffen. Indeed, it is increasingly clear that the particular views of the world that a group holds are a reflection both of reality and of their position in the world. The cultural studies tradition referred to in the first chapter, especially postcolonial theory, has addressed these ideas more fully. For the moment our concern is: If not continental divisions, then what?

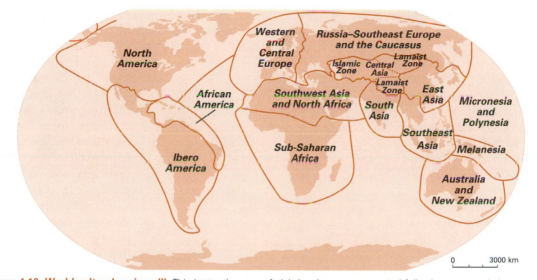

Figure 4.16: World cultural regions III. This interesting map of global regions was presented following an extended account that stressed the problems of regionalizing at this scale, but that also identified the value of regionalizing at this scale. It was acknowledged that 'all human geographical entities are conventional rather than natural' and that there is accordingly a need to use regional names that are generally understood rather than names that have different meanings in different contexts (Lewis and Wigen 1997:196).

The fourteen regions recognized are those of East Asia, Southeast Asia, South Asia, Central Asia (divided into Islamic and Lamaist zones), Southwest Asia and North Africa, Sub-Saharan Africa, Ibero America, African America, North America, Western and Central Europe, Russia-Southeast Europe and the Caucasus, Australia and New Zealand, Melanesia, and Micronesia and Polynesia. It is acknowledged that any attempt at regionalization is necessarily problematic.

Source: M.W. Lewis and K.E. Wigen, *The Myth of Continents: A Critique of Metageography* (Berkeley: University of California Press, 1997):187.

- First, it is helpful to think of Africa, Europe, and Asia as comprising the supercontinent of Afro-Eurasia.
- Second, note that denying that Europe is a continent does not prevent it from being treated as a cultural region because, of course, an area does not need to be a continent in order to merit cultural region status. Effectively, Europe can be thought of as Western Eurasia or as Northwest Afro-Eurasia.
- Third, the regionalization of Asia is reconceived, being divided into five regions, an explicit acknowledgment of both size and regional complexity. The inclusion of a Central Asian region is a distinctive addition to most regionalizations, as

is the reference to Lamaism (that is, Mahayana Buddhism).
- Fourth, the regionalization of the Americas is similarly reconceived. Latin America is rejected as a regional name, and the central American area is named African America.

Concluding Comments

A regional approach to cultural geography derives both from the established tradition of geographic regional study and from the specific incorporation of regions in the landscape school. Notwithstanding some terminological confusion regarding the words 'region', 'area', 'landscape', and 'place', cultural geographers have embraced the approach.

Your Opinion 4.8

What is your reaction to the three maps of global regions and the accompanying text comments? Do you feel that world cultural regionalizations of this type are suggestive of the value of regionalization as a device for obtaining a clearer understanding of the world? Do you feel that they demonstrate the difficulties, indeed dangers, of oversimplification? Given these questions, it is probably not surprising that most cultural geographic regionalizations are explicitly presented as devices to aid general world comprehension rather than as definitive statements.

Landscape and region are the preferred terms in this chapter, with landscape as a general term referring essentially to the visible scene, and region as the more specific term referring to a part of the surface of the earth that has a similar landscape. Application of the concept at the subnational scale is the favored approach with most studies centering on American regions. Application of the regional concept at a global scale, to aid the identification and mapping of cultural realms, has a venerable and controversial tradition and is currently experiencing something of a resurgence. There are two different, perhaps contradictory, approaches to global regionalization, namely the civilizational and the world system approaches. The resurgence of interest in global issues is related to the dramatic changes occurring in the contemporary world, especially concerning globalization.

There is a key outstanding issue that has been raised on more than one occasion in this chapter, but that has not been addressed directly. Recall that some regions can be described as homelands, a term that is employed to reflect an especially close link between people and both the physical and built landscape of their region. This idea has sparked some controversy and is a useful entry point into a larger debate about the legitimacy of distinguishing distinct groups of people and related distinct regions.

Homeland, as the term is used in this chapter, refers to such requirements as a people, a place, a sense of place, and control of place, or, alternatively, to such group characteristics as the length of time a people have been present in an area and the extent to which they are a majority of the population. As such, homelands are cultural regions that are especially closely linked to a distinctive group. But such a seemingly straightforward idea is not without critics. For example, both the identification of the Hispano group and the reference to a Hispano homeland have been questioned with the argument that there are two myths involved in these ideas, namely the myth that there is a distinctive Hispano group, and the myth that there is a real and discrete region occupied by the supposed group, a homeland. 'It is hard to imagine anything more powerful by way of an existence–statement about a culture than to depict its map location, its boundaries, and its internal subregions' (Blaut 1984:159). (See Your Opinion 4.9.)

Notwithstanding some methodological concerns, and as the discussion of regional studies in this chapter makes clear, many cultural geographers both favor and continue to work in a regional tradition that does not place priority on matters of social production. The success of this continuing and continually changing regional perspective, as demonstrated in this chapter, is certainly undeniable.

Your Opinion 4.9

There is little doubt that questions of group identity are complex, elusive, and certainly changing. Are we really able to divide people into neat parcels such as cultures or ethnic groups or are these groupings really arbitrary creations? There is no clear-cut response to this question and it is sufficient to raise these concerns in the conclusion to this chapter without attempting any resolution. Indeed, in addition to the elusiveness of group identity, it is now increasingly acknowledged that both human identity and human landscapes, regions, and homelands, are socially constructed or produced. Although there is debate on the merits of this perspective, the popularity of the idea in much contemporary cultural geography is undeniable. This chapter on cultural regions has largely ignored both this perspective and the debate surrounding it, as such matters are more appropriately introduced in chapters 7 and 8 as part of a more socially informed cultural geography.

Further Reading

The following are useful sources for further reading on specific issues.

Mason (1895) and Wissler (1917, 1923) are early anthropological applications of the regional concept, a tradition that culminated with the publication of a major work on the cultural and natural areas of North America (Kroeber 1939).

According to Carter (1948) the anthropological contributions attracted much geographic attention. A survey of the culture area concept as used in anthropology and a comparison with geographic applications was provided by Mikesell (1967:621–4, 626–8).

Hartshorne (1939, 1959) on geography as the study of regions.

Butzer (1989a) on the methodological conflict between Sauer and Hartshorne.

Harvey (1969) on regions as an example of classification procedures.

Gilbert (1988), Pudup (1988), Johnston (1990), and Lee (1990) on the new regional geography.

There are few examples of cultural region formation and transformation outside of North America that have employed core, domain, and sphere concepts notwithstanding the generic flavor of the model. For Wales, two principal cultural regions can be delimited according to the relative strength of the English and Welsh languages, and a bilingual zone between the two can be equated with the Meinig domain (see Carter and Thomas 1969; Pryce 1975). Also see Withers (1988) for an account of the Scottish Highlands cultural region.

Harris (1978) on the historical geography of North American cultural regions.

Uses of the first effective settlement concept include: Gastil (1975) in an account of areas of cultural homogeneity; Elazar (1984, 1994) on the United States political scene with reference to the westward migration of cultural traits but without any explicit use of the work of cultural geographers; Wacker (1968) on the Musconetcong Valley of New Jersey and Wacker (1975) and Wacker and Clemens (1995) on the larger area of New Jersey.

Aschmann (1965) used the preadaptation concept, without using the particular term, in a discussion of five Apache cultural traits that made possible their occupation of the American Southwest.

Jordan and Kaups (1989) on the linking of specific ethnic groups with specific environments and Garrison (1990) for a critical commentary on this idea.

West and Augelli (1966:11–16) identified cultural regions and related landscapes in Central America, empha-sizing the interplay of physical geography and history as reinforced by isolation. Jordan (1996) mapped regions in Europe based on particular variables, such as language and religion, while the attempt at a broadly defined cultural regionalization of the continent was in terms of North and South, East and West, and core and periphery.

Luebke (1984) and Shortridge (1995) explained many developments in the Great Plains in terms of distinct incoming cultures.

Meinig (1969) on Texas as a cultural region; Hilliard (1972) and Nostrand and Hilliard (1988) on the American South; Cobb (1992) on the Mississippi Delta; Wyckoff (1999) on Colorado.

Jordan, Kilpinen, Gritzner (1997) on the folk landscape of the North American mountain West.

Rose (1972) and Jeans (1981) on the possibility of delimiting cultural regions in Australia.

Christopher (1984:193) noted that throughout much of Africa 'New European societies were formed which were able to impress their ideas upon the landscape and create an image of France or England overseas'; nevertheless, there has been little interest in identifying any resultant cultural regions.

Zdorkowski and Carney (1985) identified past vernacular regions through the inclusion of a question on name changes through time; Lamme and Oldakowski (1982) reported a close link between vernacular regions and an established cultural divide in Florida; Miller (1968) studied the Ozark region with reference to folk materials, identifying the self-sufficient family farm as the principal way of life; Stein and Thompson (1993) discussed the distinctive identity of Oklahoma, as viewed from both inside and outside, in terms of what they called, psychogeography, with the emphasis on attitudes rather than material landscapes.

Trépanier (1991) and Estaville (1993) on French Louisiana.

Meinig (1965) is the pioneering cultural geographic analysis of the Mormons although the presence of the region was generally acknowledged prior to that analysis (Zelinsky 1961); Francaviglia (1978) provided the seminal landscape analysis (see also Jackson and Layton 1976, Jackson 1978); the symbolic importance of the region to Mormons has been described by Jackson and Henrie (1983).

Meinig (1971), Carlson (1990), and Nostrand (1992) on aspects of the formation of the Hispano cultural region and the characteristic landscape; Smith (1999) on cultural change occurring at the northern edge of the Hispano homeland.

Kollmorgen (1941, 1943), Augelli (1958), Lehr (1973), Raitz (1973b), Gerlach (1976), and Gade (1997) on ethnicity and cultural islands.

Melko (1969), Sanderson (1995), and Wilkinson (1987) on the writings of civilizationists.

Wolf (1982), Gills (1995), and Sanderson and Hall (1995) on world systems.

The principal contributions to world systems theory are by Wallerstein (1974a, 1974b, 1980, 1989).

Hugill (1997:348) noted that the 'the most vital debate in social science' is the attempt to develop a theory of the history of the capitalist world, and possibly even a theory for all human history. Political and economic geographers have made major contributions to this debate (Dicken 1992; Taylor 1996).

Mosely and Asher (1994) is a comprehensive atlas of language; Renfrew (1988) stressed the relationships between language families; the role of language as a basis for group delimitation and communication between group members has been noted especially by Wagner (1958b, 1974, 1975); cultural geographic overviews of religion include those by Sopher (1967, 1981), Gay (1971), and Park (1994).

Sauer (1940) outlined a global regionalization that stressed the historical origins of regions in early civilizations.

Ó Tuathail (1996:240–9) on possible failings of the Huntington world regionalization and Kaplan (1994, 1996) on suggestions that conflicts might be based on differences in qualities of life and access to resources.

Ecology and Landscape

That cultural geography has a considerable and sustained interest in ecological analyses—the study of organisms in their homes—is hardly surprising. As noted in the opening chapter, cultural geography, and for some even the larger discipline of geography, can be broadly interpreted as the study of humans and land. Of all the various approaches to the study of humans and land discussed in Chapter 2, it is several versions of the ecological approach that have proven most useful and that have been attracting increasing interest in recent years. Indeed, in addition to the continuing importance of what might now be thought of as traditional cultural ecological concerns (evident in some nineteenth-century geography and in the Sauerian school), there are other areas of more recent focus. For example, there is an increasing concern with human impacts on environment, a concern that has given rise to a plethora of literature in physical and social sciences. There is also a concern based on recent advances in biology that explicitly acknowledges the variability, rather than the stability, of nature. Further, and not surprisingly, developments in social theory have prompted more sophisticated versions of ecological thought, especially those of ecofeminism and political ecology. This chapter considers both the traditional and the newer interests. The sequence of material in this chapter is as follows.

- There is a discussion of ecology, stressing the nineteenth-century origins in physical science, the geographic interpretations made by Barrows and Sauer, the interests shown by other social sciences, and the use made of systems concepts. As this lengthy opening section makes clear, ecology is a multifaceted arena that, although sometimes referred to as a discipline, is more appropriately thought of as an approach open to multiple interpretations and applications.
- There is an account of cultural ecological analysis that includes brief overviews of resources as cultural appraisals and of population and technological changes through time. The key concept of adaptation, discussed in Chapter 3, is reintroduced and placed into a larger social science context. There is a discussion of the 'new ecology' that acknowledges the ever-changing natural world and a related brief comment on the origins of agriculture, one of the topics that has been frequently addressed in ecological terms. The section concludes with an account of cultural ecological analyses in a global environmental context, an account that, in the grand tradition of ecology, stresses the links between environmental and human matters.
- The human and nature relationship, previously discussed in Chapter 2, is reintroduced with emphasis on the two topics of environmental ethics, including the attitudes and values that relate to our behavior in environment, and of ecofeminism. Both are areas of increasing concern, with philosophers, psychologists, and many other scholars all having something important to say about the way in which humans perceive and use environments.
- There is a discussion of political ecology as this approach came to the fore in the 1980s in the context of attempts to understand problems of soil degradation in the less developed world. Political ecology can be broadly described as an integration of political economy and cultural ecology.

The many applications attest to both the popularity and success of the approach, especially in less developed world settings.

- The chapter concludes with an acknowledgment of the changing character of ecological approaches and a brief discussion of the prospects for ecological approaches as used by cultural geographers.

The earliest ecological analyses were physical science studies of the ways that plants and animals related to their environments, but studies of humans soon followed and, at first thought, it seems reasonable to suppose that if ecological analyses including humans are to have one logical disciplinary home, then that home should be in geography. But this has not proven to be the case. Put simply, geography, despite extending several invitations, has not proven to be an especially congenial abode for human ecology. Notably, the views of such mid-nineteenth-century geographers as Humboldt and Ritter centered on human and land relations, and these relations were also a concern in both the Vidalian and Sauerian traditions. But twentieth-century geographers tend to have identified both a physical science/human science division in their discipline, and a division between ecological approaches and other dominant approaches, first regional and later spatial analytic. The failure of geography to center on an ecological approach is at least partially a consequence of these internal identity uncertainties.

It is not difficult to understand why Zimmerer (1996:161) referred to ecological concepts as being both 'persistently foundational and yet doggedly problematic in their application' in human geography. Uncertainty concerning the merits of ecological approaches and concepts is evidenced throughout this chapter, although the principal conclusion reached is that the advantages currently perceived significantly outweigh the disadvantages.

Ecology: A Unifying Science?

Ecological approaches in cultural geography have a checkered history. Unlike the evolutionary and regional approaches, both of which have explicit foundations in the landscape school and subse-

quent histories of focused research activity, ecological approaches are not associated with a specific seminal statement, nor do they have a coherent body of empirical research. Ecology has a long scholarly heritage and specific origins in physical science, and there are versions of human ecology in several of the social sciences. As noted in Chapter 2, the term 'ecology' is derived from two Greek words, *oikos* meaning 'place to live' or 'house', and *logos* meaning the 'study of'. Accordingly, the central concern of ecology as a science is with the ways in which living things interact with each other and with their environments or homes.

Nineteenth-Century Origins

TWO MISCONCEPTIONS

It is helpful to begin this section by dispelling two popular misconceptions.

- It is a mistake to equate ecology with concern for environmental **conservation** and preservation; ecology was not introduced in response to concerns about human use of land. Indeed, in nineteenth-century thought there were numerous interpretations of the human and nature relationship, including two diametrically opposed general views of nature; from one perspective, nature was there to be exploited for the benefit of humans while from another perspective, nature needed to be conserved in order to minimize the damage being inflicted by humans. Ecology came to the fore in this contradictory intellectual environment, with some scholars regarding ecology as a means of facilitating the continuing exploitation of nature by pointing out ways in which the resultant damage could be minimized.
- Notwithstanding some attempts in this direction and given the current organization of knowledge, it is also a mistake to think of ecology as an academic discipline. The origins of ecology as a scientific approach can be traced to the third chapter of Darwin's *The Origin of Species*, which considered the various adaptations and interrelationships of organic beings and environment, although the more general context was biologists' attempt to understand how the distribution of species is related to environment. Few of the early ecolo-

gists were interested in questions of species evolution. The first use of the term 'ecology' is generally considered to be by Haeckel in the 1860s, and by the end of the nineteenth-century ecological approaches were well established in both plant and animal science. There has never been a single discipline of ecology; rather, ecological concepts, notably the concept of a plant or animal community, were applied to a number of existing disciplines: 'many scientists could do "ecology" while retaining their primary disciplinary loyalty elsewhere' (Bowler 1992:365). This multidisciplinary characteristic of ecology is a recurring component of the discussion in this section.

ECOLOGIES IN PHYSICAL SCIENCE

Ecology is, then, best thought of as an approach to the study of relationships between organisms and their environments and not as a particular body of subject matter, and it is for this reason that there are ecological approaches in many of the physical and human sciences. This is precisely how ecological concepts, as applied to humans, were introduced in Chapter 2 along with alternative approaches such as environmental determinism and possibilism.

Early advances in ecological thought are inextricably bound up with various arguments put forward by Spencer. Recall from Chapter 2 that Spencer, in his articulation of what is usually labeled Social Darwinism, argued for an analogy between animal organisms and human groups with regard to the struggle to survive in a physical environment. The logic of this Social Darwinism, a form of naturalism, is straightforward, with cultural groups evolving in accord with their ability to adjust to particular physical environments. Indeed, it was Spencer who introduced the oft-quoted phrase, 'survival of the fittest'. These ideas are related to views about biological evolutionism as an inexorable form of social progress and, of course, to the conception of society as a superorganic entity.

Following introduction of the term 'ecology', the first significant advances were made by Clements, an American botanist, who developed the idea that a plant community was a distinctive superorganism that had a life of its own. The intel-

lectual inspiration for this idea was the work of Spencer, and there are parallels with the superorganic concept of culture. Thus, a plant community was seen as something more than the sum of the individual species and as responding to laws that operated only at the community level. The first textbook to describe the new methodology of ecology as applied to plants was *Research Methods in Ecology* (Clements 1905). The Spencerian idea of social progress was reflected in the concept of climax vegetation, which is the logical outcome of a process of evolution.

By 1900, ecology was an established approach in biology, botany, and marine science—the British Ecological Society was founded in 1913 and the Ecological Society of America in 1915. Six principal related concepts were employed in early physical science ecological analyses.

- *Community* refers to the idea of a group of related organisms in a particular area.
- *Competition* refers to the struggle between member species of a community.
- *Invasion* refers to the ability of some species to take over areas occupied by another species.
- *Succession* refers to the sequence of change as a community moves toward the finished or climax situation.
- *Dominance* refers to the fact that one or more species in a community is normally able to seek out and occupy the most favorable environment.
- *Segregation* refers to the spatial patterning of species.

INTRODUCING HUMANS

Although, overall, ecology has been concerned with the study of nature without humans, or certainly without privileging humans, versions of human ecology soon appeared. The first direct reference to humans was made by Darwin in *The Descent of Man*, published in 1871, but it was Spencer who was the principal intellectual inspiration for the early advances and it was plant ecology that provided the basic concepts. Stoddart (1986:168) noted that 'from about 1910 "human ecology" was used for the study of man and environment, not in a deterministic sense, but for man's place in the "web of life" or the "economy of nature"'.

The principal concerns of ecological analyses today reflect these nineteenth-century origins with a continued focus on the relationship of a given organism both to other organisms and to surroundings. But, as discussed later in this chapter, the understanding of surroundings in contemporary ecological analyses centered on humans is much broader than in the past as it is less closely tied to concepts developed in physical science and more culturally informed. Indeed, of all the various ecologies, it is the several versions of human ecology that have the most tenuous links to the parent physical science disciplines.

Human Ecology and Geography

Ecology is an approach applied in several academic disciplines, and what is sometimes called human ecology is itself an approach applied in a number of academic disciplines. Thus, Young (1983:2) described human ecology as 'a strange nonfield, a hybrid attempt to understand the ecology of one species, the being *Homo sapiens*'.

The first geographic use of the term 'human ecology' was in 1907 by Goode, a geographer at the University of Chicago. Following Clements and other plant scientists, Goode envisaged a human ecology as an organizing concept for the discipline of geography. Such an idea was popular at the time with several ecologically inspired papers presented at meetings of geographers, while Huntington became the second president of the Ecological Society of America. The culmination of this interest in a human ecology was the seminal methodological statement by Barrows (1923), 'Geography as Human Ecology', initially delivered as a presidential address to the Association of American geographers in 1922. It is useful to think of this statement as another in the series of attempts to claim a particular identity for the discipline of geography. (See Your Opinion 5.1.)

THE CONTRIBUTION OF BARROWS

- The specific aim of the 1923 statement was to define the field of geography as the study of mutual relations between man and his natural environment and thus as human ecology.
- More generally, the intention was to carve a niche for geography separate from other academic disciplines: 'Geography finds in human ecology, then, a field cultivated but little by any or all of the other natural and social sciences. Thus limited in scope it has a unity otherwise lacking, and a point of view unique among the sciences which deal with humanity' (Barrows 1923:7).
- Emphasis was placed on adjustment to, as opposed to influence of, physical environment. Geography was distinguished from geology and other physical sciences in that it was not concerned with the origins and distribution of physical features, such as landforms, climates, vegetation types, and soil types, but only with human adjustment to those features.
- An explicit distinction was also made between geography and history: the historian began 'with what our remote ancestors saw' while the geographer began 'with what we ourselves see' (Barrows 1923:6).
- In sum, geography as a discipline needed to concentrate on human ecology interpreted as human adjustment to physical environment.

Overall, geographic contemporaries of Barrows had good reasons to be somewhat perplexed at the thrust of this statement. Given the established concerns of geography at the time, some objected to the explicit exclusion of the traditional geographic content of physical geogra-

Your Opinion 5.1

Is it surprising—given the long history of interest in human and land relations and the specific geographic precedent set by Humboldt and Ritter—that none of the pioneering definitions of geography by leading late nineteenth-century practitioners chose to identify geography as human ecology? The closest to the major statements were Ratzel's claims concerning environmental determinism and the idea that political states could be analyzed like organisms. Most of the other statements focused on the region as the core idea of geography. It might be argued that the twentieth-century history of geography would have been very different had one of the late nineteenth-century practitioners chosen to stress an ecological focus.

phy, others were concerned at the lack of reference to the traditional regional focus, while still others were dismayed at the rejection of environmental determinism. Overall the statement was not well received and it is not surprising, therefore, that Barrows created little positive impact on the discipline of geography. In retrospect, it can be seen as an idiosyncratic attempt to forge an identity for the discipline.

Certainly, the interest in human ecology as an approach to geography following Barrows can be best described as uneven and uncertain. Goode (1926) discussed the geography of Chicago with reference to a human ecological perspective but without any clear basis in the statement from Barrows. One impact on university instruction was the textbooks, *Geography: An Introduction to Human Ecology* (White and Renner 1936) and *Human Geography: An Ecological Study of Society* by the same authors (White and Renner 1948). From this perspective, geography was the study of human society in relation to the earth (Box 5.1). Overall, however, most subsequent statements about geography as human ecology were not sympathetic to the Barrows interpretation.

LINKS TO THE LANDSCAPE SCHOOL

The fact that the landscape school incorporated ecological content—ideas about human and land relationships—is, of course, clear from a consider-

ation of the discussion at the beginning of Chapter 3 and especially from the landscape school diagram (see Figure 3.1). Given this earlier discussion, it is not necessary to belabor the links between the two at this time. Hence, this account is a brief one.

Although it is usual to acknowledge Sauer as one of the geographers who first expressed substantive concern about human impacts on earth, such acknowledgment is usually based on his writings from the 1950s and later. But Sauer was from the very beginning expressing such concerns, at least partly because of his strenuous opposition to environmental determinism.

- Thus, Sauer (1924:33) noted: 'Land is passing out of economic use in this country more rapidly than new land is being occupied. Timber devastation, soil erosion, overgrazing, soil depletion, failure of irrigation, dry farming, and drainage, here and there causing a pitiless revaluation of areas. Land is increasingly becoming a restricted economic good.'
- Again, Sauer (1927:192) claimed: 'To what extent is man as a terrestrial agent, that is by his areal expressions of culture, living harmoniously in nature (symbiotically), and to what extent is he setting narrowing limits for future generations by living beyond the means of the sites that he occupies? Man . . . appears, periodically, to effect his

Box 5.1: Adjusting to Environment—a Human Ecological Theory

White and Renner (1948:635–7) outlined a human ecological theory for geography that built on the ideas of Barrows. The theory can be summarized as follows.

- Human societies are living organisms with a particular social structure and patterns of activities.
- On encountering a new environment, a group adapts that environment to their needs and habits so far as is feasible, but also adjusts their previous institutions and behaviors as necessary.
- Thus, a human group in an environment is always some compromise between past practices and environmentally inspired responses.
- The adjustments to environment comprise three types of relationship, namely use relationships, control relationships, and ideological relationships, all of

which are continually changing in response to technological change.
- The outcome of these relationships is the creation of a cultural landscape.
- 'It is the geographer's contention, therefore, that human society can be understood only when its culture, make-up, and behavior are viewed against the background of its location, the space it occupies, and the resources which it utilizes' (White and Renner 1948:636).

This interesting formulation has not been significantly developed or applied by other geographers, although it is clearly similar in general outline to much of the cultural geography discussed in both chapters 3 and 4.

own ruin. . . . He has certainly been an engine of unparalleled destruction'.

Certainly, a third principal approach initiated within the Sauerian tradition, along with landscape evolution and regionalization of landscape, was ecology. Nevertheless, the ecological content is not as readily acknowledged as are the other components of the landscape school, at least partly because Sauer rarely used the terms 'ecology' or 'human ecology'. Possibly this is because Sauer took exception to the statement by Barrows on the grounds that it involved an inappropriate limitation on the discipline of geography, rendering it simply the study of humans in relation to natural environment. In common with the majority of geographers at the time, Sauer was less than enthused with the human ecology proposed by Barrows, and it may be that the failure to integrate the two approaches was partly a result of Sauer objecting to Barrows's exclusion of physical geography. Interestingly, several of Sauer's students completed major ecological analyses, including some studies of vegetation and the exemplary cultural geographic studies by Wagner (1958a) and Mikesell (1961). For Mathewson (1999:268), ecology's place in many of the cultural landscape studies conducted by landscape school geographers is 'implicit and pervasive' and indeed there does not seem 'to be any end in sight for this tradition, despite the emergence of new varieties of cultural geography whose practitioners jejunely and routinely issue

---

Your Opinion 5.2

Why was the landscape school statement by Sauer so influential while the human ecology statement by Barrows had a relatively minimal impact? A brief comparison may be instructive. Thus, Sauer not only proposed a geographic approach, he also restated and clarified his position on several occasions, he accomplished analyses based on the approach, and he attracted students to work on related projects. Barrows was not similarly successful in any of these areas. With regard to intellectual antecedents, the landscape school, while original in an American context, built on a rich tradition of European work and incorporated physical geography and regional study, whereas Barrows was arguing for something that was, relatively speaking, totally new. Given these differences, the human ecology proposed by Barrows was not readily endorsed by other geographers. Indeed, Barrows was criticized by some contemporaries for advocating little more than a diluted version of environmental determinism, namely adjustment, and by other contemporaries for the exclusion of physical geography. Sauer, on the other hand, incorporated the physical environment in his conceptual framework, explicitly rejected environmental determinism, and also related the landscape school ideas directly to other geographic writings.

---

uninformed dismissals of what has preceded them'. (See Your Opinion 5.2.)

SOME OTHER PROPOSALS

Most of the suggestions that geography adopt a human ecological perspective have come from geographers sympathetic to an integrated physical and human geography.

- A second landmark geographic statement, after Barrows, was by Thornthwaite (1940:343), who was a student of Sauer: 'human ecology must transcend all of the present academic disciplines, and . . . the development of a science of human ecology must involve cooperation of geography, sociology, demography, anthropology, social psychology, economics, and many of the natural sciences as well'. Such a view of a unifying human ecology, a cooperative enterprise, bears little resemblance to either the earlier statement by Barrows or the Sauerian cultural tradition.

- A later statement focused on the community concept, drawing parallels between biological and human communities. This account acknowledged that any form of organic analogy in the Spencerian tradition was subject to criticism, but argued the case for viewing human communities in two ways: '1) Ecologically, as aggregates of different species focusing on man, including all the animals and plants which depend on man and all those on which man in turn depends. 2) Socially, as aggregates of one species, consisting of individuals with a variety of interests and functions which, joined together, make possible one social and economic unit' (Morgan and Moss 1965:348). In both cases, the key feature is the uniting of individual community members.

- There have been other statements, such as that by Chorley (1973), that aimed to show how human ecology provided a unifying link between physical and human geography, but these have not typically resulted in significant empirical work.
- There have also been occasional statements that are quite different in tone, such as that by Norwine and Anderson (1980:vii), which built on the idea of the environment as a factor 'impacting man's destiny'. Again, such a statement was an exception that has had little impact on the discipline of geography.

Overall, it seems clear that human ecology has been characterized more by proposals of method than by demonstrations of method. Reviewing human ecology in geography between the years 1954 and 1978, Porter (1978:15) referred to 'a decade of progress in a quarter century'. However, since about 1978 there have been five principal advances in ecological thought in geography—namely the continuing application of systems concepts, analyses using the concept of adaptation, acceptance of the new ecology of instability from biology, analyses reflecting the views contained within new philosophies of nature, and the argument that spatial economies at all scales are a reflection of political economy. Discussion of each of these is delayed until later in this chapter.

Ecologies in Social Science

The geographic encounter with human ecology was not atypical. Other social sciences similarly enjoyed some acquaintance with the approach as earlier developed in physical science, but in all cases the contact was confused and uncertain. Certainly, human ecology was a fragmented field, multidisciplinary rather than interdisciplinary. Ecological approaches in sociology, anthropology, history, and psychology are now noted, especially as these aid the understanding of human ecological approaches in geography.

Sociology

Notwithstanding the fact that geography might be regarded as the most convivial disciplinary context for human ecology, given the traditional geographic concerns with both physical and human worlds, it was sociology that was the first of the social science disciplines to develop a coherent human ecological approach. This development took place at the University of Chicago where several members of the geography department, including both Goode and Barrows, were also expressing interest in human ecology. The key sociological figure was Park, one of the leaders of early American sociology. In conjunction with others—notably Burgess, McKenzie, and Wirth—Park successfully articulated and employed a human ecological approach to the analysis of urban areas.

For Park and his colleagues, usually described as the Chicago school of sociology, human ecology centered on the distribution of humans and of human institutions and their interactions. This sociological work involved the mapping of these distributions and explanation by reference to the various concepts developed within plant ecology. As previously noted, these are community, competition, invasion, succession, dominance, and segregation. Application of the ecological concepts was straightforward.

- Park employed the concept of an animal or plant community as virtually synonymous with society, the basic disciplinary concept of sociology.
- Humans were seen as competing for space, a competition that was reflected in differential urban land values and therefore varied land uses.
- Invasion and succession referred to land use changes through time.
- Dominance referred to the fact that particular land uses were able to locate in the most desirable area; within the city this area was the central zone and the dominant land uses were those belonging to the financial and retailing sectors.
- Segregation, the presence of urban areas usually distinguished by land use or ethnic identity, was one outcome of competition.
- These six related concepts were applied especially in the Chicago context, allowing the sociologists

to suggest a concentric zone model of intraurban land uses.

What Park accomplished was a human ecology explicitly modeled on plant and animal ecology—a human science derived from physical science. But these conceptual advances were stimulated not only by physical science ecologies, but also by personal experiences. 'I expect that I have actually covered more ground tramping about in cities in different parts of the world, than any other living man. Out of all this I gained, among other things, a conception of the city, the community, and the region, not as a geographical phenomenon merely, but as a kind of social organism' (Park 1952:5). Humans were regarded as organic creatures affected by the laws of the organic world.

It might be suggested that this is precisely the type of approach that geographers might have been expected to have adopted, but their interests typically lay elsewhere—in environmental determinism, in the regional approach, and, of course, in the emerging landscape school. Indeed, the sociologists placed great emphasis on the intellectual separation of sociology and geography, especially stressing that geography was an **idiographic** science whereas sociology was **nomothetic**. For Park and his colleagues, then, the role of geography was a quite limited one, namely to supply facts that the human ecologist could explain. Park also regarded geography as a discipline concerned with humans and their physical environments, whereas sociology was concerned with humans and their social environments.

This early human ecological approach in sociology was much more successful than were any of the early attempts to promote a geographic human ecology. The conceptual and empirical initiatives, mostly in midwestern American cities, prompted considerable interest, an interest that geographers did not share until the rise of a spatial analytic approach in the 1960s, by which time the approach was largely out of favor in sociology.

Despite this early work, the history of ecological analyses in sociology has proven uncertain. A principal criticism of this human ecology from within sociology, readily understandable in the context of the first of the two philosophical debates

introduced in Box 2.2, was of the interpretation of a human group in physical science terms. Hence, a major revision was the introduction of the idea that various symbolic associations and sentimental attachments could become identified with city areas and that these variables were able to counter the ecological processes. Most notably, Ericksen (1980) criticized previous approaches because of their de-emphasis of humans, because of the naturalism involved, and because most sociological human ecology thoroughly misrepresented society. Rather than treating the ecological process as dominant, emphasis was placed on humans themselves as leaders, not followers, of the ecological process. The conceptual inspiration for these ideas is the approach to sociology known as symbolic interactionism (Box 5.2).

ANTHROPOLOGY

The sociological interest in human ecology has had little impact on the work of cultural geographers, but the anthropological interest is another matter, a clear reflection of the close ties between these two disciplines since at least the 1920s. Although the origins of a human ecological approach in anthropology rely on the same physical science foundations as are appropriate for both geography and sociology, anthropologists have developed their contributions more fully. This is notwithstanding the fact that their initial interest came a little later.

Steward, who had a first degree in zoology and took graduate work with Kroeber, provided the seminal ecological statements in anthropology in a series of empirical analyses in the 1930s. Thus, ecology was introduced into anthropology by application rather than programmatically. A functional relationship between culture and ecology was considered similar to a cultural law.

The first explicit outline of an ecological approach came much later and Steward (1955:30) is generally credited with introducing a new term to describe the approach: 'In order to distinguish the present purpose and method from those implied in the concepts of biological, human, and social ecology, the term cultural ecology is used.' This approach of **cultural ecology** as both articulated and employed by Steward includes some notable characteristics.

- The failings of the various human ecologies result from an insistence that a new subdiscipline is being defined, whereas an ecological approach is better seen as precisely that—an approach—a means to an end and not an end in itself.
- The aim of the approach is to explain particular cultural features, not to derive general laws.
- Culture is included as a superorganic factor, with a distinction drawn between the behavior of humans compared to the behavior of other organisms. This is not, however, the extreme

superorganic concept—Steward (1955:36) also referred to the 'fruitless assumption that culture comes from culture'.
- The concepts of community, competition, invasion, succession, dominance, and segregation, as introduced by physical scientists and as adopted by Park and others, are rejected as being inappropriate for humans. Rather, it is culture that is seen as the key to understanding human relationships with nature. Competition, for example, may be present, but it can only be interpreted in

Box 5.2: Symbolic Interactionism

Symbolic interactionism refers to a group of related social psychological theories that were prompted by the ideas of the social philosopher, George Herbert Mead, and subsequently developed especially by Herbert Blumer. There are close links to the philosophy of **pragmatism**. The three fundamental premises are:

- Individual humans learn the meaning of things primarily through interactions with others, such that we learn how to define the world through our experiences of social interaction.
- Although social interaction presents meanings of things to us, we do not simply accept those meanings, rather they serve as a basis for our understanding.
- Human behavior results from our interpretations of the meanings presented to us.

In addition to these three fundamental premises, symbolic interactionists generally agree on an additional seven assumptions:

- Humans are unique in their ability to use language.
- Other unique human traits arise because of language.
- Human development is dependent on social interaction.
- Humans are purposive and rational animals.
- Humans seek rewards and avoid costs.
- Humans are conscious actors.
- Humans are active.

All major versions of symbolic interactionism accept the three fundamental premises and the seven additional assumptions, although there is often some disagreement concerning additional conceptual content.

It is not difficult to see that a human ecology built on these ideas is rather different from the other versions advocated by sociologists. Specifically, symbolic interactionism stresses that humans are different from other animals in some very important ways, whereas many

of the other versions of ecology, explicitly derived as they are from physical science, emphasize similarities between humans and other animals rather than differences. The most important difference identified by symbolic interactionists concerns the efficiency of human language and the implications for what it means to be human given that effective communication is the key to complex behavior. Other important differences include the interaction with others that characterizes human life, the purposiveness of human behavior, and the high degree of awareness that humans possess. It is also not difficult to see why Ericksen (1980) opted to employ a symbolic interactionist interpretation of human ecology. The key reason is the emphasis that the approach places on interaction, which is of course central to any ecological approach.

Although the human ecology proposed by Ericksen (1980) has not been adopted by cultural geographers, there has been discussion and application of the symbolic interactionist perspective. Most of the relevant work is associated with the cultural geography that criticized the perceived superorganic concept of culture employed by Sauerians, and the larger context for this work is contained in Chapter 8. As Duncan (1978:269) noted: 'The self is largely a product of the opinions and actions of others as these are expressed in interaction with the developing self.' One major acceptance of the idea that spatial and social behavior are entwined was a social geography textbook, the central concern of which was 'human spatial behavior and the derived geographical patterns from the point of view of society: the summation of a population's symbolic interactions' (Jakle, Brunn, and Roseman 1976:7). Note also that the definition of culture offered by Jackson and Smith (1984:205), and included in Box 1.2, is based on this approach.

cultural terms and needs to be approached through an understanding of culture and not as some independent process. Further, human culture extends beyond the range of local communities. It is the rejection of the physical science concepts and the emphasis on culture that explain the choice of name for the approach—cultural ecology.

- The role of physical environment is limited. Implicit in the approach is the assumption that 'cultural and natural areas are generally coterminous because the culture represents an adjustment to the particular environment', and also the assumption that 'different patterns may exist in any natural area and that unlike cultures may exist in similar environments' (Steward 1955:35).
- The approach is one of environmental adaptation that takes into account the character of the culture. These ideas are comparable to the possibilism introduced by Vidal.
- Three fundamental procedures are evident in this cultural ecology, namely analysis of technology and environment relations, analysis of behavior patterns affecting environment, and determination of relations between the behavior patterns and other aspects of culture.

Numerous other anthropologists have continued this pioneering work, often in amended form, but there have also been some substantive criticisms of the approach. Understandably, given some of the larger changes in social theory, the superorganic implications of the approach have been subject to particular criticism. There have also been arguments for a unified science of ecology rather than for a specific cultural ecology, thus reintroducing the idea that human and non-human behavior could be studied in similar ways. There are various other related developments in anthropology, three of which are especially noteworthy.

- The approach of cultural materialism proposed a theory of sociocultural evolution based on the doctrine of natural selection and focusing on the demographic, economic, technological, and environmental causes of cultural evolution (Box 5.3).

- Barth (1956, 1969) introduced and used the idea of an ecological **niche** in an analysis of ethnic group distribution.
- An approach labeled ecological anthropology argued for 'numerous unifying concepts and principles' (Hardesty 1977:vii), including those of adaptation, human energetics, and the ecological niche. This third interpretation is pursued in greater detail in the account of adaptation later in this chapter.

Notwithstanding the varied anthropological contributions, the principal concern was that introduced by Steward. Certainly, there are evident distinctions between this most substantive anthropological cultural ecology and the earlier derivation of human ecology in sociology. In turn, both of these ecologies are different from the proposals that emanated from geographers. Both the sociological and anthropological ecologies have intellectual origins in plant and animal ecology, and both added humans to the ecological equation, but the specific concerns, both conceptually and empirically, are quite different. Reflecting their parent disciplines, the sociological human ecology was urban and Western in orientation, while the anthropological cultural ecology was rural and preindustrial. It is perhaps significant to an understanding of the differences between the two to note that Steward was also influenced by a legacy of earlier anthropological work by Kroeber and others, whereas Park created a human ecology with minimal inspiration from within larger sociology.

HISTORY

In history, the introduction of ecological approaches was related to the belated appreciation that human behavior and environment are closely related—as recently as 1984 Worster (1984:2) found it necessary to propose 'the development of an ecological perspective in history', while, according to Crosby (1995:1180), historians of earlier generations 'could not see what they were not ready to see'. Thus, they did not see the significance of the ecological analyses conducted by geographers such as Humboldt and Marsh, nor did they see the relevance of population growth and technological change as factors basic to historical analyses. For most historians, the significance of environment was merely as a backdrop to human affairs rather than as a meaningful player in human affairs. Even at the global scale, both Toynbee and Spengler discussed civilizations with only minimal reference to environment.

But there were two major contributions in the 1930s and 1940s concerned with the settlement of the Great Plains that focused explicitly on physical environment and culture (Malin 1947; Webb 1931). Both of these studies treated Great Plains settlement in the context of human and land relations through time. Natural resources such as soil and water were discussed in the context of technology and culture. The fact that these important American studies largely failed to generate substantive

analyses for other times and places was a reflection of historical scholarship at the time. It was not until the 1970s, following the more general interest being shown in environmental matters, that historians began to conduct ecological analyses.

Ecological history is now inspired by the geographic work of Réclus, Ratzel, Semple, and Huntington, by the anthropological work of Kroeber, Wissler, Steward, Rappaport, and Harris among others, and by the historical work of Webb, Malin, and the German historian, Wittfogel. Following Wittfogel, and the argument that the 'fundamental relation underlying all social arrangements . . . is the one between humans and nature', Worster (1984:4) provided a powerful argument for the inclusion of more nature in the study of history. Despite the rel–

Box 5.3: Cultural Materialism

Cultural materialism is an ecological research strategy that is 'based on the simple premise that human social life is a response to the practical problems of earthly existence' and that 'opposes strategies that deny the legitimacy or the feasibility of scientific accounts of human behavior' (Harris 1979:ix). At the outset, cultural materialism distinguishes between:

- thoughts and behavior, acknowledging that both can be studied from the perspective of either the observers or the observed
- emic and etic operations, with the former allowing those observed to determine the legitimacy of any analysis and the latter allowing the observer to so determine

While humanists favor an emic perspective, cultural materialists emphasize an etic perspective: 'The starting point of all sociocultural analysis for cultural materialists is simply the existence of an etic human population in etic time and space' (Harris 1979:47). Analyses of such populations recognize an infrastructure that includes modes of production, such as technologies of food production, and modes of reproduction, such as technologies of population change, a structure that organizes production and reproduction at both domestic and political social scales, and a superstructure that includes activities such as rituals and science.

With these points as a basic framework, cultural materialism asserts that 'the etic behavioral modes of production and reproduction probabilistically deter-

mine the etic behavioral domestic and political economy, which in turn probabilistically determine the behavioral and mental emic superstructures' (Harris 1979:55–6). This is the key principle of infrastructural determinism, claiming the priority of etic and behavioral conditions and processes over their emic and mental equivalents, and claiming the priority of infrastructural conditions and processes over their structural and superstructural equivalents. This is not a form of monocausal determinism as infrastructure comprises a host of demographic, technological, economic, and environmental variables.

Human landscape-making behavior can be treated as an instance of human adaptation to natural and social environments and much of that behavior is directed at ensuring successful adaptation, survival. Human survival is dependent on certain kinds of behavior. Cultural practices, group behaviors, that aid the survival of the group are themselves likely to survive. Infrastructure, as defined by cultural materialism, is the principal interface between culture and nature. This means that an understanding of human behavior can be achieved by reference to its natural consequences. Put differently, cultural practices result from material causes.

Because cultural materialism argues that behaviors, as responses to environmental variables, precede mental rationalizations as to the reasons for responses, it is subject to criticism from many contemporary social scientists—including many cultural geographers—who prefer to think of cognition as a more appropriate approach to the understanding of human behavior.

atively late introduction into history, ecology is now a thriving approach with two rather different thrusts, namely concerns with natural history and with the history of ideas about nature.

PSYCHOLOGY

In psychology an ecological approach was initiated by Lewin (1944:22–3) as one part of the concern with the role of non–psychological inputs into human behavior: 'Any type of group life occurs in a setting of certain limitations to what is and what is not possible, what might or might not happen. The non-psychological factors of climate, of communication, of the law of the country or the organization are a frequent part of these "outside limitations"'. Lewin (1951) thus referred to behavior as a function of individuals and their psychological environment, which together make up the life space, and of the larger non–psychological environment. Together, these three components comprise what is called a field. This theory is more fully discussed in the context of behavioral topics in Chapter 6 (see also Figure 6.2).

This approach is not clearly related to comparable developments in geography, sociology, and anthropology. The basic principle is similar, namely that humans are parts of environments and not objects in and separate from environments, but the psychological contribution is very clearly a part of that larger discipline. Perhaps the principal sharing of ideas concerns the psychological interest in the

---

Your Opinion 5.3

A principal conclusion of this overview of human ecologies is that there have been several and varied interpretations of the concepts initially developed in physical science. Clearly, there is no one human ecology, and it is not appropriate to think of ecology as a coherent approach or a distinctive subject matter, but there is one additional development that merits note. In a small number of North American universities an attempt has been made to claim the title of human ecology as descriptive of an academic discipline concerned with the improvement of humankind: 'Human ecology is a new academic discipline which seeks, through a study of the interaction of man with his environment, to improve the near environment and the quality of life. It focuses knowledge of technical advances and improved coping skills on strengthening the ability of people to cope with the environment and to improve human–human and human-environment relations' (Edwards, Brabble, Cole, and Westney 1991:3). Most of the content of this proposed discipline is taken from more established disciplines, including those discussed earlier in this section, but also from the area of home economics, and it appears unlikely that the interpretation will be well received in the larger scholarly community. It is significant, however, that such a claim can be made—certainly, the meaning of human ecology continues to be unclear such that new claims on the term can be made. Do you agree with this assessment?

---

way that humans adapt to their surroundings, an interest that includes consideration of culture as a 'group's adaptation to the recurrent problems it faces in interaction with its environmental setting' (Berry 1984:87). This approach in psychology is related to cultural ecology in anthropology, and the idea is pursued further in the following section. (See Your Opinion 5.3.)

Ecology and Systems Analysis

The idea that ecology is a unifying science has not been widely adopted. For most interested scholars, usually approaching problems from a specific disciplinary perspective, the challenges are simply too great. Such a unified science would need to integrate both physical and human phenomena and be genuinely interdisciplinary. It has not even proven possible to integrate the various human ecologies, as the discussion so far in this section makes clear, and, indeed, the evidence points to increasing diversity of approaches rather than to integration. But there is one principal exception to these statements—the application of systems analysis in an ecological context. Many of the ecological analyses in various disciplines have turned to systems concepts to support their work, such that these concepts represent the closest that researchers have come to a genuine integration of ecological approaches.

Rather surprisingly, the term **ecosystem** was not introduced until as recently as 1935, when it was coined by the English botanist, Tansley. Sometimes called an ecological system, an ecosystem is any self-sustaining collection, an interacting system, of living organisms and their environment. Although this is a simple–sounding definition, ecosystems are usually both complex and dynamic. It is these characteristics of an ecosystem and the related difficulties of conducting analyses using traditional methods that prompted the use of the procedures of systems analysis as a means of identifying key variables and regulating factors. Ecosystems have four principal properties.

• An ecosystem is monistic, meaning that it integrates living things and their environment, allowing for recognition of the interactions between elements. In a general sense, this property is in accord with the ambitious aims of geographers such as Humboldt and Ritter.
• An ecosystem has a structure that can, in principle, be investigated.
• An ecosystem functions, in the sense that there are interactions occurring within the system.
• An ecosystem can be viewed as an example of a general system, meaning that all of the properties of a general system apply also to a particular ecosystem.

Attractive features of the concept include the recognition that ecosystems exist at a multitude of scales, from the world to perhaps a single pond, the emphasis on complexity of interrelationships between all member organisms, and the acknowledgment that an ecosystem is in a dynamic balance with the environment, constantly adapting in response to changing conditions.

Systems analysis is a scientific approach to complex phenomena that proceeds through five stages (Figure 5.1).

• In the first stage, systems measurement, objectives are outlined and data collected.
• The second stage, data analysis, involves calculating relationships between variables in order to determine the key variables given the objectives of the study.
• Systems modeling is the third stage, during which models are built to provide for a theoretical interpretation of the system.

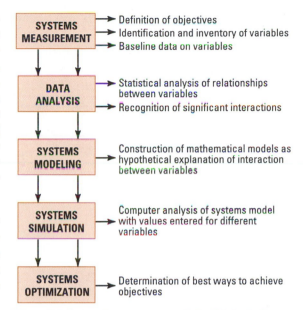

Figure 5.1: Stages in a systems·analysis. This basic diagram outlines five stages through which a typical systems analysis of an ecosystem might proceed. Details of each stage necessarily vary according to the specific aims of the analysis; for example, the fifth stage is appropriate only if the aim is to solve a particular problem. The procedure can be modified if the objective is to identify the consequences of a particular ecosystem change, such as the location of an industrial plant.

Source: C.H. Southwick, *Global Ecology in Human Perspective* (New York: Oxford University Press, 1996):45.

• The fourth stage, systems simulation, involves manipulating the theoretical models in order to assess the consequences of specific changes.
• Finally, during the stage of systems optimization, the aim is to develop strategies for achieving desired objectives.

The use of modeling and simulation is often essential given the size and complexity of many of the systems analyzed. As an obvious example, consider attempts to study the effects of an increase in humans' carbon dioxide emissions on our global ecosystem. Similarly, one of the reasons why so many attempts to predict future population growth have been so spectacularly unsuccessful is because the complexity of the factors involved have rarely been adequately acknowledged. In many cases, futures are predicted on the basis of assuming that relevant current trends will continue without sig–

nificant change, which is rarely a reasonable assumption. In brief, systems analysis, which does at least allow for the recognition of complexity, is often the only feasible way to gain some understanding of ecosystems and their changing character.

In principle, analysis of an ecosystem that includes humans focuses on human modification of environment, whether accidental or deliberate, and also on the consequences of modification. Typically, human involvement in an ecosystem leads to species destruction and loss of equilibrium. Most such analyses have been conducted from within the physical sciences, with humans seen as but one component of the system. A typical systems analysis with a social science emphasis employs the ecosystem concept in broad terms, rarely following the stages outlined in Figure 5.1.

There are a number of favorable overviews and applications of the approach by both geographers and anthropologists. Certainly, the attraction of a systems framework for the cultural geographer seems clear, with the explicit recognition of human and land relations and of humans as a part of nature. Interestingly, following an outline of the approach, Foote and Greer-Wootten (1968:89) noted that they had 'presented "old wine in new bottles"', with the vintage being Sauer (1925). (See Your Opinion 5.4.)

Cultural Ecological Analysis

The extended discussion of ecologies in the preceding section emphasized the attractiveness of the approach in both physical and human sciences, but

Your Opinion 5.4

Although there is at least superficially a compelling logic to the systems approach, it has not been widely employed by cultural geographers. Why is this the case? Are analyses of complex systems, employing sophisticated analytical skills and typically conducted by teams of scientists (not by individuals) largely outside of the cultural geographic tradition? Is there an underlying concern for many cultural geographers, regardless of their particular philosophical persuasion, concerning the legitimacy of viewing humans as but a part of a system? Does it seem logical to say that, almost of necessity, cultural geographers view humans as a distinctive component of a system? Certainly, the systems approach was most widely advocated during the 1960s, a time when geographers in general were especially attracted to theoretical approaches, with most of the applications during the following two decades. Currently, the approach is less popular, with some of the decline in interest associated with the rise of a new ecology as discussed in the section on cultural ecological analysis.

also stressed the fact that ecology has not proven to be the unifying approach that might perhaps be expected—there is not a single coherent ecological approach. Further, practitioners in particular disciplines not only use their own often distinctive versions of the approach, they also sometimes cross disciplinary boundaries to share concepts with other disciplines—the adaptation concept is a prime example. Further, many of the problems studied in this way are genuinely interdisciplinary—the question of agricultural origins is a prime example. Cultural geographers certainly work within their own traditions, but also have been attracted as well to both the sociological and anthropological ecological traditions.

Resources and Population

Chapter 2 included discussions of nature as a resource for humans, and of both technological and population change through time. Following the basic ecological argument included in this chapter, these themes are now briefly reviewed as they relate to each other.

Resources are cultural appraisals in that they are defined not only by their physical presence but also by human awareness, by technological availability, by economic feasibility, and by human acceptability. The presence of a particular **resource** is therefore no guarantee that it will be used by humans because both human ability and need are prerequisites for resource use. The process of cultural change through time typically increases the resources available, especially because of advances in technology and increasing scientific knowledge. These simple ecological statements are, of course, a compelling argument against any simplistic environmental determinism.

Thus, the resources used by humans vary according to cultural background, but regardless of specific cultural interpretations, it is usual to divide resources into two discrete categories. **Stock resources**, such as land and minerals, are fixed in their supply as they cannot be replaced except over the long period of geological time. **Renewable resources** are replaced sufficiently rapidly after their use by humans so that they can effectively be regarded as always available. However, such a basic distinction is misleading and it is better to think of a use renewability continuum (Figure 5.2).

These statements about resources stress the obvious point that resource use is dependent on, among other things, technology. As defined in the Glossary, technology is the ability to convert energy into forms useful to humans, and one of the most fundamental features of cultural evolution is the increasing capacity to use energy sources more effectively. Prior to the nineteenth century, most groups relied almost exclusively on solar power, and the new ability to tap the energy stored in fossil fuels combined with advances in scientific understanding resulted in huge increases in our ability to use resources. One consequence of our massive use of fossil fuels is the ongoing search for new forms of energy to replace the decreasing supplies of fossil fuels.

Also linked to the use of fossil fuels was an exponential increase in population numbers involving a decline of death rates without a corresponding decline of birth rates, a situation that characterized the industrial world of the nineteenth century. *It is, of course, the particular interpretation of the combination of these three—resources, technology, and population—that results in the huge differences of opinion concerning our human future.*

Some commentators from Malthus onwards have envisioned continuing population growth, related resource depletion, and a human future fraught with problems. This **limits to growth** thesis was resurrected in the 1960s and has continued to dominate both scholarly and public opinion since then. The alternative **cornucopian thesis** sees our human future in rosier terms, with less human crowding, less pollution, and fewer ecological problems. Contemporary neo-Malthusian claims such as those made by Kaplan (1996) may be open to criticism because of their geopolitical underpinnings and their apparent suggestion of abandoning some parts of the less developed world through a Western policy of containment and exclusion. This debate is important but, as Smil (1987:341) noted, the 'most fundamental difference is the initial mind-set'. The debate resurfaces in the Chapter 7 discussion of global inequalities.

Central to the ecological cultural geographic work is the concept of **carrying capacity**—the

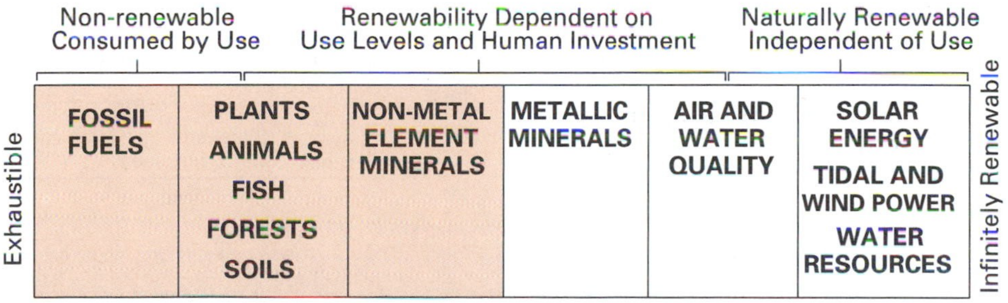

Figure 5.2: The natural resources continuum. Rather than conceiving of resources as either replaceable or irreplaceable, it is appropriate to recognize a continuum, with all resources replaceable and the difference between resources as the length of time needed for replacement to occur. At one end of the continuum are those resources that are renewable over a very short time period and the availability of which is therefore unrelated to use. Included in this group are solar energy, wind and tidal power, and water resources. At the other end of the continuum are those resources that are consumed when used and which are effectively non-renewable. Fossil fuels are the principal example. Between these two extremes is a wide range of resources, the renewability of which is a function of the way in which humans use them.

Source: J. Rees, 'Natural Resources, Economy and Society', in *Horizons in Human Geography*, edited by D. Gregory and R. Walford (New York and London: MacMillan, 1989):369.

maximum human population that can be supported in a given area with a particular level of technology. A particularly useful amendment of this concept is the Boserupian thesis, which argues that increases in population serve to stimulate technological advances, hence increasing the carrying capacity of an area (Boserup 1965).

Butzer described the concerns of cultural ecology as follows:

> It focuses upon how people live, doing what, how well, for how long, and with what environmental and social constraints. It emphasizes that human behavior has a cognitive dimension and is dependent upon information flow, values, and goals. Finally, cultural ecologists recognize that actions are conceived and taken by individuals, but that such actions must be examined and approved by the community, in the light of tradition and the prevailing patterns of institutions and power, before decisions can be implemented (Butzer 1989b:192–3).

Note that, in this tradition, culture is not superorganic; rather, there is a concern with cognition and individuals are viewed as members of groups. Conceptually, there are closer links with symbolic interactionism (Box 5.2) than with cultural materialism (Box 5.3). See Box 5.4, which summarizes one example of an analysis of the evolution of landscape.

Adaptation

DEFINITION

The single most useful and yet a much debated concept employed in cultural ecological analyses, including those with a systems analysis focus, is probably that of **adaptation**. Loosely defined as changes in behavior that are designed to improve quality of life, or rather more formally as the strategies adopted to achieve ecological success, the meaning of the concept is clouded especially by the recurring issue of environmental as opposed to cultural control and by the choice of social scale of analysis, such that the meaning of adaptation is necessarily imprecise.

Box 5.5 offers a particular interpretation, one based on the idea that all human behavior is social and therefore learned such that adaptation is

appropriately viewed as one part of the matrix of a society. According to this argument, 'explanations about variations in adaptive processes and about adaptive innovations . . . are to be looked for in variations of other elements of sociological systems, and not in changes of either ecological or biological systems' (Mogey 1971:81). Regardless of the specific interpretation, the central question addressed using this concept concerns why humans behave as they do in the environments they occupy. Not surprisingly, then, adaptation is a concept introduced, interpreted, and applied in a number of disciplines, including psychology, anthropology, and cultural geography.

ADAPTATION AND SOCIAL SCALE

In *psychology*, there is concern with the way in which individuals adapt, and an ecocultural psychology centers on adaptation to the recurrent problems posed by environments with the argument being that individual behavior must be considered in some complete ecological and cultural context. From this perspective, adaptation is interpreted in terms of the reduction of dissonance within a system. Ecological balance or harmony is increased through three types of adaptation labeled adjustment, reaction, and withdrawal.

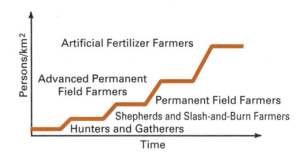

Figure 5.3: Hypothesized relationship between population size and technology. The logic behind this diagram is the idea that humans increase the carrying capacity and therefore the population of a given area through technological advances. These increases occur in a steplike fashion as technology changes. In this example, five levels of technology are identified: hunters and gatherers, shepherds and slash-and-burn farmers, permanent field farmers, advanced permanent field farmers, and artificial fertilizer farmers.

Source: U. Emanuelsson, 'A Model for Describing the Development of the Cultural Landscape', in *The Cultural Landscape: Past, Present and Future*, edited by H.H. Birks, H.J.B. Birks, P.E. Kaland, and D. Moe (New York: Cambridge University Press, 1988):112.

- Adjustment is the most usual form of adaptation, involving changes in behavior designed to reduce conflict with environment.
- Reaction involves a form of retaliation against the environment to make the environment adjust.

- Withdrawal involves conflict reduction through human removal from the source of conflict.

Recognition of these three versions of adaptation is comparable to earlier work in psychology

Box 5.4: Modeling the Evolution of the Cultural Landscape

It is quite usual to use the term 'cultural ecology' to refer to a series of stages of human and nature interactions. In a northern European analysis, Emanuelsson (1988) focused on changes in vegetation and stressed the roles played by both physical and human factors. A model to study human impacts on nature that centered on the interaction of both factors was outlined. There are three components to the model.

- The *first* component is based on the following argument. Human history is characterized by a series of inventions that allow humans to support ever-increasing population numbers with a fixed resource base; our discussion of cultural evolution since the late Pleistocene in Chapter 2 addressed this general point. Each of these inventions allows for a stepwise change in the human impact on landscape (Figure 5.3). Four such steps between five levels of technology are proposed. These are outlined in terms of specific impacts on landscape as reflected by population density characteristics for southern Sweden, although the principles are argued to apply generally. The *first* level is that of a foraging economy. This can typically provide food for population densities between 0.5 to 2.0 people per square kilometer (0.2–0.8 people per square mile); the vegetation in the immediate vicinity of the group is affected. The *second* level is that of shifting cultivation or of pastoralism. This supports about twenty people per square kilometer (eight people per square mile). In the case of shifting cultivators, vegetation is substantially affected in the areas close to temporary settlements while in the case of pastoralism, vegetation over a larger area is affected. The *third* level is that of permanent farming. This supports about fifty people per square kilometer (nineteen people per square mile) with substantial and permanent vegetation impacts in the areas occupied. The *fourth* level is that of a more efficient application of permanent farming, reflecting advances in natural fertilizer use, in cropping rotations, and in the selection of crops. This allows population densities to increase to 200 people per square kilometer (seventy people per square mile) and again causes permanent vegetation change. The *fifth* level is that of permanent farming

that is dependent on artificial fertilizers. This supports 1,000 people per square kilometer (386 people per square mile) and results in permanent vegetation change.

- The *second* component of the model recognizes that the development of a cultural landscape does not necessarily reflect a smooth transition through these five stages, for there are periods during which population densities decline. This occurs because of overexploitation of soil or because of catastrophic events such as wars or disease. Each of these events is likely to result in a loss of ecological control or what might also be called a loss of the established equilibrium between humans and nature.

- The *third* component of the model incorporates the fact that specific land uses and related human impacts are different in different parts of the world. The explanation offered concerns variations in physical geography. Most important, spatial variations of land use within a given technological level are related to climate. For example, in northern Europe the climatic gradient, which moves from a maritime to a continental climate from west to east, is very significant. In some other areas an extreme climate may make the transition from one stage to the next particularly difficult. In addition, within any given climate region, specific land use activities are related to variations in other aspects of physical geography, notably soils and landforms.

Clearly, although this is described as an evolutionary model formulation, the primary concern is the significant ecological content. It is a formulation that deals with long spans of time and, further, that says little about the details of the human landscape, emphasizing rather the consequences of human and physical geographies for vegetation change. The model provides a helpful example of one aspect of the basic Sauer dictum—humans operate on physical landscapes through time to affect change in those landscapes. In this case, the changes considered are not those with which cultural geographers are usually concerned, namely human additions to the landscape, but with specific changes in the physical landscape.

that recognized movement toward, against, or away from a stimulus. This perspective sees all adaptations as individual solutions to particular environmental circumstances.

This psychological concern with individual behavior highlights questions about the social scale at which adaptation occurs, with anthropologists and cultural geographers usually favoring a group scale (see Box 5.5), although acknowledging the need for a variety of scales. Both individual and group behaviors can be seen as adaptations, although the details of the processes are different according to social scale. What may be a sound adaptive response for an individual may not be so for the group, and, further, sound adaptive responses for both individual and group may not

Box 5.5: Adaptation as Social Adjustment

It may sound surprising to discover that the question of the relative importance of the physical environment and of humans to an understanding of the human and land relationship continues to be discussed by both cultural geographers and anthropologists in the specific context of adaptation. While the extreme of environmental determinism is no longer seriously argued, the complexity of the relationship and the variations from place to place encourage analyses that consider a variety of particular situations. Both Malin and Webb emphasized human adaptation to fixed and stable environments, rarely recognizing that the human mind was a crucial variable, while an analysis of the intermountain Mormon landscape by Speth (1967) argued that an adaptation focus lay midway between the two extremes of environmental and cultural determinism. Similarly, Porter (1965:419–20) concluded a study of subsistence activities in Kenya by noting: 'A model which seeks to describe culture as adaptive, through subsistence, to environmental potential cannot ignore man himself as a causative agent of environmental change.' Certainly, discussion of adaptation has renewed debate about determinism, with Mogey (1971), in a penetrating but infrequently referenced argument, proposing that all human behavior is social.

In order to maintain themselves, all societies have social mechanisms enabling them to adapt to environment, mechanisms that include both techniques and objectives and that are learned through time. There are different sets of social mechanisms in specific societies and are subject to ongoing change. With these claims in mind, Mogey (1971:80) explicitly argued against any form of environmental determinism, noting that even major natural disasters do not lead to social change with many societies adapted to specific disasters. From this perspective, all adaptation is social. Community was favored as the appropriate social unit for analysis because it is at this social scale that adaptation occurs,

with different communities employing different adaptive strategies. The specific differences between communities are based on two factors, technology and values. Following this argument, six propositions were advanced.

- Community is the unit of social structure through which environments affect human behavior. It is this process that is labeled adaption or adaptation.
- Adaptive innovations require modification of the value system and of the techniques of the community if they are to be acceptable.
- In equalitarian communities, rates for the appearance of adaptive innovations are low.
- The mere presence of a status system that supports a set of leaders does not itself guarantee the ready acceptance of innovations.
- Adaptive innovations are frequent and acceptable in those communities with bureaucratic organizations because they have values based on achievement and a status hierarchy that demands performance.
- But bureaucratic organizational systems have their own internal rigidities, which may impede their performance in the process of adaptation. Hence, more work on the interrelations of value systems and techniques in political, religious, and family systems is needed.

Note that these are propositions and not conclusions, and they do not imply a version of sociological determinism. As noted, these propositions have not been widely tested or adopted by geographers in either cultural or social traditions.

The final point made by Mogey implies that adaptation does not occur simply because it is needed. Certainly, many of the institutions of human society are constructed precisely because they facilitate cultural continuity—religion is a prime example—and such institutions are a 'prison de longue durée' (Braudel, quoted in Tarrow 1992:179).

be so for the environment. Group adaptation may be interpreted as the state of management of physical resources.

In *anthropology*, adaptation is closely linked to cultural evolution, with three levels distinguished—behavioral, physiological, and genetic. Anthropologists are primarily concerned with the behavioral level. According to Hardesty (1977:23), behavior 'is the most rapid response that an organism can make and, if based upon learning rather than genetic inheritance, is also the most flexible'.

From this perspective, two types of adaptive behavior may be distinguished, namely idiosyncratic (that is, unique individual responses as studied by psychologists) and cultural (that is, shared responses as studied by anthropologists and cultural geographers). Group cultural adaptation occurs because of changes in technology, organization, and ideology, all of which are aspects of culture. How do these changes foster adaptation? The usual response is to assert that they provide solutions to the problems posed by environment. However, three other processes may be relevant in that changes in cultural variables improve the effectiveness of solutions, provide adaptability, and provide awareness of environmental problems. Adaptation is thus characterized by continual changes in human and environment relations. A simple adaptation model building on these ideas is as follows (Hardesty 1986):

- the focus is on individuals and their fitness to adapt, with fitness linked to cultural baggage and learning
- space is culturally meaningful, with ideology as the key determinant as to what is meaningful
- adaptation itself frequently occurs because of some revolutionary, as opposed to evolutionary, change in ideology

In *cultural geography*, some recent work has modified the adaptation concept as traditionally employed. Denevan (1983) and Butzer (1989b) both argued that a version of cultural adaptation could be a key explanatory procedure for the cultural geographer, with human survival dependent on an available supply of adaptive responses, the appropriate scale of analysis being the ecological popu-

lation, and the research aim being that of accounting for the sources of variation and the processes of selection. There is also increasing recognition of the importance of ethnicity and of social and political power. A specific concern with adaptation as human behavior at the individual scale was favored by Knapp (1991) in a study of pre–European wetland agriculture in Ecuador, with emphasis on coping behaviors such as decision making and problem solving. These modifications of the adaptation concept are closely related to a revised ecological approach known as the 'new ecology'.

The 'New Ecology'

Recall from several earlier comments in this chapter that a basic idea behind much ecological thought, including the systems approach, is that nature strives to be in balance and indeed undergoes a process of evolution as it proceeds to a state of balance. The rise of the new ecology can be understood in the larger context of reactions against the intellectual traditions within which the initial ecological approach arose, especially the scientific tradition of mechanistic analysis and the Enlightenment tradition of continuing progress—an intellectual context that was detailed in Chapter 2. It is these fundamental ideas that are being challenged by the new ecology, the central feature of which is the proposal of a non-equilibrium view of nature. A focus on the volatility of the natural world challenges the conventional ecological wisdom that depicts nature steadily progressing toward equilibrium. Originating in biology, this new ecology is especially critical of systems analysis that is seen as implying that nature can be understood in mechanical terms.

- Zimmerer (1994:109) noted that 'this new perspective calls attention to the instability, disequilibria, and chaotic fluctuations that characterize many environmental systems as it challenges the primordial assumption of systems ecology, namely that nature tends towards equilibrium and homeostasis'.
- Referring to the concepts of vegetational succession and to an eventual climax vegetation as expounded by Clements, Blumler (1996:32) noted: 'I have my doubts, given the highly dualistic

nature of feminist writings, but it does seem possible that Clements' succession model represents a case of unconscious male psychosexual wish fulfilment: vegetation proceeds from small and flaccid to hard woody and upthrusting, becoming ever bigger and better until the ideal climax is reached.'

New ecology thus adds a revised conception of nature to earlier ideas about cultural adaptation, a revision that involves a thorough reinterpretation of biophysical environments, and that has implications both for our understanding of nature and for the study of human impacts, both matters that are raised later in this chapter. (See Your Opinion 5.5.)

ORIGINS OF AGRICULTURE

One particular research theme that has benefited from the application of ecological approaches, especially that of adaptation and, more recently, of new ecology, is the question of the origins of agriculture. Scholars from many disciplines have addressed this problem at least partly because it was the first major cultural, as opposed to biological, revolution. Certainly, ecological questions are at the heart of the agricultural origin problem, specifically the relative roles played by environment and culture and by individuals and groups. Several of the key ideas about origins are introduced in Chapter 2, and likely centers of origin are mapped in Figure 2.4.

As discussed in Chapter 2, the questions of where, when, how, and why humans first domesticated plants and animals are crucial to an understanding of the history of humans and their relationships with nature. The introduction of agriculture as the dominant human means of subsistence revolutionized human impacts on environment and the human way of life. The agricultural origins problem is addressed by both physical and social scientists, with two of the pioneering contributions from a botanist, Vavilov, in 1926, who introduced the idea that there were centers of origin, and an archeologist, Childe, in 1928, who popularized the idea that there was an agricultural revolution. Multidisciplinary research teams have used various methods and concepts, including the method of systems analysis and the concept of adaptation, to help answer questions about agricultural origins. But acceptance of the new ecological argument that nature is inherently unstable necessitates rethinking some of the basic concepts used to facilitate understanding of agricultural origins.

Most notably, if, as suggested by the new ecology, changes in nature are usual and not only the result of human activity and of extreme natural occurrences, then such concepts as those of plant communities and of climax vegetation are suspect. The implication is that it was individual plant species that changed locations because of their particular requirements and adaptations rather than whole communities changing locations in response to climatic change. The idea that vegetational change proceeds through a succession to a climax state can also be questioned on the grounds that it is a form of evolutionary thought, in the Spencerian tradition, that simply presupposes that change is linear, developmental, and progressive. The par-

Your Opinion 5.5

Interestingly, although the implications of this new ecology are as yet barely touched upon, the basic logic is attractive to cultural geographers because it can be seen as akin to the ideas of Sauer in one key respect. Specifically, one aspect of the argument is that many of the interpretations of the humans and nature relationship made by social scientists inevitably lean toward some form of cultural determinism, at least partly because few social scientists have a sophisticated appreciation of nature. For cultural geographers, this is not a difficult idea to accept. In this book, much of Chapter 2 is concerned with identifying and discussing the many and varied interpretations of humans, of nature, and of relationships between these two. Blumler (1996:26) wrote: 'I am more inclined to the Sauerian viewpoint, which although strongly opposed to environmental determinism, nevertheless recognizes that nature and culture were equal players, interacting with each other in non-linear fashion. . . . In this sense, the Sauerian approach is consonant with the new ecological paradigm.' Do you agree?

allel here is with criticisms of the models of evolutionary cultural change (see Box 2.6).

Ecology and the Global Environment

Perhaps the most significant contemporary applications of cultural ecology are those concerned with the global environment. Although human impacts on environment are studied in many disciplines, both physical and human, geography plays an important, often integrating, role. Within geography, contributions from cultural geographers are increasing in both number and significance. Particularly significant are a number of thoughtful proposals from Wagner, who has long demonstrated a concern for the study of human impacts both from a scholarly and from a practical and concerned perspective. In the 1960 volume, *The Human Use of the Earth*, the aim was to show how humans were continually remaking landscapes (Wagner 1960), but by 1972, in *Environments and Peoples*, it was realized that 'the cultural propensities of mankind in *themselves* tend to bring about a kind of "dislocation" of communities of men in respect of their habitats. That, I think, is the kernel of the modern ecological dilemma' (Wagner 1972:x). Recall also the ambitious statement by Wagner (1990:41) included at the beginning of Chapter 1.

The fact that the origins of ecology were largely unrelated to what is now described as the environmental or green movement was stressed at the beginning of this chapter. Today, however, the two are closely related—at least in the public imagination—such that ecology is often interpreted as a concern with environment. Certainly, two things are becoming increasingly obvious to many people, especially in the more developed world:

Farmers use slash-and-burn agriculture—that is, clearing an area of forest by burning—to transfer nutrients from the biomass to the soil to increase agricultural yield. They cultivate this area for a few years, then abandon it when soil fertility and crop yields decline. This method leads to ecological degradation and is a main cause of deforestation in the Amazon (*Digital imagery ® copyright 1999 PhotoDisc, Inc.*).

- our human activities pose ever-increasing threats to environments
- human activities in one part of the world can have impacts elsewhere, even globally

Both of these changes in our thinking relate to broad social issues and to the physical science acknowledgment of population and environment relationships. Scientists now recognize that human populations are able to destroy their own habitat. Strictly speaking, however, ecology refers to the relationships between organisms and their environment and is not exclusively associated with a particular value system. There are also important philosophical and practical implications arising from the ideas of the new ecology as previously discussed—the recognition that nature changes regardless of human activity is important to any understanding of human impacts, but does not undermine concerns about what impacts humans are having on environments.

It is the interest in human impacts that has been responsible for pushing ecology to the fore, such that it is usual today to equate ecology with concern for the environment, notwithstanding the fact that ecology is a term referring to the study of relationships and not, as noted, a term with any specific ideological implications. Tansley, the botanist who introduced the term 'ecosystem' in the 1930s, was one of those who argued against the idea that ecology necessarily involved questioning human exploitation, recognizing instead that human activity was one component of an ecosystem such that it was not easy to identify those human actions that were destructive. This argument was in opposition also to the claims of Clements concerning the supposed superiority of natural systems, claims that had their origins in Darwinian evolutionary ideas.

ORIGINS OF ENVIRONMENTALISM

A classic argument proposes that current environmental problems are rooted in the Jewish and Christian traditions because the Christian God was seen as giving the earth to humans for them to use as they chose, including the choice of destroying. Some commentators reject this argument, claiming that there is a substantive difference between the biblical concept of dominion and the concept of environmental domination. The subsequent debate around this issue includes discussions of religions other than Judaism and Christianity to suggest that, even where religions traditionally see humans as a part of nature, there is evidence of ruthless exploitation of nature. Major challenges to the Western tradition that saw humans as different from and superior to nature generally and other animals in particular came from several and different directions.

- Malthus envisaged disastrous social consequences because of the excessive population growth, relative to the increase in resources, that humans seemed unable to control.
- John Stuart Mill (1806–73) anticipated that increases in population and in resource consumption could not simply continue.
- Henry David Thoreau (1817–62) viewed all creatures as a part of the community of nature.
- John Muir (1838–1914), the founder of the Sierra Club, urged the preservation of selected natural environments.
- Darwin, of course, provided scientific support for the idea that humans were but a part of nature.
- Geographers have also long been concerned with deleterious human impacts, and the contributions made by such major scholars as Humboldt, Ritter, and Marsh are especially noteworthy.

Henry David Thoreau, naturalist, philosopher, and writer, was born in Concord, Massachusetts in 1817. In 1845, to prove to himself and to others that it was possible to live off the land, he built a hut near Walden Pond, where he lived in solitude for two years. *Walden; or Life in the Woods*, published in 1854, records this experience (*Private collection*).

- The Sauerian interest in such matters was a principal spur to a 1955 conference and to the subsequent publication of the landmark volume, *Man's Role in Changing the Face of the Earth* (Thomas, Sauer, Bates, and Mumford 1956).

In their different ways, each of these contributed to a revised view of nature and of the human place in the natural world.

THE AGE OF ENVIRONMENTALISM

Aldo Leopold (1886–1948) urged society to adopt a new land ethic aimed at rejecting the idea of land as mere property and at protecting the environment, and it is frequently claimed that this argument represents the true philosophical beginnings of the contemporary environmental movement—certainly Leopold (1949) outlined a non-anthropocentric, holistic environmental ethic. But widespread popular interest in the environment dates from the 1960s, and especially from the 1962 publication of *Silent Spring* (Carson 1962)—a powerfully critical account of the effects of pesticides on human life. It was this work that ushered in the age of environmentalism, loosely defined as a concern with changing human attitudes and behavior regarding environment. (See Your Opinion 5.6.)

As one part of a larger discussion of group identities, Castells (1997: 113) identified five principal types of environmental movements 'as they have manifested themselves in observed practices in the past two decades'. The five, noted in Table 5.1, are each defined by some combination of three characteristics that are seen as defining social

movements generally, namely identity, adversary, and goal.

- The traditional type of movement is that concerned with the conservation of nature; the Sierra Club is a leading example.
- The fastest-growing type is that of local community involvement, sometimes seen as the 'Not in my back yard' syndrome, although the concerns of many such movements often extend beyond the purely parochial.
- Concern about environment has also been associated with the emergence of groups that expressly oppose some traditional social beliefs and practices; the ecofeminism discussed in the following section is an example of this type.
  - Greenpeace, which was founded in Vancouver in 1971 and is now an international organization, is the leading example of the save-the-planet type of environmental movement.
  - Finally, the rise of political green parties, most notably in some European countries, represents a fifth type of environmental social movement.

The interweaving of the physical and human—cultural, political, and economic—aspects of this topic is undeniable. 'It is now taken for granted that the *global environmental crisis* and a renewed concern with *global demography* (the return of the Malthusian specter) are inseparable from the terrifying map of *global economic inequality*' (Peet and Watts 1996:1). Certainly, the current concern about environment cannot be detached from concern about people, as the United

---

Your Opinion 5.6

Has the typical Western attitude toward human impacts on environment changed in recent years? A useful distinction that many commentators make is that between an earlier anthropocentric or human-centered world view, and a more current (but by no means dominant) ecocentric or environment-centered world view (see Glossary). This is seen as a transition from an overwhelming emphasis on material well-being (part of the anthropocentric view) toward greater emphasis on the quality of life (part of the ecocentric view). Pinkerton (1997) described the ecocentric view as enviromanticism, noting that it may prove to be powerful politics. Certainly, contemporary environmentalist and green movements are complex social and political forces, incorporating many different viewpoints, ranging from those who favor preservation of particular areas to those who adopt more radical ideas, perhaps favoring a dismantling of capitalist society and economy. It is not difficult to see how such environmentalist ideas can be translated into formal political movements.

Table 5.1: Typology of Environmental Movements

Type	Identity	Adversary	Goal
Conservation of nature	Nature lovers	Uncontrolled development	Wilderness
Defense of own space	Local community	Polluters	Quality of life/health
Counterculture, deep ecology	The green self	Industrialism, technocracy, patriarchialism	Ecotopia
Save the planet	Internationalist ecowarriors	Unfettered global development	Sustainability
Green politics	Concerned citizens	Political establishment	Counterpower

Source: Adapted from M. Castells, *The Power of Identity* (Malden, MA: Blackwell, 1997):112.

Nations sponsored–conferences on Environment and Development in Rio de Janeiro in 1992, on Population and Development in Cairo in 1994, and the World Conference on Women in Beijing in 1996 have made abundantly clear. Today, it is usual to portray concern for the environment as being in opposition to progress, science, technology, and population growth, an attitude that might be explained by reference to the close association between the expansion of European empires and related environmental impacts. (See Your Opinion 5.7.)

THE STATE OF THE EARTH

Notwithstanding the recent rise of an ecocentric world view, much human use of the earth continues to reflect an anthropocentric view. Briefly, our use of the earth involves pollution of land, water, and air, changes to global climate, loss of species, loss of land, loss of wilderness, loss of cultural diversity, and dramatically uneven patterns of consumption and quality of life. In this sense, many observers judge that there is a crisis of environment, with environment including humans and nature—effectively a crisis of contemporary culture. Discussions of the specific environmental issues are commonplace in both academic and other circles,

and it is not appropriate to include any detailed account in this textbook.

Certainly, there are many sources of information that are geographic in character. Three important volumes merit mention. First, as noted earlier, *Man's Role in Changing the Face of the Earth* (Thomas, Sauer, Bates, and Mumford 1956) provided a substantive account of human impacts, but this is necessarily dated both conceptually and factually.

Second, a more recent volume in a similar vein, *The Earth as Transformed by Human Action: Global and Regional Changes in the Biosphere in the Past 300 Years* (Turner et al. 1990) focused on human impacts globally during the period since the onset of industrialization. This work included discussions of changes in population, technology, culture, urbanization, and human perception of these and other changes; accounts of impacts on environment both generally and in specific regional studies, including such critical issues as **acid rain** and the human–induced **greenhouse effect**; and also thoughts about humans and nature and **maladaptation**, gender issues, and the relevance of ecological approaches.

Third, *Regions at Risk: Comparisons of Threatened Environments*, by Kasperson, Kasperson, and Turner (1995) emphasized that the causes and consequences of human-induced environmental change are unevenly distributed over the surface of the earth and proceeded to identify some critical environmental regions. Understanding such critical regions requires acknowledgment of the following factors:

- criticality is interactive, relating to the type and rate of environmental change, the characteristics of the ecosystem, and the culture in the area
- criticality needs to be studied in both temporal and spatial contexts
- criticality arises if the relations between culture and environment involve a situation susceptible to changes in economy or environment
- criticality refers to a reduction in environmental quality as this relates to human occupance
- future situations of criticality are often difficult to predict
- substantial human impacts usually result in economic gains
- current human activities have different degrees of sustainability, and there is usually a range of other options available
- there are driving forces such as population growth, technological capacity, affluence, poverty, state policy, character of economy, and beliefs and attitudes

Revisiting Humans and Nature

The preceding discussions of the new ecology and of ecology and the environment have referred to some of the ideas contained in Chapter 2, and it is appropriate now to reconsider some of that earlier

---

Your Opinion 5.7

It is interesting and possibly somewhat disturbing to acknowledge the considerable scholarly, popular, and governmental attention being paid to the question of human impacts on environment, and to compare this attention to that being paid to the question of human impacts on other humans. It is commonplace today to suggest that the measure of concern that a society has for their environment is a reflection of the human condition, but do we similarly assess ourselves according to our treatment of others? For cultural geographers, and for practitioners in many other disciplines, the answer seems uncertain. Notwithstanding many areas of concern, such as the ecofeminist arguments discussed below and discussions of the role played by the division of labor in our understandings of nature, it is possible that more popular and scholarly attention is paid to the way we treat the earth than to the way we treat other humans. To paraphrase Wagner (1990:41), a principal aim of cultural geography should be that of developing a theoretically well-grounded, intellectually vigorous, and practically effective discipline that might well assume, in time, a major role in guiding and guarding our attitudes toward and treatment of ourselves.

---

content. Recall that the discussions of humans and nature in Chapter 2 included brief references to some of the more recent interpretations of this complex theme, particularly the idea that nature may not be quite so natural after all, but rather may be more correctly interpreted as a social construction, meaning that when we talk about nature we are saying as much about ourselves as we are about the thing we are describing. This idea is now revisited, especially in light of the additional context provided by the rise of the new ecology. The context for revisiting these ideas is a discussion of environmental ethics.

The Need for Reinterpretation

Although the discussion of new ecology focused on the implications for ecological analyses, such as the concern with the origins of agriculture, there is another larger issue involved in the idea that nature is inherently unstable, namely our human image of nature. Cronon noted:

Recent scholarship has demonstrated that the natural world is far more dynamic, far more changeable, and far more entangled with human history than popular beliefs about the 'balance of nature' have typically acknowledged. Many popular ideas about the environment are premised on the conviction that nature is a stable, holistic, homeostatic community capable of preserving its natural balance more or less indefinitely if only humans can avoid 'disturbing' it. This is in fact a deeply problematic assumption (Cronon 1995:24).

It is especially problematic because abandoning our belief in an essentially unchanging nature is psychologically disturbing. How are we to judge our actions without such a benchmark? We have a double difficulty here.

- First, nature is socially constructed. Thus, our descriptions of it reflect not just nature itself, but also ourselves, our values, and our assumptions, implying that we cannot in any objective sense 'know' the natural world.
- Second, nature changes regardless of our human activities.

What does all this mean for us when we ask ethical questions of ourselves about how we should treat nature? There are not, of course, any clear responses to such a question, but the double difficulty noted does provide a useful backdrop for considerations both of contemporary environmental ethics and of ecofeminism.

Environmental Ethics

How are we to determine what is ethically right when we consider our treatment of the natural world? Human attitudes toward environments and non-human animals have undergone some significant changes in emphasis since the dominant early nineteenth-century view that supported the idea that it was right and proper for humans to use environments as they saw fit and also the idea that humans were different from all other animal species. Today, there is considerable philosophical debate concerning the proper assessment of humans as a part of nature. The complex issues introduced in this discussion do not offer any resolutions, but they all encourage consideration of the question whether we should be behaving in some environmentally friendly manner because we need to for our own sake, or rather because it is ethically right for us to do so in terms of the natural world itself. Seven different lines of thought are noted, and there are further contributions on this theme in the accounts of postmodernism and inequality that are contained in Chapter 7.

RELIGION AND ENVIRONMENT

The earlier reference to the debate about the relation between Christian ideas and human impacts is but one aspect of the larger question of religion and environment. Most religions identify what is important to humans beyond everyday concerns and also offer rules of behavior that are said to derive from some spiritual source. Inevitably, then, religion has something to say about human activities as these affect other humans and the environment that humans occupy. Overall, most religions provide guidelines for human activities and for an understanding of nature.

Many traditional religions were not as separated from the rest of human life as the Christian religion has been. In traditional Hawaiian culture, god, humans, and nature were related to one another and formed a single interacting community with all species, not only humans, considered sentient. Thus, the entire world was seen as alive, capable of knowing and acting, in the sense that Western culture sees only humans as alive. Such beliefs allowed for a genuine kinship that is lacking in the Christian tradition, and also for a quite different perception of the world. These comments apply generally to a great many of the traditional belief systems that were replaced or substantially weakened as Europeans spread overseas, but they apply also to several of the major world religions.

In Islam, respect for the natural world is founded on the principle that God created all things, and there is an acknowledgment that humans are not the sole community living in the world. Use of the natural world to support humans is appropriate, but needless destruction is not sanctioned. Both Hinduism and Buddhism traditionally viewed humans as but one part of nature such that abuse of the natural world is also abuse of people. However, most observers suggest that the materialistic orientations of Western culture have affected these traditions with a corresponding decrease in concern for the natural world.

EXTENDING THE PRINCIPLE OF EQUALITY

Environmental ethics breaks new ground from a philosophical perspective. All ethics is concerned with seeking an appropriate respect for life, but only environmental ethics asks whether non-human life should be included. There are several important philosophical arguments today that propose viewing the natural world, including non-human animals, from an ethical perspective. The question here is whether we should extend the principle of equality beyond our own species. Further, if we do extend the principle of equality to include all sentient life, then should we also extend it to insentient life? Note that asking these questions would not be necessary in some traditional cultures, as the discussion of religion and environment made clear.

Referring to the principle of equal consideration of interests, which allows us to claim that other humans should not be exploited because, for example, they have lesser intelligence or a different culture, it might be argued that it is similarly wrong for humans to exploit members of other species. Following this logic it is, for example, wrong to eat meat, wrong to use animals in experiments, and wrong to display animals in zoos. These are very complex ethical issues that are at the heart of our human relationship with parts of the natural world.

What are some of the basic arguments against keeping animals in zoos? Most obviously, the zoo environment is an alien one that restricts the animals' freedom of movement, limits their ability to seek out food, and, more generally, prevents them from behaving in ways that are natural to them. Further, it has been argued that many animals suffer in zoos, often dying young, suffering injury, or developing deformities. But there is another rather more difficult argument against zoos that focuses on the question of why we wish to display animals in zoos. In addition to the conventional justifications concerning human entertainment, education, scientific analyses, and species preservation, it can be argued that they serve to reinforce our human sense of superiority. The fact that we are able to confine animals means that they are there for our use and pleasure. It is in this sense that zoos can be viewed as a reflection of an outmoded environmental ethic.

Zoos are the last refuge for the Siberian tiger, an endangered species whose numbers were estimated in 1994 at between 150 to 200 in the Amur region of Russia. The Siberian tiger population, like that of other wildlife in the region, is dwindling rapidly due to intense poaching and habitat loss (*Toronto Zoo*).

ECOLOGY, ECONOMICS, ETHICS

The growing field of environmental economics begins from the fundamental premise that much economic activity is unecological, and that there is therefore a need to bring economic concepts of value more into line with ecological concepts of value. This may not be a difficult task given the fundamental unity of ecology and economics, for both are concerned with the management of the household or home. What is needed is to adjust the traditional economic conception of a household as a given group of people to an ecological scale where that group is the entire world population and not some subset of that population. The issue is, then, one of social scale as Hardin (1968) noted in the classic account of 'The Tragedy of the Commons'. Do we behave appropriately for ourselves only, or for ourselves and some others, or for ourselves and all others, with this final possibility implying also a respect for the natural world?

One major stumbling block to the widespread acceptance of environmental economics is that it does not accept the conventional economic wisdom that growth, measured in some monetary terms, is an end in itself. A possible resolution of this difficulty lies in the acceptance of the adaptive strategy of **sustainability**.

BEHAVING ETHICALLY

It may be helpful to think in terms of environmental problems resulting from the actions of individuals and organizations, and to seek understanding of why we behave as we do from a psychological perspective. There is a contradiction between the environmental behaviors that many of us practise and the environmental attitudes that we hold—many people may have pro-environmental attitudes and yet participate in environmentally destructive behaviors. How are we to understand and change this attitude-behavior gap? Rather than emphasizing additional environmental education, one approach might be to focus on the psychology behind the inconsistency. Four general conclusions emerge from such a focus.

- Environmentally damaging behavior is often not a result of indifference to the environment; rather, it occurs notwithstanding legitimate concern for the environment. Thus, it is possible that requiring organizations to adhere to certain environmental standards acts in a counterproductive fashion, and also that the decision not to recycle occurs despite both liking the idea and believing that it is good for the environment.
- Inappropriate behavior may result from faulty mental models of the world; in other words, we are not fully aware of the environmental problems that we are encountering, such that our behaviors are less than ideal.
- There are real difficulties in assessing the implications of environmental change for human life and welfare.
- It is extremely difficult to place an economic value on environmental goods and outcomes.

LIVING WITH DISCORDANT HARMONIES

Can harmony with nature only be achieved through an acceptance of the basic logic of the new ecological recognition that nature is ever changing, combined with proper use of our technologies to aid understanding of nature? Because humans are but one part of a larger living changing system that is global in scale, perhaps our aim should be to accept natural changes—that is, the discordant harmonies—and strive to make the earth as comfortable as possible for ourselves. How can we expect to use the natural world wisely if we do not understand that natural world? For example, knowing that nature changes means that we need to rethink ideas about valuing wilderness for its intrinsic characteristics or conserving particular natural landscapes as we think they were before human impacts. If nature changes of its own accord, how can we possibly conserve it? Of course, these questions are not arguments against conservation and preservation, but rather arguments about the reasons for such actions.

THE 'GOLDEN RULE'

Another way to approach the question of environmental ethics is to think in terms of how we wish to be treated. We might approach this matter by first questioning two of the frequently stated bases for developing an appropriate ethic, namely the related claims that humans are permanently damaging the earth and that we need to learn how to behave as stewards for the earth we are threatening. Both of these propositions reflect an exaggerated view of our human role because humans are but one of millions of species and we have, geologically speaking, only appeared on the earth recently. But to think in such grand terms is, of course, misleading.

To help develop an appropriate ethic, we need to think in more local terms, namely that we are capable of the significant impacts of eliminating both the natural world upon which we depend and also ourselves. Following this logic, Gould (1990) proposed that an appropriate environmental ethic is one that can convince people of the need for clean air, clean water, reforestation, and so forth as being the best solutions for ourselves and our environment here and now, rather than one that stresses long-term consequences. *The ethic in question is simply that of our treating others as we wish to be treated by others.*

Transforming Culture

The previous brief references to the rise of an ecocentric world view obscured the many and varied attempts in this direction. One of the principal conceptual arguments is that favoring a deep ecology or what is often called sustainable development. Thus, the ecophilosopher, Naess (1989), proposed that humans need to develop a new world view that sees nature as having value in its own right, that sees all humans as connected, that sees a need to work with rather than against nature, and that sees ecosystem preservation as a primary goal. In addition, such a deep ecology views humans as part of nature, emphasizes the idea that every life form has in principle a right to live, is concerned with the feelings of all living things, and is concerned about resources for all living species. Thus, deep ecology is both an environmental ethic and a spiritual basis for rethinking what it means to be human. Several of the basic concepts employed by Naess are derived from Hinduism.

In a related fashion, but pursuing an original argument, one writer has proposed that we need to move away from the oppression and emphasis on wealth that are characteristic of modernity through a reintroduction of the specific basic qualities that we once possessed, namely a connection with nature, a sense of belonging, and an egalitarian community (Box 5.6).

Ecofeminism

Although the discussion of **feminism** is delayed until Chapter 7, it is helpful to note briefly the basic argument at this point in order to facilitate the discussion of ecofeminism. Although there are many versions of feminism, there is general agreement among feminists that sexism prevails and is wrong and needs to be eliminated. In the broadest sense, something is a feminist issue if an understanding of it helps one understand the oppression or subordination of women in any place and at any time.

In addition to this basic argument, academic feminism has increasingly recognized that the liberation of women cannot occur until all women are freed from the many oppressions that structure our gendered identities. Included in these multiple oppressions are those based on social class, skin color, income, religion, age, culture, and geographic location. But, despite these statements, there is not one feminism. Nor is there one environmental philosophy and, accordingly, there is not one ecofeminism or one ecofeminist philosophy.

Definition

Given the above, an explicit statement of interest is as follows:

According to ecological feminists ('ecofeminists') important connections exist between the treatment of women, people of color, and the underclass on one hand and how one treats the nonhuman natural environment on the other. Ecological feminists claim that any feminism, environmentalism, or environmental ethic that fails to take these connections seriously is grossly inadequate (Warren 1997:3).

These ideas first appeared in the mid-1970s through an integration of some feminist ideas and the green movement—from feminism comes the idea that humanity is gendered such that women are subordinate and exploited, and from the green movement comes the idea that humans are damaging the natural world. There are two fundamental versions of ecofeminism, both of which were noted in Chapter 2, but without direct reference to ecofeminism at that time.

- *Cultural ecofeminism* chooses to emphasize and focus on the links traditionally made between women and nature and the related circumstance that men oppress both. Thus, cultural ecofeminism is the response to the feminist claim that Western culture has identified women and nature as belonging together and has devalued women and nature. Human nature is seen as grounded in human biology in that sex/gender relations have resulted in different power bases. As such, cultural ecofeminists see the world as dominated by technologies and ideologies developed and controlled by men, and they seek to elevate and liberate women and nature, usually by means of direct political action. The relations between women and the environment are seen as essential.

- *Social ecofeminism* rejects the biological determinism implied in the previous position, choosing instead to emphasize and focus on nature as a social and political construct and not as a natural construct. Thus, social ecofeminism advocates liberating women by overcoming the constraints imposed by marriage, the nuclear family, patriarchal religion, and capitalism. There is a rejection of the essentialist idea that the linking of women and nature results from biology in favor of the idea that the relations between women and the environment are constructed. Because they are constructed, these relations are not universal; rather, they vary from place to place and through time.

Box 5.6: Social Evolution and the Environmental Crisis

For Earley (1997), it is the continuation of modernity that poses the greatest threat to human survival in the sense that our current environmental crisis results from the way our world is structured and our related view of nature and our place in nature. To move away from this modernity requires changes to our social structures and institutions. These points are clarified by means of a model of social evolution (Earley 1997:2–4).

At the outset of social evolution, humans enjoyed three *basic qualities*:

- connection with nature
- sense of belonging and richness of experience
- egalitarian community

Through time, populations grew, resources were limited, and conflict occurred, resulting in three *emergent qualities* that provided for more conscious choice, increased freedom from environment, additional population growth and resource exploitation, and a new understanding of ourselves and of nature. These are:

- technology
- reflexive consciousness
- social structure

The difference between these two sets of qualities is striking. The initial ground qualities—what we were—reflect vitality and organic wholeness, while the later emergent qualities—what we have become—reflect power and differentiated organization. Put simply, it is the emergent qualities, especially the resultant economic system emphasizing material growth, that are responsible for our alienation from nature, from others, and, indeed, from ourselves. The result is that:

- we have gained technology, but destabilized the environment
- we understand more, but empathize less
- we have developed social structure, but lost equality and community

Only in recent years has it been possible to grasp fully the implications of this process of social evolution and to appreciate that it is insufficient and, of course, impossible, to return to the ground qualities. What is needed is a rediscovery of the ground qualities and an integration of ground and emergent qualities, resulting in:

- ecological technology
- integrated mind and heart
- a social structure that promotes equality and community

In summary: '*we must now consciously chose to regain our wholeness and vitality in conjunction with our complexity and autonomy, as individuals and communities, as organizations, as a world society, and as a living planet*' (Earley 1997:4).

Thus, a fundamental area of disagreement between these two positions is what we might describe as the debate between essentialism and **constructionism** (both terms are listed in the Glossary).

IMPLICATIONS

The fundamental implication of ecofeminism is best described in the larger context of feminism in general. Thus, since about the 1960s, the various forms of feminism have effectively called into question virtually all of the established and cherished belief systems and institutions of the dominant patriarchal cultures. The implications of ecofeminism for the larger ecological perspective are enormous. For many, ecocentric views, such as deep ecology, are seen as inadequate correctives, although, for others, ecofeminism is a union of deep ecology and feminism. Ecofeminism confirms the view that the dominant development paradigm is one that sees the earth as a resource to be exploited and developed for human benefit, confirms that there are compelling connections between the treatment of nature and women, and confirms that there is a larger domination of the world by the values of a Western cultural tradition. Accordingly, ecofeminists are striving to move beyond the various dualisms, such as nature/culture and female/male, in order to create a new consciousness of people and the earth in harmony. One way to demonstrate some of these matters, and also to clarify the difference between cultural and social versions of ecofeminism, is to review the example of the Chipko movement, the most often discussed example of the many struggles that symbolize the relationship between women and environment.

Chipko, meaning embrace or hug in Hindi, is a grassroots, non-violent, women–initiated movement that began in 1974 when women in an area of northern India expressed their concern about the removal of broad–leafed indigenous trees. Since then it has spread remarkably and is credited with saving perhaps 4,634 sq. mi. (12,000^2 km) of trees. Why the concern? Because these trees are a critical resource for the rural poor, providing fuel, food, fodder, building materials, household utensils, and also limiting soil erosion. Why was it women who initiated the movement? The straightforward factual answer is that it is because women are more

dependent than men on trees, and hence it is women who are the principal sufferers if trees are removed; it is also women who collect fuel and fodder and perform household tasks. Replacement of the indigenous trees with pine or eucalyptus (as was planned by the government) provides employment for men. A cultural ecofeminist answer focuses on the idea that ancient Indian cultures worshipped tree goddesses and also recognizes that movements similar to the Chipko movement were in place over 300 years ago. A social ecofeminist answer argues that it is because women are marginalized members of society and have consequently created a link with nature.

Thus, this example can be interpreted as being supportive of the cultural ecofeminist essentialist claim that women are closer to nature and are, therefore, necessarily more committed to caring for the environment. On the other hand, it can be interpreted rather differently following the social ecofeminist constructionist argument that women in the less developed world view their relations with nature differently from women in the more developed world. To summarize this discussion of ecofeminism is not easy because generalizations are difficult, but the following statements reflect a moderate ecofeminist position:

- nature does not need to be ruled over by humans
- humans need to recognize their intimate relationship with nature
- from a religious perspective, God is neither male nor anthropocentric
- mutual interdependency replaces the traditional hierarchies of domination that have God above humans above nature, or perhaps have God above men above women above children above animals above plants
- prevailing patterns of interdependency based on sex, gender, skin color, class, culture, income, religion, age, and geographic location need to be reconstructed to create more equitable situations. (See Your Opinion 5.8.)

Political Ecological Analysis

The introduction of the new ecology, and the myriad implications of recent conceptions of humans and nature, are not the only major revisions of traditional ecological approaches. There is also a related set of changes that are labeled political ecology because they involve some integration of the relatively well-established approaches of cultural ecology and ecological anthropology with the newer Marxist–influenced political economy and peasant analyses.

As previously discussed, both human and cultural ecologies built on ideas developed in physical science and typically viewed cultures as adaptive systems. Political ecology advocates a quite different approach. Specifically: 'The phrase "political ecology" combines the concerns of ecology and a broadly defined political economy. Together this encompasses the constantly shifting dialectic between society and land-based resources, and also within classes and groups within society itself' (Blaikie and Brookfield 1987:17).

Political ecological analyses recognize and incorporate a variety of social and spatial scales, and stress the need to understand a problem in a broad cultural, social, political, and economic context. Thus, an analysis may be of a local area and of a small group, but understanding is sought at various scales; further, the problem may be specifically economic, but other human and physical considerations are incorporated as well in the understanding.

The Approach in Context

Suggestions that political and economic processes need to be considered in human or cultural ecolog-ical analyses began to appear in the 1970s. Arguments focused on a socialist human ecology as the best means of practice, with such a human ecology focusing on environmental histories, human adjustments, system stability, and the fate of folk ecologies. There were also analyses of population, food supplies, and famine in parts of the less developed world that incorporated some of the characteristics of this emerging approach. But political ecology as an essentially Marxist–inspired analysis of the human use of resources and related impacts on environment is usually linked to two important works, namely those by Blaikie (1985) and by Blaikie and Brookfield (1987).

For example, with reference to soil erosion in the less developed world, the conventional explanation that referred to such factors as overpopulation, inappropriate use of resources, environmental difficulties, and the collapse of markets was rejected. Rather, emphasis was placed on social causes, specifically on land managers, or households that were actually using local resources and that were obliged to extract surpluses thus leading to land degradation. A chain of circumstances was identified: first, the land managers have direct contact with the land; second, land managers relate to each other, to other land users, and to other groups in society; third, they are linked also to the state and to the world economy. The attractiveness of this political ecology needs to be understood within a number of specific larger contexts, in addition to those provided by the ecological traditions discussed in this chapter.

Your Opinion 5.8

The argument that nature may not be quite so natural after all, but rather may be more correctly interpreted as a social construction is precisely that, an argument. The principal advocates are postmodern philosophers for whom the argument is but one component of the larger claim that insists on the lack of objective givens in virtually all areas of enquiry. A contrary argument about the value of nature claimed that: 'Social construction is necessary but not sufficient for our being. Some values on earth are not species-specific to homo sapiens *(Rolston 1997:62). As evident in much of the content of chapters 7 and 8, the general social constructionist argument has been well received by many contemporary cultural geographers and is evident in many recent analyses, especially those focusing on the identity of both people and place. It is important to remember, however, that philosophers do debate these issues, and it is always appropriate to query, or at least be aware of, the underpinnings of any analysis. Just as it needs to be understood that some cultural geographic studies, including many of those referred to in chapters 3 and 4, are premised on an essentialist view of culture, and that some other studies are premised on a constructionist view, so it needs to be understood that the relative merits of these two philosophical assumptions are debated.*

- First, there is the context of political economy. For geographers, this is not an easy term to define as it is really an umbrella phrase to cover a variety of radical emphases mostly inspired by Marxism. The essence of the approach can be summarized as the idea that the political and the economic are irrevocably entwined such that any study of human activities must incorporate both of these. Central to this argument is the idea that much human activity is bound up with social and political struggles and with competing claims to such things as ownership and right of access. Political ecology is a radical perspective, emphasizing social justice and equity issues. Political ecology is, of course, also about politics, specifically the unequal power relations between those involved in affecting the environment.

- Second, there is the larger context of environmental and social concern that has been evident since about the late 1970s, especially in the less developed world context. Unlike the emergence of such concern in the early 1960s, the later concern was bound up with the larger questions of economic and cultural globalization and with the increased awareness that humans were affecting a global environment. Political ecology is critical of environmental policies that do not consider the impacts of global capitalism, typically favoring major changes to both local and global political economy.

- Third, it is possible to identify links with **structuration theory** as outlined in a series of studies by the sociologist Giddens (for example, 1984). Key features of this theory concern the links between human agents and the social structures within which they function, with individuals viewed as operating within both local social systems and larger social structures; indeed, there is a substantial and continuing debate in social science—the structure and agency debate—centered around this issue. Structuration is a complex set of ideas that seeks to integrate the work of a wide variety of thinkers, including Durkheim, Freud, Marx, and Weber. Individuals are viewed as agents operating within both the contexts of local social systems, sometimes called **locales**, and the larger social structures of which they are a part. The key characteristic of structuration is the identification of the dualities associated with

social structure and human agency. Thus, social structure enables human behavior, while at the same time behavior can also influence and reconstitute culture. Further, the rules of any social structure are both constraining and enabling—they are constraining because they limit the actions available to individuals, but they are also enabling as the rules do not determine behavior. Certainly, both political ecology and structuration theory consider the ways in which global, national, or regional processes articulate with local scale processes through time. Related to structuration theory is the context provided by the socially informed new regional geography referred to in Chapter 4. This regional geography sees regions and regional change as bound up with social processes. 'Human activities do not cause regional change; rather human activities shape, and are shaped by, place and history. Human identities and activities constitute the economic, political and ideological processes which form and transform regions. In turn, regions shape human activities due to particular contextual details of place' (Jarosz 1993:367).

- Fourth, in a discussion of 'liberation ecologies' Peet and Watts (1996) incorporated aspects of development theory, various poststructural critiques and related discourse theory, and social movements and related political forms; these additional conceptual insights are not addressed in any detail at this time as they are more usefully considered in chapters 7 and 8.

It is clear from these brief comments that political ecology is a complex and multifaceted approach that shares conceptual and empirical concerns with several other interests. As is the case with some of the other topics raised in this chapter—for example, the new ecology and ecofeminism—political ecology is an approach that continues to evolve as the underlying theoretical concepts continue to be developed and clarified.

Analyses

This section briefly reviews two examples of political ecological research (for other examples, see especially the various articles in a Special Issue of *Economic Geography*, 69, no. 4, 1993).

PEASANT–HERDER CONFLICTS

In west Africa since the early 1970s, large numbers of Fulani pastoralists have moved south from semi-arid environments in the Sahel, especially from Mali and Burkina Faso, into the northern savanna region of the Ivory Coast. Their movement has been welcomed by the Ivory Coast government because of the contribution made to beef production, but one unfortunate consequence has been a series of conflicts between the incoming pastoralists and the local Senufo peasant farmers that they encountered. The principal reason for conflict concerns the damage inflicted by the cattle on crops grown by the Senufo and the subsequent lack of any compensation. Traditional geographic explanations for such movements and conflicts rely on the circumstances of a declining resource base relative to population, while a political ecological focus stresses the need to consider human activi-

ties as responses in context, and looks at the chain of causality leading to conflict.

'It is at the intersection of Ivorian political economy and the human ecology of agricultural systems in the savanna region that one can begin to identify the key processes and decision-making conditions behind the current conflict' (Bassett 1988:469). Specifically, the peasant farmers are aggrieved at the consequences of pastoral intrusions into their cropping areas because of the impact of uncompensated crop damage on their already marginal standard of living. Although it might be suggested that this is the cause of conflict, it is not seen as a sufficient cause of the peasant uprising that included the murder of about eighty herders in 1986. Table 5.2 indicates the key determinants of conflict, distinguishing between ultimate causes and proximate causes, and identifying stressors and counter-risks.

Throughout Africa, the grazing of livestock has altered wildlands thus affecting many native species. Regulations on overgrazing (if any) tend to be relaxed during drought years when the effects of grazing are most pronounced (*CIDA Photo/Roger Lemoyne*).

Table 5.2: Conflict Determinants, Northern Ivory Coast

Ultimate Causes	Proximate Causes	Stressors	Counter-risks
Ivorian development model: • (surplus appropriation by foreign agribusiness and the state • (livestock development policies	Low incomes and beef consumption	Uncompensated crop damage	Compensation
	Insecure land rights	Political campaigns	Fulani expulsion
	Intersection of Senufo agriculture and Fulani semi-transhumant pastoralism	Theft of village cattle	Crop and cattle surveillance
Savanna ecology			Corralling animals at night
Fulani immigration			

Source: T.J. Bassett, 'The Political Ecology of Peasant-Herder Conflicts in the Northern Ivory Coast', *Annals of the Association of American Geographers* 78 (1988):456.

As with any ecological approach, political ecology thus acknowledges complexity and interrelationships. Although some key factors may be evident, the principal concern is to understand an issue in context. Thus, three ultimate causes are noted, one of which is the Ivorian development model, which a traditional cultural ecological approach would not seriously consider. There are also more specific proximate causes that relate to the particular circumstances of the groups involved. Note specifically that poverty, as expressed in terms of low incomes, is only a proximate and not an ultimate cause. It is necessary to know how poverty comes about, rather than simply to use it as an explanatory factor.

BANANA EXPORTS AND LOCAL FOOD PRODUCTION

Colonial powers encouraged banana production in colonies with the appropriate physical environment; thus, bananas are imported by France from Martinique and Guadeloupe, by Spain from the Canaries and the Azores, and by Britain from the Windward Islands in the eastern Caribbean. Indeed, for the islands of the eastern Caribbean, bananas have been the principal export crop since the early 1900s, with most production on small (less than 5–acre/2–ha) plots and by farmers who were often owner occupiers.

In many islands of the eastern Caribbean during the early 1990s, there was evidence of increasing banana production accompanied by decreasing production of local foodstuffs and increasing food imports. The usual explanation for such circumstances assumes that it is the increase in export crop production that is the cause of the other changes. But in a political ecological analysis of this situation in the eastern Caribbean island of St Vincent, Grossman (1993:348) noted that such explanations do not 'specify the precise means by which banana production is supposed to interfere with local food production', nor do they 'analyze the significance of the role of the state and capital in influencing these relationships'. An important political economic factor is that marketing policies and state intervention result in banana production being advantaged over local food production, especially because of the guaranteed market and fixed prices for bananas, while an important local factor is the reduced possibility of theft of bananas. Grossman (1993) also considered the question of potential conflicts in the productive sphere by reference to any seasonal conflicts of time allocation, to spatial patterns of land use, and to issues of intercropping. It was concluded that such conflicts were not the principal cause of local food production decline. Rather, the explanation lies in the different political economic contexts of banana and

Bananas, which comprise over 70 per cent of St Lucia's exports, are vulnerable to hurricanes and crop diseases (*CIDA Photo/Dilip Mehta*).

local food production. The banana market is attractive for reasons related to the state and to foreign capital.

Interestingly, the value of this approach is further evidenced by changes in world trade regulations. In accord with protocols agreed to in 1957, European countries were able to afford preferential treatment to their banana-producing areas. In 1993 the European Union amalgamated the various protocols into a single preferential trading structure. This system of protectionism for specific banana-producing areas has, however, been condemned by the new World Trade Organization following a challenge initiated by the United States and the banana-producing countries of Ecuador, Mexico, Guatemala, and Honduras. This global economic change, only one aspect of the movement toward free trade, is a major blow to those banana-producing countries, such as St Vincent, that previously received preferential treatment. As another example of the complex relations between banana production and larger policies, the European Union issued a statement of standard in 1994 requiring bananas to be at least 1 inch (27 mm) wide, 5.5 inches (14 cm) long, and without abnormal curvature. The bananas grown in the islands of the eastern Caribbean do not conform to such standards.

Concluding Comments

'Human ecology means different things to scholars in different disciplines' (Porter 1987:414). Although it is evident that there is no consensus in the social sciences as to what comprises an ecological approach, it is equally evident that much useful research has been conducted under the general ecological umbrella. Conceptual plurality is accompanied by substantial and quite varied research achievements.

As introduced in Chapter 2, ecology is one means of thinking about human/human and human/nature relations. Accordingly, it is not central to any one social science, but rather it is of interest to geographers, sociologists, anthropologists, psychologists, historians, and others. Geography might appear to be the most natural home of ecology because of the physical and human content and the long-standing idea that the discipline can serve as some sort of bridge between physical and human sciences, but this has not proven to be the case, and the geographic use of the ecological approach has been erratic. There was limited acceptance of the pioneering call from Barrows, while later attempts to define the ecosystem as a geographical model have not won many converts: Stoddart (1967:538) proposed the idea of a geosystem as a replacement for ecosystem, arguing that systems analysis 'at last provides geography with a unifying methodology, and using it geography no longer stands apart from the mainstream of scientific progress'—an appeal that has gone largely unanswered. Such a failure may reflect the fact that geographers do not seek a unifying methodology, especially given the diverse subject matter of the discipline. Perhaps the most successful uses of ecology in geography are the current linkage between ecology and environmental concern, the development of a political ecological focus, and the uses of ecology in analyses of population change as it relates to technology.

Geographers have adopted and applied ecology as developed in other disciplines, especially in sociology and anthropology. For example, the Chicago school of sociology was a major inspiration for urban geography in the 1960s, despite such limitations as the omission of a cross–cultural perspective and of any reference to symbolism. In some cases, concentric patterns of urban social areas were identified and such processes as invasion and succession noted. Cultural geographers have preferred to turn to anthropological cultural ecological analyses, especially to adaptation ideas, in their analyses. More recently, geographers have been attracted to the new ecology in biology and to political economy as developed in politics and economics especially. In addition, in a discussion of

Australian prehistoric cultural landscapes, Head (1993) employed some of the ideas associated with the new cultural geography in a study of landscape as a symbolic expression.

Ecological analyses by geographers have a rich and complex heritage. Criticisms of that heritage and of the many early studies are inevitable given recent advances in social theory, especially feminist concerns, but much current evidence strongly suggests that ecology is an increasingly popular and even more diverse tradition today than previously. There seems every indication that cultural geographers will continue their fascination with this approach and that conceptual revisions and additions will be accompanied by increasing numbers of empirical analyses. It seems likely that ecological analyses, in all their variety, will continue to inform us about ourselves and the world in which we live.

Further Reading

The following are useful sources for further reading on specific issues.

Stoddart (1965, 1966) on geography and ecology in general terms and Martin (1987b) on details of the ecological tradition in geography.

Koelsch (1969) on Barrows and historical geography.

Schnore (1961:209) described the article by Barrows as 'little more than a piece of intellectual history'.

Leighly (1987) on Sauer and human ecology.

Clarkson (1970) and Grossman (1977) on geographic versions of human ecology.

The traditional sociological view was that geography was principally concerned with factual information about humans and their physical environments rather than with explanations of human distributions. Thus, Quinn (1950:339) suggested that geographers were concerned with direct human and land relations whereas ecologists were concerned with interrelationships; Hawley (1950:72) identified a similar distinction, but also stressed that geography, specifically regional geography, was atemporal while human ecology was not. Schnore (1961) conveyed a more positive view in an enlightened interpretation of then current human geography in ecological terms—work in economic, urban, and population geography was cited for ecological content.

In sociology, some initial human ecological ideas were outlined by Park (1915) with the first textbook account offered by Park and Burgess (1921).

Entrikin (1980) on a discussion of human ecology in early sociology.

Hawley (1950, 1968) built on the human ecological work of Park to propose closer links with biology and a de-emphasis of humans.

Firey (1945, 1947) argued, with specific reference to Boston, for the importance of symbolic variables rather than physical science-based human ecological variables in any account of urban areas.

Kroeber (1928) noted a neglect of ecological considerations in then current ethnological studies, but chose not to pursue the theme any further.

Steward (1936, 1938) related group characteristics to ecological circumstances.

Bennett (1976:48) and Vayda and Rappaport (1968:492) on criticisms of the cultural ecology proposed by Steward.

The plethora of ecological approaches in anthropology prompted Meggers (1954) to discuss the limits set on cultural evolution by environments, even to the extent of arguing for cases of determinism, and resulted in Netting (1977) suggesting that any theoretical or methodological overview of the approaches was not practicable.

Boulding (1950) on economics and ecology.

Crosby (1995:1186) chose 'for the sake of convenience' to identify the work of Sauer and the 1955 symposium, which produced the seminal volume *Man's Role in Changing the Face of the Earth* (Thomas, Sauer, Bates, and Mumford 1956), as the scientific debut of the ecological approach in history.

Headland (1997) proposed that anthropology adopt a historical ecology and identified a number of myths that such an approach could address.

Reviews of and proposals for ecological history are provided by White (1990), Worster (1988, 1990), Cronon

(1992), and Merchant (1990). Some examples of ecological studies by historians include those by Cronon (1983) on environmental changes caused by different cultural groups in New England, by Crosby (1986) on the biological expansion of Europe, by Merchant (1989) on changing ideas of nature, gender, and science in New England, and by Arnold (1996) on environment, culture, and European expansion. Somewhat separately, explanations of the rise of the Western world also focus on the relevance of physical geography as it relates to human activities—recall the references to the works by Landes (1998) and Diamond (1997) in Chapter 3. Articles on the study of past environments by historical geographers and environmental historians are included in a Special Issue of *The Geographical Review* (April 1999).

The ecological psychology outlined by Lewin was more fully developed into a substantive research area by Barker (1968) with the term 'behavior settings' used to describe the larger context within which behavior occurs. The relative significance of the individual and of the environment to an understanding of behavior is related to the character of the environment, with the individual being more important if the environment is relatively stable and the environment being more important if it is varied and changing. A related advance is that of environmental psychology (Ittelson, Proshansky, Rivlin, and Winkel 1974).

Chapman (1977), Huggett (1980), and Wilson (1981) provided broad-based geographic accounts of the systems approach. The most substantial applications of a systems approach by geographers are studies in the natural hazards tradition (see Kates 1971), while there are also studies of subsistence (Nietschmann 1973) and of civilizations as ecosystems (Butzer 1980).

Using a cultural ecology framework, the anthropologist, Geertz (1963), studied land use change in Indonesia in generic systems terms, while Rappaport (1963) analyzed human involvement in island ecosystems. A key assertion evident in much of the geographic and anthropological work is that human populations can be treated in much the same way as any other population. Rappaport

(1963:70) noted that the 'study of man, the culture bearer, cannot be separated from the study of man, a species among other species'.

Kaplan (1996) for an interesting global argument adopting a limits to growth perspective.

Simon and Kahn (1984) for a forceful argument in favor of the cornucopian thesis.

Carneiro (1960) on the concept of carrying capacity. There are numerous studies, especially of agricultural areas in the less developed world, which employ this concept (see, for example, Bernard and Thom 1981).

Some ecological analyses employ the term 'cultural landscape' to refer to altered nature (Rowntree 1996:129)—a not uncommon use of the term in British and European literature (see also Birks et al. 1988; Simmons 1988; Svobodová 1990). A larger context for what is sometimes called historical ecology is contained in Crumley (1994).

Concerning terminology, Goldschmidt (1965:402) noted: 'Our investigation is a study in cultural adaptation, in ecological analysis, in the character of economic influence on culture and behavior in social micro-evolution—depending upon which of the currently fashionable terminologies one prefers'.

Berry (1984) on ecocultural psychology and Berry (1997) on use of the approach to a study of immigration and subsequent acculturation.

Bennett (1993) on adaptation and the social scale in psychology.

Worster (1993) on the new ecology.

For some, the introduction of a new ecology is but one component of a larger scientific change that includes a non-material paradigm of the universe and the emergence of chaos theory. 'Chaos begins where classical science stops. Disorder, or "noise" in the atmosphere, the sea, rivers, population fluctuations and biological functions, is the discontinuous and erratic side of nature' (Conacher 1992:178).

Smith (1995) and Harris (1996) on the agricultural revolution.

Adams (1997) on some philosophical implications of the new ecology.

Zimmerer and Young (1998) provide examples of analyses that focus on the complex character of environmental change in the less developed world.

White (1967) on links between the Christian tradition and environmental abuse; Gore (1992) on problems with this argument; Kinsley (1994) on aspects of both points of view.

de Steiguer (1997) on the age of environmentalism.

Inglehart (1990) on changing attitudes to environment.

Crosby (1986) and Griffiths (1997) on links between use of environment and political power.

Dickens (1996) on links between the division of labor and our understanding of nature.

Mungall and McLaren (1990) and Simmons (1997) on human impacts on environment.

Botkin (1990) on the psychological difficulties involved in the idea that nature is constantly changing.

Singer (1993) and Jamieson (1985, 1997) on questions relating to our human treatment of animals, including animals in zoos.

Diesendorf and Hamilton (1997) on environmental economics.

Tenbrunsel, Wade-Benzoni, Messick, and Bazerman (1997:2) on the discrepancy between environmental attitudes and behaviors.

Botkin (1990) on living with discordant harmonies.

Plummer (1993) on how nature, women, and other subordinated groups have been manipulated as a result especially of the two processes of backgrounding (that is, denying value) and instrumentalism (that is, service without recognition).

Shiva (1988) on ecofeminism.

Dwivedi (1990), Merchant (1995), Mellor (1997), Rose, Kinnaird, Morris, and Nash (1997), and Warren (1997) on the Chipko movement.

Baker (1997) identified the Landcare movement, a community-based approach to the land degradation crisis in Australia that has the potential to tackle larger issues of ecological sustainability.

Wisner (1978) developed an early argument for a socialist human ecology (see also Bennett 1976; Ellen 1982).

Kjekshus (1977), Porter (1979), Watts (1983), and Grossman (1984) are early studies in the political ecology tradition; more recent studies include those by Keil, Bell, Penz, and Fawcett (1998), Low and Gleeson (1998), and O'Connor (1998)

Fairhead and Peach (1996:10–15), Mayer (1996), and Bryant (1997) provide accounts of the larger intellectual context supporting political ecology.

Hawley (1998) on developments in human ecology.

Behavior and Landscape

It is in this chapter that the substantive philosophical debate introduced in Box 2.2—concerning whether or not the study of humans can be modeled on approaches developed in physical science—surfaces most clearly and most controversially. Indeed, the question of naturalism is effectively a recurring theme in this fourth of the six chapters that identifies and discusses an approach to the study of cultural geography.

In some respects the behavioral approach is the least clearly *cultural* approach, although cultural geographers played a prominent role in its popularization beginning in the late 1960s. The approach includes a relatively traditional perspective, **perception**, but also includes a major departure for cultural geographers, namely a questioning of the value of culture as a central concept. In this approach the preference is to focus on human behavior often at the scale of the individual and as it relates to landscape. Of the approaches discussed so far, this consideration of behavior necessarily brings cultural geography closer to social science in general, and hence embroils cultural geography in ongoing debates about the character of social science. The direction that this approach to cultural geography might take in the future seems especially uncertain.

The sequence of material in this chapter is as follows.

- There is an overview, necessarily in broad and general terms, of major psychological traditions to the understanding of human behavior. This overview serves the important function of placing much of what follows into an appropriate larger intellectual context. The principal traditions noted in this essentially chronological account are psychoanalysis, behaviorism, Gestalt psychology, and cognition, as well as ecological and environmental psychology.

- There is an account of the conceptual bases for a cultural geography centered on behavior. The account commences with references to some relevant early statements, followed by a consideration of the model of humans implied by the spatial analytic approach and an identification of the related behavioral geography. There is also a consideration of the model of humans implied by humanistic approaches and the related behavioral geographic interest in perception is identified. (Note that the term 'behavioral geography' is employed within geography to refer to the study of human behavior regardless of the conceptual inspiration for that study. Indeed, there are two quite different interests within this behavioral geography—some behavioral geography is informed by spatial analysis and some is informed by humanism. This chapter follows convention by using the term 'behavioral geography' to refer to both of these approaches.) Together, the first two sections build on the preceding chapter opening comments and offer two principal and divergent interpretations of the study of human behavior.

- The third section employs both the scientific and humanistic concepts in a series of empirical discussions of landscape change. The intentional inclusion of dramatically different concepts in a single chapter section is intended to encourage critical evaluation of the material presented. The landscape examples described include the Mormon landscape, the southeastern Australian pastoral frontier, colonial western Australia, the Great Plains, and Prairie Canada.

- The conclusion stresses the commitment that most contemporary cultural geographers have toward the more subjective approaches to the study of behavior and landscape.

In addition to the fundamental philosophical issue of naturalism, understanding the material in this chapter requires appreciation of some important events in the twentieth-century history of the discipline of geography. The relevant events are as follows.

- During the period from about 1900 to the late 1950s, as outlined in Box 6.1, the foremost approach to geography was regional geography as favored by Hartshorne and others, a regional geography that did not place any priority on culture, but that can be baldly characterized as empirical, descriptive, atemporal, and often implicitly environmental determinist.
- By the late 1950s, the Sauerian cultural geography that was first introduced in the 1920s was an important component of the larger discipline of geography—as it was concerned with questions of landscape evolution and regionalization and with a growing focus in ecology—but it was not the dominant geographic interest in North America. Recall also that this approach might be interpreted as being superorganic because of the concern with culture as cause and the related lack of focus on individual human qualities.

- A growing discontent with regional geography specifically was expressed in a landmark article that argued for a more scientific approach to geographic subject matter—for a nomothetic rather than an idiographic emphasis (see Glossary) (Schaefer 1953). This discontent was most fully expressed in what can be described as the quantitative revolution, or the **spatial turn**, that ushered in the approach to geography known as spatial analysis (see Box 6.1).
- Thus, the 1960s was a distinctive period in the history of geography as it was characterized by an approach that claimed to be equally applicable to both physical and human geographic facts and that, at least informally, embraced the philosophy of positivism, more specifically **logical positivism**, thus introducing what some commentators have described as a dehumanization of human geography. Box 6.2 is a discussion of the impact on the Sauerian school of this spatial turn.
- Reaction to the rise of the spatial analytic approach was swift and seemingly decisive, with some human geographers, especially cultural geogra-

Box 6.1: Regional Geography and Spatial Analysis

Regional Geography 1900 to Late 1950s	Spatial Analysis Late 1950s to 1970
Asks what and where?	Asks what, where, and why?
Aims to describe	Aims to both describe and explain
Atemporal	Atemporal
Focuses on facts	Includes both factual and conceptual content
Essentially a classification	Uses theories, hypotheses, and quantitative methods
Implicitly environmental deterministic	Rejected environmental determinism
Linked to empiricist philosophy	Linked to positivist philosophy
Declined about 1955 because not scientific	Declined about 1970 because dehumanizing

The above is a basic summary of some differences between the traditional regional and the spatial analytic approaches as practised in twentieth-century geography. The dates are approximate.

phers, arguing for a quite different approach to humans, particularly favoring a behavioral geography inspired by a variety of humanistic philosophies, and some other human geographers, who were more committed to the spatial approach, developing a behavioral geography within an essentially positivist framework. (There were also other geographers, including cultural geographers, promoting a more radical human geography, and still others maintaining the Sauerian tradition, but these two interests need not concern us at this time as the radical approach is central to Chapter 7 and the continuing Sauerian interests were discussed in chapters 3, 4, and 5.)

The critical point in the present context is that, by about 1970, some cultural geographers were initiating a behavioral geography based on humanistic concepts while some other cultural geographers were at least indirectly involved in the emerging behavioral geography that was linked to spatial analysis. It is these two concerns, both interested in human behavior and yet based on quite different philosophical positions, that provide the material for this chapter. The debate about how to study humans was thus present at the c.1970 origins of the cultural geographic concern with human behavior, hence the importance of the debate throughout this discussion.

Box 6.2: Impact of the Spatial Turn on Cultural Geography

Why was it that cultural geography was essentially unaffected by the spatial turn? The general answer is that cultural geography was separate from the dominant geographic tradition that was challenged by the spatial turn, namely regional geography. The cultural geographic interest in evolution and in the study of process through time was especially significant in distinguishing between cultural and regional geography. The challenge to the regional concept was aimed explicitly at regional geography rather than at regions as the outcome of evolutionary processes as typically studied by cultural geographers. Hence, the loss of the region as the core concept was of major significance to such interests as economic geography, which lacked a substantive conceptual basis other than that provided by regional geography, but was of little significance to cultural geography because of the firm foundations contained within the landscape school. Also, because the spatial turn maintained the atemporal emphasis of the regional school, cultural geographers saw little to attract them in the new spatial emphasis.

Further, those geographers who advocated the new spatial analysis directed their critical attention especially to the areas of economic and physical geography because it was there that the potential for theory construction and the use of quantitative methods, as stimulated by a positivist philosophy, seemed most appropriate. The deductive procedures involved in theory creation were certainly unattractive to cultural geographers who were often intent on analyzing the particular and not the general. For a research tradition primarily concerned with questions of history and culture, a positivistic perspective was necessarily of limited appeal arguing, as it did, for objective research and for research strategies adopted essentially from the physical sciences. This is notwithstanding the fact that criticisms of the superorganic content of the landscape school included a concern with the 'behaviorist claim that habit should be construed not as thought but as activity' (Duncan 1980:194–5).

In one respect the failure of the spatial turn to affect cultural geography is surprising because the discipline of history was undergoing some methodological debate along with an increasing positivist concern at about the same time. Further, both the French *Annales* school of historical method and the mostly North American new economic history embraced some theoretical and quantitative content. Although these developments in history are comparable to the spatial turn in geography, cultural historical geographers made little conceptual connection. Interestingly, toward the end of the 1960s when the spatial turn was being challenged by other approaches, the deficiency of excluding historical work was recognized and a process to form methodology proposed. Process to form ideas are explicitly historical, but they have not been welcomed by most cultural historical geographers.

The failure of the spatial turn to affect cultural geography is regrettable in one respect. Thus, although spatial analysis was quickly challenged by other approaches, it can be suggested that geography, but not including cultural geography, benefited from the spatial turn in that new ideas surfaced and new methods of analysis were investigated. Cultural geography has not been able to retrospectively single out any aspects of spatial analysis that might be of value.

Understanding Human Behavior

One theme that is evident throughout this book concerns the idea that knowledge and learning are divided into disciplines that change through time such that the contemporary social sciences are social constructions with boundaries that are neither necessary nor unchanging. Nowhere is this more evident than in attempts to understand human behavior. Two points are relevant here.

- First, although long an area of scholarly interest, it is only in the twentieth century that the study of human behavior has become identified particularly with the disciplines of psychology and sociology. Prior to this time, such work was conducted primarily by philosophers and historians.
- Second, the current division of interest between psychology and sociology may be an unfortunate distinction given the shared interest in human behavior. The division reflects trends in nineteenth-century disciplinary formation, especially the desire of Comte and later of Durkheim to

establish sociology as an independent science of society. Hence, while any full understanding of human behavior requires consideration of both individuals and collectivities, most theoretical formulations are derived from disciplinary origins and are limited to one or the other. Box 6.3 provides a brief overview of this question of social scale.

This section stresses the concepts developed in psychology as these are the most relevant given the work accomplished by cultural geographers. There is certainly an abiding tension to these discussions as psychologists have often desired the objectivity of natural sciences but have been concerned that they do not and cannot possess such objectivity because some of what they study, such as mental states, seems necessarily subjective and incapable of measurement.

The content of this section is important both because there is a need to appreciate the sophisticated, albeit varied, psychological emphases *per se*, and also because there is a need to understand

Box 6.3: The Question of Social Scale

Individuals or groups? A recurring issue in social science. Because individuals live in societies, and because societies comprise individuals, there is no clear distinction between the two and it can be claimed that they must be considered together. Indeed, it was only in the nineteenth century that social scientists established disciplines distinguished on this rather flimsy basis. Certainly, psychology as the study of individuals, and both sociology and anthropology as the study of groups, suffer from poorly defined key terms; recall, for example, the discussions of culture and society in Chapter 1. Even today, the distinction between individual and group is less than clear with university courses on social psychology often included in both psychology and sociology departments. Psychological social psychology typically studies social relations from the individual outwards by reference to such individual characteristics as attitudes and the approach is often derived from naturalism. Sociological social psychology typically employs group identities such as class, community, and culture, and is more receptive to methods related to the idealist perspective.

What is the appropriate social scale for analyzing the human behavior of interest to cultural geographers?

There is no simple answer to this question, with different approaches favoring different scales. Indeed, the question itself is misleading as an interest in groups can range anywhere from two individuals to the global population. Of course, Sauerian cultural geography focused explicitly on culture: 'Human geography, then, unlike psychology and history, is a science that has nothing to do with individuals but only with human institutions, or cultures' (Sauer 1941a:7). Further, most contemporary cultural geography accepts the need for a group scale. For example, the radical approaches that are discussed more fully in Chapter 7 contend that humans are more than individuals and that there is a need to consider the intersubjectivity of social life. From this perspective, the question becomes one of who does something—an individual or a class, an institution or a culture? The radical answer is that individuals need to be defined within the wider cultural context. Focusing on individuals is demonstrating a concern for their beliefs or ideas, which need to be defined culturally. Humanistic geographers, because of their concern for subjectivity and free will, employ philosophies that typically encourage an individual scale of analysis (see Box 6.6).

geographic interests in the light of relevant psychological material. It is not possible to do full justice to the diversity of psychological interest in human behavior, but the principal schools of thought are readily identifiable and those to which geographers have turned for inspiration are relatively few. Psychologists have introduced many general theories of behavior and other more specific theories concerned with such issues as perception, rote memory, and discrimination. This discussion is restricted to some of the general theories and introduces some of the difficult issues that these theories have needed to address.

Some Key Questions

- Is it appropriate to regard human behavior as possessing what is often labeled purposive or teleological qualities; that is to say, are some phenomena best explained in terms of ends (what they have become or what they achieve) rather than in terms of causes? Certainly, some theories view goal striving and purpose as integral components of individual behavior while other theories see these factors as of marginal relevance and argue that they accompany behavior but are not causes of behavior.
- What is the relative importance of conscious determinants of behavior of which the individual is aware, and unconscious determinants of which the individual is unaware? There is general agreement in psychology that there are unconscious determinants, but there is much debate about their importance.
- How important are reward and pleasure as causes of behavior? Some argue for cause and effect while others argue for association.

Among the many other issues that are debated are the importance of learning to behavior, the question of individual uniqueness, the relatedness of acts of behavior, and the degree of relationship between behavioral acts and environmental contexts. Certainly, all of these are complex and unresolved issues that psychological theories continue to address.

Geographers, who have understandably been most attracted to those psychological concepts that appear to relate directly to their specific concerns,

have largely ignored these difficult issues. This is an important point, and one that refers back to the way in which academic disciplines are socially constructed. What cultural and other human geographers have been attracted to are those theories that appear to speak directly to the traditional geographic concerns with space and environment. Hence, the attraction for geographers of theories that employ such concepts as cognitive or mental maps and perceived environments. Typically, such theories focus on the importance of the psychological environment, or what might be called the subjective frame of reference. This focus on subjective environments involves recognition that the physical world can only affect individuals in so far as it is perceived or experienced. Thus, in much of the geographic work, objective reality is not seen as a cause of behavior, but rather it is objective reality as perceived or assigned meaning by the individual that is judged important.

During the nineteenth century, psychologists introduced two approaches to the study of the human mind—the original Greek tradition of psychology. Some argued that the mind could best be understood through analysis of its parts or structures (see structuralism in Glossary), with the principal research method being that of introspection. Others argued that the mind was to be understood in terms of ongoing thought processes that are responsible for human learning; this approach, known as **functionalism**, was influenced by Darwinian ideas of evolution and adaptation.

From these bases, twentieth-century psychology introduced the approaches of psychoanalysis, behaviorism, Gestalt psychology, humanism, and cognition. One of the major areas of disagreement between these approaches concerned the subject matter of psychology. Was it the study of the human mind or of human behavior, or indeed was it the study of both of these? The five approaches are now considered, although it is important to emphasize that, as noted, geographers, including cultural geographers, have not made great use of these complex and divergent theoretical formulations. Rather, they have been attracted particularly to those approaches, or parts of those approaches, that appear to be in accord with prevailing geographic perspectives or that appear to be potentially geographic in content. Of course, this may not be such a bad thing if Hall and Lindzey (1978:69) were correct in stating: 'The fact of the matter is that all theories of behavior are pretty poor theories and all of them leave much to be desired in the way of scientific proof.'

Psychoanalysis

There are several psychodynamic theories, of which the most important is psychoanalysis as introduced and developed by Sigmund Freud (1856–1939). According to Freud, the human mind is like an iceberg as it is mostly hidden from view and thus cannot be observed directly (Figure 6.1). This deterministic approach was in accord with the prevailing scientific perspective of the late nineteenth century in that it claimed that human behavior is beyond the control of the individual such that the role of free will is minimal. Employing an analogy from physical science, Freud envisaged mental activity as a form of psychic energy as opposed to the better understood physical energy.

Because it is the presence of a dynamic unconscious that explains why people behave as they do, the task of a psychoanalyst is to uncover the forces that operate in the unconscious. But this unconscious cannot be accessed directly and there is accordingly an emphasis on uncovering this hidden world through other means such as the analysis of dreams. It is further claimed that the character of the adult unconscious is largely determined by drives, particularly sexual, that are acquired in early childhood.

Freudian theory continues to be an important part of discussions of human behavior, the human mind, and personality in contemporary psychology. Although several of the key concepts, such as those of psychic energy and the topographical mind, are not generally accepted, much of what Freud introduced is now accepted wisdom. Certainly, there is general agreement concerning the importance of childhood years to the development of personality. More broadly, the picture of individuals functioning in a personal world that is part reality and part make–believe is one that many psychologists find appealing.

Overall, the approaches to understanding human behavior proposed by Freud, by those who have refined his ideas, and by those who have out-

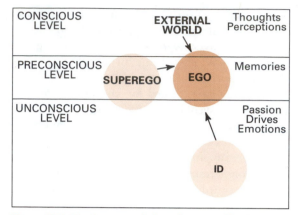

Figure 6.1: The human mind and personality structure according to Freud. There are three levels of the human mind. There is only a small area, the conscious mind, that is visible, with the much larger subconscious mind and the intermediary preconscious mind remaining unseen.

- The conscious mind includes current thoughts and perceptions.
- The preconscious mind includes memories that are not a part of current thoughts but that can be accessed.
- The unseen subconscious mind is effectively an underworld of feeling, passions, drives, and emotions that controls the conscious mind and therefore largely determines human behavior. In this sense, human behavior is essentially both biological and instinctual.

In addition to asserting the existence of three levels of the mind, the mind is seen topographically as comprising of three parts. The three components of personality—id, ego, and superego—are not to be thought of as things; rather, they are terms for psychological processes. Nor are they to be thought of as separate parts; rather, personality comprises an integration of the three. Notice that the ego is affected by both the id and superego in that it strives to satisfy their different demands, and also that it is affected by the external world. Understanding human behavior is complicated by these several impacts.

The id operates only at the level of the unconscious and is driven by primitive biological urgings directed at the pursuit of pleasure. The ego operates at all three levels—conscious, preconscious, and unconscious—and functions according to a reality principle rather than the pleasure principle of the id. Because the ego takes the real world into account, it strives for appropriate adaptations to environments. The third part of the mind, the superego, also operates at all three levels and comprises moral standards; it is loosely equivalent to the idea of a conscience. In a very general sense, the id can be seen as the biological part of personality, the superego as the social part, and the ego as the psychological part. Freud explained the role played by each of these three parts of the mind in a theory of personality development that stressed the weak development of the ego in infants and its subsequent development with growth.

lined alternative psychoanalytic theories have proven of limited interest to cultural geographers and, indeed, to human geographers generally. Certainly, prior to the 1990s there was little explicit use of this body of work, although it might be referred to in setting the psychological backdrop for accounts of behavioral geography. This characteristic reluctance of geographers to engage meaningfully with psychoanalysis is understandable given the controversial nature of the ideas within psychology, the complexity of the concepts, and the implicit determinism. The most substantial attempt to employ psychoanalysis to facilitate understanding of matters of interest to cultural geographers has focused on the ways in which space and self intertwine. These emerging ideas resurface in chapters 7 and 8.

Behaviorism

John B. Watson (1878–1958) introduced a behaviorist philosophy to psychology in 1913. According to this view, the subject matter of psychology was behavior, not the study of the human mind, and behavior was to be studied by means of the objective procedures practised in physical sciences; conscious experience and mental processes were not considered because they could not be observed. The approach was both deterministic and stimulus–response in character, with the concept of reflex (a stimulus that elicits a response) as the basis of an understanding of behavior. This first behaviorist movement, known as methodological behaviorism, was short-lived, but stimulated several other approaches, the most important of which is the radical behaviorism introduced by Skinner. As behaviorist psychology developed, a distinction was drawn between respondent (sometimes called Pavlovian or classical) conditioning, which involves the environmental conditioning of involuntary behavior known as the reflex concept, and operant (sometimes called instrumental) conditioning, which involves the learning of voluntary behaviors as well as an active organism operating on the environment.

Radical behaviorism is not a stimulus-response approach to the study of behavior—the basic concept employed is that of operant, not respondent, conditioning. This philosophy has given rise to a research tradition known as behavior analysis, which involves a set of concepts and techniques used to study behavior in its environmental context. The basic ideas are few and straightforward.

- Operants are behaviors that are maintained or changed by their immediate consequences. Consequences are events that affect the likelihood that an operant behavior will be repeated. Events that succeed a behavior and lead to an increased frequency of the behavior are known as reinforcers, while those that lead to a decreased frequency of the behavior are known as punishers.
- Events that are consistently present when behaviors are reinforced or punished are also relevant because they determine under what circumstances particular behaviors are likely to occur. These are known as antecedent or discriminative stimuli because they serve to discriminate in favor of particular behaviors.
- Thus, both the antecedent stimuli that precede a reinforced behavior and the consequences of that behavior serve to control the behavior.
- This sequence of antecedent stimulus, operant, and consequence is known as a contingency statement. Operant conditioning is a process that involves the environment reinforcing those behaviors that are the most adaptive and effective in achieving reinforcers and avoiding or escaping from aversive stimuli. The key idea, the law of effect, is that consequences influence behavior.

Although these few simple ideas have been highly influential in psychological research, they have not proven adequate for many investigations of behavior, and additional concepts have gradually been added. Most important, the concept of rule-governed behavior addresses the challenge posed by the fact that some contingencies have outcomes that are delayed and thus might not be efficient reinforcers. A rule is a verbal description of a contingency that provides a mechanism for understanding how thoughts or self-talk might control goal-oriented behavior; in this sense, rules function as antecedent stimuli. Thus rules, instructions, or guidelines that specify the outcomes of actions can control behavior. Indeed, we live in a world in which rules, as instructions or advice, as laws or social norms, regularly serve to provide guidelines for effective behavior. It is this rule-governed behavior that encourages humans to optimize delayed outcomes. Similarly, the behavior of others can serve as antecedent stimuli with behavior being learned through observational, vicarious, operant conditioning; that is, through watching others perform the behavior. (See Your Opinion 6.1.)

In their commentaries on behaviorist approaches, it has not been unusual for geographers to assume that all behaviorist approaches share the deficiencies of the original methodological behaviorism, specifically the limitation to behavior as respondent conditioning. Thus, Gold and Goodey (1984:544–5) stated, 'behaviorism viewed human behaviour in terms of stimulus–response relationships in which specific responses could be attached to given antecedent conditions'. Further, there has been some related confusion concerning the extent to which geographers have employed a behaviorist approach. Although the behavioral geography of the late 1960s that evolved in association with spatial analysis was behavioristic in broad outline, there have been very few attempts to apply behavior analytic ideas to problems in cultural geography. Two efforts in this direction are included later in this chapter.

Your Opinion 6.1

Might it be appropriate to employ the working assumptions, constructs, and methods of behavior analysis not only in analyses of individual behavior but also in the context of group behavior? Obviously, this is a deep question for which there is no simple answer, but it is an important question for cultural geographers to consider for two reasons. First, behavior analysis has proven influential in some other social science disciplines with the argument that the propositions of behavioral psychology are the general explanatory propositions of all the social sciences. Second, there are close links between radical behaviorism and cultural materialism (see Box 5.3). Thus, radical behaviorism identifies the principles of individual behavior in terms of reinforcers and punishers, while cultural materialism focuses on group behavior in terms of benefits and costs. Both approaches consider that behavioral responses to environmental variables precede mental rationalizations as to the reasons for responses. Cultural geographers have not typically followed either of these precedents.

---

field within which the specific object is contained. The human mind and perception are one whole and not a number of parts. This way of analyzing conscious experience is different from those proposed by both psychoanalysis and behaviorism. The emphasis is on how the world is perceived rather than on how individuals behave because perception determines behavior. Thus, behavior can be understood by uncovering the perceived world of individuals rather than by studying the actual objective world.

The principal method of analysis employed by Gestalt psychologists is that of **phenomenology**, involving the study of individual perception and subjectivity. Phenomenology emphasizes the need for holistic descriptions; hence, the idea that the whole is greater than the sum of the parts succinctly summarizes the Gestalt approach. Although this basic idea is appealing to many psychologists and cultural geographers, the Gestalt approach has not maintained a prominent place in contemporary psychology. Field theory, one extension of Gestalt psychology, proved to be of particular interest to geographers and continues to be of importance. This field theory is an application of field concepts as developed in physical science, with behavior interpreted as a function of the field that exists at the time behavior takes place.

- A first step is to formally separate the individual person from the rest of the world; this is represented diagrammatically in Figure 6.2(a).
- The psychological field is diagrammed in Figure 6.2(b), with the person (P) within a psychological

Gestalt Psychology

The Gestalt movement in psychology developed in the early twentieth century, initially arising in opposition to the structuralist tendency to divide things into component parts. The key idea is that human behavior is determined by a psychophysical field that is analogous to a gravitational or electromagnetic field as described in physical science. For example, human perception of a particular object is to be explained by reference to the larger

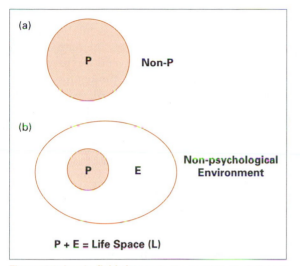

P + E = Life Space (L)

Figure 6.2: Lewin field theory. In Figure 6.2(a), the individual person, P, is formally separated from the rest of the world, non-P. Lewin represents this distinction in graphic form, and not simply in writing, because a spatial representation is susceptible to mathematical manipulation. The shape and size of the figure enclosing P are irrelevant; all that matters is that there is a bounded area completely surrounded by a larger area.

In Figure 6.2(b), a second step is taken in the representation of psychological reality. In this case, a second bounded figure is drawn around the bounded figure that encloses P. The area between the boundary of P and the boundary of the second figure is E, the psychological environment. The total area of both P and E is labeled the life space. Finally, the area outside of the boundary of the second figure is the non-psychological environment that includes physical, cultural, and social facts.

environment (E) and the sum of P and E defined as the life space or L. Beyond E is the non-psychological environment, which comprises physical, cultural, and social facts. The significance of L is that it contains all the facts that directly influence individual behavior—thus, behavior is a function of the life space.

- But Lewin also claimed that those physical, cultural, and social facts that are outside of E can influence behavior indirectly through their impact on E. Psychological ecology was thus proposed as the study of these physical, cultural, and social facts.
- Further, Lewin saw the psychological world as influencing the physical, cultural, and social worlds, and thus identified a two-way interaction.
- Finally, there is also a two-way interaction between P and E. Thus, both boundaries in Figure 6.2(b) are permeable.

These few basic ideas developed by Lewin have proven highly influential, and it is from these that the ecological and environmental psychologies discussed in Chapter 5 developed. Both of those approaches stress the role of the environment external to the individual as a cause of behavior. Behavioral geographers have been attracted to the Lewin field theory because of the emphasis on perception and subjective environments, the concern with individuals as members of groups, and the explicit acknowledgment of the role played by external non-psychological environments. Further, the positivistic overtones were attractive in the context of 1960s geography.

Humanism

Notwithstanding that there are numerous varieties of humanistic psychology, some clear differences are evident between humanism and the other approaches discussed in this section. Most important, a deterministic logic is rejected; thus, the psychoanalytic idea that behavior is the result of compelling internal forces is rejected, as well as is the behaviorist idea that behavior results from a conditioning process. 'Humanistic psychology opposes what it regards as the bleak pessimism and despair evident in the psychoanalytic view of humans on the one hand, and the robot conception of humans portrayed in behaviorism on the other hand' (Hall and Lindzey 1978:279). Accordingly, humanism has been described as the third force in American psychology.

As employed by psychologists, humanism rejects the scientific tradition of conducting research that is replicable and verifiable by other researchers, seeking instead an understanding of the world as it is appreciated by the individual or group being studied. This humanism centers on human perception, asserting that behavior results from subjective understandings of the world. Although there is a central concern with human experience and personal growth in all versions of humanistic psychology, there are numerous specific variations. Further:

- there is some sharing of interest with Gestalt psychology concerning phenomenology—the method that emphasizes that what is perceived by an individual is in effect his or her personal reality when it comes to understanding the person's behavior

- major differences between a Gestalt approach and humanism concern the Gestalt focus on cause and effect explanation and also the group membership considerations
- there is also some sharing of interest with cognitive psychology as discussed below, especially concerning the importance of mental imagery

The most well-developed humanistic approach in psychology is probably **existentialism**, an approach that employs the phenomenological method as already discussed in the context of Gestalt psychology. Existentialism has nineteenth-century philosophical and literary origins, although the movement did not create a major impact in social science until the mid-twentieth century, with Martin Heidegger (1889–1976) typically seen as a key link between the original philosophy and later social science developments. A key idea in the psychological interpretation is that each individual is a being-in-the-world, an idea that emphasizes the unity of humans and the world. Specifically, it is argued that there are three components to the world of human existence, namely the physical surroundings (*umwelt*), the human environment (*mitwelt*), and the individual (*eigenwelt*). Causality is not considered in this approach; rather, attention is on motivation as the means of understanding behavior. Existentialism forcibly opposes any dehumanization of humans.

Given the unifying thrust, the opposition to causation, and the explicitly human focus, it is not surprising that these ideas have appealed to some cultural geographers. Rather unfortunately, however, the quite varied and often nebulous content has limited application of the ideas within psychology. Indeed, the future of humanistic psychology seems particularly uncertain. According to Wertz (1998:61), it is in crisis because of a relatively small number of interested researchers and practitioners and because of the absence of any supporting institutional framework.

Cognition

Cognitive psychology is an approach that argues that cognitions, especially thinking, are the princi-

pal causes of behavior. **Cognition** is a broad term referring both to a group of mental processes including perception, memory, attention, pattern recognition, problem solving, and the psychology of language and also to all those processes by which sensory input is transformed, reduced, elaborated, stored, recovered, and used. Thus, cognitive psychology explicitly addresses two principal related criticisms of behavior analysis, namely:

- the claim that the presence of mental factors means that any attempt to explain all human behavior in terms of physical laws is necessarily inadequate
- the claim that, while it may be the case that behavior can be shaped by consequences, it is still the case that cognition must be studied in its own right

Cognitive psychologists pay particular attention to memory, thinking, language, and consciousness. The concept of memory, defined as the persistence of experience and learning over time, involves the acquisition, storage, and recall of information. Some memories are stored verbally and some as images. Thinking can be defined as the processing of information through the internal manipulation of symbols, often involving visual and spatial images. Consciousness involves awareness, defined as the ability to express something symbolically.

Given the emphasis placed on information processing, it is not surprising that a principal related approach is that of cognitive science, which aims to explain cognitive processes in terms of the means by which information is handled by the mind, usually proposing that the mind works in a series of logical steps rather like a computer. Some cognitive psychologists have borrowed concepts from such areas as computer science, communications science, and neuroscience and there has been a related geographic interest as well in the computational process modeling of spatial cognition and behavior.

These advances in cognitive psychology have proved to be attractive to some geographers concerned with understanding behavior, especially as they support related ideas from humanism and from the environmental psychology discussed in

Chapter 5. Most notably, some cognitive concepts link well with some of the concerns of a humanistic approach, for example, the emphasis placed on such emotional characteristics as attitudes, feelings, beliefs, and values. Further, the geographic analysis of mental maps and of environmental cognition generally was encouraged. However, the concepts associated with the cognitive science component of cognitive psychology are not viewed sympathetically by humanists.

Debating the Merits of the Approaches

All too clearly, psychologists do not agree concerning the best approach to the study of human behavior. Each of the approaches noted has some merit, and it is not possible to definitively state that one is in some way inherently superior to the others. A balanced response to the above overview may be to suggest that different approaches are suited to different specific problems. Unfortunately, however, even this limited conclusion is complicated by the fact that distinguishing between seemingly different approaches is often not quite as straightforward as might be expected. Take, for example, the distinction between radical behaviorism and cognition.

COGNITIVE BEHAVIORISM?

As noted, cognitive psychology emerged as an identifiable field beginning in the 1960s and is a major focus in contemporary psychology, competing with versions of behaviorism, especially radical behaviorism, as the dominant approach to the study of human behavior. The distinction with radical behaviorism is clear in principle as cognitive psychology stresses the need to study internal mental processes rather than observable behaviors on the grounds that it is cognitions, particularly thinking, that control behavior. However, the distinction between behaviorism and cognition is not quite so clear in practice, as one school of thought, sometimes called cognitive behaviorism, claims that cognitions are best thought of as behaviors that occur inside the body. Following this logic, the fact that a cognition, such as a thought, cannot be seen does not make it any different from an observable behavior, such as waving a hand. For the cognitive behaviorist, both types of behavior are subject to the same underlying principles. This

important integration of seemingly quite different schools of thought merits some elaboration. The possibly fading distinction between behaviorism and cognition in contemporary psychology is evidenced by:

- Slocum and Butterfield (1994:59), who observed that 'scholars from the two perspectives have invented similar concepts to describe behavior'
- Schnaitter (1986:303), who asserted that 'something rather like behaviorism, and something rather like cognitivism, may eventually evolve to fit felicitously one with the other'
- Phares (1991:323), who noted that 'a new language has begun to intrude into the world of behaviorism. Such terms as *cognitive behavior modification, cognitive relabeling, stress inoculation,* and *rational restructuring* have started to infiltrate the language. What this signals is a growing cognitive emphasis in the field of behaviorism.'

Thus, some cognitivists do accept the general principles of operant conditioning while they also see mental processes as playing an active and indeed critical role between the antecedent stimuli and the behavior, affecting how one particular behavior can be selected from a range of possibilities. Similarly, many contemporary radical behaviorists consider that unobservable (private) behaviors —that is, mental states such as thoughts and feelings—are no different in principle from observable (public) behaviors. Many of the discussions in the emerging field of cognitive behaviorism employ techniques that rely largely on the concept of rule-governed behavior.

This example of the overlap and integration between two approaches that are often thought of as opposites is but one instance of the uncertain status of the contemporary conceptual landscape of the understanding of behavior. There are also some interests shared between cognitivists and humanists, as both object to the behaviorist exclusion of mental life, focus on perceived meaning as the basis of behavior, and argue that humans are quite unlike animals.

Links to Cultural Geography

As the above account suggests, psychology is a pluralistic discipline, and it is not uncommon to find adherents to particular approaches complaining that their approach is misunderstood and misrepresented by adherents to other approaches and, indeed, by historians of the discipline and introductory textbook authors. Further, it seems clear that psychologists and geographers are usually attracted to an approach that is in accord with their larger philosophical preference. Certainly, cultural geographers might profit from approaching this wealth of ideas with a flexible mindset, although, unfortunately, application of a specific approach is often bedeviled by internal differences of opinion.

Each of the various psychological traditions has something to offer the cultural geographer concerned with behavior and landscape; certainly, all have been considered and employed to some degree, but none of the traditions can be said to have been wholeheartedly embraced, or indeed especially favored, by cultural geographers. It is possible that the general lack of explicit integration of psychological and geographic traditions results from the presence of distinctly different parent disciplines. Generally speaking, geographers interested in the study of human behavior were attracted first (that is, in the 1960s heyday of spatial analysis) to ideas with positivistic overtones such as behaviorism and Gestalt psychology and, second, in the 1970s reaction to spatial analysis, to ideas with humanistic and cognitive content.

- The possible relevance of psychoanalytic theories has proven somewhat difficult to appreciate until quite recently, an unsurprising fact given the prevailing interpretations of cultural geography. The present interest in these theories is one part of the cultural turn in cultural geography and is discussed in Chapter 8.
- Although behaviorism is generally considered to have been a principal stimulus for much of the behavioral geography that appeared in the late 1960s, there is good reason to suggest that there were some misinterpretations and resultant confusions about precisely what is involved in a behaviorist explanation.
- Ecological and environmental psychology as developed from the larger Gestalt tradition certainly informed much behavioral geography with both shared research interests and techniques,

although there has been limited evidence of genuine integration. An early use of Gestalt ideas was by Kirk (1963:365), asserting that the environment has 'shape, cohesiveness and meaning added to it by the act of human perception'. Implicit in this idea was the existence of both individually perceived worlds and group views.
- Humanistic traditions in psychology have served to reinforce the post–1970 cultural geographic interests in phenomenology and existentialism especially. The humanistic interest that permeated geography during the 1970s was attractive for essentially the same reason that made field theory and its outgrowths attractive, namely the emphasis placed on the subjective environment.
- Cognitive psychology has encouraged geographers to pursue interest in the humanistic approach to behavioral geography and also to develop a more analytic behavioral geography.

Certainly, much work accomplished by cultural geographers has been at least partially informed by the various psychological traditions, and the next two sections review a selection of cultural and related geographic work, respectively conceptual and empirical, concerned with behavior and landscape in which the work accomplished by psychologists plays an important role.

Behavioral Cultural Geography

As discussed at the beginning of this chapter, geographic studies of human behavior generally belong to one of two broad interests, both of which have borrowed from psychological traditions.

- There is some continued interest in aspects of positivist thought. The principal concern here is with analytic behavioral geography, much of which is cognitive in emphasis. However, most of this work lies outside of the cultural geographic tradition. In addition, there have been some limited attempts to apply behaviorist ideas in a cultural geographic context.
- There is a focus on humanistic thought. The principal concern here is with perception, mental maps, and images. This is the dominant tradition in cultural geography.

This section provides an overview of behavioral cultural geography as it was informed by both spatial analysis and by humanism, beginning with some of the early statements and progressing to a discussion of the two relatively distinct views of humans that are associated with the two interests identified above. Discussions of examples of relevant research are included in the following section.

Early Statements

Although it can be argued that geographers have had a sustained interest in seeking causes of human behavior—witness the logic of environmental determinism, possibilism, probabilism, and the landscape school, each of which assumes a particular cause of human activities and landscape change—it is appropriate to note that, prior to the 1960s, geographers only occasionally acknowledged the importance of behavior in geographic studies, and that there were few theoretical or empirical advances specifically concerned with behavior. (See Your Opinion 6.2.)

The suggestions that there are objective and subjective worlds and that both are of importance were not incorporated into mainstream geographic thought, although they were implicit in the writings of Vidal and Sauer

with both these geographers incorporating a strong visual content into some of their geographic descriptions. Indeed, humanistic cultural geographers have frequently referred to Vidal and Sauer as principal early inspirations. Other than these general contributions by European geographers

and by the Sauerian school, there were two specific proposals for a geographic study of behavior before the 1960s, although, in both cases, their impact was minimal at the time and their influence is debated even today.

- The first of these was not clearly related to other geographic literature and nor did it adequately signpost future research directions, but Wright did explicitly distinguish between the subjective and the objective:

Objectivity . . . is a mental disposition to conceive of things realistically. . . . The opposite of objectivity would, then, be a predisposition to conceive of things unrealistically; but, clearly, this is not an adequate definition of subjectivity. As generally understood, subjectivity implies, rather, a mental disposition to conceive of things with reference to oneself. . . . While such a disposition often does, in fact, lead to error, illusion, or deliberate deception, it is entirely possible to conceive of things not only with reference to oneself but also realistically (Wright 1947:5).

Thus, a subjective view of the world might range anywhere along a continuum from complete error to complete accord with reality—a discipline named geosophy and centering on these true and false geographical ideas was proposed. However, these original ideas were not related to other disciplines, such as psychology, nor to other developments in geography, and there was no substantive empirical attempt to pursue the ideas.

Your Opinion 6.2

In some respects, the lack of explicit cultural geographic concern with behavior may be surprising given the following pioneering statement by Humboldt: 'in order to comprehend nature in all its vast sublimity, it would be necessary to present it under a twofold aspect, first objectively, as an actual phenomenon, and next subjectively, as it is reflected in the feelings of mankind' (quoted in Saarinen 1974:255–6). Indeed, Humboldt went much further than merely advocating such a concern with both objective and subjective worlds, effectively stimulating an aesthetic tradition in geography, a tradition that emerged from Humboldt's ability to combine intuition and science. For example, Humboldt (quoted in Bunkse 1981:138) wrote: 'In the uniform plain bounded only by a distant horizon, where the lowly heather, the cistus, or waving grasses, deck the soil; on the ocean shore, where the waves, softly rippling over the beach, leave a track, green with the weeds of the sea; everywhere, the mind is penetrated by the same sense of the grandeur and vast expanse of nature, revealing to the soul, by a mysterious inspiration, the existence of laws that regulate the forces of the universe.' Are you surprised by the failure of cultural geographers to build on this early concern with both objective and subjective worlds and also on the tradition of aesthetic writing?

• The second specific contribution, by Kirk (1951), was similarly uninfluential at the time of publication, at least partially because it was published in a journal that was relatively inaccessible to North Americans such that the work was not generally known until the 1960s. This was a more explicit attempt to incorporate behavior into geographic analyses with a distinction made between the world of facts, both physical and human, and the environment in which these facts were culturally structured and in which they acquired cultural values. The world of facts was labeled the Phenomenal Environment, and the world in which they received meaning was labeled the Behavioral Environment. This is a distinction that is comparable to the work of Gestalt psychologists as diagrammed in Figure 6.2. Kirk (1963) further developed these ideas to include the suggestion that human behavior in the Behavioral Environment was rational. These original ideas, for geography, have not become central to subsequent geographic perception research.

Both Wright and Kirk identified the presence and importance of subjective worlds, as Humboldt had a century earlier, but these contributions did not form a continuous development of ideas, nor did they generate geographic research into human behavior. Prior to the 1960s, then, the geographic interest in behavior was quite limited. This failure by geographers was despite the various advances in psychology, as discussed in the previous section, but may be understandable given the academic isolation of geography within the social sciences prior to the 1960s along with the importance of the regional approach and the landscape school. Further, anthropology, the discipline to which cultural geography was undoubtedly closest, was not successful in developing a behavioral anthropology, at least partly because of the prevalence of a cultural determinist perspective.

The Model of Humans

Understanding the ways in which geographers have approached the study of human behavior, including whether humans are viewed as passive objects or as active subjects, is aided by an appreciation of the character of geography as an academic discipline from about 1960 onwards. Recall from the opening remarks of this chapter that the 1960s rise of spatial analysis encouraged a behavioral geography that incorporated a positivistic emphasis, while the subsequent rapid demise of spatial analysis after about 1970 encouraged the rise of a more humanistic behavioral geography. Indeed, that demise was partly attributable to the interest shown in humanism by cultural geographers who were never attracted to spatial analysis. These two related but different approaches to the study of human behavior in geography involved two principal and quite different conceptions of the model of humans.

BEHAVIORAL GEOGRAPHY INFORMED BY SPATIAL ANALYSIS

Consider the following statements:

• All social sciences study social life or 'patterns of conduct that are common to groups of people' such that it 'is not their subject matter but their approach to it that differentiates the various social sciences' (Blau and Moore 1970:1).
• Geography is included in every 'list of the social sciences' with each discipline having 'its own subject matter, its own part of the study of man' (Senn 1971:60).

During the 1960s especially, with spatial analysis as an important approach to geography, many human geographers concurred with these statements, viewing their discipline as a social science concerned with some aspect of behavior, with behavior often used as an umbrella term to refer to human activities in general.

• A leading cultural geographer asserted: 'Regardless of the tradition they follow or the methodology employed in the definition of their role, most human geographers have no hesitation in identifying themselves as social scientists' (Mikesell 1969:227).
• Similarly, Ginsburg (1970:293) noted that human geography was concerned with 'questions of human behavior to the same degree, though not necessarily in the same way, that the other social sciences are'.

It was in this intellectual context, with human geography viewed by many as a social science, that this interest in behavioralist geography surfaced. *This behavioral geography subscribed to a model of humans that had close links to the then geographic interest in spatial analysis and, accordingly, there was little room for human creativity and understanding.* It is incorrect, however, to suggest as some have that this behavioral geography was behaviorist in the psychological sense of that term. Two instances of this misunderstanding are as follows.

- First, Relph (1984:209) stated: 'Since I have never been able to establish just what "behavioral geography" is and how it distinguishes itself from other sorts of geography, I have assumed it to be a version of B.F. Skinner's behaviorism somehow transferred from psychology to geography.'
- Second, Pile (1996:36) described this behavioral geography as behaviorist and also incorrectly described both the Watsonian and Skinnerian versions of behaviorism as premised on the logic of stimulus and response.

Certainly, it is the case that some behaviorist work in psychology was regarded favorably such that this development of behavioral geography involved an informal acceptance of some of that work, but it is not the case that this behavioral geography was behaviorist in the psychological sense of the term.

Thus, this behavioral geography reflected neither the stimulus–response methodological behaviorism advocated by Watson, nor the more sophisticated operant conditioning approach of radical behaviorism advocated by Skinner. Regardless, for many subsequent geographic commentators, the spatial analytically informed behavioral geography is seen as a particularly unfortunate episode in the history of the discipline, an episode too closely tied to a spatial analysis that 'had no time for humanism, and instead supposed that a narrowly conceived science could help us to learn everything we wanted to know about human activity in geographical space' and that 'restricted itself to a fairly narrow conception of how human beings think and act, and in so doing subscribed to versions of *behaviourism*—admittedly not always

the simple "stimulus–response" behaviorism that critics have suggested' (Cloke, Philo, and Sadler 1991:66–7).

However, as stated above, it is correct to note that the model of humans employed in this behavioral geographic work was one that involved humans responding to particular stimuli and one that paid minimal attention to questions of human freedom and dignity. In addition to this rather narrow view of humans as rational decision makers, borrowed from economics and largely employed in economic geographic analyses, there was a more flexible view of humans as optimizers. This idea was in accord with some work in economics that viewed humans as boundedly rational animals, and major geographic studies included work on natural hazards. A classic study of farming practices in a part of Sweden further established that humans adopted strategies that aimed to find a satisfactory outcome, and the terms 'satisficer' and **satisficing behavior** were introduced (Wolpert 1964). Many of the studies conducted in this tradition studied human behavior and inferred cognitive characteristics such as perception, a research framework that is described below as the inference problem.

Before proceeding to a discussion of the humanistic approach to behavioral geography, two distinctive, tentative, and quite controversial cultural geographic contributions are briefly noted. Neither have close conceptual affinities with behavioralist work in that they do not stress the importance of learning; rather they imply that some aspects of human behavior result from spontaneous responses to stimuli. Specifically, both approaches incorporate suggestions that humans are innately programmed to behave in a certain way. These two are prospect and refuge theory and the Geltung hypothesis and they are introduced in Box 6.4 and Box 6.5 respectively.

BEHAVIORAL GEOGRAPHY INFORMED BY HUMANISM

For many cultural geographers, the fundamental problem with the behavioralist approach concerned its inability to accommodate individual human characteristics, accordingly, a humanistic approach to the study of behavior was favored. Three key assumptions of the humanist philosophy from which this approach is derived are relevant to this discussion.

- First, humans are ontologically and epistemologically irreducible. This means that claims about knowledge cannot be derived from physical science, and involves emphasizing such human phenomena as creativity and understanding; 'human beings present a unique reality, irreducible to the order of animal behavior' (Wertz 1998:53).
- Second, the focus is on human experiences and symbolic expression, not abstract principles, and there is an acknowledgment of many different truths.
- Third, individual freedom and dignity are respected.

Following the ideas of humanistic philosophers, especially Dilthey and Edmund Husserl (1859–1938), it is recognized that the subject matters of physical and human science necessitate different concepts and methods. Most obviously, this is the case because humans are free to act as they choose. But it is also because, unlike physical phenomena, the human phenomena that are the subject matter of the human sciences are internal, not external, to the experience of the scientist. This is a critical observation and Wertz (1998:49) expressed it this way as follows: 'Because physical phenomena and their laws of interconnection are not directly given, they must be explained by constructed models, theoretically derived hypotheses, and experiments revealing functional relationships among isolated factors. Human phenomena,

Box 6.4: Prospect and Refuge

Appleton (1975a, 1990) argued that some aspects of human behavior in landscape can be profitably studied with reference to animal behavior, with the emphasis on biological drives and a corresponding denial of the relevance of human imagination and creativity. Thus, understanding human behavior in landscape is considered comparable to understanding why humans mate, protect their young, and eat. This essentially Darwinian approach employs concepts from the study of animal behavior not to seek direct parallels in humans, but to locate clues to the understanding of human behavior. There are two principal hypotheses.

- *Habitat theory*—relying directly on studies of animal behavior that show that species seek optimal environmental conditions—'asserts that aesthetic satisfaction, experienced in the contemplation of landscape, stems from the immediate perception of landscape features which, in their shapes, colours, spatial arrangements and other visual attributes, act as sign-stimuli indicative of environmental conditions favourable to survival' (Appleton 1975b:2). Thus, this hypothesis stresses the idea of spontaneous human response to rather than rational appraisal of landscapes that are encountered. Learned patterns of behavior are considered to be of secondary status to inner needs. In those circumstances where there is a correspondence between human habitat and human inner needs, an aesthetic sensibility of landscape results; if such a correspondence does not prevail,

then anxiety is evident. Simply put, humans seek a natural habitat.
- *Prospect-refuge theory* moves beyond the first hypothesis in an effort to identify those innate sign-stimuli that continue to be important for humans. The basic argument is that the ideal environment is one that humans are able to retreat to in safety, meaning that it is a refuge in which they cannot be seen, and also one that provides the opportunity to observe surroundings, meaning that it serves as a prospect. 'It is a part of our nature to wish to be able to see and to hide when the occasion demands, and the capacity of an environment to furnish the opportunity to do these things therefore becomes a source of pleasure' (Appleton 1975b:3). Prospect has to do with perceiving, with the acquisition of information, while refuge has to do with shelter and seeking safety. Simply put, humans have an innate desire to see without being seen.

Using these two hypotheses to inform analyses of landscapes thus involves classifying both physical and built landscape features and situations that serve as habitats in terms of their prospect and refuge characteristics. Appleton (1975b) offered a comparison of British and Australian symbolic landscapes in these terms that stressed the importance of such physical characteristics as light, landforms, vegetation, along with such elements of the built landscape as houses and settlements.

because they are immediately given to knowledge and internally related to each other, need not be hypothesized but are to be described and understood in their meaningful connections.'

Humanists do acknowledge the need for hypothetical systems in physical science, but reject their application in the human sciences, arguing that the reason why the human sciences often fail in their tasks is because of their adoption of physical sciences methods that are totally unsuited to their subject matter. Expressed more simply, humanists demand a quite different model of humans from that which results from the use of physical science methods.

In geography, humanist ideas were introduced primarily by those who saw themselves as cultural or historical geographers, and humanism was a powerful critique of spatial analysis. Box 6.6 outlines the three principal humanistic interests advocated by cultural geographers—phenomenology, existentialism, and **idealism**—while humanistic geography is further considered in Chapter 8 in the context of discussions of landscapes as places, of identity and landscape, and of symbolic landscapes. Intriguingly, however, as discussed in Box 6.7, some scholars interpret the larger humanistic tradition quite differently.

Box 6.5: The Geltung Hypothesis

Why do we behave as we do? More specifically, how might we begin to behave more appropriately, both toward each other and toward the environment? Various proposed answers to these questions were noted in the Chapter 5 discussion of environmental ethics, and the admittedly speculative Geltung hypothesis might have been introduced in that context. Undoubtedly, this is one of the most thoughtful and novel advances toward an understanding of behavior involving the original claim that 'human beings are innately programmed to persistently and skillfully cultivate attention, acceptance, respect, esteem, and trust from their fellows' (Wagner 1996:1). In short, we are born to show off, to strive for what is labeled Geltung. The hypothesis was developed from ideas about diffusion and communication with the recognition that the 'opportunity for transmitting an (always innovative) message with success depends upon the Geltung of its sender compared with that of the recipient' (Wagner 1988:190). In this respect, recall the reference in the Chapter 3 discussion of cultural diffusion to variations in innovativeness, especially as this involves the presence of opinion leaders.

Given the basic argument that humans are genetically programmed to behave in a particular way, this hypothesis is quite different from most current social science approaches to the understanding of human behavior. The idea that humans put on displays and that other people are important to us is a well-established one in some social psychological theory, but the suggestion that such behavior is instinctual, a part of our animal heritage, is relatively novel. 'I shall contend that human behavior responds universally to the interpersonal effec-

tiveness in communication that I have termed Geltung and that the latter reflects a special feature acquired through evolutionary selection' (Wagner 1996:12).

It is personal Geltung that is the principal explanation for our social relationships, integrating individuals as members of larger groups. It is personal Geltung that explains our activities in environment. It is personal Geltung that allows humans to both cooperate and compete with others in order to attain positions of influence and power. Conceptually most interesting, it is personal Geltung that clarifies the meaning of culture as the dynamic consequence of a series of imitative diffusions.

Further, elaboration of the hypothesis permits a series of prescriptive recommendations, including the need to respect the importance of place, respect the importance of people, expose and deconstruct vanities, challenge spatial monopolies of power, oppose war, enhance personal development, respect environment, and modulate population processes. The Geltung hypothesis introduces an argument that merits the attention of cultural geographers, not simply because it has been proposed by a distinguished practitioner but because it permits some stimulating suggestions about, first, how humans behave toward each other and how humans treat the earth and, second, how humans *ought to* behave toward each other and how they *ought to* treat the earth. As this book consistently notes, these are central issues in cultural geography. According to the Geltung hypothesis, it is not so much culture that matters but rather human behavior as it relates to communication that is designed to enhance personal Geltung.

Most work in humanistic geography is associated with phenomenology and stresses human subjectivity and free will, the relevance of participant rather than observer research, and the need for a hermeneutic or interpretive focus. For Bunkse (1996:361) the key humanistic element in cultural landscape creation is imagination—'the unique human ability to step outside the immediate context of life, to, as it were, absent ourselves from the here and now and to enter into other realities or to create entirely new realities'. Perhaps surprisingly, the attraction of humanistic ideas in the 1970s for those geographers, including many cultural geographers who were disillusioned with the positivistic spatial analytic tradition, did not involve close links with psychology. Rather, as noted in Box

Box 6.6: Phenomenology, Existentialism, and Idealism

Much of the support for the humanistic movement in cultural geography as it emerged in the early 1970s was the opposition to the model of humans embraced by positivistic spatial analysis. Humanism contends that nature can be explained, but that humans, social life, and individual behavior, need to be understood. Necessarily, then, the scientific method is seen as inappropriate for research into human behavior because humans have intentions; this is the principle of subjectivity. Further, it is claimed that social phenomena are not entirely external to the researcher, and thus a hermeneutic tradition has developed, a tradition that aims to reveal expressions of the inner life of humans by *verstehen (see* Glossary). A 'principal aim of modern humanism in geography is the reconciliation of social science and man, to accommodate understanding and wisdom, objectivity and subjectivity, and materialism and idealism' (Ley and Samuels 1978:9).

The three humanistic approaches referred to in this box are well established in the larger philosophical literature. Although it has been usual to employ this tripartite classification in geography, there is much overlap and integration of the three approaches given the common interest in subjectivity and individuals.

- *Phenomenology,* as introduced especially by Tuan and Relph, has prompted numerous variants. For example, although initially articulated as a study of individuals, phenomenology has been adapted to the social scale by Alfred Schutz. A key concern of the perspective is the contention that there is not an objective world independent of human experience, and hence there is a focus on the individual lived world of experience and the understanding of meaning and value. The implications for cultural geography are varied, but certainly a major task becomes that of reconstructing individual worlds.
- *Existentialism* was proposed especially by Samuels, although 'the boundary between existentialists and phenomenologists cannot be drawn precisely ... exis-

tentialists concern themselves with the question of the nature of "being" and understanding human existence' (Entrikin 1976:621). Again, there is a concern with individuals and their relationship both with the world of things and the world of others. There is little evidence to suggest that this philosophy is emerging as central to cultural geography.

- *Idealism as* advocated by Guelke is a specific version of humanism and one component of the larger idealist alternative to naturalism discussed in Box 2.2. The Guelke version derives specifically from historical idealism as proposed by the historian, Collingwood. Idealism insists that phenomena are only significant when they are a part of human consciousness; to comprehend the world, it becomes necessary to rethink the thoughts behind actions in order to discover what decision makers believed rather than why they believed it. Although this approach was cogently advocated, and although there have been several applications, most notably by Guelke (especially 1982), idealism conceived in this manner has not proven to be especially attractive to other cultural geographers.

Each of these three approaches is humanistic in character, stressing human subjectivity and free will. Each shares the view that a spatial analytic geography is dehumanizing because humans are treated as objects not subjects. But the claim that the appropriate scale of analysis is that of the individual is not accepted by most contemporary cultural geographers as it is recognized that the behavior of the individual is subject to many cultural constraints. A focus on individuals can be criticized as psychological reductionism, explaining behavior in terms of individual mental processes. For example, Jackson and Smith (1984:21) noted that one problem of humanistic philosophies was that of 'building an effective bridge between individual cognition, perception, and behavior, on the one hand, and an appreciation of man's place in society, on the other'.

6.6, humanistic geographers turned directly to some of the philosophical sources. Regardless, the interpretation of humanism made by psychologists and by cultural geographers was fundamentally similar.

Interestingly, notwithstanding the many supportive arguments, one leading humanistic geographer argued that no real humanistic tradition in geography has developed and even some ten years after being introduced, it was claimed that the approach consisted of 'little more than a few expressions of possibilities' (Relph 1981:134). If this claim is correct, humanistic geography can be unfavorably compared with corresponding interests in sociology and psychology where scholars have been able to develop ideas on the basis of substantive research activities.

Nevertheless, and despite some difficulties—especially that relating to the preferred individual social scale of analysis—and also some uncertainties about a specific identity, humanistic approaches that anticipated a cultural geography centered on an intersubjective world of lived experience and shared meanings have generated some important concepts. Together, such concepts are employed to discuss parts of the surface of the earth as these are understood, perceived, lived in, and experienced.

- The need to focus on **place** and not space was a key feature of early humanistic geography and continues to be a paramount concern.
- There are the related ideas of **sense of place** and **placelessness.**
- The term **topophilia** refers to love of place.
- More generally, it is usual to talk about the **taken-for-granted world** of ordinary, everyday experience.

TWO MODELS OF HUMANS—ONE OR TWO GEOGRAPHIES OF BEHAVIOR?

Behavioral geography as it developed within and from the spatial analytic tradition was not closely associated with work in cultural geography, unlike

Box 6.7: Modern Humanism as a World View

So far in this book it has been usual to draw a distinction between humanism and science, a fair reflection of the prevailing cultural geographic and larger social science understanding of these terms. But there is another way to approach humanism and science.

In a sensitive series of writings, Hutcheon has outlined a twenty-six century history of humanist thought that has one great idea at heart—namely, the philosophical premise of naturalism. This idea that humans are grounded in nature is seen as compatible with the recognition of humans as a distinctive species—distinctive because we have developed critical consciousness and culture. Thus: 'Humanists believe that we humans are the only species thus far to have evolved the capacity for constructing reliable knowledge of our surroundings, and about ourselves. Therefore, we no longer need to resort to myths of revelation from on high, or to fictions about mysterious intuitive messages from unknowable forces beyond what is accessible to human experience' (Hutcheon 1995:31). This modern humanism also has an overriding focus on morality and social activism, but is philosophically rather than politically motivated. Most critically:

- it is an interpretation that is committed to the scientific method, including the operation of cause and effect in human behavior
- it is an interpretation that argues that the longstanding foundations of this humanism are occasionally attacked from within by those who become attracted to some version of subjectivism; existentialism, as discussed in this chapter, and postmodernism, as discussed in Chapter 8, both fall into this category

Because this interpretation has not been adopted by either humanistic psychologists or cultural geographers—who as we know have been attracted instead to versions of humanism that are subjectivist, opposed to the scientific method, and unsympathetic to naturalism—it is not considered further. This failure to elaborate is not intended as a criticism of the interpretation but rather is necessary in a textbook that is designed to reflect the practice of cultural geography. If this interpretation were accepted, then the basic arguments at the opening of this chapter concerning the debate about naturalism would be inappropriate.

the humanistically flavored behavioral geography that was promoted by cultural geographers. But this clear distinction in principle, specifically involving different inspirations and different models of humans, was much less clear in practice as geographers struggled to introduce behavioral analyses into the discipline. The principal research activities focused on the presumed links between human perceptions, behavior, and the creation of landscape, and often built on the assumption that perception caused the behavior, which in turn caused landscape.

Regardless of which of the two quite different emphases was favored, the emergence of a behavioral geography was initially focused on the idea of perception, with a key distinction made between real and perceived environments. The idea—discussed through examples in the following section—that humans imaged environments as something other than reality emerged as a fundamental basis for research. The classic early study along these lines was one that intentionally and effectively integrated positivist and humanist geography to explain and understand American inner city life with specific reference to the Black area of Philadelphia (Ley 1974). For some contemporary cultural geographers, such an integration of approaches is necessarily flawed, although Ley insisted that measurement and interpretation were both quite compatible:

> For while one may conduct a multilocale ethnography of more than one place, an upper limit is quickly reached. In moving from two sites to 20 some standardization of accounts—usually measurement—becomes necessary. I see nothing but intellectual gain from incorporating the harvest of an intensive interpretation of one or a few sites with an extensive analysis of a large number (Ley 1998:79).

In addition to the substantive body of work that centered on perceived environments, one geographer adopted the transcendental phenomenology of Husserl. According to this view, humans have misunderstood the character of the world as they have come to believe that a scientific, mathematical, and objective view of the world is also the world itself. Rejecting such a view, Husserl claimed that it was not appropriate to interpret events in our life-world scientifically, but rather to seek to understand the general structures of meaning of the life-world. To achieve such an understanding, it is necessary to suspend our preconceptions; that is, to put aside any suggestion that the real world is naturally ordered. With these ideas as a basis, Hufferd (1980:19) argued that: 'In order to explain why an object (human or not) acts or reacts in whatever fashion, it is necessary to understand it in relation to the perspective resulting in its action.' According to this argument, any attempt to understand human behavior needs to proceed as follows.

- First, the accounts offered by numerous participants need to be recorded as precisely as possible until major themes become evident and no additional new themes are forthcoming.
- The researcher then lists the frequently mentioned topics, classifies them, and compares the different interpretations.
- Finally, topics are summarized with key features identified to allow for the creation of as authentic as possible reconstruction of the perception held by the group.

Arguably, this reconstruction forms the basis for an understanding of human activities. (See Your Opinion 6.3.)

THE INFERENCE PROBLEM

One additional point needs to be raised before proceeding to the discussion of empirical work. Thus, it is not satisfactory to approach an understanding of the cultural landscape solely from the rather naive proposition that perception causes behavior causes landscape. All too clearly, factors other than perception influence behavior and factors other than behavior influence landscape. Nevertheless, much useful and innovative work has centered on the role played by perception. But should cultural geographers organize their research around the logic that perception causes behavior, or should they infer perceived environments on the basis of actual behavior? Both of these research frameworks involve difficulties.

- On the one hand, the use of images to explain behavior is difficult in that it can be argued that perceived environments cannot even be objectively measured. Further, a close correspondence between perception and behavior is often difficult to demonstrate.
- On the other hand, inferring images from behavior also poses logical problems as it is one example of the inference problem; that is, assuming the identity of causes on the basis of an analysis of effects.

On balance, the second concern is the more problematic. Inference is an uncertain form of discovery for there are no guarantees that the inferred cause is indeed a true cause. To demonstrate the dangers of inference in this context, consider the following simplified circumstance. Farmer A perceives that either crop x or crop y will produce a satisfactory income, while farmer B perceives that either crop y or crop z will produce a satisfactory income; both choose to cultivate crop y, and therefore their two landscape decisions are identical, despite being based on different perceptions of reality.

standable that behavioral geography includes a wide range of topics and utilizes varying conceptual frameworks and related methods of analysis. Indeed, much behavioral geography does not clearly qualify as cultural in character, and the material included in this section is but a sample of the larger body of behavioral work.

Cultural geographers have focused their attention especially on the various humanist traditions, and three examples of the importance of environmental perception and related behavior are discussed. But this section on empirical analyses first briefly discusses two examples of a more controversial approach that opt to build on radical behaviorist ideas as developed in psychology. It is stressed that this approach is detached from mainstream cultural geographic work, although the examples are explicitly cultural.

A Radical Behaviorist Approach

At the outset, it is acknowledged that both Skinner and the approach that he advocated have proved to be controversial.

Your Opinion 6.3

Although several geographers writing in the 1960s anticipated that a behavioral approach might prove to be a panacea for the ills of the discipline, this has not proven to be the case. Why not? One reason sometimes cited for this failure is the close correspondence between spatial analysis and some behavioral geography, although a more general reason may be the convoluted nature of the larger behavioral geography that did appear. In addition to the two quite different versions of behavioral geography, increasing awareness of the psychological literature also emphasized the more general idea of cognition such that the emerging behavioral geography was a less than clear mixture of a physical science approach, humanistic interests, and cognitive psychology. There has also been some confusion concerning precisely what is involved in using positivism or humanism—what are commonly called approaches are actually different epistemological systems that need to be appraised in quite different ways.

Walmsley and Lewis (1984:4) identified the rather uncertain identity of behavioral geography by referring to such characteristics as the focus on individual decision-making units, the relevance of both acted-out and mental behavior, the emphasis on the world as it is rather than as it should be under some theoretical constraints, the interest in economic locational topics, and the use of models. As you will appreciate, this list is very suggestive as to the mixed inspirations and interests of behavioral geography.

Spatial Understanding and Spatial Perception

Given the diverse geographic and psychological background to studies of behavior, it is under-

- Catania (1984:473) observed that: 'Of all contemporary psychologists, B.F. Skinner is perhaps the most honored and the most maligned, the most widely recognized and the most misrepresented, the most cited and the most misunderstood.'

- Smith (1997:985) explained some of the rejection of Skinner's ideas and work by noting that, 'There is little doubt that Skinner was often treated as emblematic of the faults of positivist science and not read.'

    Nevertheless, the argument here is that cultural geography might benefit from a reconsideration of the philosophy of radical behaviorism because that philosophy has led to the development of a well-established set of concepts and principles (known as behavior analysis) that have been applied in several of the social sciences. This is, of course, an argument that many cultural geographers reject immediately because of the concern that such an approach lacks any meaningful cognitive or humanistic content and, hence, is necessarily inadequate as a basis for understanding human behavior. But it does seem possible that behavior analysis might be more appealing with the introduction of two concepts that greatly extend the explanatory power of the approach, namely the concepts of rule–governed behavior and establishing conditions.

- Recall that, for the radical behaviorist employing the concepts and principles of behavior analysis, the principal factors that establish and control behavior are the antecedent conditions that precede behavior and the consequences that follow behavior rather than thoughts, feelings, or other mental processes. But, as previously noted, most contemporary behavior analysts do recognize cognition as important and incorporate thoughts and feelings into analyses as unobservable behaviors. Specifically, behavior analysts address the problem of delayed consequences, not easily explained by reference to the basic concepts of behavior analysis, using the concept of rule–governed behavior. Rules are able to serve as antecedent conditions because they provide a mechanism for understanding how thoughts or self-talk, which are traditionally concerns of cognitive psychology, might control goal–oriented behavior. Thus, some behavior, especially that which has an immediate consequence, is shaped directly by consequences, whereas other behavior, especially that where the consequence is delayed, is under the control of a rule. As already noted, examples of rules include cultural norms, religious beliefs, and advice from various respected others.

- A second addition to the basic concepts described earlier involves recognizing two types of antecedent events, namely those that are discriminative stimuli, the traditional meaning of the term, and those that are establishing operations. This distinction is based on function—antecedent events that have discriminative functions are discriminative stimuli while antecedent events that have motivative functions are establishing operations. Specifically, an establishing operation is an event or stimulus that momentarily changes both the reinforcing effectiveness of other events and the frequency of occurrence of behaviors that have had those other events as consequences. There is a relation between rule–governed behavior and establishing operations, such that a rule can be viewed as a conditioned establishing operation rather than as a discriminative stimulus; this is because the rule changes the momentary effectiveness of consequences as reinforcers.

    Both of these additional concepts may be helpful in applying this approach to problems addressed by cultural geographers, such as those of landscape change. Two examples are offered.

REVISITING THE MORMON LANDSCAPE

Recall the brief description of the Mormon church and the discussion of the Mormon intermountain region as a homeland in Chapter 4. This account is in a quite different vein, focusing on a behavior analytic interpretation of Mormon settlement and related landscape behavior that employs the concepts of antecedent conditions, operant behavior, consequences, rule–governed behavior, and establishing operations. There are two sets of antecedent conditions.

- Priesthood authority with a hierarchical administrative structure that involves all members governs the church; leaders are divinely called to lead and members typically accept requests or instructions, known as callings, from those leaders. Although there is debate about this matter, the consensus is that the submission of individual will to larger group interests was characteristic of Mormonism during the frontier settlement period.

- Mormons practised cooperative effort and support for other members in accord with the principle of community.

Together, these two sets of antecedent conditions, properly described as establishing operations because they function to motivate particular behaviors, prompted Mormons to formulate rules to guide behavior. Being a faithful Mormon meant acceding to requests from leaders and also living and working in a cooperative manner. Group success was valued above individual success.

There were also delayed consequences in place that encouraged settlers to behave in accord with requests from leaders and as members of a community. The two most important were:

- the collective belief that being a church member in mortal life is only one stage in a process of eternal human relationships
- the collective goal of economic independence

Further, there were direct-acting contingencies related to the statement of rules: for example, the fact that commitment to the Mormon faith was shared with others reinforced decisions. Following rules was reinforced through the daily activities of living as a faithful Mormon, especially the regular group expression of gratitude. Overall, direct-acting contingencies encouraged Mormon settlers to conform to accepted Mormon behavior until the reinforcing consequences specified by the rules came into play.

Several previous studies of cultural groups in landscape implicitly recognized the role played by adherence to cultural norms; that is, rule-governed behavior was occurring. Much previous work on Mormon settlement falls into this category, as does work on members of the Dutch-Reformed church in southwestern Michigan. Bjorklund noted:

The basic principles followed by the adherents to Dutch-Reformed ideology can be stated simply as follows: (1) there are particular rules governing the conduct of life which must be obeyed literally; (2) man is obliged by these rules to perform both physical and spiritual work; and (3) opposition or intrusion of conflicting rules of conduct cannot be tolerated, because life after death depends upon the literal conduct of life on earth on a principled basis and is not subject to individual interpretation (Bjorklund 1964:228).

Given these three principles, it is not difficult to envisage a behavior analysis of the Dutch-Reformed settlement experience in southwestern Michigan.

THE SOUTHEASTERN AUSTRALIAN WHEAT FRONTIER

The research question addressed in this discussion is that of the transition from a pastoral landscape to an arable landscape in nineteenth-century southeastern Australia. Specifically, why did British settlers move away from the established coastal region to initiate an inland pastoral economy beginning in the 1820s? From a behavior analytic perspective, a first answer to the question about the transition from pastoral to arable activity is that the antecedent conditions changed such that the physical, economic, social, and political environments encouraged wheat farming. Thus:

- there were changes in the British economy that decreased the demand for wool and increased the demand for wheat
- there were changes in world prices for the two products to the advantage of wheat
- there were increases in local markets
- some of the pastoral area was suitable for wheat cultivation, especially given developments in technology and railroad construction
- there was a need to feed the rapidly increasing mining population
- there was a powerful political impetus to provide land for the poor and to take land out of the hands of a few wealthy pastoralists

These considerable changes in the antecedent conditions combined to produce a different contingency, more properly described as a metacon-

Grain cultivation is combined with sheep raising over large tracts of the coastal ranges' inland slopes in Australia (*Victor Last*).

tingency as the behavior under consideration is that of a group and not an individual.

The physical environment serves as a discriminative stimulus, a cue informing farmers what to do in order to get the economic success that they already want, while the economic and social environments are establishing operations, motivating the behavior of wheat cultivation. Specific examples of establishing operations include the presence of neighbors cultivating wheat, the presence of a railroad, and high market prices received for wheat.

Wheat farming is an operant that was reinforced and hence maintained by its consequences. Although the key reinforcer of economic success is not direct-acting, in that there is a long interval between sowing seed and receiving income for the crop, there are direct-acting contingencies related to the statement of rules that specify the consequences of wheat farming. Examples of rules include the reasoned expectation, a prevailing view held by most members of the group, that wheat farming was a profitable activity because of such

factors as the railroad network and the availability of markets. The principal delayed consequence was that of economic success. There were also direct-acting contingencies related to both the statement of rules and to the following of rules that served to support arable activity until that activity paid off with the consequences noted. For example, the relevance of rules favoring wheat cultivation was evident to a new farmer who was able to observe and imitate existing selectors already experiencing the reinforcing consequences relating to economic success and social stability. These direct-acting contingencies encouraged wheat farmers to continue arable behavior until the reinforcing consequences specified by the rules came into play.

The two examples of Mormon settlement and landscape change in southeastern Australia are suggestive of the value of applying the concepts and principles of behavior analysis to cultural geographic problems concerning such topics as settlement behavior and landscape change. Although the underlying philosophy—radical behaviorism—is not in accord with most current thought, which favors subjectivist rather than objectivist emphases, the examples may be sufficiently suggestive as to encourage further explorations in this direction.

Subjective Environments

Most cultural geography concerned with human behavior favors approaches that focus more explicitly on cognition than does behavior analysis and especially favors approaches that address humanistic concerns such as that of subjectivity. Work in this more important tradition is now considered.

Images, Mental Maps, and Cognitive Maps

Probably the single most important concept in the behavioral geography literature as it developed beginning in the late 1960s is that of environmental perception in conjunction with the related concepts of images, cognitive maps, and mental maps—terms that were not clearly distinguished at the time. Cultural geographers were fully involved in these interests, with Mikesell (1978:6) noting that, for cultural geographers, the 'recent development of greatest potential interest has been the proliferation of work on environmental perception'. The essential logic of perceived environments is certainly most attractive to the cultural geographer, with the clear focus on subjectivity and on the role played by both individual and cultural appraisals.

The concept of a subjective, imaged environment different from the objective, real environment was fundamental to the development of a humanistically inspired behavioral cultural geography, and cultural geographers were encouraged to turn to such books as *The Image* by Boulding (1956) for conceptual content and to *The Image of the City* by Lynch (1960) for both conceptual and empirical content.

*The key idea supporting this type of research is that humans do not respond directly to the environment but rather to their mental image of the environment, with the result that human landscape-making activity is related to the **image** held of the environment.* Although it is possible to conceive of more complex relationships and interactions, the basic logic of this approach is straightforward (Figure 6.3). Certainly, the idea that people represent environmental, including spatial, information in the form of an image, specifically a **mental map**, in their heads has proven very attractive to cultural geographers. It is also well established that mental maps err in certain consistent ways. For example, local areas are usually known in relative detail while more distant areas are less known.

The related concept of a **cognitive map** was derived from earlier work in psychology. Downs and Stea (1973:7) defined cognitive mapping as a 'process composed of a series of psychological transformations by which an individual acquires,

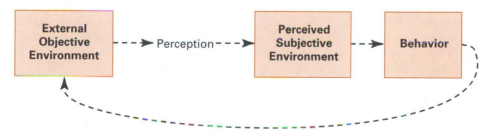

Figure 6.3: Perception, behavior, and landscape. This is a basic schematic diagram that outlines the essential logic underlying many of the studies of perception in cultural geography (see also Haynes 1980:2). There is an external objective environment that is perceived by humans. The environment that is perceived—the subjective environment—is necessarily different from the objective environment. Further, the perceived environment is different for each individual, although individuals who belong to a group, such as a culture, are likely to have perceptions that are comparable. This is because perception is affected by both individual characteristics and group characteristics. Behavior in landscape is a response to the environment as perceived rather than a response to the real environment. Behavior changes landscape and thus changes both the objective and subjective environments through time. Perception analyses in cultural geography typically adopted a framework of this type.

Several attempts have been made to conceptualize these ideas more explicitly, and Kitchin (1996) reproduced nine rather more sophisticated formulations centered on the theme of cognitive mapping. There has been special emphasis on trying to understand what is going on in the human mind as environments are perceived, with most such work built on the ideas of cognitive psychology and including perception as one component of cognition.

stores, recalls, and decodes information about the relative locations and attributes of the phenomena in his everyday spatial environment'. This concept raises some important questions, especially concerning whether or not people represent their environments as maps. Evidence from psychology is contradictory on this question, with descriptions of human spatial memory as both propositional and analogical in form; propositional refers to the idea that place knowledge is stored in lists, and analogical refers to the idea that the storing is directly comparable to the depicted objects.

In geography, Bunting and Guelke (1979:453) advanced important criticisms of two of the fundamental claims of this type of behavioral geography, namely:

• the claim that *'identifiable environmental images exist that can be measured accurately'*
• the claim that *'there are strong relationships between revealed images and preferences and actual (real-world) behavior'*

It seems fair to say that the current status of this humanistically inspired behavioral geography within geography is uncertain. Certainly, in cultural geography there has been much interest in this more humanistic version of behavioral geography, although most of the principal analyses date from the 1970s and 1980s rather than from more recent years.

The focus in the following discussion of subjective environments is, first, on an example of cognitive mapping applied to one American region, Appalachia, and, second, on selected work in the historical cultural geography of areas of European overseas expansion. This second focus builds on some of the content of Chapter 3. There is a general account of the image–making process, followed by specific examples of the Swan River Colony in Australia, the American Great Plains, and the Canadian Prairies. These historical examples are of subjective environments that, at the time under discussion, were significant departures from reality.

APPALACHIAN COGNITIVE MAPS
Recall the definition of vernacular cultural regions in Box 4.1 and the related examples of such regions

in Chapter 4. These are areas that are presumed to have a regional identity and that are the product of the spatial perception of average people. One attempt to delimit an Appalachian vernacular region employed the strategy of acquiring individual cognitive maps and subsequently producing generalized maps for each of a number of groups (Raitz and Ulack 1981b). The use of the term 'cognitive map' in this research is analogous to the term 'mental map' as this is used in much other research in that it is referring essentially to a place–preference surface. This example is not concerned with any detailed investigation of the psychological concept of cognition as defined earlier, although it is very suggestive concerning the importance of location as a factor affecting spatial cognition.

Cognitive maps were acquired from 2,397 students at sixty-three colleges in and close to the Appalachian region. Respondents were asked to outline Appalachia on a base map that included only state boundaries and names. Composite maps were then produced for each of two groups in each state. One group comprised those who lived in Appalachia and the other those who lived outside Appalachia; in the case of West Virginia, there was only one group as the entire state is within the region. The regional delimitation on which these distinctions were made was a conventional physiographic demarcation.

For each group, the composite map produced showed the core area, defined as that area identified by 80 per cent or more respondents, and the larger Appalachian region, with the regional boundary being the 20 per cent isoline. Most of the composite maps indicated that the groups shifted both the core and the regional boundary toward their home area, and that there was a consistent decline in recognition of the region with increasing distance—what might be called cognitive distance decay. Respondents from the central area of Appalachia tended to increase the size of the core and, relatively, to reduce the total area. Figure 6.4 shows the core of Appalachia (defined as the most frequently occurring location identified as being in Appalachia) for a selection of the groups identified in the analysis.

Overall, the results of this research suggest that proximity affects cognition and that groups resid-

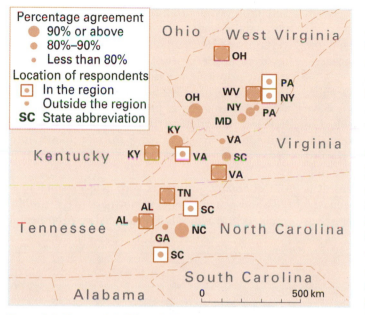

Figure 6.4: The core(s) of Appalachia. This figure shows the location of the Appalachian core for each of twenty groups. Note the differences in core location as identified by those groups comprising respondents resident in the region and those groups comprising respondents from outside the region, and also the tendency for the core to shift toward the state in question. These results suggest that most individuals relate to an environment that is too complex to permit any full understanding, that near areas are better appreciated than distant areas, and that individual members of relatively cohesive groups share similar cognitive maps. Overall, cognitive map distortion reflects the preferences, attitudes, and values of group members.

Source: Adapted from K.B. Raitz and R. Ulack, 'Cognitive Maps of Appalachia', *Geographical Review* 71 (1981):209.

ing in a region are likely to have a significantly different view of a place than those groups that live outside the region. To understand the basis behind these types of cognitive maps, it is clearly helpful to distinguish between the perceived environment that is accessible and known, and the presumed or inferred environment, which is some larger area about which detailed knowledge is lacking. In this context, it is useful to distinguish between those familiar areas where individuals can rely on cognitive representation and less familiar areas where external information is needed. Two of the most important issues that arise from an analysis of this type concern:

- the question of the origin of regional stereotypes
- the implication that differing cognitive maps may have for behavior

For Appalachia, the first issue was addressed by Hsiung (1996) using a theoretical framework to examine the internal connectedness of the region and external links to American society more generally. The second issue is discussed in the examples of imagined environments below.

CREATING IMAGES

Notwithstanding some concerns about an approach that stresses the links between perception and behavior, it seems clear that much of the behavior that was critical for landscape formation occurred in situations of uncertainty and that in these uncertain environments, what happened may have been more closely related to views developed from myth than to the realities of the environment. Indeed, myth may be a 'prime generating factor in the decision to migrate. Though the mythical goal is seldom realized, the processes of mobility and discovery are just as real as if the myth had been realized' (Salter 1971b:18). Although inappropriate behavior often resulted from perceptions that were significant distortions of reality, it was usual for errors not to be repeated (Box 6.8).

Historical cultural geographers have shown much interest in understanding the differences between objective and subjective environments—especially in areas of European overseas expansion during the periods of exploration and early European settlement—because of the realization that any comprehensive understanding of the evolution of settlement and the transformation of regional landscapes over large areas of the globe cannot be accomplished without reference to images and image construction. Three principal issues have been addressed:

- How were images that were different from reality created?

- What was the extent of the difference between image and reality?
- Is it possible to understand behavior and the changing landscape by reference to the image?

These are not easy questions to answer, either conceptually (as the previous discussions have made clear) or empirically, given the difficulties of collecting information on subjective environments in a historical context. Attempts to answer the first of these three questions have stressed the role played by those who provided descriptions. It is often appropriate to distinguish between the accounts of those who saw the area for themselves and those who produced second–hand descriptions, and it is also recognized that the image conveyed relates to the motivations of those who are involved in the process of image creation. In the context of much European overseas activity, it may be useful to identify the different motivation of such image makers as promoters, officials, settlers, travelers, and natural historians—each group of individuals brought their particular biases into play as they produced their assessments.

- Most notably, promoters, such as leaders of immigration schemes and shipowners, were prolific writers and typically exaggerated environmental

Box 6.8: Learning in Landscape

Learning often occurs in environments of uncertainty where there is a discrepancy between objective and subjective worlds such that adjustments to behavior were usual.

Colonization invariably resulted in the confrontation of imported cultural systems with new, strange and often inhospitable environments. Rarely were the particulars of these environments consistent with colonists' perceptions of them, and even more rarely were they totally amenable to the resolution of colonists' aims. Perceptions and aims had therefore to be modified and this initiated a process of active, conscious adjustment of learning which continued until such time as perceptions were consistent and aims became realistic and attainable (Cameron 1977:1).

Further developing this idea, Cameron (1977) utilized a model of adaptive learning that is acknowledged to be linked to Gestalt concepts but that also has behaviorist overtones. The model provides a framework for an analysis of settlement in pre-1850 western Australia (Figure 6.5). A colonization process, such as that outlined in Figure 6.5, involves the establishment of behavior patterns in a new environment, and these patterns are a consequence of a complex interplay of factors. The model discussed is a useful one, although it does not focus explicitly on such factors as the prior experience of the colonizer, including any preadaptive traits, and the attitudes held by individuals or group.

Clearly, much behavior in landscape is affected by prior experience and the relative importance of indi-vidual identity and group membership, as in the example of Irish settlers in eastern Canada mentioned in Chapter 3.

- Prior experience may be a key consideration if the goals established in the new area are similar to those pursued previously, if the new environment is similar to the old, if there are limited links with other groups, if the image held of the new environment is such that repetition of prior behavior is judged appropriate, if institutional considerations do not play a major role, and if group membership is more important than individual characteristics.
- The characteristics of each individual are likely to be an important determinant if group membership patterns are weak, if institutional considerations are lacking, if goals are unclear and if the environment is relatively benign.
- Group membership will be important if individual characteristics are invariable, if institutions do not dominate, if links with other groups are limited, and if there is a shared environmental image.

As one example of the importance of perception in this larger context, consider the reaction of European settlers to North American forests. In southern Ontario, civilization and progress were deemed incompatible with the continuing presence of forest, with the result that forests were attacked 'with a savagery greater than that justified by the need to clear the land for cultivation' (Kelly 1974:64). This was despite the fact that the agricultural value of retaining some forest was understood. A lack of prior experience was also important in

quality in order to encourage settlement. Immigrant expectations were largely derived from the writings of promoters.

- Literature produced by officials was more variable, sometimes misleading, and sometimes relatively objective.
- The usually optimistic—especially during the initial phase of settlement—settler literature was taken seriously by potential settlers, but was limited in quantity.
- Travelers and natural historians produced the least useful accounts because they often focused on topics of very limited interest to prospective settlers.

One of the more notorious instances of a subjective environment concerned the supposed existence of a navigable North West Passage across the north of North America. Proposed both by explorers and armchair geographers in Britain, this incorrect image influenced the history of the area:

The whole enterprise was founded on a misapprehension, a geographical fiction, a fairy tale, springing out of the kind of stories sailors tell to amaze landsmen or to delude other sailors, to which were soon added the inferences, speculations and downright inventions that scholars manufactured to amaze themselves (Thomson 1975:1).

This illustration from *The British Farmer's Guide to Ontario* illustrates the back breaking labour to hew a homestead out of the wilderness (*National Archives of Canada C44633*).

this situation because most incoming settlers were ill-prepared for the vastness of North American forested areas. The goal of establishing commercial farming required some land clearing, but not on the scale employed. In this example, the behavior of settlers largely resulted from their lack of prior experience with the type of environment that they encountered, from the behavior of others, from the perception of the forest as some sort of threat to advancement, and from related attitudes.

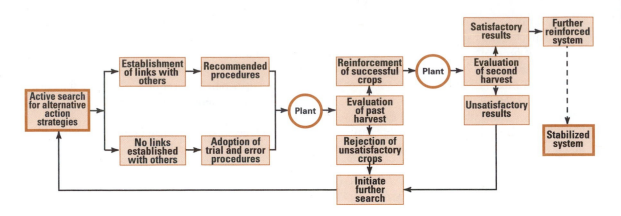

Figure 6.5: A simplified model of colonizer behavior.

- First, movement into a new environment prompts a search for those actions that will satisfy preconceived goals. Hopefully, a colonizer can simply imitate others who are already achieving the equivalent goals, but, if this is not possible, then actions need to be determined subjectively on the basis of the image held of the area.
- Second, actions are taken.
- Third, when the consequences of actions are known, a process of evaluation is undertaken. Actions that were successful in achieving the desired goals are repeated, and those that were unsuccessful are rejected and a new search procedure begun.

   This sequence—search, operate, evaluate—is repeated as many times as needed until the desired goals are achieved. Through this process, the colonizer learns a set of actions that are repeated on a regular basis.

Source: Adapted from J.M.R. Cameron, *Coming to Terms: The Development of Agriculture in Pre-Convict Western Australia* (Nedlands: University of Western Australia, Department of Geography, Geowest 11, 1977):3. Reprinted by permission.

The three examples that follow are not quite so dramatic and colorful as the example of the North West Passage, although they are certainly indicative of the important role played by the distortion of environmental images, a distortion that was often intentional.

SWAN RIVER COLONY

'Between the "place" and the "people" themselves, however, lay the "media" by which the so-called facts of the one were made available to the audience of the other' (Heathcote 1972:81). An excellent example of the creation of an incorrect image resulting from the way the media presented material to the public concerns the Swan River Colony of western Australia in the period from 1827 to 1830.

   An 1827 exploratory voyage by Captain Stirling and Fraser (a botanist) was the sole substantive source of information about the Swan River area during this period. Although their visit to the area was brief, the subsequent report to the British Colonial Office supported the idea that Swan River

was a suitable location for a colony and settlement because of fertile soils, a suitable climate, and adequate water. This report was made available to the public only in the form of an article published in a journal, the *Quarterly Review* (1829), and this article was then the basis for later journal articles in the *Mirror* (1829), *New Monthly Magazine* (1829), and *Westminster Review* (1830).

- The first article, in the *Quarterly Review*, was written by Barrow, the acknowledged expert on the southern hemisphere, and was aimed at an audience of shipowners, merchants, and wealthy potential settlers. The generally favorable Stirling and Fraser report was the basis for the article, but Barrow employed various strategies, such as deleting unfavorable comments and reporting optimistic conjecture as fact, in order to promote the area.
- The three later articles continued this process of image distortion to such a degree that the *Westminster Review* article showed only superficial resemblance to the original report.

Two examples of the extent of the distortion are shown in Table 6.1: the two most important features of the area for potential settlers, namely the size of the cultivable area and the quality of the climate, were distorted almost beyond recognition. Conditions favored distortion in this example because of the short time period involved, the existence of only one basic source of information, and the pro-colony mood of the British public at the time, but the sequence by which the information was disseminated differed but slightly from other colonial areas. Undoubtedly, this is a compelling instance of the misleading details that were made available to those considering investing in or moving to an overseas area. Consider the following two statements.

- First, the *New Monthly Magazine* reported that 'the soil is a fine brown loam, alternating with broad valleys of the finest alluvial soil; the hills appeared finely timbered; the valleys produced an immense luxuriance' (quoted in Cameron 1974:65).
- Second, one early settler complained: 'Not a blade of grass to be seen—nothing but sand, scrub, shrubs, and stunted trees, from the verge of the river to the top of the hills. . . . I may say with certainty, that the soil is such, on which no human being can possibly exist' (quoted in Cameron 1974:74).

The image presented to settlers by the media and the reality—or, more correctly, the perception of reality—that the settlers encountered were far apart.

GREAT PLAINS OR GREAT AMERICAN DESERT?

Although the Great Plains is probably the most studied example of a subjective environment in the historical cultural geographic literature, it is also one of the most confusing. It has often been claimed that the popular image of the plains in the period from about 1820 to about 1870 was that the area was a desert; indeed, the area was known as the Great American Desert. But this scholarly perception by both historians and geographers that the area was perceived as a desert has itself been challenged as an incorrect perception.

The dominant scholarly view, largely based on a consideration of the reports of explorers and educational texts, is that the area was perceived as desert, an idea that was initiated by early nineteenth-century European explorers. Three principal reasons have been offered to explain the emergence of this perception and its continuance for a period of about fifty years.

- First, individuals tended to record what impressed them most and, for explorers from the eastern United States, the small areas of desert in the plains merited substantial description.
- Second, relative to the eastern region, the treeless plains could be described as desert.
- Third, there was a powerful political impetus to move west to the Pacific Coast as quickly as possible, and a desert perception of the extensive plains region accorded with that goal.

In these three ways, the plains were being defined in eastern terms. Removal of the desert perception occurred after about 1870 with improving knowledge of the area. But a major challenge to this dominant scholarly view claimed that 'the desert belief was far from universally held in America in the mid-nineteenth century' (Bowden 1976:119–20). According to this argument, the

Table 6.1: **Image Distortion in Swan River Colony**

	Stirling and Fraser, 1827	New Monthly Magazine, 1829
Cultivable area (sq. mi.)	100	22,000
Conjecture, as % of total description of climate	9.7	51.1

Source: J.M.R. Cameron, 'Information Distortion in Colonial Promotion: The Case of Swan River Colony', *Australian Geographical Studies* 12 (1974):66, 68. Copyright © Institute of Australian Geographers Inc.

incorrect idea that the plains were perceived as desert can be understood in terms of the variable climate of the area and the powerful impacts of a small number of historians, particularly, Turner, Webb, and Malin.

'A NEW AND NAKED LAND'

Settlement of the Canadian Prairies by Europeans between about 1895 and 1914 was not simply the next logical move in the process of overseas movement and a response to the demand for wheat, it was also prompted and sustained by an organized campaign of image distortion. 'The more distant and inhospitable the land, the greater the blandishments necessary to attract settlers to it. Cold, dry, treeless plains half a continent and—for Europeans—an ocean away were hardly alluring' (Rees 1988:4).

In the mid–nineteenth century, the region was viewed as either a northern continuation of the Great American Desert or a southern continuation of the Arctic— the area was indeed *The Great Lone Land* (Butler 1872). But this was an unsatisfactory state of affairs for both British and Canadian officials who were concerned about the lack of settlement between Ontario and the Pacific Coast, especially given the possibility of American expansion. Accordingly, there was an 1857 British expedition under the leadership of Captain John Palliser to the area and also an 1858 Canadian expedition under the leadership of Henry Youle Hind. The reports of these expeditions introduced the idea that the Prairies had agricultural potential, with the exception of the grasslands area in the southwest portion, which became known as Palliser's Triangle. A process of image change was initiated.

During the 1860s, 1870s, and 1880s, numerous favorable reports of the area appeared, often written with a limited factual basis. Thus, A.J. Russell, the inspector of Crown agencies in Canada, stated in 1869 (without visiting the Prairies) that both of the

expedition reports probably underestimated the agricultural potential of the grassland region, while John Macoun, a government botanist and author of *Manitoba and the Great North-West*, declared in 1881 that the area was 'all equally good land' (in Rees 1988:7). These and other largely unjustified claims about the physical environment encouraged massive campaigns by the two principal promoters of the area, the Canadian Pacific Railway and the government. But despite this ongoing process of image change, the results were disappointing, with little immigration from the United States, Britain, or western Europe. Something else was needed to attract immigrants to the Prairies, a fact that was recognized by Clifford Sifton, who was placed in charge of the Ministry of the Interior in 1896. Sifton initiated three principal policy changes.

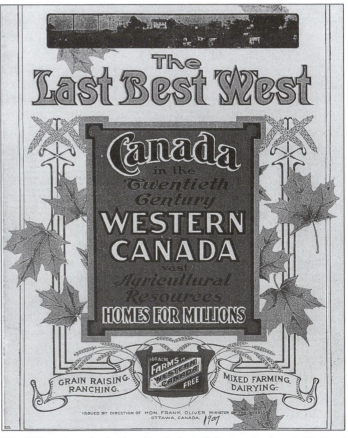

This 1907 poster exhorted immigrants to settle in western Canada (*National Archives of Canada C-30621*).

- He extended the range of advertising into non-traditional immigrant source areas, notably to eastern and southeastern Europe and Ukraine.
- The quantity of the promotional literature was increased substantially.
- Most important, given our current concern with images, Sifton began to make major changes to the content of the promotional literature. Thus, Sifton disallowed overseas publication of Manitoba temperatures, required that the words 'snow' and 'cold' not be used in government publications, that snow not be included in any illustrations, that the often inadequate rainfall in the growing season be hidden through the use of annual averages, and that the great open expanses of the Prairies be softened through descriptions that focused on enclosures. In this example, image distortion was intentional and substantial.

It is clear that the promoters of settlement were operating in a context of high emotion and high expectations. There was a perceived need to settle the West as quickly as possible and this appeared to justify the employment of dubious strategies to entice immigrants to the area. For many settlers, the reality on arrival was quite different from the expectations generated by the image makers: an unliterary child from Ontario described the Prairies as 'a new and naked land' (in Rees 1988:36).

The conclusion reached by Christopher with reference to southern Africa applies equally well to the Canadian Prairies and to many other areas of European overseas movement:

To a large extent men believed what they wanted to believe or that which they were given to believe, rather than the truth; and there can be little doubt that large tracts of southern Africa would never have

Sod houses were built on the Prairies where sod was the only building material available. Sod was cut into lengths, placed with the grass side down, and used like bricks to build walls, which were plastered and sometimes whitewashed. The roof was made by putting poles across the top of the walls, covering them with hay and then with a layer of sod (*National Archives of Canada C4743*).

been settled by Europeans or would have been set-tled in different circumstances had the true state of affairs been appreciated. Immigrants' illusions were sometimes shattered soon after they reached southern Africa, but illusion had brought them, and they had to make the best of it (Christopher 1973:20).

The discussion of subjective environments, particularly the examples of historical cultural geo-graphic research, demon-strates that the subjective environment is often a significant distortion of the objective environment and that, as a consequence, much behavior occurs in situations of uncertainty. Certainly, humans know little of the consequence of behaviors that are not repetitive or that are not being imitated. Given an initial uncertainty, it is reasonable to suggest that a learning process usually proceeds and involves increasingly more appro-priate behaviors through time (as described in Fig-ure 6.5). (See Your Opin-ion 6.4.)

Your Opinion 6.4

If it is reasonable to suggest that much human behavior during the critical early stages of settle-ment in a new environment had a rather flimsy rationale, does this mean that alternative behav-iors might well have occurred? The suggestion that different decisions might easily have been taken and then might have had different impli-cations for landscape does seem a reasonable one. This is a stimulating, if controversial, thought for the cultural geographer to consider (Box 6.9).

Concluding Comments

This chapter opened with a reference to the debate about naturalism that is included in Box 2.2 and, certainly, questions about the possibility of achieving a science of behavior have been evident throughout the chapter. Unfor-tunately, there has been and continues to be much

---

Box 6.9: What Might Have Been? Part I

One of the most interesting yet rarely pursued ques-tions that historical cultural geographers might con-sider raising is 'What might have been?' Given that much landscape-making behavior occurred (and indeed continues to occur) in situations of uncertainty, it is readily apparent that behaviors different from those that took place *might* have taken place. Logically, it seems sensible to suggest that any such different behaviors might result in different landscapes. In this sense, the basic logic behind **counterfactual** queries is simple, and their use can be justified as follows: 'Every statement of causation implies the counterfactual proposition that in the absence of the causative factor the event would not have occurred' (North 1977:189). A counterfactual is any statement that is untrue in that it typically queries, 'What would have happened if...?'

Because the usual object of scholarly enquiry is what is real, there is an implicit assumption that this reality was in some sense a necessary consequence of what went before; this can be described as the bias of hind-sight. To think counterfactually is to acknowledge that things might have happened differently. If things had occurred differently and a different outcome resulted, then this would suggest that the presumed causes were indeed important. If, however, things had occurred differently and the outcome was similar to the reality, then this would suggest that the presumed causes might not be critical to the outcome. One of the dangers of *not* thinking counterfactually is that we tend to adjust our thoughts such that the expected proba-bility of something occurring that has in fact occurred is increased, while the expected probability of some-thing occurring that has not occurred is decreased. Contingency may be ignored if counterfactual logic is not employed.

The principal debate about causation and counter-factuals takes place in philosophy, although historians have long taken an interest in such matters. Typically, historical analyses in this vein belong to two types.

- Some counterfactual histories may be largely specu-lative, involving reasoned assumptions about what might have happened had some part of the past been different; military historians have been attracted to this type of argument because of the recognition that many critical events depended on individual decision making or on conditions that might very well have been different, such as weather conditions.

confusion about this matter in the social science literature:

> At one and the same time, authors sought a law–like science of society and upheld belief in individual freedom and dignity. To put it bluntly, they denied *and* asserted individual freedom of action. This contradiction correlated with the subject–object distinction—the split between the knowing mind, the sphere of reason, and the object known, the sphere of physical events—embedded in Cartesian dualism (Smith 1997:261).

Viewed retrospectively, as evidenced by reviewing surveys of twentieth-century human geography, it seems clear that the study of behavior has not been regarded as core subject matter in the discipline. For example, the encyclopedic *Dictionary of Human Geography* (Johnston, Gregory, and Smith 1994) includes only incidental references to behavior. Further, most of the accounts of the subdiscipline of behavioral geography interpret it, in retrospect, as an episode in the history of the discipline rather than as an approach that has strengthened through time and become established as a key subdiscipline of human geography. This apparent failing of behavioral geography is understandable given that, in a general sense, all human geography involves the study of behavior. Certainly, as evidenced by the empirical discussions in this chapter, thoughtful analyses focusing on human behavior in landscape, especially those concerned with perception, images, and mental maps, have been major contributions to the body of cultural geography literature.

In this respect, a behavioral cultural geography has developed significantly since the 1960s and the initial associations with spatial analysis and humanism. There have been some substantial conceptual advances, with increasing recognition that behavior is complex in origin. Cultural and other human geographers continue to turn to psy-

- Other counterfactual histories are designed to specifically test ideas about the past, for example, the idea that North American westward expansion was closely related to railroads; most of these studies are in the tradition of the new economic history that developed in the 1960s and involve formal procedures of hypothesis testing by means of quantitative methods.

Historical cultural geographers have taken limited interest in conducting counterfactual analyses despite being offered two very suggestive examples.

- First, a leading North American historical geographer thoughtfully described some of the consequences for the emerging political unit of Canada that might have resulted from minor changes in the exploratory activities of Champlain during the years 1605–8 (Clark 1975).
- Second, an Australian historian whose many works have inspired geographers because of their recurring concern with spatial issues, compellingly argued for counterfactual analyses as follows:

> We easily forget that every statement we make of why events happened is in part speculation. In looking at history we have to ask: what might have happened? Every time we affirm the profound importance of a particular event—whether the finding of gold in 1851 or the failure of the North Queensland separation movement in the 1880s—we unmistakably imply that such an event was a turning point and that society—but for that event—would not have changed direction. There can be no discussion of a powerful event without realizing that it is like a traffic juncture where a society is capable suddenly of changing direction. In writing history we concentrate more on what did happen, but many of the crucial events are those which almost happened. The born and the unborn may seem completely different but essentially they have to be analyzed and discussed in the same way (Blainey 1983:202–3).

Given the increasing emphasis on uncertainty and the role of chance, it is possible that cultural geographers might pursue counterfactual analyses if it seems reasonable to suggest that the details of the process that is presumed to have affected landscape creation might have been different.

chological and philosophical literature, especially cognitive and humanistic ideas, but also to psychoanalysis and radical behaviorism. It seems probable that cultural geographers will continue to be interested particularly in work that facilitates understanding of perception and related matters. Certainly, there has been a real advance from the early and rather uncompromising view that perception is the cause of behavior, which is in turn the cause of landscape change. (See Your Opinion 6.5.)

This chapter on behavior and landscape has included a selection of material that focuses essentially on questions relating to why humans behave as they do in landscape, although it is recognized that questions of behavior arise in all chapters of this textbook—hence the frequent references to other chapters. Indeed, it might be suggested that the principal subject matter of cultural geography, and of the social sciences in general, is that of human behavior. But what is distinctive about this chapter, within the context of cultural geography, is the explicit attention paid to behavior. In other chapters, discussions of behavior are usually less explicit.

A final comment. A diverse set of ideas about human behavior has appeared as part of the larger postmodernist interest discussed in Chapter 7. As expressed by various contemporary thinkers, especially by a number of French social theorists, and as implicit in much of the new cultural geography, behavior is related to difference. Thus, groups of people, defined by such socially constructed categories as ethnicity, gender, sexuality, age, and class, have sets of behaviors that are in some way deemed appropriate for them. Simply put, groups of people—cultures as the term is used in the cul-

---

Your Opinion 6.5

The need for cultural geographers to focus on human behavior was well stated by Saarinen (1974:252): 'Every day, all over the world, men are making decisions which lead to transformations of the earth environment. Although the impact of an individual decision may be small, the cumulative effect of all such decisions is enormous, for both the number of people and the technological power at the command of each is greater than ever before and is growing rapidly.' Add a time dimension to this quote and the situation becomes even more complex, such that a cultural landscape can be studied as the dynamic product of many individual decisions made at different times within differing cultural and physical environments. Although cultural geographers may debate the merits of different approaches to this subject matter, the relevance of the basic ideas seems clear. Do you agree?

---

tural studies and related literature—are centers of meaning and behavior. Such an approach typically recognizes that behavior is linked especially to matters of power and desire, which affect the way the world is interpreted. Further, the constructed categories are spatially configured and subject to change and negotiation. These ideas are already sufficiently important in cultural geography that they resurface as part of the concern with geographies of difference in Chapter 7 and as part of the concern with cultural identity and place identity in Chapter 8.

Further Reading

The following are useful sources for further reading on specific issues.

Sopher (1972) on cultural geography and positivism.

Amedeo and Golledge (1975), Chisholm (1975), and Norton (1984) on process to form methodology.

Gold (1980:7–15) and Martin (1991:6–11) for brief accounts of psychological approaches to the study of behavior.

Sayer (1979, 1982) on radical approaches and social scale.

Watson (1913) on behaviorism and Skinner (1969) on radical behaviorism.

Homans (1987) and Lamal (1991) on behavior analysis and groups.

Neisser (1967, 1976) on cognitive psychology.

Smith, Pellegrino, and Golledge (1982) on the geographic modeling of spatial cognition and behavior; Golledge (1987) on environmental cognition.

Jensen and Burgess (1997) on misrepresentations of behaviorism; Wertz (1998) on misrepresentations of humanism.

Norton (1997a) on various geographic confusions about behaviorism.

Werlen (1993:15) outlined an original argument for a human geography that was oriented to society and action on the grounds that 'no activity can be explained by reference to psychological factors only, but must be explained in relation to social context'.

Daniels (1985:144) on Vidal and Sauer as inspirations for humanistic cultural geography.

Christensen (1982) on the positivist–humanist split.

Golledge (1969) and Kitchin, Blades, and Golledge (1997) on relations between geography and psychology.

Rose (1981) on Dilthey and humanistic geography.

Pioneering contributions in humanistic geography are those by Relph (1970) and Tuan (1971), both of whom identified phenomenology as a legitimate approach.

There have been partially successful attempts to base humanistic geography on an idealist philosophy (see, for example, Guelke 1974), on an existential philosophy (see Samuels 1978), and on a critical encounter with historical materialism (see Cosgrove 1978, 1983).

Tuan (for example, 1972, 1975) and Pickles (1985) on phenomenology and geography; Samuels (1978, 1979, 1981) on existentialism and geography; Guelke (1971, 1975) on idealism and geography. For critical commentaries on idealism, see Watts and Watts (1978) and Harrison and Livingstone (1979).

Tuan (1974), Relph (1976), Ley (1977), and Seamon (1979) on some humanistic concepts introduced by geographers.

Burton (1963) and Brookfield (1969) on the merits of behavioral geography as a corrective for perceived flaws.

Cullen (1976) and Bunting and Guelke (1979) on the unfortunate links between spatial analysis and behavioral geography.

Norton (1997b) on a behavior analysis of wheat farming in southeastern Australia.

With reference to mental maps, Downs and Stea (1973) reported that there was a tendency to straighten curves and to generally simplify reality; Gould (1963) proposed that each individual has an image of environments that serves as a preference surface, with different locations being varyingly attractive; for various examples, see Gould and White (1986); the most comprehensive account of these and related ideas is in Golledge and Stimson (1997).

Downs (1979), Rushton (1979), and Saarinen (1979) for responses to some of the criticisms of research on images and mental maps.

Spencer and Blades (1986) on some links between psychology and geography.

Powell (1977) on the role of images in the transformation of global environments.

Merrens (1969), Heathcote (1972), Christopher (1973), and Johnston (1979, 1981) on images and image makers during European overseas expansion.

Francis (1988) on the image of the Canadian West during the settlement period.

Dicken and Dicken (1979) on historical perceptions of Oregon, and Peters (1972) on surveyor perceptions in Michigan in the early nineteenth century.

There are discussions of counterfactuals in historical geography (Norton 1984:14–15; also see Norton 1995 for an example), in world politics (Tetlock and Belkin 1996), and in history (Ferguson 1997).

Fogel (1964) on counterfactuals in the new economic history.

Unequal Groups, Unequal Landscapes

Much of the content of this and the following chapter reflects a range of theoretical contributions including Marxism, feminism, and what is often described as the cultural turn, with most of the concepts discussed in these two chapters having been introduced into cultural geography quite recently in scholarly terms—usually no earlier than about 1970 and more often during the 1980s or 1990s. Returning again to the ideas introduced in Box 2.2 and the related discussions that have appeared throughout this book, the cultural turn in particular is explicitly opposed to any suggestion that the study of humans should be based on concepts and methods as developed in physical science. It is certainly fair to say that the Sauerian approach continues to be a vitally important component of cultural geography, a situation that is clearly evident from much of the material incorporated in chapters 3, 4, and 5, and that is also evident in both this and the following chapter. However, it is similarly fair to say that the cultural geography associated with Marxism, feminism, and the cultural turn—what was broadly referred to in Chapter 1 as the new cultural geography—is an integral component of the contemporary subdiscipline.

The material included in chapters 7 and 8 is based on a useful but certainly by no means absolute distinction.

- The approach that lies at the heart of this chapter is the interest taken in questions of **identity** and difference (related particularly to the idea of race and ethnicity) as these are reflected in inequalities. There are different groups of people and different landscapes and, crucially, it is often the case that a principal basis for distinguishing one group from another is in terms of relative **power**—groups are unequal, and the landscapes that they occupy are unequal. This basic idea has been stimulated by, and in turn has stimulated, a substantial body of concepts and analyses, with the three most significant bodies of supporting ideas being those of Marxism, feminism, and postmodernism.

- Chapter 8 is more clearly informed by humanism, feminism, and postmodernism, with central themes being the construction of human identities, the geography of others, the idea that landscapes can be studied as texts, and ideas about the symbolic interpretation of landscape.

The content of both chapters 7 and 8 is characterized by being both spatial and social, both historical and contemporary, and both urban and rural.

This chapter is structured as follows.

- There is an account of a diverse conceptual background. This conceptual background prompts and supports the concern in this chapter with human identity, difference, and inequality, as well as the discussions of places, identity construction, others, landscape interpretation, and symbolism in Chapter 8. The introduction of culture into traditional social geography is noted, and there are overviews of Marxism, feminism, and the cultural turn—the latter seen as including cultural studies, poststructuralism, and postmodernism. The perspectives and approaches introduced through these various bodies of thought have had a considerable impact on the concepts and practice of all areas of social science, including cultural geography. All are complex, interrelated, and even contested bodies of thought that are continually evolving. Accordingly, the summaries contained do not attempt to reflect the full richness and diversity of these ideas. Rather, what is attempted here is a series of capsule accounts with emphasis on material that is especially relevant to the cultural geographic studies discussed in this and the following chapter.

- The conceptual material is followed by three closely related discussions concerned with the way in which human groups and their landscapes exhibit the characteristics of what James (1964) called 'One World Divided'. The idea of race and related matters of supposed racial differences are the subject matter of the first of these three discussions. The principal empirical topic discussed is that of the apartheid era in South Africa.
- There is an account of ethnicity and ethnic identity that includes consideration of the roles played by language, religion, and both national and local identities.
- There is a discussion of both the evidence for and the significance of the current tendency for ever-increasing globalization—politically, economically, and most significantly, culturally.
- There is a brief concluding section that raises the issue of human rights in the light of the earlier chapter discussions.

Note that it is helpful to appreciate that the ideas covered in the first section inform subsequent analyses in much the same way that the section on the landscape school at the beginning of Chapter 3 informed much of the empirical content of chapters 3, 4, and 5. But there is, of course, one principal and significant difference between the two bodies of ideas—the landscape school concepts were developed within the subdiscipline of cultural geography specifically to facilitate cultural geographic analyses, whereas the concepts introduced in this section were developed elsewhere and have been imported into cultural geography.

Three Challenges

There are three potentially confusing issues (perhaps a better term is challenges) that arise in this and the following chapter and that need to be raised, but by no means resolved, immediately.

- First, as stated in Chapter 1, the new cultural geography is a welcome addition to, rather than a replacement for, the Sauerian approach. Indeed, although the cultural turn particularly is most correctly seen as a set of new developments, the Sauerian antecedents of some aspects of these new developments are evident from the contents of earlier chapters. Further, there is no clearly definable new cultural geography because the term does not refer to an approach but rather to several and varied conceptual advances that may have some common content but that also demonstrate substantive differences. This is an important point to stress at this stage of the book—much of the content of chapters 7 and 8 reflects the new cultural geography, but this does not mean that this content reflects some monolithic approach. None of this should be surprising, of course—the Sauerian approach that has been such an important conceptual underpinning for much of the material contained in earlier chapters was not and is not a single monolithic approach.
- Second, because of the relative newness of much of the conceptual material in this and the following chapter, there is less certainty about what is and what is not important—the consensus judgments made today are unlikely to be identical to those made tomorrow. Any academic discipline travels many conceptual pathways, all interesting at the time perhaps, but not all equally profitable in terms of the empirical analyses prompted and the contribution made to knowledge. Indeed, several geographers have expressed concern about the seemingly indiscriminate, and perhaps even self-indulgent, borrowing of ideas from other disciplines and also about the short lifespan of many of the concepts introduced. More generally, other geographers worry that 'human geography is to be subsumed within a broadly defined social science' with a resultant loss of disciplinary identity (Unwin 1992:210). It is certainly the case that some of the recent conceptual advances embraced by some human geographers have further added to what was already a rather uncertain discipline,

seeming to further remove the *geo* from human geography.

- Third, many of the recent conceptual advances in cultural geography are more difficult to understand and apply than is the Sauerian approach. This is because they are often associated with complex and disputed philosophical and/or social scientific origins—the brief remarks about feminism in Chapter 5 are indicative of this point—with the result that there are typically multiple versions of each of the newer conceptual advances. It is also because many of these advances are linked to postmodernism, a term that refers to a varied corpus of ideas but that characteristically insists upon the need to take into account non-traditional sources and the need to question established wisdom. Finally, it is because the cultural turn involves writing styles and vocabulary that makes it seem exclusionary to some and pretentious to others.

Together, these three issues combine to make cultural geography one of the most exciting, most challenging, and, yes, sometimes rather difficult areas of contemporary scholarly analysis. Added to and sometimes building on Sauerian landscape foundations, these conceptual advances and related empirical work are making major contributions to both academic and practical knowledge.

Conceptual Underpinnings

Cultural geographers have, of course, always been concerned with human identities and related inequalities, especially as these are expressed through cultural differences. Thus, there are long-standing interests in such identifying characteristics as way of life, language, religion, and ethnicity—interests that have been reflected in earlier chapters. What makes the current concerns with identity and inequality distinctive is both the explicit use of theoretical work and the broadening of the subject matter beyond the more traditional view of culture—as employed within the Sauerian school—to encompass a more meaningful view of human identity. One component of this broadening is contemporary cultural geographers' willingness to incorporate social geographic sub-

ject matter into their analyses, a circumstance that has at least partially removed the rather artificial distinction between the two subdisciplines. Indeed, social geographers have also taken an interest in various identifying characteristics, especially those of class and ethnicity.

Geography and Society

SPACE AND SOCIAL THEORY

Until the 1980s, social theory was almost exclusively the property of the discipline of sociology and, accordingly, there was little meaningful consideration of spatial matters. However, consider these statements from a sociologist:

- Human geography is a social science as 'there are no logical or methodological differences between human geography and sociology' (Giddens 1984:368).
- 'The social sciences developed as a family of disciplines such that the different "areas" of human behavior that are covered by the various social sciences form an intellectual division of labour which can be justified in only a very general way' (Giddens 1987:9).

And these statements from geographers:

- 'Space is a social construct—yes. But social relations are also constructed over space, and that makes a difference' (Massey 1985:12).
- There is a 'recent revival of interest in a social theory that takes place and space seriously' (Agnew and Duncan 1989:1).

These claims concerning the status of human geography as a social science and the related importance of place in social theory are similar to those noted in Chapter 5 where the rationale for such claims was based on the perceived shared concern with human behavior, although these more recent claims also identify a shared conceptual base. Certainly, the view that human geography is a social science continues to be strenuously voiced while statements of human geographic subject matter usually emphasize the spatial organization of societies and the relationships between people and their environments. The inclusion of place or space in social theory, previously almost unheard of, is now quite usual.

- The first substantial work in this area was the structuration theory of Giddens. As noted in Chapter 5, key features of this theory concern the links between human agents and the social structures within which they function, with individuals viewed as operating within both local social systems and larger social structures. Further, structuration theory explicitly recognizes the links between social relations and spatial structures, with space as a medium through which social relations are produced and reproduced rather than as a mere backdrop against which social relations unfold. The term 'locale' has been used to refer to the settings in which everyday social interactions occur (see Glossary). Although several geographers have endorsed and used structuration theory, others have questioned its value and there is little evidence that it is to play a major role in contemporary cultural geography. Sympathetic accounts by human geographers tend to be critical of geography for viewing place as inert when it ought to be viewed as a historically contingent process, as a transformation of space and nature that cannot be detached from the transformation of society, but cultural geographers might well resist any suggestion that they have viewed place as inert. The many regional studies by Meinig may not be conceptually sophisticated by structuration standards, but they are explicit analyses of related spatial and social change.
- Another body of work to which geographers have been attracted is that concerned with what might be called the production of space, a phrase that refers to the fact that the spaces within which social life occurs are socially produced—a rational and scientific interpretation of space is replaced with an interpretation that sees space as experienced or imagined.
- But the most important advances in contemporary social theory for the cultural geographer are those that strive to build on ideas that are explicitly opposed to the Enlightenment concern with an empiricist epistemology and with the search for universal truths. Accordingly, there has been

much interest shown in the works of some critical social philosophers. These theoretical contributions are considered later in this section in the context of a discussion of the importance of cultural studies as a source of concepts for cultural geographic analyses of both identity and difference.

SOCIAL GEOGRAPHY

But what of social geography? Until the 1980s, social geography could be regarded as the British counterpart to North American Sauerian cultural geography. As discussed in Chapter 1, the different experiences are explained in terms of the different intellectual origins of the two subdisciplines. Thus, although cultural geography was stimulated by leading European geographers such as Humboldt, Ratzel, and Schlüter, it was really initiated by Sauer and never fully incorporated into geography curricula outside of North America, while the British social geographic tradition has origins in French sociology, in British human geography, and in the Chicago school of human ecology. Box 7.1 is an overview of these developments.

One principal North American argument was for a social geography built on the ideas of symbolic interactionism (see Box 5.2), with social geography defined as the study of human spatial behavior and related human landscapes from the point of view of society as the totality of a population's symbolic interactions. The label 'social geography' was favored over that of cultural geography because of the implied emphasis on communication processes through which symbolic interactions take place and the related claim that individuals are constrained in their behavior by the character of the groups to which they belong. The cultural geographer, Wagner (1972, 1974), likewise saw distinct landscapes and societies evolving in response to communication processes often generated by social institutions, but this interest has not proven to be a dominant focus.

Early British social geography built especially on the ideas of Le Play to become a concern with the identification of different regions as these could be delimited according to the association of social phenomena. However, during the 1960s and 1970s three transformations occurred: namely the introduction of a behavioral perspective; the use of quantitative methods; and a focus on social problems often stimulated by Marxist thought. But as the first three of the following four quotes indicate, social geography continued as an uncertain subdiscipline with much relevant material regarded as belonging to other compositional subdivisions such as population geography and urban geography. (See Your Opinion 7.1.)

- Dennis and Clout (1980:1) began their textbook with the following concern: 'Of all the varieties of geography, "social geography" is probably the most elusive and poorly defined.'

- Similarly, Cater and Jones (1989:viii) began their textbook with a discussion of 'social geography's identity crisis'.
- A few years earlier, Eyles and Smith (1978:55) concluded their review of the status of social geography on a cautionary note: 'Social geography's greatest achievement may yet turn out to be the acceleration of its own destruction.'
- Hamnett (1996:3) is more certain as to the identity of social geography: 'The approach I take is that social geography is primarily concerned with the study of the geography of social structures, social activities and social groups across a wide range of human societies.'

Box 7.1: Cultural Geography and Social Geography

The intellectual origins and history of *cultural* geography have been detailed earlier in this book, especially at the beginning of Chapter 3. The term '*social* geography' was used by French sociologists of the Le Play school in the 1880s, by Réclus in the 1890s, and a detailed proposal was made by Hoke (1907). However, none of these developments had significant impacts on geography, and it was not until the 1930s that social geography emerged as a subdiscipline following a programmatic statement by Roxby (1930), the publication of social geography books by members of the French school, and some of the sociological human ecological work referred to in Chapter 5 that emphasized social values. The result was a British social geography essentially concerned with the regional distribution of population, occupation, and religion.

By the 1960s, this British social geography was an uncertain field because of the varied intellectual stimuli, although Pahl (1965:81) described the subdiscipline as a concern with the 'processes and patterns involved in an understanding of socially defined populations in their spatial setting'. At the same time North American cultural geography had a much clearer identity, with a single dominant school of thought, namely the Sauerian school that focused on evolution, regions, and ecology. Certainly, British social geographers, in common with sociologists, tended to see culture as the concern of the anthropologist.

The differences between the two were further emphasized during the 1960s with cultural geography remaining outside of, and social geography being involved in, the spatial analytic turn. However, a convergence of the two interests was initiated in the 1970s when aspects of both subdisciplines were inspired by the perspectives of humanism and Marxism, and, according to some observers, this convergence has proceeded apace, especially since the mid-1980s, with shared interests in the materials covered in this and the following chapter (Figure 7.1).

Nevertheless, it is misleading to suggest that the two have in some way become one. Indeed, the simplified form of Figure 7.1 is readily apparent given the material presented in Chapter 1 of this book. Specifically, the extent to which cultural and social geography have experienced a genuine integration is open to much debate, as evidenced by both the form and content of the various Progress Reports written for the journal, *Progress in Human Geography*. Certainly, the content of these reports indicates that the viability of a meaningful cultural and social integration is questionable.

- Gregson (1995:139), for example, noted 'the ascendance of cultural, as opposed to social, theory in social geography (and in human geography more generally)', and wondered whether a discussion of progress in social geography was really an 'obituary'.
- Hamnett (1996:4), on the other hand, claimed that the 'revival of the "new" cultural geography in Anglo-America has been very influential in the revival of social geography'.
- Finally, Peach (1999:286) maintained that work in social geography continued 'to have more in common with the spatial side of sociology than with the cultural side of geography'.

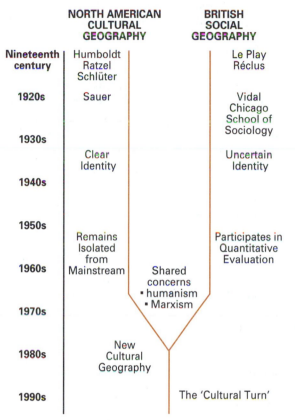

NORTH AMERICAN CULTURAL GEOGRAPHY	BRITISH SOCIAL GEOGRAPHY
Nineteenth century — Humboldt Ratzel Schlüter	Le Play Réclus
1920s — Sauer	Vidal Chicago School of Sociology
1930s	
Clear Identity	Uncertain Identity
1940s	
1950s	
Remains Isolated from Mainstream	Participates in Quantitative Evaluation
1960s — Shared concerns • humanism • Marxism	
1970s	
1980s — New Cultural Geography	
1990s	The 'Cultural Turn'

Figure 7.1: Cultural geography and social geography: origins, concerns, and convergence. This is a schematic diagram that highlights the separate origins, the different concerns, and the convergence, which began about 1970, of cultural and social geographic traditions.

Marxism

Marxism, the first of the conceptual underpinnings discussed in this chapter, is also the oldest (having been introduced to human geography about 1970) and is not associated with the cultural turn of the 1980s. Indeed, in some respects, Marxism was one of the approaches that the cultural turn reacted against because of the emphasis that Marxism placed on grand theory—the claim that there is necessarily some right answer to a problem—and also because

of the essentialist character of the approach. Nevertheless, it is appropriately discussed at this time because the study of unequal groups and unequal landscapes was largely initiated within this research tradition. Marxism, as developed by Marx and Friedrich Engels (1820–95), is both a body of social theory and a political doctrine. During the twentieth century the political doctrine was implemented, in a modified form, in a number of countries, notably the former Soviet Union and China. The concern in the present context is with the social theory.

MARXIST SOCIAL THEORY

It is often suggested that in developing a social theory Marx followed Hegel and employed a **dialectic** as an alternative to formal logic, although it seems more appropriate to suggest that the dialectic used by Marx was ontological (see Glossary), referring to the fact that societies are organic wholes that evolve only to eventually collapse as a consequence of their internal contradictions.

The three components of Marxist social theory are as follows.

- First, principal types of human society were identified in a historical context. Examples of societies are slavery, feudalism, capitalism, and socialism (all four terms are included in the Glossary). Each society was discussed as a **mode of production**, comprising both **forces of production**—that is, the raw materials, implements, and workers that produce goods—and the **relations of production**—that is, the economic structures of society. It is the forces of production that produce goods, while the relations of production determine the way in which the production process is organized. The most important relation concerns matters of ownership and control. Two insights emerge from this component of Marxist social theory. First, there is a historical transition from one mode of production to another. Second, social classes play a key role in social formation and social change. In addition to these distinctions, society can be differentiated into **infrastructure (base)**—another term for relations of production—and **superstructure**—the legal and political system and forms of consciousness. These terms are helpful because infrastructure can be interpreted as a determinant of superstructure, meaning that human thinking results from material conditions. This is effectively a form of economic determinism.

- Second, the transition from one type of society to another was explained in terms of two related processes, namely technological change and class struggle. Emphasis was placed on the emergence of new classes and on the fact that an existing dominant class was able to limit the opportunity for further social change. Because a dominant class typically favored maintaining things as they were, Marx recognized the need for class struggle as the basis for the replacement of one mode of production with another mode that comprised different classes.

- Third, particular attention was paid to the analysis of nineteenth-century capitalism and to the anticipated transition to socialism. Marx was especially critical of capitalism because the dominant class (owners) were concerned with profits—or what can be called the ceaseless accumulation of capital. Further, owners were seen as exploiting the dominated class (workers). Although such exploitation of one class by another was by no means new, Marx recognized that what was previously open in modes of production, such as slavery and feudalism, was disguised in the capitalist mode of production. Marx envisaged capitalism as the last of the class-based societies. Much of the political doctrine of Marxism is concerned with the various mechanisms by which classless socialist societies would eventually replace capitalism.

THE IMPORTANCE OF DIVISION OF LABOR

Marxist social theory is concerned with the material basis of society and aims to understand society and social change by referring to historical changes in social relations. This approach is termed **historical materialism**. Marx conceived of his-

tory in materialistic terms, and a central feature of Marxist social theory concerns the division of labor and the formation of classes.

For Marx, the first human historical act is the production of means to satisfy material needs, and fulfillment of these needs leads, as population numbers increase, to other needs. The first form of social relation, the family, which is associated with the fulfillment of the initial material needs, proves unable to satisfy these new needs and thus a new social formation develops. In this sense, it is appropriate to identify each mode of production with a particular social formation. This is what is intended by saying that society has a materialist basis. Thus, in tribal societies there is an elementary division of labor based on age and gender, but the division becomes increasingly complex with the development of agriculture, civilization, and industry. In the industrial world of the nineteenth century, the capitalist mode of production involved a separation and conflict of industrial and agricultural interests, a separation and conflict of capitalist and worker interests, and a separation and conflict of individual and community interests. The implications of these divisions of labor, from the elementary forms found in the family to the much more complex forms found in capitalism, are enormous.

- There are unequal distributions of labor, of the products of labor, and of private property.
- The distinction between individual and community interests leads to the creation of states that serve the community interest.
- There is an estrangement or **alienation** of social activity. Alienation refers to the reduction of human relations to commercial relations—a form of commodification.

Thus, Marx was concerned both with the material conditions of workers and with the way in which their lives had assumed an alien form.

VERSIONS OF MARXISM

Marx was a prolific writer whose works included numerous uncertainties and contradictions as ideas developed and transformed through time and, accordingly, there have been several interpretations of both the social theory and the political doctrine.

Certainly, there is no one correct reading of Marx. Two principal versions of Western Marxism appeared during the twentieth century.

- The Frankfurt school of critical theory emphasized the humanistic aspects of Marxist thought and added elements of social psychology. From this perspective, Marx was a voluntarist, seeing human history as reflecting human intentional activity. This school criticized bourgeois society for being dominated by a form of technological rationality and for the prevailing scientific orientation of social science.
- A second version rejected the humanistic aspects and focused on the structuralist character of Marxism. Structuralism was a popular approach during the 1960s, emphasizing that the understanding of a social system required an appreciation of the structural relations between the component parts. From this perspective, human problems were essentially structural in character as they were rooted in some specific economic system, and humans were not considered as autonomous active subjects—refer to the first component of Marxist social theory as discussed above.

In addition to these two versions, there have also been attempts to read Marx in terms of rational choice theory in economics—this is known as analytic Marxism—while Marxist theory has also been influential in some feminisms, in postmodernism, and in cultural studies generally.

Regardless of the specific version, Marxism is always concerned with analyzing inequality, oppression, and subordination, and with identifying some means to overcome these. The basic solution for Marx was, of course, that of the creation of a socialist society involving the abolition of class, of the private ownership of the means of production, and of commodity production.

MARXISM IN GEOGRAPHY

As the account of Marxist social theory has indicated, there is no single Marxist tradition to which geographers could turn and because of this contested situation, any Marxist approach adopted has been challenged not only by those geographers who are unsympathetic to the approach generally but also by other Marxist geographers who favor an alternative version. The most significant debate, not surprisingly, has been that between Marxist geographers inclined to the humanistically oriented Frankfurt school of thought and those inclined to structural Marxism. The disagreement between these two is one part of the larger structure and agency debate; that is, the concern about the capabilities of human beings as they function given the constraints imposed on their behavior by social structures. We have encountered this debate in previous chapters under the guise of what was called the social scale of analysis.

- Not surprisingly, humanistic geographers especially have objected to structural Marxism on the grounds that, like positivism, it adopts a passive view of humans.
- But some other human geographers have favored structural Marxism because of the emphasis placed on the presumed crucial role played by the infrastructure—meaning the economic structure of capitalist society—on the superstructure of the human geographic world, which includes the landscape.
- Some other geographers have acknowledged the importance of both the more humanistic Frankfurt school and of structural Marxism, noting with favor that Marxist thought is open to a variety of interpretations ranging from an active view of people as the makers of their own history to a more passive conception of human development as the determined product of relatively autonomous structures.

Within cultural geography specifically, there are two important arguments as follows.

- It has been argued that a Marxist cultural geography corrects perceived deficiencies such as the failure to recognize that culture has a political component and the failure to recognize that cultures are divided into classes. Further, a Marxist view is seen as one that avoids the humanistic tendency to oppose individuals and culture and that tends to neglect external constraints from society at large (Blaut 1980).

- It has been claimed that 'Marxism and cultural geography share important basic presuppositions concerning the significance of culture, but in different ways and for different reasons both have failed to sustain those presuppositions in their practice and have not developed a dialogue with each other' (Cosgrove 1983:1). Cosgrove's attempt to correct this failing recognized a common focus on the historical aspect of human and land relations and stressed conceptual similarities such as a shared Marxist and Vidalian view of the relationship. (See Your Opinion 7.2.)

Your Opinion 7.2

Do not forget that cultural geography is concerned with both humans and landscapes, with both people and place. Surely it is of great importance that there are massive differences in terms of quality of life between different groups of people and also between different parts of the surface of the earth. Traditional Sauerian cultural geography rarely considered these differences, and it is certainly one of the major contributions of Marxism to cultural geography that it forces our attention to this critical matter. Admittedly, the many internal squabbles within the Marxist community result in some confusion, but the effort needed to understand Marxism is more than repaid by the insights that it affords. Agreed?

Feminism

Recall that feminism was introduced in Chapter 5 in the specific context of ecofeminism when it was noted that, although there are many versions of feminism, all concur that **sexism** prevails, is wrong, and needs to be eliminated. **Patriarchy** is usually seen as the fundamental wrong. Feminists are concerned with the oppression or subordination of women, and also with other oppressions. Two versions of ecofeminism were identified in Chapter 5, namely cultural ecofeminism and social ecofeminism. This discussion of feminism broadens the earlier account in order to provide an appropriate basis for the discussions of the social construction of **gender** included in this chapter and the discussions of landscapes of sexuality included in Chapter 8.

THREE VERSIONS

- The basic tradition of feminist thought, dating back to the late eighteenth century and known as the first wave or liberal feminism, aimed to obtain equal rights and opportunities for women. Note that from some later perspectives, such a seemingly innocent approach can be seen as funda-

mentally flawed because it maintains the male as the norm.

- The feminisms that appeared during the 1960s and continued into the 1980s are often described as the second wave of feminism thought. These tended to assume that it was possible to identify a specific cause of the oppression of women. Suggested causes included men's ability to control women's fertility, and the capitalist requirement of a pliable workforce. Cultural ecofeminism as discussed in Chapter 5, with a focus on the technologies and ideologies developed by men and the idea of essential relationships, belongs in this second wave category. These more radical feminisms often have the general intent of moving beyond the redistribution of rights and resources as in liberal feminism in order to initiate changes to the structure of society. Further, many of these radical feminisms contend that the oppression of women is deeply rooted in both psychic and cultural processes such that much more than superficial change is necessary.

- During the 1980s, the third and current wave of feminist thought appeared. This involved a critique of the earlier assumption that it was appropriate to attempt to universalize the experiences of women (usually White and middle class) in the more developed world. As noted in Chapter 5, feminists thus developed a fuller concern for the multiple oppressions that structure our gendered identities. These multiple oppressions are based on social class, skin color, income, religion, age, culture, and geographic location. It also involved an expanding interest in sexual difference as well as in gender, an interest stimulated by psychoanalytic theory. There is increasingly a realization that to insist on equality is to deny the importance of sexual difference. Finally, it involved an engagement with aspects of poststructuralism and postmodernism that includes a concern with the social construction of gender. Most of the current work in feminist geography is identified with this third wave.

Note that, for most versions of contemporary feminist theory, sex is a natural category based on biological differences between women and men (but see the discussion of sex and gender in Chapter 8), while gender, although based upon the natural category of sex, is a social construction that is developed over time and that refers to what it means to be female or male. Social ecofeminism as discussed in Chapter 5, with a focus on the constraints imposed by institutions such as marriage and the idea of socially constructed rather than essential relationships, belongs in this category.

LOCATING FEMINISM IN GEOGRAPHY

It is clear that despite a considerable body of both conceptual and empirical literature by geographers and others, feminism does not merit any significant explicit mention in many standard reviews of geographic approaches. The contrast with the receptions given to Marxism and to postmodernism is striking. It does appear that the discipline of geography is less than certain about how to accommodate a body of thought that not only includes new perspectives, new approaches, and implies new subject matter, but also questions the discipline of geography as a legitimate enterprise. Cloke, Philo, and Sadler (1991:xi) explained their decision not to include a chapter on feminism in their account of approaches to human geography as follows: because 'the issues involved are more important than this "ghettoising" might imply', because of the belief that feminism is something 'more than just "another" approach', and because 'such discussion is most ably pursued by women'.

In addition to this uncertainty, there are also debates about feminisms and feminist geographies within the feminist geography literature itself. Some have identified areas of shared feminist and geographic interest and argued for integration, while others express concern about the fundamental masculinism of geography that might effectively prohibit meaningful integration. The term 'masculinist' refers to work that claims to be exhaustive, but that ignores the existence of women and that accordingly is concerned only with the position of men. Another important debate was evident in the writing of a feminist geography textbook aimed at introductory level students (Women and Geography Study Group of the Royal Geographical Society with the Institute of British Geographers 1997). This book was written by a collective that made the decision ('not an easy one') to exclude men from the writing team (Rose et al. 1997:3). (See Your Opinion 7.3.)

Your Opinion 7.3

How do you react so far to these comments about the approach being taken to feminism? If you are like the discipline of geography generally, you may be experiencing some qualms. Indeed discussion of feminism in geography is proving to be much more controversial than might be expected given the relative ease with which other new approaches have been discussed. The reason? It may be that feminism provides much more of a challenge than do other new concerns because feminism strikes at the very core of the geographic enterprise. Domosh (1996:411) noted: 'Unlike many of the other conceptual terms that have entered geographical discourse in the past twenty years (Marxism, structuralism, postmodernism), feminism provides neither a methodology or a theory for human geography, although methodological and theoretical stances have been derived from it. Rather, feminism as it was originally introduced to geography had a far more radical goal—it provided a political basis for a critique of the practices of geography, practices that had allowed for the invisibility of women as both practitioners in the field and objects of enquiry.' Feminism thus challenges established epistemologies as to what can be regarded as legitimate knowledge and is accordingly seen by some as a real challenge to the discipline in general.

FEMINIST CHALLENGES TO GEOGRAPHY

Feminist geography has been concerned with various issues, which have changed through time in general accord with changes in the larger feminist context.

- In the early 1970s and in accord with the logic of liberal feminism, there was concern expressed about women in geography, especially since the discipline of geography, like many other disciplines, was dominated by men.
- There has been a concern with the geography of women involving an incorporation of gender as a variable into geographic studies, along with the more usual variables of culture, ethnicity, and class. Studies of oppression included considerations of oppression within the discipline and within the larger world and many of these studies adopted a Marxist orientation.
- The early concern about the lack of women in the discipline of geography expanded to include concerns about the essential masculinism of geography, especially because of the traditional link with exploration and the continuing involvement with fieldwork.
- In a more radical tradition and with links to postmodernism, there have been attempts to construct theory from the perspective of women in response to the idea that geography is a masculinist discipline that claims to speak for everyone but that only speaks for White, middle-class, heterosexual males.
- Feminist geographers and others have begun to further explore the social construction of human gendered identities, both femininity as well as masculinity.

Probably the single most important concept in feminist geography is that of gender, along with the derivative ideas of **gender roles**, **gender relations**, and patriarchy (see Glossary for all four terms). Gender roles are those sets of behaviors that are deemed to be socially appropriate along with another set that are deemed inappropriate, solely on the basis of sex. Few, if any, of these behaviors rely on inborn sexual differences. Most important, gender roles vary culturally and they are, accordingly, best thought of as learned differences in

behavior. The concept of gender relations recognizes that gender roles are not necessarily accepted; rather, they are sometimes contested and such roles change through time. Both of these characteristics of gender roles refer to the idea that gender involves power relations between women and men, specifically the relationship of patriarchy whereby men dominate women. A similar argument about sexual identity is included in Chapter 8.

It is clear that feminism poses several challenges, perhaps better thought of as invitations, for geographers to consider. Most dramatic of all, since about the 1960s, forms of feminism have challenged virtually all of the belief systems and institutions of dominant patriarchal cultures, including geography. Contexts for knowledge are questioned, as are basic concepts such as reason and logic. For geography this has involved questioning the viability of approaches such as positivism, humanism, and Marxism because of what is seen as their sexist bias. Indeed, feminists were among the first to explicitly recognize the importance of the social construction of knowledge, meaning that those who create knowledge are responsible for determining the identification of research problems, the data to be used, and the methods to be employed—the earlier discussion of Marxism clearly demonstrated the validity of this observation, but it has also been implicit in all of the material covered in this textbook following the first introduction of the idea in Chapter 1. Similarly, then, feminists exposed the importance of positionality and the situated character of knowledge—ideas that are central also to postmodernism.

The Cultural Turn

A CONTEXT

It is usual to suggest that both the social sciences and humanities have undergone a cultural turn (see Glossary), an increasing importance placed on culture—and the related downplaying of other considerations such as economy and politics—in discussions of humans and human activities. An important component of this turn is the emphasis on language, such that there is also a linguistic turn, a term that is sometimes used as an alternative to cultural turn and sometimes seen as one component of the cultural turn. Specifically, there is an increasing realization that language is important in that it enables the communication of meanings but, at the same time, it is constraining because of the limitations resulting from the meanings attached to words—we do not know the world for what it is, but rather only as it is mediated through language and other symbolic systems. Expressed simply, the linguistic turn challenges the ability of language to function as a neutral conveyor of information.

This discussion considers the cultural turn from three related perspectives, namely those of cultural studies, poststructuralism, and postmodernism. It is recognized that this is a less than perfect way to categorize material that is by definition fluid and dynamic, but it is considered that the imposition of some order—however artificial it might appear to be—is important to the student of cultural geography at this time because, as noted at the outset of this chapter, many of the ideas being raised are new to the subdiscipline, are of uncertain status, and are contested. However, it is appropriate to stress that the very spirit of the cultural turn is unsympathetic to classification and the imposition of order. This is an idea that resurfaces on several occasions in this discussion, but it is also an idea that proves difficult to maintain at times.

Another point of information is in order. It is not difficult to understand that there might be a cultural turn in some other disciplines, but it may sound confusing to claim that the turn is affecting cultural geography—how can there be a *cultural* turn in *cultural* geography? The short answer is that the meanings of culture involved in the cultural turn are different from the meaning incorporated within Sauerian cultural geography—refer back to the various definitions of culture included in Box 1.2, and recall the deterministic implications of the classic Sauerian definition in contrast to the understanding of culture as a negotiated intersubjectivity implied in some, but by no means all, of the more recent definitions.

CULTURAL STUDIES

As noted in Chapter 1, cultural studies is perhaps most appropriately seen as an umbrella term incorporating a wide range of philosophical and social theoretical ideas including feminism, poststructuralism, and postmodernism, and also aspects of Marxism and humanism. Central to cultural studies is the claim that established disciplines are unable to cope with the complex circumstances of the cultural world, such that the advantages of an interdisciplinary approach are stressed. But this is a claim that is difficult to honor in many instances, as the publication of cultural studies readers or textbooks makes clear—somewhat ironically, reference has been made to 'the delineation both of a tradition and a future trajectory for cultural studies as the Academy's most upwardly mobile discipline' (Carter, Donald, and Squires 1995:vii).

Box 7.2 provides an overview of some of the more important authors and ideas that are proving to be influential in cultural geography. The information in this box aims to facilitate the understanding of cultural studies, specifically of poststructuralism and postmodernism.

The cultural studies literature typically rejects modernism in addition to the Enlightenment concerns with an empiricist epistemology and with a search for universal truths. Together, the multiple and diverse bodies of thought that are labeled cultural studies comprise a key feature of the cultural turn in the social sciences and humanities generally. Principal components of this work include several British studies of working-class culture (for example, Thompson 1968), the concept of culture introduced by Geertz (1973), and the Frankfurt school of Marxist critical theory. Also important in the current context are the several advances in social theory associated with such writers as Gramsci, Derrida, Foucault, Baudrillard, and Deleuze, as these are discussed in Box 7.2, along with the several traditions—such as feminism, postcolonialism, subaltern studies, and studies of racism—that are linked by their concern with exposing the Eurocentrism associated with the production of knowledge in the Western world since the Enlightenment.

Understandably, different bodies of thought that are gathered under the cultural studies umbrella emphasize different matters: feminism, for example, is critical of much of the gender blindness of traditional Marxism, while Marxism is at the same time a major source of inspiration for many postmodern social theorists. But together these works stress the importance of culture as a dynamic and primary force that is not necessarily always predictable from political, economic, and social forces. Further, all see culture as a site of negotiation and conflict in societies that are dominated by power and that are splintered especially in terms of ethnic identity, gender, and class. Much of this work can be described not only as part of the cultural turn, but also as part of the linguistic turn because of the emphasis on the role played by language.

POSTSTRUCTURALISM

Recall from Chapter 2 that structuralism is one of several bodies of thought that view culture as a system of ideas, being based on advances in linguistic theory that see a sound and the object it represents as entirely arbitrary—that is, it is a concern with relationships rather than with those things that create and maintain the relationships. Two of the principal insights of structuralism, both antihumanist, are important, in modified form, in **poststructuralism**.

- The first insight, following the linguistic philosopher, Ferdinand de Saussure (1857–1913), is that language is a medium for understanding social organization. Thus, meaning is not reflected through language, but rather produced within language. Simply put, the facts do not speak for themselves.

Box 7.2: Cultural Studies: Scholars and Ideas

Although many of the scholars and ideas associated with cultural studies are also closely identified with feminism, postmodernism, or poststructuralism, they are discussed together in this box for reasons of convenience. Specifically, because much of the empirical material in both chapters 7 and 8 relies on the ideas presented in this box, it aims to serve as a reference informing many of those studies.

Antonio Gramsci (1891–1937) was a Marxist theorist and political activist whose most influential conceptual contribution involved the elaboration of the idea of **hegemony**. According to Gramsci, any class that achieved a position of economic importance and that desired to attain power in a larger society first needed to achieve a degree of cultural and intellectual hegemony. This hegemony allowed for the expression of the class's world view, for the structuring of social and other institutions such that they accorded with the class's aims, and most generally for the creation of a context sympathetic to the class.

Gramsci also introduced the term 'subaltern' (see Glossary) to refer to those socially subordinated groups that lacked both the unity and the organization of more dominant groups that are able to exercise **authority** and control. The term has been popularized by Indian Marxists and generally refers to those groups who are considered to be in some way socially inferior to other groups.

The work of Michel Foucault (1926–84) both crosses and challenges disciplinary boundaries and is, accordingly, especially difficult to classify. One important contribution concerns the ideas of **discourse** and **episteme**: Foucault sees subjectivity as constructed within and through discourses, with a discourse defined as a system, comparable to a language that enables the world to be made intelligible. All terms, then, need to be discussed within the specific context of a given discourse as the meaning of a term varies from one discourse to another. Discourses are important because they serve to legitimize a particular view of the world that then becomes part of the taken-for-granted world (see Glossary). Discourses define our world for us: 'the exercise of power perpetually creates knowledge and, conversely, knowledge constantly induces effects of power' (Foucault 1980:52). It can be argued, for example, that the discipline of geography evolved within the late nineteenth-century discourse of imperialism. The understanding contributed by Foucault concerns the necessarily partial and situated character of any particular discourse. In short, language cannot 'serve as a perfectly transparent medium of representation' (Duncan and Ley 1993:5). Expressed simply, language can be seen as constructing rather than conveying meaning. What Foucault achieved in this respect was a reversal of the idea that knowledge is power, with the claim that only those in power have the right to say what is knowledge. To quote a few simple examples, consider what is involved in renaming rape as sexual assault, jungle as rainforest, and swamp as wetland—in each case the identical subject is being placed in a different context for evaluation.

Jean Baudrillard (1929–) is a particularly controver-

- The second insight, following the Marxist philosopher, Louis Althusser (1918–90), concerns the interpretation of subjectivity. Specifically, notions of individuality are seen as ideological constructs; thus, humans cannot act autonomously, rather their actions are constrained by various structures.

Both of these ideas are also important in postmodernism. Poststructuralism is a broad term, reflecting an expansion of, rather than a break with, structuralism. Many of the most important ideas, including those modified from structuralism, are often included as one component of the cultural studies tradition and are, as noted, also identified with postmodernism.

POSTMODERNISM

Box 7.3 provides a basis for outlining the key arguments and content of postmodernism by offering distinctions between four terms, namely modernity, modernism, postmodernity, and postmodernism. According to many (but by no means all) observers, the contemporary world is increasingly displaying the characteristics of postmodernism. Although this discussion offers a series of generalizations about postmodernism, it is not intended to convey the impression that there is some clearly agreed upon set of ideas. Rather, postmodernism is itself a highly contested arena with numerous competing ideas. Certainly, postmodernism includes a range of ideas that together constitute a real chal-

sial postmodernist writer. As is the case with many other twentieth-century theorists, Baudrillard challenged Marxism both for being deterministic and for reducing culture to a secondary status at the expense of the political and the economic especially. Contemporary society was described as a **consumer culture** based on consumption and not on production. Baudrillard also introduced the idea of the simulacrum, arguing that the contemporary world is dominated by signs, images, and representations, obliterating the real and rendering any search for truth and objectivity fruitless.

The French philosopher, Jacques Derrida (1930–), has demonstrated a continuing concern with the taken-for-granted character of much philosophical thought and, according to some postmodernists, has effectively shown that philosophy has no privileged access to meaning and truth. Derrida is best known for the introduction of **deconstruction** and the related idea of **text**. Essentially a school of philosophy and literary criticism, deconstruction has had a significant impact on many of the traditional social science and humanities disciplines. However, it is not by any means universally accepted, especially by philosophers, at least partly because it is such a radical set of ideas. Most generally, it has been used to challenge some of the various structures of the dominant European cultural tradition.

Gilles Deleuze (1925–95), a philosopher, and his occasional collaborator, Felix Guattari (1936–92), a political theorist and psychoanalyst, are best known for their attempts to integrate the ideas of Marx and Freud. One of their joint efforts has been described as a 'vast,

chaotic rag-bag of a book' (in Honderich 1995:183), and although they have been welcomed into cultural geography, especially by Shurmer-Smith and Hannam (1994), the long-term significance of their work remains to be determined.

Edward Said (1935–), a literary and cultural theorist, has contributed an influential body of ideas—**Orientalism**—concerning the idea that the Orient is really an invention of those who study it from outside, and that America and Europe employ ideas from within their cultures, ideas such as freedom and individualism, to facilitate their conquest and domination of other regions. Thus, the Orient is a necessary European image of the **other**, a construct of a dominant European discourse, as that term is discussed above. For Said, culture is best seen as a hegemonic environment with specific prevailing modes of thought. The contemporary importance of these ideas is evident in the claim that the United States sees the Arabic world in dehumanized terms, a tendency that is exacerbated by the ongoing conflict between Arabs and Israelis.

Much of the work that is considered part of the cultural studies tradition, including feminism and studies of racism, qualifies as critical theory—a term that refers to the conviction that both the social sciences and humanities need to be emancipatory and not ideological instruments. Critical theories are fundamentally different, epistemologically, from theories in physical science in that they are reflective and not objectifying. There are origins in the Frankfurt school of critical theory, especially the work of the German philosopher and sociologist, Jürgen Habermas (1929–).

lenge to established wisdoms and the significance of postmodernism for social theory is considerable.

- *It effectively rejects all other attempts at social theorizing because of the emphasis that those attempts place on establishing foundations for knowledge.* Social theories developed within modernism (for example, Marxism) are seen as involving a privileging of science

and of the related rise of foundational bodies of knowledge that discuss matters in terms of truth and falsity. Postmodernism, on the other hand, is antifoundational, opposing attempts at grand all-embracing theories, and favoring instead more local and particularistic studies that emphasize difference. This aspect of postmodernism is sometimes rejected as something of a distraction.

Box 7.3: Modernity and Postmodernity, Modernism and Postmodernism

Terms can be very confusing and it is helpful to begin, following Duncan (1996:429), by distinguishing postmodernity and postmodernism (see Glossary). Postmodernity refers to a new phase of history and a new cultural system, while postmodernism is a narrower term referring to a set of ideas, a scholarly movement that opposes the intellectual logic of modernism. Much of the academic literature describes both the historical phase and the scholarly movement employing the one term 'postmodernism', but it is certainly useful to distinguish the two in principle, although it does prove to be rather more difficult in practice. By extension, modernity and modernism can be similarly distinguished from each other.

- Modernity, the modern period, was initiated by the eighteenth- and nineteenth-century processes of increasing industrialization, urbanization, secularization, and nation state creation. Durkheim identified the increasing division of labor as the key to the onset of the modern period, while Tönnies identified the transition from *Gemeinschaft* to *Gesellschaft* (see Glossary). In broad terms, the contemporary world continues to display the basic characteristics of modernity. However, some observers see an emerging phase of postmodernity that is gradually replacing modernity.
- The principal characteristics of postmodernity are those of a postindustrial economy accompanied by globalization processes, a loss of national identities, and a corresponding rise of more local ethnic identities. Harvey (1989) analyzed postmodernity as a condition of society. For many geographers, one appeal of postmodernity is the interest shown in spatial difference as compared to the emphasis on progress through time. Jameson (1991) described postmodernity, using Marxist terms, as a mode of production, a period of **late (taken-for-granted) capitalism** that is

characterized by multinational capitalism and by the commodification of culture.
- Modernism, as a scholarly movement, is associated with the Enlightenment ideals of progress and rationality, with an empiricist epistemology, and with advances in scientific knowledge (see Glossary and Box 2.1). It was in this intellectual context that the various academic disciplines established their place in universities. This is the process of institutionalization described in Chapter 1.
- Postmodernism is a scholarly movement based on a loss of faith in the values and ideas that support the modernist project. Postmodernism emphasizes a blurring of disciplinary boundaries and the rise of new interdisciplinary movements and, in some instances, even argues for a postdisciplinary academic context.

Conflating these ideas, as many scholars do, it can be argued that there is occurring today a significant set of changes such that the modern Western world is in a state of crisis, as reflected by such circumstances as the general weakening of Western political traditions, the lessening of authority of some social institutions, and the resurgence of religious fundamentalism.

Note that, unfortunately, the terms 'modernity' and 'modernism' are inherently flawed because what is modern at one period of time is not necessarily modern at some later period. It is for this reason that the rather clumsy terms 'postmodernity' and 'postmodernism' can be described as following, subsequent to, and even in opposition to modernity and modernism respectively. Note also that the terms 'postmodernity' and 'postmodernism' are problematic because they appear to preclude the possibility of a coexistence of modernity with postmodernity and modernism with postmodernism. The use of the prefix 'post', meaning after or behind, is usually interpreted to mean that one circumstance is replaced by another.

- *Further, postmodernism questions—even denies—the ability of science and theory to establish truth, and also rejects the claim that any such truths, were it possible to attain them, can be liberating.* These are challenging ideas based on the argument that all knowledge, including truth claims, is socially created. Postmodernism proposes that the myths and ideologies of modernism serve only to promote ethnocentrism, specifically Eurocentrism. This postmodernist way of looking at human identity encourages studies of disadvantaged groups such as women, the working class, ethnic minorities, and gays—groups that were previously marginalized or excluded because of their lack of power. In addition to encouraging studies of the disadvantaged and oppressed, postmodernism also asserts that the voices of such groups are as important as the voices of so-called authorities. Authors and authority, as identified in modernism, are not privileged. For a postmodernist, individuals are repositories of cultural systems and therefore unable to speak for themselves; rather, they speak for the cultural system. It is clear from these comments that any adoption of postmodernist claims has some significant implications for cultural geography.
- *Emphasis is thus placed on the idea of social construction—a recurring phrase in much contemporary cultural geography that derives from both feminism, as already discussed, and postmodernism.* Postmodernists share the feminist concern with gender, class, and ethnic identity, recognizing that these are social constructions rather than givens. Accordingly, they can be changed. Like feminists, they thus stress the importance of positionality and the situated character of knowledge. Postmodernists further acknowledge, with feminists, that academic disciplines such as geography are also social constructions that only exist because they were created (and that can therefore be dismantled) by society. Note that this general idea, without the accompanying philosophical argument, was central to the account of the discipline of geography included in Chapter 1.
- The claim that language is at the heart of all knowledge, in the sense that social constructions are reflected in and often maintained through language, means that it is necessary to uncover

the limitations that language places on our thinking and understanding. For this reason, a key feature of the postmodern agenda is the philosophy and method of deconstruction (see Box 7.2). Deconstruction seeks to uncover what is not said in a text and what cannot be said because of the constraints imposed by language. Further, all aspects of a text are necessarily intertextual in the sense that meaning is produced from one text to another rather than being produced between the world and any given text. More generally, this **intertextuality** refers to the idea that all things are related. Any reading of a text, meaning also any view of reality, is as good as any other reading. Deconstruction is preferred to theory construction and, as applied in the context of social science generally, a key purpose of the approach is to reveal the interests served by theories.

Closely related to postmodernism is the **postcolonial studies** tradition, which is concerned with theories of identity, the role played by power, and the production of culture, especially in the less developed world.

Evaluating the Concepts

In addition to the stimulation and challenges implied in both Marxism and feminism, there has been a cultural turn in cultural geography, and together these three philosophical movements have offered new research norms for cultural geography. Two principal new directions are noted.

- First, Marxism, feminism, and the cultural studies tradition have encouraged studies of human identity, human difference, and the politics related to these matters. Many contemporary cultural geographic studies are characterized by a persistent questioning of concepts, classifications, and categories, a situation best described as one of **constructionism**. Identity, for example, is no longer taken to be something that is pre-given but, rather, it is acknowledged to be socially constructed and therefore open to change. Logically then, cultural geographers have developed an increasing sensitivity to culture, especially concerning relations between dominant groups and other groups, and also an awareness of the politics of difference. The role played by power relations is a central feature of all of the approaches discussed in this section. There is also much evidence, contained later in this chapter and in Chapter 8, that contemporary cultural geography is including, even welcoming, the voices of those previously excluded. A humanistic justification for a constructionist view is included in Chapter 8.

- Second, any discussion of ideas and facts is now interpreted with a recognition of the specific discourse being employed; this is the idea of **contextualism**. There is, then, increased awareness of and sensitivity to how knowledge is constructed, by whom, and for what purpose. It is for this reason that much contemporary cultural geography, especially that informed by feminism and postmodernism, explicitly acknowledges the positionality and situatedness of the author.

Your Opinion 7.4

Is there an intriguing ambiguity in the postmodern emphasis on social construction on the one hand and the continuing concern with academic disciplines and disciplinary status on the other hand? Certainly, it was noted earlier that, for some practitioners at least, cultural studies aspires to disciplinary status. More significantly for cultural geography, it is evident that much of the human geographic literature continues to work happily within the idea of the socially constructed human geography discipline as traditionally constituted, while at the same time challenging other social constructions such as those of human identity. This tendency is especially clear in some discussions of race and nation. With reference to the impact of postmodernism, Benko and Strohmayer (1997:xiii) noted: 'more than any other disciplines, with the possible exception of Cultural Studies, it was Human Geography that came away from this encounter with a renewed sense of mission, vindication, even of pride'; the continuing existence of disciplines is explicit and even reinforced by the capitalization used.

There have also been more general expressions of concern about the seemingly ceaseless introduction of new ideas into geography, a process that might be explained in terms of the commodification of knowledge with competition resulting in flexible specialization. Berry (1995:95) referred to a 'combination of self-defeating trendiness and the calculating acceptance of a mainstream agenda by academics more concerned about creating successful niches than about principle', and suggested that postmodernism in particular is fundamentally flawed because a geography without a rational and science-based core cannot accommodate specialties. (See Your Opinion 7.4.)

Regardless, it is the case that the current philosophical and theoretical background that underpins cultural geographic research is multifaceted, complex, and dynamic, perhaps even uncertain and unstable. It has certainly been responsible for a wealth of empirical analyses as discussed below and in Chapter 8.

But these developments, especially those associated with the cultural turn, are not without their critics. Indeed, according to Duncan (1994a), a 'civil war' occurred within cultural geography between Sauerians and new cultural geographers during the 1980s and early 1990s. Further, concerning the cultural turn in social geography, Gregson (1993:527) observed that 'on the one hand social geographers move to deconstruct the socially constructed categories which form the basis for social differentiation (in so doing, frequently criticizing these categories as essentialist), on the other they find themselves having to recognize that these categories are the very ones through which people make sense of themselves and of social life'.

A Divided World: The Idea of Race

A principal concern in the remainder of this chapter is with the relations between concepts of ethnicity, related inequalities, and links to landscape and way of life. In Chapter 8 there is a further discussion of ethnicity, but with an emphasis on identity and symbolic landscape. First, however, in order to inform these discussions effectively, it is necessary to clarify the idea of race and to identify what is meant by such terms as racism, prejudice, and stereotyping, as these are all aspects of the identities that some individuals and groups impose on other individuals and groups.

The Unity of the Human Species

A single and simple fact to begin—there is only one species of humans in the world, namely *Homo sapiens sapiens*.

- It is correct to say that this human species displays variation among its members, but it is biologically incorrect to say that these differences are sufficient to merit making **race**—that is, subspecies—distinctions.
- The biological differences that do exist are the result of human groups being isolated from each other for a period of time—during the global human movements between about 100,000 years ago and about 10,000 years ago as discussed in Chapter 2—such that some selective breeding and some adaptation to physical environments occurred allowing the formation of the groups that are sometimes labeled Negroid, Mongoloid, Caucasoid, and Australoid. This period of time was not, however, sufficiently long to allow the formation of races in the biological sense of genetically distinct subspecies.
- Thus, humans are a biologically variable species, with variability resulting from the same processes that have produced variability in other living things, namely natural selection in different physical environments, but humans cannot be divided into biological subspecies or races.
- Indeed, there is typically more genetic variation within one supposed race than there is between supposed races.
- The statement that all humans are members of the same species is biologically confirmed by the fact that any human male is able to copulate and procreate with any human female.
- In the context of humans, race, then, is not an objective fact.

A History of the Idea of Race

Notwithstanding the fact that there are not distinct human races, it is clear that the supposed existence of such races is an important idea in the contemporary world, and it is this idea—this social construction—that is important in the present

discussion (Figure 7.2). It is also usual to claim that the idea of race is a long-standing one, such that many groups in the past perceived themselves and others as having racial status. But it can be argued that there was no significant racial thought in the European world prior to the late seventeenth century and that the claim that there was such thought is itself largely a product of the nineteenth century. Thus, there is no word resembling that of race in Jewish, Greek, or Roman writings, although there is much evidence of divisions between groups, especially the political division between the civic and the barbarous. For Hannaford (1996:12), the Greek 'political idea involved a disposition to see people not in terms of where they came from and what they looked like but in terms of membership of a public arena' and it therefore 'inhibited the holding of racial or ethnic categories as we have come to understand them in the modern world'. The European idea of race, as opposed to this political idea, was initiated because of four developments.

- First, polygenetic theories of human origin. These were in opposition to the previously dominant Augustinian view of a single human origin and appeared following the Reformation.
- Second, the process of European overseas expansion. It seems likely that—because some of the people encountered by Europeans in the context of overseas movement really did look different— the idea that races existed seemed logical.
- Third, the rise of natural history and related attempts to classify humans. Thus, there were major advances in the scholarly classification of humans, such as those by Carolus Linnaeus (1707–78), Johann Friedrich Blumenbach (1752– 1840), and Buffon, but without any values necessarily being attached.
- Fourth, Enlightenment philosophers such as Montesquieu paved the way for environmental racial theories. Overall, the period from the late seventeenth to the early nineteenth century witnessed the rise of the idea that civilization advanced not through the 'public debate of speech-gifted men and the reconciliation of differing claims and interests in law but through the genius and character of the *Völker* naturally and

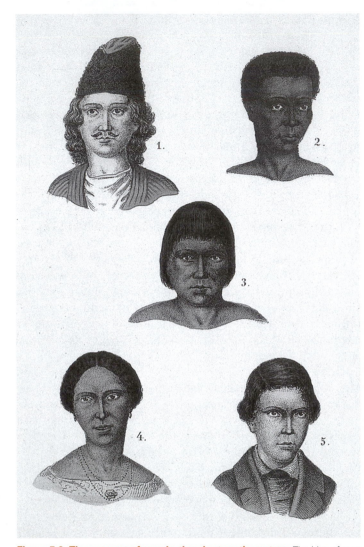

Figure 7.2: The concept of race in the nineteenth century. The idea of race was routinely accepted by scientists in nineteenth-century Europe and this figure shows the frontispiece of one typical volume on natural history, *Martin's Natural History*, First Series, published in 1862. Five principal races and several mixed races are identified in the book. The illustration shows three of the principal races and two of the mixed races as follows: (1) the White or Caucasian race, (2) the Black or Negro race, (3) the American or Red race, (4) Mulatto, and (5) Mestizza. The other two principal races identified in the text are the Yellow or Mongolian race and the Brown or Malayan race.

Source: Sarah A. Myers, trans., *Martin's Natural History*, First Series (New York: Blakeman & Mason, 1862):frontispiece.

But it was nineteenth-century Europeans who created a history of **racism** through a reinterpretation of the past using the new language—discourse—of national identity, progress, and Social Darwinism. Barthold Georg Niebuhr (1776–1831) rewrote the history of Greece and Rome using racial rather than political classes, while Augustin Thierry (1795–1856) performed a similar task for French history. Together, these works prompted the idea of an original European Aryan culture superior to both Greece and Rome. More generally, the tendency for most cultures at most times to dislike and even fear strangers and those who are in some way different, as noted in Chapter 2 and as discussed in the following section, was confused with racism. Critically, the new doctrine of nationalism was supported by the idea of race, with group membership contingent upon such characteristics as language and skin color. Nations, whether real or imagined, were replacing religion as the basis for group identity. Racial classifications, now including hierarchies, were proposed by many scholars.

- Count Arthur de Gobineau (1816–82), in his *Essay on the Inequality of the Human Races* (1853–5), classified Europe into races with Germanic people as the superior group, doubted the ability of Africans and Native Americans to become civilized, and generally saw race as the sole explanation for progress.
- But the most influential and widely accepted work was the 1899 study of civilization, *The Foundations of the Nineteenth Century*, by Houston Stewart Chamberlain (1855–1927), a work that argued that race explains everything and that saw civilization as the sole prerogative of the German race. A typ-

biologically working as an energetic and formative force in the blood of races and expressing themselves as *Kultur'* (Hannaford 1996:233).

ical sliding scale of human groups that equated ability and culture with physical characteristics saw Nordic or Teutonic peoples at the top, followed by Alpine, Mediterranean, Slavic, Asiatic, and African groups.

One component of these classifications and rankings was the explicit distinction made between Germans as Christians and Jews. Jews in the West were a well-defined group traditionally excluded from the state, but political anti-Semitism developed only in the late nineteenth century in relation to these race studies. Certainly, by the early twentieth century race was a dominant organizing idea and it was generally understood that humans could be appropriately classified and ranked according to appearance. In Germany, these ideas contributed to the program of the National Socialist Party, which declared that only members of the German race could be citizens of the German state, and also to the activities of that party before and during the Second World War.

But racist ideas, with roots in Enlightenment and nineteenth-century thought, were not influential only in Germany. Rather, they were widely accepted and implemented in various ways, both practical and scholarly, throughout much of Europe and North America. Nationalism, **colonialism**, and **imperialism** were all supported by the idea of race, as **neocolonialism** continues to be at the present time. Thus, racist immigration policies were introduced in the United States and in many of the European colonies, while the study of eugenics was widely pursued.

It is also not insignificant that the various social science disciplines were institutionalized at a time when race was the organizing idea, claiming precedence over all previous formulations of nation and state. Certainly, geography, anthropology, sociology, and psychology were all caught up with the idea of race. In geography, for example, Ratzel was a founder of a German journal devoted to the study of race, and also popularized the idea of **lebensraum**, a concept referring to the living space required by a political state.

Correcting a Misunderstanding?

The suggestion that the idea of race was unscientific was stated by the anthropologist Boas (1928) and more fully detailed by Huxley and Haddon (1936), who argued that the concept was fundamentally flawed and should not be used. The most important such statements were produced under the auspices of the United Nations after the Second World War. But these studies exposing the falsity of the idea of race did not prevent the rise of scholarly race relation studies that typically were critical of racism, but at the same time accepted the idea of race as a legitimate basis for analysis. Thus, sociological analyses were initiated in the 1920s by Park, and by the Swedish economist Gunnar Myrdal (1898–1987) who produced a major study of race relations in the United States (Myrdal 1944). In this way, the study of race relations was established as a legitimate component of the social studies curriculum, including human geography, and it has remained so through to the present. These and similar studies also contributed to the race relations legislation that was produced in many Western countries and to the racial labels used for various official purposes. (See Your Opinion 7.5.)

Race may not be a fact, but racism most certainly is. Racism is a particular form of prejudice, typically incorporating a prejudgment of others. **Prejudice** refers to the holding of negative opinions about another group, negative opinions that are based on inaccurate information and that are extremely resistant to change. Racism similarly incorporates a schema or blueprint about others, and a **stereotype** is a set of beliefs and expectations about another group that is difficult to change even where there is clear evidence that the stereotype is incorrect.

Racist Geography: Apartheid in South Africa

Before commencing this account of the racist geography of South Africa prior to the early 1990s, it is worth noting that there have been few discussions by geographers of the most notorious of racist geographies, namely the policies and activities of the National Socialist Party in Germany before and during the Second World War that culminated in the killing of some 6 million Jews and also large numbers of other people who were considered undesirable.

The concern in this discussion is with the policies introduced and practices employed by Europeans in South Africa to ensure that they were able to retain power in a country in which, unlike most other mid–latitude settler colonies, they were a minority of the total population. Although there has been considerable debate about the reasons for the emergence of **apartheid**, especially between liberal and Marxist scholars, the present account elects to situate the issue in the larger context of the discussion of the idea of race as presented above.

GROUP IDENTITY IN THE SOUTH AFRICAN CONTEXT

During the extended period of European overseas expansion from the fifteenth to nineteenth centuries, there were numerous encounters between incoming European groups and indigenous populations, a topic that was addressed in Chapter 3 in the specific context of cultural contact and change. The indigenous population of South Africa included several Khoisan groups located primarily in the western part of the country and also Bantu groups in the east and north. The Dutch East India Company established a settlement at the southern tip of South Africa in 1652, but there was little subsequent immigration. The area became British in 1806 and, although there was some British immigration, the area never proved to be as attractive a destination as the various other British mid–latitude possessions. In the late seventeenth century, Dutch settlers had begun to move inland across the western area occupied by the Khoisan group, and this expansion and related introduction of disease prompted indigenous population decline that was

---

Your Opinion 7.5

Clearly, a principal difficulty with many of the attempts to combat racism concerns the continuing use of the term 'race', a practice that can be interpreted as implying an objective reality, notwithstanding that it is now conventional to note that '"race" is a social construction rather than a natural division of humankind' (Jackson 1987:6). This practice is especially popular among more conservative scholars (see, for example, Sowell 1994). It can certainly be argued that the insistence on perpetuating the use of the term 'race'—albeit with a different meaning in mind from that of most earlier writers—has contributed to the contemporary situation whereby many groups of people elect to identify themselves in racial or ethnic terms. In this context, Bonnett (1997:198) noted that there is a tradition within the discipline of geography of 'eliding issues of "race" with "marginal" and "subjugated" peoples, most particularly non-Whites'.

---

similar to other mid–latitude areas such as the United States. But the Bantu groups in the north and east were less susceptible to the imported diseases, and a series of conflicts occurred between the Bantu and both the migrating Dutch (known as Afrikaners) and the incoming British. There were also ongoing conflicts between the Afrikaners and the British colonial authorities, especially related to the nineteenth–century formation of Afrikaner republics in the north. Further, as occurred in several other British colonial areas, indentured laborers were brought from India, most of whom worked in the sugar cane plantations in Natal colony in the northeast.

Thus, South Africa included several groups and a series of hostilities, and the creation of the Union of South Africa from four British colonies in 1910 was really a compromise between the two powerful but quite different European groups. In 1910 the population was about 5 million of whom about 1 million were European, although the proportion of Europeans has steadily declined to about 12 per cent at the end of the apartheid era in 1994. The apartheid era had begun in 1948 with the election of the Afrikaner–dominated Nationalist political party and at that time about 55 per cent of the Europeans were Afrikaners.

A key consideration in any account of racist geographies concerns the self-conception of the dominant group and the identities that were assigned by that group to subordinate groups. Hence, it is important to ask: Who were the Afrikaners, the self-declared 'White Tribe' who assumed control of the country in 1948?

As noted, Dutch settlers slowly moved inland over considerable distances, establishing an extensive farming economy, such that a distinctive Afrikaner identity is generally considered to have been initiated during this almost 250-year period of movement. Key factors involved were the process of land acquisition and settlement, opposition to both Dutch and British colonial authorities, isolation from others, a Calvinist religious tradition, the Dutch-based Afrikaans language, conflict with Bantu groups, republic formation, and war with the British (1899–1902). Lester equated the rise of the Afrikaner identity with nineteenth-century nationalist movements, noting that there was an:

> . . . implicit externally derived, but internally manifested, threat to the culture, especially religion and/or language of a body of people; a constructed but widely perceived history, the interpretation of which rests on the shared experience of 'the group' as opposed to alien 'others'; a threat to the economic fortune of most members of this historically defined group; and the creation of a set of symbols by which the group is demarcated and exhorted to act (Lester 1996:60).

For Afrikaners, it is critical to stress that the self-conception included both a distinctive culture and a pure race, although the introduction of apartheid after 1948 necessarily involved grouping all White people—Afrikaners, British, and other Europeans—together. Two quotes serve to present the Afrikaner rationale for the implementation of apartheid.

- The Minister of Justice stated in 1953: 'If a European has to sit next to a non-European at school, if on the railway station they are to use the same waiting rooms, if they are continually to travel together on the trains and sleep in the same hotels, it is evident that eventually we would have racial admixture, with the result that on the one hand one would no longer find a purely European population and on the other hand a non-

European population' (quoted in Christopher 1994:4).
- The Minister of Labour stated in 1954: 'I want to say that if we reach the stage where the Native can climb to the highest rung in our economic ladder and be appointed in a supervisory capacity over Europeans, then the other equality; namely political equality, must inevitably follow and that will mean the end of the European race' (quoted in Christopher 1994:2).

DIVIDING PEOPLE, DIVIDING SPACE

It is not easy to convey the enormity of the apartheid endeavor, this 'uniquely selfish and degrading concept' (Christopher 1994:8). Box 7.4 outlines the key legislation that was designed to ensure that both social and spatial separation was complete, to ensure that the politically and economically dominant White group was able to retain power over the country given increasing demands for political rights by the Black majority. The basic mechanisms devised to ensure the desired survival of the White race were the removal

A beach sign during the apartheid era made it plain that non-Whites would not be welcomed (*William Norton*).

of anyone who was not considered to be White from those areas deemed to be for the occupation or enjoyment of the White population and the elimination, so far as was feasible, of all contact between Whites and others.

The key first step in the implementation of apartheid was to formally classify all individuals according to their supposed racial type, hence the 1950 Population Registration Act. Inevitably, the task of assigning a racial identity was controversial in many specific cases, and Figure 7.3 shows the changes in racial description that occurred between 1983 and 1990 as a result of appeals. As the legis-

lation listed in Box 7.4 indicates, there were three levels or spatial scales.

- First, there was an ambitious attempt to redraw the political map through the removal of all Black areas from the state and the creation of independent Black homelands. An indication of the population movements involved as one stage in this process is evident from Figure 7.4. This process was begun but incomplete at the time of the collapse of apartheid in the early 1990s. When it was clear that the apartheid system was crumbling, there were several proposals to partition

Box 7.4: Segregation in South Africa: Legal Milestones

In South Africa, as in most other colonial areas, forms of spatial and social segregation in both rural and urban areas were evident from the early days of European involvement such that the development of apartheid was rooted in the colonial era. This colonial segregation was intensified and formally institutionalized by the dominant European group beginning in 1910, after the colonial period, when the four former colonies united as the Union of South Africa. The principal acts were:

- *Natives Land Act, 1913*: An act that was regarded as the first stage in the establishment of a permanent line between Europeans and the indigenous population; identified those rural areas assigned to the indigenous population.
- *Natives (Urban Areas) Act, 1923*: Required local authorities to establish separate locations for the indigenous population.
- *Natives Administration Act, 1927*: Effectively made the areas designated for the indigenous population subject to a different political regime from that of the rest of the country.
- *Slums Act, 1934*: Enabled demolition of inner city suburbs and displacement of any indigenous population residents.
- *Native Trust and Land Act, 1936*: Extended the rural area assigned to the indigenous population.

In addition to the above and numerous amendments, there were other acts for specific regions to limit the locational choices open to Asian and Coloured populations. Beginning in 1948, with the election of two allied parties that united to form the National Party in 1951, the policy of apartheid, total segrega-

tion, was introduced, building on the segregation of the colonial period and the Union period. The principal acts were:

- *Prohibition of Mixed Marriages Act, 1949*: Prohibited marriages between Whites and others.
- *Immorality Amendment Act, 1950*: Banned extramarital relations between Whites and others.
- *Population Registration Act, 1950*: The cornerstone of apartheid. Provided for compulsory classification of all individuals into what the government considered to be distinct racial categories. Initially three groups were identified—White, Black, and Coloured, with the Coloured group subdivided into Cape Malay, Griqua, Indian, Chinese, and Cape Coloured.
- *Group Areas Act, 1950*: Designed to effect the complete urban spatial segregation of the designated population groups, with respect to both residences and business.
- *Reservation of Separate Amenities Act, 1953*: Designed to create separate social environments for the various population groups, the so called petty apartheid.
- *Promotion of Bantu Self-Government Act, 1959*: Created a hierarchy of local governments for rural reserve areas.

The ending of apartheid took most observers by complete surprise. During the 1980s, there was a gradual dismantling of petty apartheid. A series of political developments were followed by a referendum in 1992 that resulted in over two-thirds of voters (all designated as Whites) accepting the principle of an undivided South Africa and rejecting apartheid. The first democratic government was elected in 1994.

South African township housing for Blacks effectively segregated entire populations and were a stark contrast to better quality housing for Whites (*William Norton*).

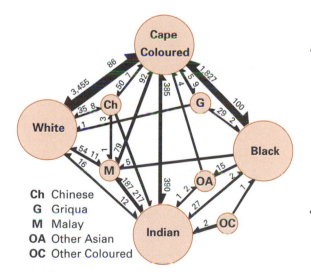

Ch Chinese
G Griqua
M Malay
OA Other Asian
OC Other Coloured

Figure 7.3: Changes in race classification, South Africa, 1983–90. Classification was based on physical appearance and social acceptability, with the general intent of ensuring that those who were definitely not White would not be classified as White. Appeals were permitted, and this figure indicates the 7,000 successful appeals between 1983 and 1990.

Source: A.J. Christopher, *The Atlas of Apartheid* (New York: Routledge, 1994):104.

South Africa, and Figure 7.5 indicates proposals from right-wing White political groups.

- The second spatial expression of apartheid concerned urban areas. The aim was to restructure these to ensure complete segregation and to restrict Blacks' access to urban areas. Certainly, the colonial period and the Union period before 1948 had witnessed the evolution of a partially segregated city, but the aims of apartheid were much more comprehensive. One proposal for the structure of a segregated urban area is shown in Figure 7.6.

- Third, personal/social apartheid included a myriad of regulations intended to minimize contact between members of the White group and others. Two examples of many suffice to make the aim of this form of apartheid clear. Thus, as evident from Figure 7.7, homes for Whites included separate quarters (usually one small room and a washroom for a Black servant), while the insistence on separating Whites and others included the identification of segregated beach areas (Figure 7.8).

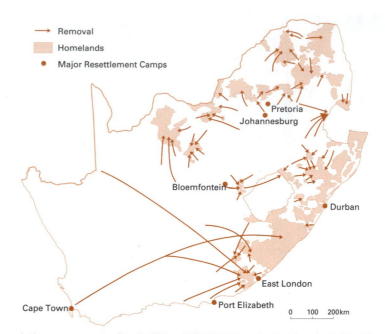

Figure 7.4: Forced population movements, South Africa, 1960–83. It is estimated that about 1.7 million people were resettled in designated homeland areas between 1960 and 1983. Most of those moved were Blacks who were regarded as surplus to the labor requirements of the White farming areas.

Source: Adapted from A.J. Christopher, *The Atlas of Apartheid* (New York: Routledge, 1994):83.

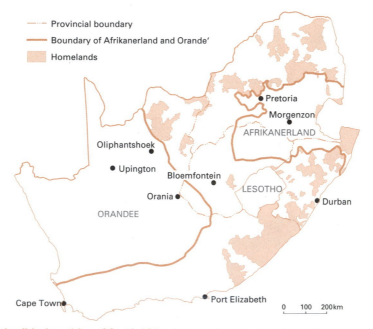

Figure 7.5: A proposed political partition of South Africa. Proposals by right-wing White political groups for the creation of specifically Afrikaner states appeared toward the end of the apartheid era. This figure shows two proposed states, Orandee, an isolated desert state, and Afrikanerland, a state that included much of the agricultural and industrial heartland.

Source: Adapted from A.J. Christopher, *The Atlas of Apartheid* (New York: Routledge, 1994):99.

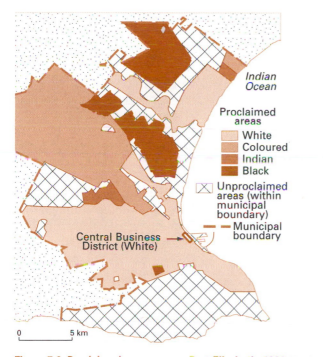

Figure 7.6: Proclaimed group areas, Port Elizabeth, 1991. The final proclamation concerning group areas in the city of Port Elizabeth. Together, figures 7.4 and 7.6 are suggestive of the enormity of the apartheid project.

Source: Adapted from A.J. Christopher, *The Atlas of Apartheid* (New York: Routledge, 1994):111.

An analysis of apartheid as it affected White and Coloured populations in the area of Cape Town stressed the interactions between space and society, with Whites manipulating space in order to manipulate the larger South African society, and with various human social relations also affecting the organization and use of space (Western 1981:6–7). This analysis detailed the manner in which apartheid policies were introduced by and for Whites, the benefits that Whites derived from measures such as the Group Areas Act, the difficulties imposed upon others by those same measures, and the hostilities that were generated between members of different groups as those groups were defined by Whites.

As a concluding comment, it should not be thought that the extreme of apartheid could only have occurred in South Africa. Box 7.5 details some provocative ideas concerning the possibility that the United States might

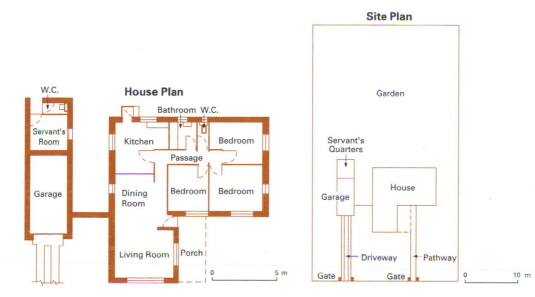

Figure 7.7: Typical South African house plan in the apartheid era. Residential segregation was enforced within the confines of each individual White residential lot.

Source: A.J. Christopher, *The Atlas of Apartheid* (New York: Routledge, 1994):142.

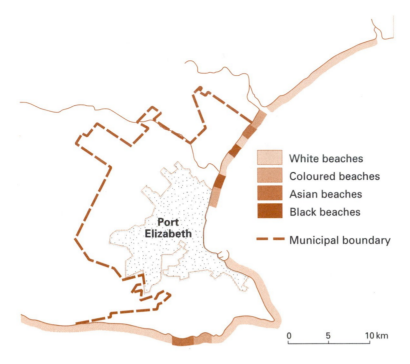

White beaches

Coloured beaches

Asian beaches

Black beaches

— — Municipal boundary

0   5   10 km

Figure 7.8: Beach zoning under apartheid, Port Elizabeth. Most of the beach areas, including the better beaches, were reserved for Whites. Together, figures 7.7 and 7.8 suggest the absurdity of the apartheid project and the humiliations it involved.

Source: Adapted from A.J. Christopher, *The Atlas of Apartheid* (New York: Routledge, 1994):146.

have adopted segregationist and apartheid policies similar to those employed in South Africa had the demographic circumstances been comparable to those in South Africa.

A Divided World: Ethnicity

The study of the interrelations of the idea of race and ethnicity has inspired a substantive body of work. In cultural and social geography, as suggested earlier, there is a considerable literature on the theme of 'race and racism' with the principal research tradition—dating from the 1960s and linked to sociological analyses of race relations—concerned with descriptions of immigration, residential segregation, **ghetto** formation, and assimilation, especially with reference to North American and British cities. Rather unfortunately, in this tradition there has been a tendency to elide the terms 'ethnicity' and 'race', and to refer to the

perceived links between culture and genetics. Certainly, notwithstanding the inherent flaws, the idea of race linked with that of ethnicity has proven to be a difficult idea for cultural geographers to ignore. Probably the most substantial area of work linking the two relates to the concept of a **plural society**. Initially developed for the study of colonial societies in southeast Asia, this concept distinguishes between societies that are culturally homogeneous with normative integration and societies that are deeply divided on an ethnic basis and held together only through strong state control.

Understanding Ethnicity

The current discussion of group identity uses the term 'ethnic' as an admittedly uncertain but assuredly more viable alternative than the term 'race'. The preceding discussion of segregation in South Africa could have been considered under the heading of ethnicity, but it was considered more appropriate to identify the example as an application of the idea of race given the specific motivations of the dominant group involved in the implementation of segregationist policies and practices.

Several of the studies that identified the unscientific status of race proposed the use of the term 'ethnic' as an alternative to race. Accordingly, the concept of ethnicity might be appropriately described as having a rather difficult scholarly origin. Add to this circumstance some considerable debate about the meaning of the term, and it is fair to say that the fact that there is not a generally accepted definition of the term 'ethnic' is both understandable and an understatement. According to Rupesinghe (1996:13); 'ask anyone to define ethnicity and the problem begins. We are left with a host of interpretations. The difficulty in defining ethnicity is that it is a dynamic concept encompassing both subjective and objective elements. It is the mixture of perception and external contextual reality which provides it with meaning.'

Given these difficulties, it is hardly surprising that an analysis of sixty-five studies of ethnicity noted that fifty-two of them offered no explicit definition (Isajiw 1974). Broadly speaking, minority groups are often defined as ethnic, especially in the North American immigrant context, and Raitz (1979:79) noted that in this capacity, 'ethnics are custodians of distinct cultural traditions'. This is the interpretation traditionally employed by some governments. In Canada, for example, a question

on ethnic ancestry, not ethnic identity, is included in the Census.

But a central consideration in much contemporary social science concerns the question of group membership, usually discussed with reference to the term 'identity'. This issue has been raised on several prior occasions in this textbook, but the tendency earlier was to treat group membership in essentialist terms; that is, as a given (but see the discussion of the Hispanic group initiated by Blaut 1984 and included in Chapter 4). The current interest in ethnic identity raises a very challenging question, representing a specific re-

Box 7.5: What Might Have Been? Part II

Separation of population groups was a feature of many areas of European overseas expansion during the colonial era and also subsequent to that era if a European group achieved political dominance. In all cases a key rationale for separation was the maintenance of European privilege and the furthering of European interests at the expense of the other groups. A brief consideration of the southern United States as it compares to South Africa is interesting in this context.

In both of these areas, the term 'segregation' came into popular usage in the early twentieth century, but the details of implementation of the idea were quite different. In the case of the southern United States, there were two principal elements to segregation before the 1960s, namely legalized social separation comparable to the rather unfortunately named petty apartheid of South Africa, and exclusion of one group from the electoral system. There was no United States equivalent to the acts identified in Box 7.4. Indeed, after the Civil War the southern United States, despite much discrimination and both *de facto* and *de jure* separation, was characterized by considerable integration. The principal reasons for the different experience in the southern United States were that the subordinate group was a minority, not an overwhelming majority, was culturally similar to the dominant group, and was a part of a democratic state and not a conquered people.

There were, however, some proposals in the southern United States to adopt a segregation system similar to that developed in South Africa. One proposal after the Civil War involved an explicit spatial separation of groups, while a second proposal favored implementing policies along the lines of the South African 1913 Native Lands Act. The circumstances of the two cases are sug-

gestive of the value of counterfactual thinking, as this line of reasoning was outlined in Box 6.9. Further, with reference to American Indians, Fredrickson wrote:

> Assume for a moment that the American Indian population had not been decimated and that the number of European colonists and immigrants had been much less than was actually the case—creating a situation where the Indians, although conquered remained a substantial majority of the total population of the United States. After the whites had seized the regions with the most fertile and exploitable resources, the indigenes were consigned to a fraction of their original domain. All one has to envision here are greatly enlarged versions of the current Indian reservations. Then suppose further that Indians were denied citizenship rights in the rest of the country but nevertheless constituted the main labor force for industry and commercial agriculture. It is hardly necessary to continue; for one immediately thinks of the kind of devices a white minority might then adopt to insure its hegemony under conditions where a majority of the Indians in fact work off the reservation and even outnumber the whites outside these designated areas. The twin objectives of white supremacy would then be, as in South Africa, to maintain direct minority rule over most of the country and some kind of indirect rule over the reservations, while at the same time providing for a controlled flow of Indian workers for industry and agriculture in the white regions (Fredrickson 1981:246–7).

In brief, what happened in South Africa might easily have happened elsewhere had the demographic circumstances been comparable.

surfacing of the debate concerning essentialism and constructionism. Recall that essentialism involves the claim that a specific identity, group membership, is basically unchanging, whereas constructionism involves the claim that identity is fluid, contested, and negotiated. Much work in contemporary cultural geography accepts that identities are socially constructed, an idea that is at the forefront of this discussion of ethnicity.

Since the 1980s there has been increased attention paid to ethnicity, especially as it relates to the political dimension of group identity—what is sometimes referred to as identity politics, a theme that is more fully addressed in Chapter 8, particularly Box 8.3. For cultural geographers, the six components of ethnicity identified by Smith (1986:21–31) appear most helpful.

• First, an ethnic group has an identifying name. This is a means by which people are able to distinguish themselves from others and to effectively confirm their existence as a group. The symbolic importance of naming is clear, suggesting as it does the reality of their identity and their difference. The related question concerning the importance of naming places as one means of indicating ownership and authority is discussed in Chapter 8.
• Second, an ethnic group has a common myth of descent. This may take the form of myths of temporal origins and spatial origins.
• Third, and closely related to the second component, there is a shared history. An ethnic group is a historical community that always relies on shared memories to unite successive generations. Each new experience is added to the previous history and interpreted and understood in terms of that history. Necessarily, such a history is but one reading of the past, a reading designed to reinforce the identity of the group.
• Fourth, an ethnic group has a distinctive and shared culture as reflected in lifestyles and values. This may take the form of a common language and religion along with, for example, particular customs, institutions, folklore, architecture, dress, food, and music.
• Fifth, an ethnic group is linked to one or more places. Most potently, there is a link to a particular territory, a place that is seen as a homeland. The group may occupy this territory, may aspire to occupy this territory, or may only have recollections of this territory. Especially meaningful places may be contained within, or may even lie outside, this homeland.
• Sixth, an ethnic group is a community that possesses a real and meaningful sense of self-identity and self-worth such that other ways of dividing people, for example, according to social class, are not able to prompt meaningful divisions within the ethnic group. Group identity also involves a belief that all members share a common fate and that individual identity is inseparable from the larger group identity—circumstances that facilitate the process of mobilizing group members and strengthening group solidarity as needed. Such ethnic solidarity is most evident during times of stress when the identity, perhaps even the very existence, of the group is challenged.

Reflecting on these six criteria, and as noted in Chapter 4 in a rather different context, Smith (1986:32) defined ethnic groups as 'named human populations with shared ancestry myths, histories and cultures, having an association with a specific territory and a sense of solidarity'. What this discussion of the meaning of ethnicity suggests is that the concept might be usefully employed by cultural geographers as a basis for group delimitation. Incorporating the conventional variables of culture as interpreted in the classic social science sense as way of life, of language, and of religion, along with a spatial component and a sense of identity, ethnicity emerges as a useful, almost catch-all, concept. Needless to say, however, the term does have some unfortunate overtones. Most obvious is the fact that it tends to be employed in many cases as a surrogate for the idea of race. Related to this interpretation is the often explicit use of the term to delimit groups in opposition to others. These difficulties are evident in the following discussions of empirical work that consider questions of ethnic group identity in conjunction with ethnic group difference and inequality in the context of, first, language and religion, and, second, national and local identities.

This extraordinarily long name of this Welsh town means 'St Mary's Church in the hollow of the white hazel near a rapid whirlpool and the Church of St Tysilio of the red cave'. It is often shortened to Llanfair P.G. (*Wales Tourist Board Photo Library*).

Language and Religion

It has been widely accepted that language and religion are two of the bases that groups employ to assert their ethnic distinctiveness and, in some instances, their national identity. Indeed, certainly in cultural geography, it has been usual to regard language and religion as the most important variables that unite groups—recall the discussions of landscapes and regions in Chapter 4 (for a brief overview of language and religion in a global context, see Box 4.5). The presumed importance of language as an indicator of cultural or ethnic or national identity is noted in Box 7.6 in the larger context of the rise of English as a global language. Language serves as a principal means by which cultures persist through time, while religion is often the repository for important beliefs, attitudes, and traditions that serve to unite a group.

LINKS TO ETHNICITY?

Interestingly, however, although language and religion have long been seen as distinguishing marks of an ethnic group, it can be argued that both are increasingly irrelevant in this respect. This is a suggestion that is, of course, in accord with the emphasis that contemporary cultural geographers place on the idea of multiple and flexible identities, but that is not in accord with the comments about Welsh and Irish languages and identities included in Box 7.6. It is also a suggestion that implies that policies favoring a minority language at the expense of some other language, as in the case of Quebec, may be misguided if the ultimate goal is retention of group identity. In brief, although traditionally they play an important role in establishing the identity of ethnic groups, both language and religion may not be critical components of ethnic identity in the contemporary world.

Indeed, language, long held to be the principal basis for defining ethnicity, may be divisive for the sense of ethnic community. In the case of Wales, for example, English-speaking people in south Wales may sense a Welsh identity similar to Welsh-speaking people in parts of north Wales. Similarly,

Scotland has a distinctive English dialect in the Lowlands, known as Lallans and sometimes considered to be a separate language, and a Gaelic-speaking Highlands, such that a continuing sense of identity may be explained in terms of the Presbyterian religion and the particular Scottish legal and educational systems rather than in terms of language.

Further, in Scotland there is census evidence to suggest that there is currently a decline of the Gaelic language. But this decline in the number of Gaelic speakers—this apparently objective indicator of ethnic identity—is accompanied by a Gaelic cultural revival as expressed by television programs, educational changes, and attendance at festivals. This seeming contradiction might be explained by reference to measures of cultural identity other than language: in other words, it is unclear as to whether language really is an appropriate indica-

tor both of the health of Gaelic culture and of identity with the cultural group.

Similarly, religion is not necessarily a good indicator of an ethnic group. Although it is important as a symbolic code of communication and as a focus for social organization, especially prior to the onset of an industrial way of life, religion often transcends conventional ethnic boundaries. However, religion may be closely related to ethnicity in at least three important ways:

- There is often a close relationship between a religious creation myth and the origin myth of an ethnic group.
- As noted in Box 4.5, there are some religions that are classed as ethnic, meaning that they are associated with a specific group of people, principal examples include Hinduism, Judaism, Taoism, Confucianism, and Shinto.

Box 7.6: English as a Global Language?

'Why a language becomes a global language has little to do with the number of people who speak it. It is much more to do with who those speakers are' (Crystal 1997:5). For a language to achieve global status, it needs to be spoken by large numbers of people as a mother tongue in some countries, to be an official language in other countries, and to be a priority in foreign language teaching in other countries. English qualifies on all three counts: it is spoken by many people as a mother tongue in Australia, Britain, Canada, Ireland, New Zealand, South Africa, the United States, and some Caribbean countries; it is an official language in the sense that it has some special status in over seventy countries (more countries than for any other language), including India, Ghana, and Nigeria; and it is taught as a foreign language in over 100 countries, again more countries than for any other language. Together, these three circumstances are resulting in increasing numbers of English speakers—a late 1990s estimate was between 1.2 and 1.5 billion.

Language, of course, does not have some existence independent of those who speak the language, such that any decline or increase in numbers of speakers has much to do with the cultural, economic, and political success of the speakers of the language. Thus, the historical importance of Latin is explained in terms of, first,

the power of the Roman empire, and, second, the power of Roman Catholicism. In the case of English, Britain was a leading world power during the critical period of the spread of capitalism, industrialization, and global migration. English, it can be suggested, was simply in the right place at the right time.

Is it possible that the rise of English as a global language is contributing to the death of some other languages? The evidence here is contradictory. It is probable that several thousand languages have appeared and disappeared, but most of the disappearances were related to the demise of a particular group of people rather than to the spread of a global language. Certainly, it has been usual to claim that retention of language is central to the survival of a distinctive cultural or ethnic group, and there has been much interest in the circumstances of a **minority language** in this context. Thus, Aitchison and Carter (1990) stressed the threat to the Welsh language posed by English. Similarly, with specific reference to Irish, Kearns (1974:86) claimed: 'That a separate language contributes to the imprint of a distinctive nationality is widely accepted.' However, a rather different interpretation of the role played by language in the preservation of a distinct identity is possible, as noted in the text discussion of Scottish Gaelic.

- The structure and organization of a religion may provide mechanisms that serve also to facilitate communication of ethnic myths and symbols.

There has been little explicit discussion of these matters in the cultural geographic literature, although Stump (1986:2) noted that religion was 'important in identifying and understanding the diffusion, distribution and character of groups defined by their social, ethnic or regional identity'.

UNDERSTANDING OTHERS

In some instances, a particularly sensitive issue concerning religion and ethnicity concerns the perception that one religious group has of another. Recall that it was suggested in Chapter 4 that the contemporary world is characterized by a clash of civilizations, especially as these are associated with religious beliefs and attitudes. Most notably, there is currently a conflict between Islam and Christianity, a conflict that assumes cultural, political, and economic dimensions.

According to Mazrui (1997:118), the Islamic world is perceived by the Christian world, especially by the United States, 'as backward-looking, oppressed by religion, and inhumanely governed', although it is in fact 'animated by a common spirit far more humane than most Westerners realize'. Recall the reference to the ideas of Said in Box 7.2 concerning the way in which dominant groups perceive other groups. It can be argued that some differences between the Christian and Islamic worlds concerning, for example, population fertility policies reflect the specific timing of cultural change rather than fundamental disagreements about quality of life. Mazrui identified numerous ways in which the Islamic world might be considered more humane than the democratic Western world: for example, Islam has always been more

protective of minority religions; it has not been hospitable to either fascist or communist governments; it has not witnessed policies of genocide such as the Holocaust; and it has been resistant to racist policies such as apartheid. Certainly, it can be argued that there is a difference between democratic principles and humane principles. (See Your Opinion 7.6.)

National and Local Identities

The above discussion of language and religion as they relate to ethnicity necessarily made reference to matters of identity, including the identity of the **nation**. Certainly, **nationalism** is now often seen as a concern with an imagined community, reflecting a constructionist rather than an essentialist view. Hudson and Bolton (1997:1) employed the term 'fabulous beast' in reference to the Australian national identity, noting that any one supposed identity was necessarily exclusionary because it ignored the reality of multiple identities, including those of groups defined ethnically and sexually.

However, it might be argued that this idea of imagined communities, while attractive, is usually obvious and often empirically unsatisfactory—obvious in the sense that all feelings of belonging are socially constructed, and empirically unsatisfactory as it appears to ignore the fact that many nationalisms, such as those of the United States and Japan, are products of shared history and shared projects. Furthermore, it may be that the **nation state** is not in a privileged position as a unit of analysis: 'The state appears to be more movable, malleable, and contestable than ever. The state consists of a set of institutions that are authoritative only because other sets of institutions recognize them; it has been socially constructed within historical and interpretive social processes and practices' (Wilmer 1997:3). Given this

---

Your Opinion 7.6

Do you agree that these comments about Islam and Christianity are important because they highlight the dangers of ethnocentrism especially clearly? One obvious danger related to the stressing of ethnic identity, especially in religious terms, is the tendency to see one group as inherently better than other groups—this is, of course, an attitude that can easily be formulated in racist terms. Do you agree with Mazrui (1997:118) that: 'In the end, the question is what path leads to the highest quality of life for the average citizen, while avoiding the worst abuses'?

debate about the identity of the nation, two further ideas might be helpfully added to the six components of ethnicity identified earlier in this discussion.

- First, the idea that groups have an identifying name implies something more than that the group has a sense of self, of identity; it also implies that the group is in some sense secure such that any threats to the sense of identity may elicit a defensive response in order to avoid psychological or physical annihilation. Some psychoanalytic ideas, including object relations theory, suggest that the self is bounded and identified with reference to positive images of who we are along with negative images of who we are not. Because this is an idea that can be applied at both individual and group scales of analysis, it is important in the context of ethnicity. Thus, national and other more local identities may be framed in terms of opposition to others. In addition to this group perception of self, there is also tension between national ethnic identity and other more local ethnic identities, a circumstance that is reflected in some accounts of English and Scottish identity.

- Second, nationalism can also be seen as a form of cultural capital; that is, the knowledge, attitudes, behaviors, and styles that characterize a group. This cultural capital occupies a space, a homeland as this term was employed in Chapter 4. It might be argued that homelands can be seen as motherlands, in patriarchal terms, because they offer group members protection and emotional security, whereas the concept of fatherland is more likely to be used with reference to matters of law and order rather than homeliness.

Your Opinion 7.7

Another way to approach this question is to adopt the argument noted above concerning the way in which group identity is formulated in opposition to some other identities. According to object relations theory, threats to the boundary of the self may be seen as threats to the existence of self. From this perspective, an understanding of ethnic conflict requires an appreciation of the complex relations involving group cultural and political and individual psychological variables. Goldhagen (1996) adopted a similar approach in an attempt to understand the Holocaust. Might it be appropriate to note, however, that attempts to understand such atrocities are attempts to comprehend the incomprehensible?

IDENTITY CONSTRUCTION AND CONFLICT

The most compelling and tragic evidence of the clash of different scales of identity—local, regional, and national—is in those parts of the world where such constructed identities have been manipulated by people in positions of power in order to seek national status and thus achieve territorial status. Knight noted:

> Ethnic or cultural integrity and uniqueness, minority status, the desire for national freedom, the natural right to independence through self-determination—these are the terms used by regional ethnic group elites as they formulate separatist political ideologies, which are, in turn, often rooted in a variety of emotional, historical (generally from a revisionist perspective), and politicoeconomic realities, as perceived and understood by the regional (ethnic) group itself (Knight 1982:523).

During the 1990s, the former Yugoslavia was the principal area that experienced the clash of national and more local identities with resultant conflict between various groups involving what has become euphemistically known as **ethnic cleansing**.

- In this context Wilmer (1997:4) asked: 'How is it that individual people are persuaded to abandon civility in favor of brutality? Why were elites *able* to so readily mobilize them to undertake acts of inhumanity on the basis of appeals to *identity*?'

- One answer to this question might note that 'history is the raw material for nationalist or ethnic or fundamentalist ideologies, as poppies are the raw material for heroin–addiction' (Hobsbawm 1993:62–3). (See Your Opinion 7.7.)

Global Power: Economics, Politics, and Culture

The preceding account of ethnicity involved consideration of politics and economics in addition to culture, and this intersection and overlapping of interests is continued in this broader account of changing global geography and what might be ambitiously described as the cultural geography of the globe. This section expands upon the Chapter 4 discussion of the shaping of the modern world and global regions, and is further informed by some of the conceptual ideas introduced at the outset of this chapter. Recall the brief account of world systems theory that was included in Chapter 4 to facilitate discussion of civilizations interpreted as world systems. In that account it was noted that world systems ideas are one way to approach long-term historical political, economic, and cultural change. Box 7.7 outlines relevant key aspects of world systems theory. Certainly, the 'notion that something fundamental is happening, or indeed has happened, to the world economy is now increasingly accepted' (Dicken 1992:1). There are three principal aspects to current economic changes:

- Economic activity is continuing to be internationalized, meaning there is a spread of activities across borders.
- It is becoming globalized, meaning there is some functional integration of dispersed activities. The increasing importance of **multinational corporations** and ongoing industrial **restructuring** are both components of these ongoing changes.
- Most critically in the present context, there are increasing inequalities evident in quality of life and in environmental impacts. The origin of the current unequal world is now considered.

Global Inequalities

In Chapter 4 there are discussions of the evolving world system and also of global regionalizations. It is now appropriate to add to those discussions with a consideration of how the contemporary unequal world, particularly as expressed in the difference between more and less developed regions, came into being. The answer provided by world systems

theory, as noted in Box 7.7, is that the rise of the more developed European world necessarily involved the creation of a dependent and less developed world, but this is not the conventional answer provided by both history and economics.

EUROPEAN MIRACLE—OR EUROPEAN MYTH?

One conventional answer to the question of why some nations are rich, while others are poor builds on the idea that Europe is in some ways different from other parts of the world, and that the particular characteristics of the European region prompted a 'European miracle' that began in the fifteenth century (Jones 1987). As noted earlier, a forceful argument along these lines has been put forward by Landes (1998), with detailed elaboration of the idea that Europe is distinctive in two respects, physical geography and culture.

- The basic physical geography logic runs as follows: mild summers permitted physical activity, cold winters reduced the dangers of disease, and adequate rainfall permitted agricultural development. In many other parts of the world climatic factors mitigated against physical activity. Further, in semiarid areas, the need to control water with vast irrigation projects involved a powerful centralized state that lessened the likelihood of private property and individual initiative. This argument—that physical geography and location do matter and are more important than any purported genetic differences—has also been proposed by Diamond (1997) in a major global historical and economic analysis of the past 13,000 years, which suggested, for example, that agricultural development proceeded especially rapidly in Europe because the continent is a region of similar climate, thus facilitating migration and diffusion.
- European culture, money, and knowledge, are also important, particularly as they explain why Europe was the first area to industrialize, to experience **development**. Thus, the British Industrial Revolution was initiated in a society with a sense of national cohesion, a competitive ideology combined with a respect for merit, along with a desire to enhance knowledge. As outlined by Weber,

Box 7.7: World Systems Theory

As noted in Chapter 4, a world system is a large social system that possesses three principal distinguishing characteristics:

- It is autonomous, meaning it can survive independently of other systems.
- It has a complex division of labor, both economically and spatially.
- It includes multiple cultures and societies.

Although historically, world systems covered only some part of the entire world, the current capitalist world system encompasses all of the world. Two types of world system can be identified, with world history characterized by the rise and fall of these systems:

- A world empire is an area that was unified in military and political terms.
- A world economy is a more loosely integrated economic system.

But beginning in about 1450, the capitalist world economy was initiated and since then it has both broadened (that is, expanded spatially) and deepened (that is, evolved into a more complex system), such that it now dominates globally. *Broadening* of the system, has involved an increasing separation into three parts. Politically and economically, the core of the system is the dominant part, while the periphery is the weakest, with the semiperiphery between these two extremes. These three parts are dependent on each other. Especially significant is the fact that the core is able and, in order to thrive, needs to maintain the underdevelopment of the periphery. As suggested by **dependency theory**, the implication of these ideas for the contemporary distribution of more and less developed countries is that, in order to become more developed, the more developed world had to create a less developed world. In Marxist terms, this process is one example of the consequences of the idea of ceaseless capital accumulation. The modern world system has broadened to cover most of the globe. There are five mechanisms involved in the *deepening* of the system. These are:

- commodification, a shift from use values to exchange values in both social and economic life
- proletarianization, the transformation of labor into employees receiving a wage
- mechanization, ever-increasing technologies applied to productive activities
- contractualization, a formalization of human relationships
- polarization, the increasing disparity between different parts of the world system

In addition to these five essentially linear evolutionary changes, there are also two types of cyclical change. There are Kondratieff cycles of economic prosperity and decline that typically last about fifty years, and there are hegemonic cycles involving dominance of the system by a single political unit. This dominance is political, economic, and cultural. A penetrating political geographic analysis argued that there have been three hegemons so far, namely the seventeenth-century Dutch, the nineteenth-century British, and the twentieth-century Americans, with the identity of any fourth hegemon as yet uncertain (Taylor 1996).

The deepening of the modern world system, like the broadening into core, periphery, and semiperiphery, can be understood in terms of the Marxist concept of ceaseless accumulation of capital. Certainly, the 'process of accumulating capital on a world scale required the continual development of the world's forces and means of production. This process was a very uneven one, and thereby continually reproduced and deepened what we call the core-periphery zonal organization of world production, the basis of the axial division and integration of labour processes' (Hopkins and Wallerstein 1996:4). This is why the modern world system is the capitalist world system and also why it cannot remain unchanging. World systems provides an explicit refutation of **developmentalism**, the idea that countries are autonomous in terms of their potential for development, and also the idea that all countries proceed through a set of similar stages as development proceeds.

A conventional criticism of world systems theory is that it overestimates the importance of the world economy, thus underestimating the role of cultural and ethnic identity. Accordingly, world systems theory has been further developed in a number of directions. Especially interesting are attempts to provide 'the beginnings of a unifying framework defining a *world* of more than political states and economic zones' with a consideration of 'people, landscapes, cultures, social institutions, economic capacities and political tenets' (Straussfogel 1997:128).

some of these characteristics were incorporated within Protestantism, the dominant religion in northern and western Europe; other religious beliefs, notably Catholicism and Islam, were seen as elevating authority and limiting individual initiative. In short, culture does indeed make a difference. This cultural argument can be extended to address the question of contemporary global differences in degree of industrialization. Poor areas are poor because of their particular cultural character that proved, and continues to prove, inimical to industrialization. In some cases the cultural character is combined with physical environments that are not conducive to physical activity.

These ideas about the singularity of Europe and European technological achievement have been challenged from several directions. Most generally, viewing history and economic change from the perspective of **multiculturalism** does not involve seeing Europe as being more advanced than other parts of the world, at least not more advanced until the Industrial Revolution, which is explained by reference to European exploitation of other areas. Thus, it is the process of European overseas expansion, colonization, exploitation of resources, the spread of European disease, slavery, and the specific character of European scholarship that are considered to be causes of the poverty in the less developed world. This line of argument is, of course, a fundamental part of world systems theory. From this perspective, ideas about the singularity of Europe are seen as reflecting Eurocentrism. (See Your Opinion 7.8.)

Your Opinion 7.8

*We are now raising a contentious issue, namely that of the legitimacy of European scholarship. It can be claimed that any argument that Europe is in some way different is itself a part of the **ideology** of colonialism and imperialism—one of the stories that Europeans like to tell about themselves. For Blaut (1993b:14–17), this fundamentally flawed European view of the world involves identifying an Inside, namely Europe, and an Outside, namely the rest of the world, with the Inside seen as naturally progressive because of specific mental factors favoring rationality and innovation, and the Outside seen as naturally backward because of a lack of these European mental factors; further, this flawed European view sees the Outside as progressing only as a consequence of contact with the Inside. Certainly, Blaut offered a powerful and well-stated critique of 'The Colonizer's Model of the World', and stressed that it is a tale about Europe by Europe and that the time has come to rethink and rewrite the tale.*

Of course, such fundamental arguments cannot be resolved in the present context; the purpose here is merely to identify what is a very important area of disagreement, both conceptually and empirically. Especially relevant for cultural geographers is the fundamental question about whether or not there are legitimate grounds for identifying weaknesses in another culture. The consensus view in much contemporary cultural geography, informed by postmodernist and related ideas, is that such judgments by one culture about another are not legitimate. It is worth noting, however, that in one important respect the disagreement on this matter is one of belief systems—a clash of conservatives and radicals. Thus, the answers to the questions of why some nations are rich while others are poor, and whether there has been a European miracle or a European myth are the way they are at least partly because of the preferred ideology, the intellectual baggage, of those who tackle such questions.

The Gap

There is a substantial debate concerning whether or not the gap between rich and poor parts of the world is widening. Again, this is a debate with powerful ideological overtones, with catastrophists viewing the contemporary world in essentially negative terms and cornucopians arguing that problems are exaggerated and that human technology will solve existing problems. Certainly, notwithstanding much ill-informed opinion on environmental matters, there is evidence to suggest that the gap between rich and poor, both between countries and within individual countries, is increasing. Cultural geographers have not focused much attention on this important topic, but there seems every likelihood that this situation will change given the increasing recognition of the need to consider economic, political, and cultural issues together rather than separately.

- Taylor (1992:20) stated: 'We need to construct a new global geography that focuses on the world map of global inequalities. We should very consciously proclaim this to be the most important map in the world and focus our Geography accordingly.'
- Klare (1996:354) argued for a new cartography that reflects the 'increased discord within states, societies, and civilizations along ethnic, racial, religious, linguistic, caste, or class lines' and noted that these stresses, along with the ever-increasing environmental damage mentioned in Chapter 5, represent the greatest threat to global security.
- Perhaps the most compelling argument along these lines is that by Kaplan (1996), although this has been criticized for making unwarranted assumptions about the maintenance of historical patterns of conflict.

The Global Present (and Future?)

As already evident, economics, politics, and culture are interwoven and cannot meaningfully be discussed as separate items. Any discussion of the contemporary world necessarily implicates all three of these concerns.

A GLOBAL CULTURE?

For cultural geographers, the single most interesting aspect of current global changes concerns the oft-stated claim that the world is becoming increasingly homogenized culturally, along with the counterclaim that it is not. Needless to say, then, there are varying assessments concerning what is happening to cultures globally. If we think of culture in terms of who we think we are—our identity—then it is helpful to note that such perceptions of identity have changed through time in accord with larger cultural changes.

Thus, it is usual to suggest that, prior to the rise of the nation state, cultures and identities were essentially local, community based, but the rise of the nation state added another dimension with

individuals being offered at least two identity options—local and national. It can be argued that the nation state acted as a type of cultural integrator with a common civil religion emerging in some cases. In some other cases, however, nation state creation resulted in much cultural uncertainty, and the fact that this circumstance has not played itself out satisfactorily in much of the world is evident from the earlier discussion of ethnicity and also, more generally, from any consideration of contemporary political events. Finally, there is the possibility that a global culture is now emerging, which necessarily adds another layer of potential identity confusion or, expressed in a more postmodern constructionist tone, contributes to the creation of multiple identities. Any discussion of this possible third stage is complicated by the presence of several quite different claims and interpretations. Three possibilities are evident.

- A *first* position focuses on the claim that there is an emerging global culture related to the erosion of local cultures. The argument that a global culture is emerging is premised on the fact that the diffusion of industrialization outwards from the West involves a diffusion also of Western culture and Western values. To date, the principal areas affected by industrialization are in Asia, and so the question as to whether parts of Asia are becoming more Western in cultural terms is often raised. Zelinsky (1992:156) discussed this general issue using the heading, 'Transnationalization of Culture'. Of course, such questions are not entirely innocent—there is little doubt that people in the European world are used to thinking of the West and the rest, and that some of the traditional distinctions that those in the West identified are becoming less and less clear. Economically and politically, Western hegemony is being challenged, especially by Asian economic growth. But what about culture? And what about values? Are Western characteristics being accepted elsewhere such that it might be proper to speak of an emerging global culture? It is not difficult to recognize that any suggestion of Western values being accepted elsewhere is a controversial suggestion. Interestingly, a rather different interpretation of the cultural consequences of the diffusion of technology was noted by James (1964:2) with the suggestion that as technology brings humans closer together, so the cultural barriers between groups become more evident.

- A *second* position sees a confrontation between the two processes of parochial ethnicity and global commerce. It can be argued that some cultures choose to be excluded from the globalization process because of the association with science and technology, and that such exclusion results in an assertion of cultural identity that is expressed through various localisms. According to Barber (1995:6), the tendencies of local ethnic affiliation, labeled Jihad, and the tendencies of global integration, labeled McWorld, are both at work, both visible sometimes in the same country at the same time and working with 'equal strength in opposite directions, the one driven by parochial hatreds, the other by universalizing markets, the one re-creating ancient subnational and ethnic borders from within, the other making national borders porous from without'. Both of these forces negatively affect the nation state.

- A *third* position involves the claim that global processes are likely to produce ambivalence or to prompt multiple identities rather than any coherence of identity. According to this argument, 'spatialized communities are the real "containers" of culture, of meaning and identity, not the "virtual" communities created by forms of electronic communication, and the networks built around flows of goods and services' (Axford 1995:164). This is essentially an argument that seeks to distinguish between the possible creation of global or world spaces in technological and economic terms on the one hand, and the possible globalization of meaning structures, of identity, on the other hand.

Clearly, the third interpretation can be regarded as one aspect of the modern to postmodern transition. It is proving particularly influential in cultural geography, and is considered further in several parts of Chapter 8, with discussions of sexuality and emerging cultures. However, it may be useful to note that most such discussions are themselves the product of a particular time and place, namely the end of the twentieth-century English-speaking academic setting.

THE DEMISE OF THE STATE

On the basis of the preceding discussion, it appears that the element of the contemporary economic, political, and cultural world that is most at risk at present is the nation state.

- The age of globalization is also an age of nationalist claims for many groups that usually base their arguments on claims of ethnic distinctiveness. In many parts of the world states are being challenged by various ethnic claims to subnational territory.
- The nation state is increasingly becoming a dysfunctional unit for organizing human activity and economic endeavor. Certainly, the rise of economic regions that transcend political boundaries, including trading blocs, confirms this trend, as do globalization trends in general.
- For Taylor, the state is challenged by current attitudes to environment:

The threat to the state comes not from the cause of globalization, an economic one world, but the consequence, the destruction of the environmental one world. It is not only the fact that pollution is no respecter of boundaries: the whole structure of the world-system is predicated on economic expansion which is ultimately unsustainable. And the states are directly implicated as 'growth machines'—it is unimaginable that a politician could win control of a state on a no-growth policy (Taylor 1994b:184).

Similarly, Wallerstein noted that ecology is now a principal political issue:

Over 500 years, the accumulation of capital has been predicated on the vast externalization of costs by enterprises. This necessarily meant socially undesirable waste and pollution. As long as there were large reserves of raw materials to be wasted, and areas to be polluted, the problem could be ignored, or more exactly considered not to be an urgent one (Wallerstein 1996:225).

Certainly, it is commonly argued that the contemporary political world is one of considerable uncertainty, a circumstance that is related to the nature and pace of change since about 1945. The period from 1945 to the 1990s witnessed a substantial expansion of the global economic system, an expansion that placed much stress on traditional political units; this occurred along with a process of decolonization that involved the creation of a great many new independent nation states; and there have also been environmental impacts linked to increasing levels of technology and population growth, issues that were identified in a larger context in Chapter 2. Further, the fundamental structure of global political power changed dramatically in the early 1990s, with the demise of the former Soviet Union and the more general transition from communism to democracy throughout eastern Europe. All of these changes have far-reaching implications for the contemporary world and there is a broad consensus that the world is entering a new epoch. Box 7.8 identifies some of the key terms and ideas related to this apparent transition.

There are no obvious answers to questions about the future of the nation state and about cultural identities as these relate to the presence of nation states. It may be that nations will become less than, but also more than, nation states, meaning that nations are of increasing importance in the world, but that they do not need to be states in order to be meaningful units. From this perspective, the two, nation and state, can be delinked without one being subordinate to the other. With reference to Canada, Penrose (1997) detailed three stages in the process of state evolution with the third stage being a delinking of nation and state:

- First, the state is constructed through the combined processes of reinforcing state structures, building a nation, and the actions of hegemonic groups.
- Second, the state is deconstructed through the impacts of both globalization and fragmentation.
- Third, it is possible that the state might be reconstructed in a form that is more appropriate to contemporary circumstances.

Interestingly, in order to achieve any understanding of the emergence and transformation of

Box 7.8: Naming the Transition

There is general agreement that the world is a changing place and Webber and Rigby (1996:1–5) identified the following new circumstances that represent relatively sharp breaks from earlier circumstances:

- industrial economies in disarray, many having huge debts
- increasing numbers of unemployed people
- more people in unstable, part-time, or contractual work
- transition from manufacturing to service employment
- major corporations are global
- the very value of growth is being questioned in light of damage to the environment
- increasing amounts of money are transferred globally
- rise of new industrial countries, especially in south and east Asia
- collapse of communism in the former Soviet Union and eastern Europe

Recall that in Box 7.3 Jameson (1991) described postmodernity, using Marxist terms, as a mode of production, a period of late (taken-for-granted) capitalism; that is, characterized by multinational capitalism and the commodification of culture. There are indeed numerous terms that aim to identify the current global transition, economic, cultural, and political. The most general terms introduced so far in this chapter are those of modernity and postmodernity (see Glossary). But it might be proposed that the transition from modernity to postmodernity is also a shift from:

- **Fordism** to **post-Fordism**
- **organized capitalism** to **disorganized capitalism**
- mass production to **flexible accumulation**
- national economies to a single global economy
- growth to stagnation
- nation states to region states or to ethnic territories
- local cultures to a global village
- the golden age to the less-than-golden age

Of course, labeling in this way increases the danger of creating a reality, the golden age, which never really existed. It is with such linguistic concerns in mind that, as noted in Box 7.2, poststructuralist philosophers, such as Derrida, Deleuze, and Foucault emphasize the heterogeneous and plural character of reality.

global systems, the value of counterfactual thinking, of proposing alternative scenarios, is being increasingly recognized. The increasing complexity emerging with the transformation of world–systems through time means that there was a great deal of chance involved in what types of structures eventually emerged. That is to say, just because a particular sequence of historical events DID happen, does not mean it HAD to happen' (Straussfogel 1997:128). This argument is similar to the ideas presented in Box 6.9 and Box 7.5.

Concluding Comments

This chapter has introduced a series of conceptual advances that are linked by a shared suspicion—in some cases an outright rejection—of science and scientific method. More particularly, and with explicit reference to cultural geography, there is a general concern about the Sauerian tradition because it is now seen by many as viewing culture and cultures in an inappropriate manner. A popular view of culture today involves emphasizing that culture is, as noted in Box 1.2, 'a domain, no less than the political and the economic, in which social relations of dominance and subordination are negotiated and resisted, where meanings are not just imposed, but contested' (Jackson 1989:ix). That such an interpretation is not universally accepted is evidenced by the body of ongoing cultural geographic literature (some of which is referred to in chapters 3, 4, and 5), a literature that continues to be informed by a Saucrian or modified Sauerian view. This interpretation is also evidenced by a large body of ongoing histor-

ical and social science literature. For example, with regard to the relative merits of a postmodern concern and the work of Landes (1998), Eichengreen (1998:133) claimed 'that the great questions of economic history that occupy Landes tend to be regarded as fair game for pundits rather than scholars is one of the intellectual tragedies of our times. . . . The postmodernism and multiculturalism that run rampant in history departments are fundamentally incompatible with the approach taken by Landes.' Sowell (1994:225) offered a similar view: 'The a priori dogma that all cultures are equal ignores the plain fact that cultures do not present a static tableau of differences, but rather a dynamic process of competition.'

It is also noted that, although the emphasis on identity that is at the heart of the discussions in this chapter is a key component of contemporary cultural geography, it is not without criticism. In addition to those cultural geographers who prefer to conduct research along more established Sauerian lines, there has also been concern expressed from contemporary social geography. Thus, Gregson (1995:139) worried about the cultural turn, specifically the preoccupation with matters of meaning and identity and the related lessening of interest in the facts of social inequality.

Notwithstanding these essentially ideological disagreements, it is evident that humans are both individual and social beings. It is also evident that humans are social with regard to a plurality of groups such that no one group is the sole determinant of our identity. Certainly, humans have allegiance to more than one group, and it is this key

> ### Your Opinion 7.9
>
> *For some, there is a real difficulty in asserting ethnic or other group rights in that such rights may be seen as legally and morally superior to individual rights. Further, it can be argued that making such distinctions between groups serves to increase frictions between groups. For others, it is appropriate to assign different rights and privileges to different groups, especially when there is a perception that there have been historical injustices that need to be addressed and when it is thought that groups might suffer some discrimination without the additional protection afforded by different status. It is this latter view that dominates in the scholarly world at the present time. However, it can, of course, be argued that a multicultural policy does not include either an appreciation of the ongoing construction and reconstruction of cultures nor an awareness of the possibility of multiple identities. Because of this circumstance, the question of individual rights and group rights resurfaces in Chapter 8 following a more substantial account of relevant conceptual and empirical issues.*

general idea that carries over into the following chapter with a shift in empirical focus from matters of inequality to matters of symbolic identity.

A final observation is in order. Any discussions about human identity and inequality raise the challenging issue of whether or not it is appropriate to talk in terms of group rights in addition to individual rights, an issue that typically arises in any multicultural society. This is an exceptionally complex matter for which there is no easy resolution at this time. (See Your Opinion 7.9.)

Further Reading

The following are useful sources for further reading on specific issues.

Duncan (1980), Ley (1981), Berry (1995:95), and Dunn (1997) on some concerns about cultural geography, borrowing ideas from other disciplines.

Gregson (1993, 1995) and Badcock (1996) on some concerns about writing styles associated with the cultural turn.

Soja (1989:Chapter 1) on space and social theory; Gregory (1994) used the work of Lefebvre (1991) to facilitate discussion of the historical production of modern capitalist spatiality; other instances of the incorporation of space into social theory include the ideas of habitus (Bourdieu 1977) and of normalizing enclosures (Foucault 1970).

Gregory (1981), Pred (1984, 1985), and Gregson (1987) on geography and structuration theory.

Dunbar (1977) and Watson (1951) on the history of social geography.

Dennis and Clout (1980) and Jakle, Brunn, and Roseman (1976) are examples of social geography textbooks that employed a symbolic interactionist focus.

Schmitt (1987) on Marxism generally and Sayer (1992b) on the importance of the division of labor.

Habermas (1972) on the Frankfurt school of critical Marxism and Althusser (1969) on structural Marxism.

There are accessible discussions of Marxism in geography in Cloke, Philo, and Sadler (1991), Unwin (1992), Peet (1996a), R.J. Johnston (1997), and Robinson (1998), and a book-length treatment by Quani (1982). As discussed by Peet (1977a, 1977b), Marxist social theory was introduced into human geography as a part of an emerging radical geography during the 1960s—the radical journal, *Antipode*, began publishing in 1969—with the first major Marxist-inspired work being the book, *Social Justice and the City* (Harvey 1973).

Peet and Thrift (1989) on the structure and agency debate.

Gregson et al. (1997:67–8) on a feminist geographic analysis of the term 'patriarchy'.

Overviews that provide a relatively accessible introduction to feminist geography include Hanson (1992), Rose (1993), Domosh (1996), Monk (1996), McDowell (1997), Women and Geography Study Group of the Royal Geographical Society with the Institute of British Geographers (1997), and Robinson (1998).

Hanson (1992) on areas of shared geographic and feminist interest; Bondi and Domosh (1992) and Longhurst (1994) on the masculinism inherent in most geography.

England (1994) and Rose (1997) on feminism and the situated character of knowledge.

Dowling and McGuirk (1998) on some of the principal themes reflected in some research on gendered geographies.

Barnett (1998a, 1998b) on the cultural turn in human geography.

Rosenau (1992) on differences within postmodernism; Berg (1993) on the claims that even the differences between modernism and postmodernism are based on false logic.

Accessible accounts of postmodernism by geographers include those of Dear (1988), Webb (1990–1), Cloke, Philo, and Sadler (1991), Unwin (1992), Duncan (1996), Edwards (1996), R.J. Johnston (1997), and Robinson (1998).

Jackson and Penrose (1994) for discussions of race and nation that challenge the social construction of these identities while at the same time working uncritically within the socially constructed discipline of geography.

Mohan (1994) on the commodification of knowledge.

Arendt (1951) and Steiman (1998) on the nineteenth-century rise of anti-Semitism.

Montagu (1997) on the fallacy of race.

Charlesworth (1992) and Cole and Smith (1995) are geographic analyses of the Holocaust.

Lester (1996:1–14) on the debate between liberal and Marxist scholars concerning the rise of apartheid.

Rogerson and Gloyer (1995) on Gaelic culture.

Juergensmeyer (1995) on the possible dangers of religious nationalism.

Anderson (1983) on nations as imagined communities; McLeay (1997a) on the construction of the Australian national identity; Castells (1997:29) for a rather different interpretation.

Taylor (1991) on English identity, Withers (1995) on Scottish identity, and Graham (1997) on Irish identity.

Hage (1996) on homelands as motherlands. One of the ways in which one group establishes authority over another group during times of conflict is by assaulting women in order to demonstrate the inability of the fatherland to protect its daughters (also see Ó Tuathail 1995:260).

Hopkins and Wallerstein (1996:1–10) on the recognition of six 'institutional domains "vectors" of the world-system' that are 'distinguishable but not separable'. These are the interstate system, the structure of world production, the structure of the world labor force, the patterns of world human welfare, the social cohesion of the states, and the structures of knowledge.

Smil (1993:5) on the all too common ill-informed reporting of environmental issues.

Renner (1996) on evidence for the increasing gap between rich and poor countries.

Dalby (1996) on some criticisms of Kaplan (1996).

Turner (1994), Axford (1995), Albrow (1997), and Lee and Wills (1997) on the need to consider economic, political, and cultural issues as one.

Zelinsky (1992:149–53) on the rise of a common civil religion.

Radcliffe (1997:289), in an analysis of the dispute over the Ecuador–Peru frontier, found 'multifaceted and complex affiliations to places within and beyond the nation, in addition to national identifications, which suggests that geographies of identity are not fixed, but neither are they, as some postmodern approaches might suggest, infinitely in play and flux'.

Ohmae (1993) on economic challenges to the nation state.

Eller (1999) on ethnic conflict from an anthropological perspective.

Kofman and Young (1996) on aspects of globalization including the development of theory, the role of the state, and gender implications.

Sowell (1996) on the historical migrations of selected culture groups.

Kobayashi (1993) and Hutcheon (1994) on the official policy of multiculturalism in Canada.

Crick (1996), Henderson (1996), and Pulvirenti (1997) on the question of individual and group rights.

8

Landscape, Identity, Symbol

As evident in the preceding chapters, contemporary cultural geographers employ a diverse set of concepts to facilitate the exploration of a wide range of issues. This chapter extends both the conceptual and empirical content of cultural geography through accounts of the ways in which people use landscapes to structure identity along with accounts of the symbolic qualities of landscape. Some of this work builds on the Sauerian tradition, which emphasizes the visible and material aspects of landscapes as a part of the physical and built environment, but most such work emphasizes the ongoing cultural construction, representation, and interpretation of landscape, place, and space. Conceptually, there are origins especially in humanism as introduced in Chapter 6 (see Box 6.6), and much of the content of this chapter is informed by the humanistic interests in place and in human experience as these were introduced into cultural geography during the 1970s. But in addition to the humanistic inspirations, this chapter content is also informed by developments in British social geography and by the cultural turn of the late 1980s as introduced in the previous chapter (see boxes 7.1 and 7.2). Certainly, the conceptual background presented in Chapter 7 informed the analyses of human identity, difference, and inequality in that chapter, but also informs the discussions of identity construction, of others, of landscape and place, and of symbolism that are contained in this chapter. Particularly important to the current discussions are the feminist and postmodern perspectives on identity creation, along with the postmodern interests in landscapes as texts and in the intertextuality of landscape interpretation.

Because the concepts employed to inform the analyses discussed in this chapter are largely of recent origin and, further, because they have often been imported from other areas of scholarship, they are sometimes uncertain and even contested. This conceptual uncertainty is a principal reason why the new cultural geography may be fragmenting. The challenges that this fragmentation implies for the student of cultural geography are a compelling reason for imposing a reasoned yet hopefully flexible organization on the contents of this chapter. The sequence of material in this chapter is as follows:

- There is a discussion of people and place that includes an overview of humanistic approaches and an account of the concept of place, including home as place. There is also a brief elaboration on the model of humans as this topic was introduced in Chapter 6 and a consideration of the way in which both human identities and the identities of places are constructed.
- There is a discussion of others and other worlds that builds on some of the material in Chapter 7. Included in this discussion are analyses based on the concepts of gender and sexuality, along with analyses of geographies of the disadvantaged and geographies of resistance. In some instances, these issues raise the idea of emergent alternative cultures, some of which are the consequences of explicit processes of resistance against the larger societies of which they are a part. Most generally, much of the content of this section can be interpreted as being concerned with geographies of exclusion rather than with the more usual geog-

raphies of belonging. The section concludes with a further consideration of the sensitive question of individual and group rights, which builds on the brief comments at the conclusion of Chapter 7.

- There is a discussion of symbolic landscapes that centers on the idea that landscapes structure identity—that cultural groups effectively write upon the world. Included in this section are accounts of the ways in which humans express their identity in landscape through the naming of places, accounts of sacred spaces, accounts of landscapes understood as texts, and accounts of landscapes as they are reflected in literature and art.
- There is a discussion of the interpretation of ordinary landscapes that stresses links to the Sauerian tradition and that offers accounts both of the personality of place and of the landscapes associated with folk and popular cultural groups and identities. The terms 'ordinary', 'folk', and 'popular', are not clearly defined in cultural geography. Ordinary and popular landscapes are the everyday, taken-for-granted landscapes that reflect ever-changing cultural circumstances, while the concept of folk refers to groups that choose to retain the traditional. Broadly speaking, ordinary and popular culture landscapes might often reflect modernity and postmodernity, while folk landscapes often retain elements of a premodern lifestyle. The account of popular culture introduces the ideas of landscape as spectacle and of landscapes of consumption.
- The concluding section includes a few brief contextual remarks.

As was the case in Chapter 7, the material discussed in this chapter is both spatial and social, both historical and contemporary, and both urban and rural. Further, this chapter is about both people and place and both identity and symbolism. The discussion of identity requires consideration of both collective and individual identity, a circumstance that moves the debate beyond culture to a more typically psychological scale of analysis. Perhaps most significantly, the bases that are acknowledged as foundations for identity construction have become increasingly varied, with the categories of gender, sexuality, the body, style, image,

and subculture added to the traditional sociological categories of class and community and the traditional cultural geographic categories of language, religion, and ethnicity. Before proceeding to the principal concerns of this chapter, two additional observations are important.

- First, one distinctive feature of Chapter 7 that is even more evident in the current chapter concerns questions about the mode of representation that cultural geographers favor. As the accounts of both feminism and postmodernism indicated, there is no one correct way of writing cultural geography, no one correct way of representing the world. The implications of this circumstance are considerable for the practice of cultural geography and are further detailed in Box 8.1.
- Second, although this is a chapter that focuses on place, it is helpful to recall the following quote about culture included in Chapter 3: 'a fascination with "origins" is both primitive and widely shared—the product, perhaps, of "the essential time bond of culture rather than its looser place bond", some rather uncomfortable words of John Leighly, which I've often pondered' (Donkin 1997:248). In other words, the material included in this chapter requires appreciation of both place and time as these relate to culture, identity, and symbolic expression.

Understanding Place and People

This section introduces a number of ideas—inspired principally by humanistic philosophies but also by postmodernist philosophies—that facilitate the understanding of places and the people who live there. Building on the rather specific ideas included in Chapter 6 as one component of the review of the model of humans employed by cultural and behavioral geographers, the current discussion includes an overview of the social constructionist argument, includes a focus on the concept of place, encourages a rethinking of the model of humans employed in cultural geographic studies, and concludes with consideration of the means by which both human identities and place identities develop.

Although it may be correct to note the absence of a substantive research tradition, there seems little doubt that humanistic approaches in geography share with humanistic traditions in other disciplines a fundamental, if less well–developed, concern with the idea of social constructionism as this is contained within phenomenology. This important idea is now pursued.

Social Constructionism

The concept of social constructionism has appeared on several prior occasions in this book as it was first introduced and defined in Chapter 1 and then discussed in philosophical terms in Box 2.2. There were also limited accounts of the social construction of nature in chapters 2 and 5, of Aboriginal groups in Chapter 3, of gender in chapters 5 and 7, of knowledge in chapters 6 and 7, and of human identity generally in Chapter 7.

In the present context, constructionism involves the idea that identities—such as gender, sexuality, and ethnicity—are fluid, contested, and negotiated. This is an idea that can be contrasted with essentialism, which involves the claim that a specific identity or group membership is relatively fixed and unchangeable.

Although it is usual in the cultural geographic literature to relate these concerns with feminist and postmodernist approaches, the concept of social construction was most fully articulated by Berger and Luckman (1966), being based on earlier ideas central to the humanistic approach of phenomenology (see Box 6.6) and to symbolic interactionism (see Box 5.2). This interest reflects a belief—shared by phenomenology, symbolic interactionism, feminism, and postmodernism—in the pervasive relevance of the socially constructed aspects of human experience. Together, these various approaches question everyday assumptions about the nature of reality. Thus, from a constructionist perspective, self, identity, community, and social reality are all creations of the human mind

Box 8.1: Modes of Representation in Cultural Geography

The task of cultural geographers is to represent the world to others, but this raises the question of what is an appropriate form for this representation. Until recently, the answer to this question was taken for granted. Today it is debated, such that there is what might be called a crisis of representation.

The previously taken-for-granted answer is that a form of representation can be used that claims to be universally valid and to result in an accurate understanding of the world. In this form of representation 'the task of writing is the mechanical one of bolting words together in the right order so that the final construction represents the thought or object modelled' (Barnes and Duncan 1992:2). Both the Sauerian approach and the occasional attempts to employ a positivistic approach are seen to belong in this category. Most important, the Sauerian approach can be seen as an attempt, usually based on observation, to accurately describe and classify the subject matter of cultural geography. It can be argued that much of the material included in earlier chapters, especially chapters 3 and 4, reflects this mode of representation.

But there are other modes of representation that reject the suggestion that it is possible to produce universally valid and accurate representations of the world because any representation is actually an interpreta-

tion. This is an idea that is central in humanistic studies, but that is also explicit in several of the more influential modes of representation introduced in the context of the new cultural geography. Thus, as discussed in the previous chapter, both feminism and postmodernism claim that accurate representations are not possible. More broadly, the hermeneutic method explicitly acknowledges the role of interpretation (see Box 2.2).

As discussed by Barnes and Duncan (1992:2–4), three critical implications arise from acceptance of the argument that it is not possible to produce accurate representations of the world, the first two of which were incorporated in the accounts of feminism and postmodernism in Chapter 7.

- *First*, if there is no preinterpreted reality for writing to reflect such that it is not possible for writing to mirror the world, then what *does* writing reflect? The general answer proposed is that it reflects earlier texts—this is the idea of intertextuality (see Glossary). From a postmodern perspective, this also implies that all truths are inside, not outside, texts.
- *Second*, all writing reflects a particular and necessarily local vantage point and is necessarily marked by its origins—a key idea in both feminism and post-

and should not be regarded as objective entities in some way separate from ourselves.

In phenomenology, these ideas were most forcefully outlined by Alfred Schutz (1899–1959) who outlined four realms of the social world, the first two of which were referred to in Chapter 6 (Schutz 1967):

- *umwelt* refers to the reality that is directly experienced, our physiological and physical surroundings
- *mitwelt* refers to the reality that is indirectly experienced, the social world of others
- *folgwelt* refers to the future
- *vorwelt* refers to the past

The social construction of reality occurs within the *umwelt*, but because people may be unreliable in their interactions with others, emphasis can be placed on the *mitwelt*, as that part of the social world involving interactions with larger social structures rather than with other individuals. Schutz further distinguished between:

- first-order processes involving the interpretation of our actions and the actions of others
- second-order processes involving social scientists' attempts to understand first-order processes

Thus, second-order processes are understandings quite unlike those produced in natural science that have no basis in our own everyday common-sense meanings, precisely because they are based on the first-order processes that humans use to make sense of the everyday world. In this way, the use of second-order processes results in a phenomenology unlike the earlier transcendental phenomenology of Husserl, which was referred to in Chapter 6 in that the key feature is the social or intersubjective component.

modernism. Bondi stated the feminist argument clearly:

According to feminist philosophers, dominant conceptions of knowledge are 'gendered'. This claim is elaborated in various ways but, of central importance, is the notion that western intellectual traditions operate through interrelated dualisms, such as reason and emotion, rationality and irrationality, objectivity and subjectivity, general and particular, abstract and concrete, mind and body, culture and nature, form and matter. In each case, the terms are defined as mutually exclusive opposites but are not equally valued: the first occupies a superior position and is positively valued; the second is subordinate and negative. And intertwined within this system of hierarchical dichotomies is a distinction between masculine and feminine. It is through this intertwining that dominant knowledge systems are 'gendered': the superior terms in the dualisms are associated with masculinity, the subordinate terms with femininity (Bondi 1997:245–6).

More generally, it is possible to claim that because all texts are a reflection of a particular and not some universal viewpoint, it becomes critical to understand the position of the author of the text and also the intent of the text. The ideas of situated knowledge and positionality assert that authors are unable to speak for themselves; rather, they speak for the cultural system of which they are a part. Thus, all knowledge is partial. Some postmodernists move beyond these claims to assert that no one text is more privileged, in some sense a better interpretation, than any other—this is clearly a claim that is especially open to debate. Ley (1998:80) noted that to abandon authority 'leads to the abandonment of responsibility, to a tentativeness that in its reluctance to achieve any closure embraces, sometimes celebrates, ambiguity and indeterminacy'. Further, it may be that the tendency for any representation to be ethnocentric need not necessarily imply that political projects ought to be excluded from academic work.

- *Third*, there is a need to employ a variety of writing strategies, tropes, and especially to use rhetorical devices to help convey meaning. In attempts to mirror the world, the principal writing strategies were those of objectivism, often involving a narrative. Metaphors are increasingly employed to describe the unknown using the language of the known.

The basic argument in this box is one component of the philosophical debates discussed in Box 2.2, and is important to an understanding of the landscape-as-text metaphor discussed later in this chapter.

Place

It is possible that the lack of a humanistic tradition in geography is related to the relative failure to build on the logic of the social construction argument. Certainly, sociology in particular developed an influential humanistic tradition during the 1960s that was linked to social constructionism while geographers largely failed to pursue this idea explicitly until additional stimuli were provided by feminism and postmodernism. Further, recall from Box 6.6 that humanistic geography explored various philosophical avenues without developing a commitment to one in particular and also became associated with positivism in the context of behavioral geography. Indeed, rather than either building upon the broad interdisciplinary idea of social constructionism or, alternatively, demonstrating some singular philosophical focus, the humanistic approach in geography as it developed after about 1970 focused primarily on a new understanding of the geographic concept of place that was informed by the idea that space and time come together in place.

Of course, geography has long been concerned with the related concepts of space, place, area, region, landscape, and location, with Sauerian cultural geography focusing especially on landscapes in a regional context and Vidalian cultural geography studying *pays*. But the phenomenologically informed concept of place adds another dimension to the more established concerns.

- Rather than using the term to refer to where something is located, as a container of things,

humanistic geographers use place to refer to a territory of meaning and to where we live.

- Place is qualitatively different from terms such as landscape, space, and region in that it involves being known and knowing others.

- Place became reinterpreted as an experiential and social phenomenon. Landscape and space 'are part of any immediate encounter with the world, and so long as I can see I cannot help but see them no matter what my purpose. This is not so with places, for they are constructed in our memories and affections through repeated encounters and complex associations' (Relph 1985: 26).

- Thus, place is intersubjective; that is, shared in the sense that the meaning of place can be communicated to others.

- How are places created? Through the human occupation of space and the use of symbols to transform that space into place.

For Entrikin (1991), however, place is best understood both scientifically and humanistically; that is, objectively as a location and subjectively in terms of experiences, a suggestion that prompts consideration of the tension that prevails in many places between local features and global features. This is an idea that appears later in this chapter in the distinction made between folk and popular cultures and related landscapes. (See Your Opinion 8.1.)

HOME

An especially useful interpretation of the concept of place concerns the idea of home, created both

---

Your Opinion 8.1

Do you agree that the interest in place represents a substantial addition to the concepts of cultural geography in the sense that 'no one lives in the world in general' (Geertz 1996:262)? Reflecting on the renewed interest in place, Agnew (1989:9) was prompted to query 'why we should have taken so long to arrive at a theoretically coherent concept of place as a defining element of geography and key idea for social science as a whole'. The answer offered was in terms of the social science focus on class and community at the expense of place, which was in turn related to the preference given to evolutionary and naturalistic emphases. Certainly, the increasingly varied use of the place concept within social science is related to the popularization of the place concept by humanistic geographers as well as to the cultural politics introduced as part of the cultural turn, which sees culture as linked to economy and politics and therefore more than an aesthetic realm. Indeed, much of the interest in place has built upon and moved beyond the various humanistic meanings toward more political meanings, with a focus on places as sites of power struggles, displacements, and contestation. Box 8.2 offers a brief statement on this theme.

symbolically and materially, as a familiar and usually welcoming setting within some larger more uncertain world. In this sense, homes may range in scale from a single room in a house through to the regional homeland concept as discussed for groups such as Hispanos and Mormons in Chapter 4.

Tuan (1991a:102) built on the popular definition of geography as the study of the earth as the home of humans to define where we live, our home as 'a unit of space organized mentally and materially to satisfy a people's real and perceived basic biosocial needs and, beyond that, their higher aesthetic–political aspirations'. Understanding place in this way invites consideration of the ideas of sense of place, of topophilia, and of placelessness as these are referred to later in this chapter (see Glossary). Although it is usual to interpret home in a positive sense, there are two issues that complicate this understanding.

- It is, of course, possible to interpret the meaning of home at the scale of the dwelling place rather differently for women and for men. Specifically, feminist geographers introduced the idea that, for women, the home is in some instances most appropriately interpreted as a center of oppression and confinement such that it may be quite unwelcoming. Women can be seen as having been constrained in their experiences and their behavior because men tried to restrict them to just the one world, home, while reserving for themselves the right to experience two worlds, home and outside home.
- More generally, the concept of place informs studies of the sense of belonging to, or perhaps being excluded from, a home or place—an idea that was raised in the Chapter 7 discussion of ethnicity without using this conceptual underpinning and that is further used later in this chapter.

The Geographical Self

The discussion that centered on human behavior in Chapter 6 introduced and used two principal and quite different conceptions of the model of humans, with the central distinction being whether humans were viewed as passive objects or as active subjects. Broadly speaking, a behavioralist perspective implied a view of humans as passive objects while a humanistic perspective implied a view of humans as active subjects. This simple distinction reflected the contested character of geography in the early 1970s as the previously dominant positivistically inspired spatial analysis lost ground to other approaches, including those related to humanistic philosophies. The discussion in the 1970s was also often framed in the context

Box 8.2: Places and Conflicts

It can be argued that there is a need to reorient human geography toward the study of places, regions, or locales because places structure 'how people tackle problems, both the small and usually trivial problems of everyday life and the large, infrequently met, problems which call for major decisions'. The proposed reorientation includes six points (Johnston 1991: 67–8):

- the creation of places is a social act; places differ because people have made them so
- places are self-reproducing entities because they are the contexts in which people learn, and they provide role models for socialization, nurturing particular belief sets and attitudes
- no regional culture exists separately from the people who remake it as they live it
- within a capitalist world economy, places are not autonomous units whose residents have independent control over their destinies
- places are not simply the unintended outcomes of economic, social and political processes, but are often the deliberate product of actions by those with power in society, who use space and create places in the pursuit of their goals
- places are potential sources of conflict

This interest is clearly concerned primarily with places as sites of conflict rather than with a humanistic understanding of place. As such the conceptual basis for these ideas as they have developed in geography and other social sciences, especially anthropology, is the postmodern concern with others. This general idea was evident in the Chapter 7 account of ethnicity and appears more fully in the discussion of landscapes of exclusion later in this chapter.

of the debate about structure (society) and agency (individual) that was referred to in the discussions of structuration theory in Chapter 5 and of Marxism in Chapter 7, and occasionally in the comparable humanistic terms of context (society) and intentionality (individual). However, the conceptual underpinnings for a discussion of the model of humans have shifted considerably in light of the emergence of both poststructuralist and postmodernist concerns as these are discussed in Chapter 7.

Indeed, for many cultural geographers there is no longer a meaningful debate between the more scientific and the more humanistic emphases as it is often argued that the former lacks credibility. Nevertheless, it would be misleading to simply conclude that the humanistic model of humans is seen as correct; rather, a revised and as yet unresolved debate has developed concerning the question of who or what is the subject.

- As noted by Philo (1991), cultural geographers might profitably consider 'the temporal and spatial variations in what selves are and in how selves conceive of them*selves*'.
- For Pile (1993:122), a 'search needs to be instigated into alternative models of the self, as a means of understanding the position of the person within the social'.

In order to seek insights into this question, cultural geographers have turned to a variety of inspirations, but especially to some recent advances in psychoanalytic theory that offer insights into the relations between the social and the individual. Most generally, Pile (1996) argued that psychoanalysis provides a basis for an understanding of the spatiality of everyday life. Essentially, the psychoanalytic argument is that identity in place is explained principally by reference to the subconscious. One popular approach is object relations theory, which, as referred to in the discussion of national and local identities in Chapter 7, seeks to unravel connections between individuals, social worlds, and material environments. A related source of inspiration from psychoanalytic theory is the work of Jacques Lacan (1901–81) which recognizes three different types of space—real, imaginary, and symbolic—and which sees the self formed in

relation to otherness. Simply expressed, this body of work sees the identities both of the self and the group as formulated in opposition to various other identities.

Constructing Place Identities and Human Identities

We now try to build upon the discussions contained so far in this section in an attempt to further clarify the ways in which the identities of places and of people, as individuals and as group members, are constructed given that there is a reciprocal relationship between the two. But, in this context, recall the difficulties involved in delimiting groups and related places that were included in the conclusion to Chapter 4. As Lewis (1991:605) noted: 'Cultures, societies, communities, ethnic groups, tribes, and nations are coming to be viewed as contingent or even arbitrary creations rather than essential givens of human existence.' Notwithstanding this circumstance, cultural geographers argue that places are indeed given meaning by people and that people are constituted through place. *Expressed simply, people interpret themselves and are also interpreted by others according to the place they live in, belong to, or originate from.*

One related theme that certainly merits increased attention from cultural geographers concerns the links between identity and place as these are understood by Aboriginal populations in areas that experienced the effects of European overseas movement. Ray (1996:1) noted: 'Many of Canada's indigenous peoples define themselves in terms of the homelands that sustained their ancestors. These are the places where their spiritual roots lie. Drawing from their natural surroundings, Native groups have developed powerful metaphors, symbols, and narrative traditions to express their religious and philosophical views . . . in effect, the land was their history book.' Thus, the concept of a homeland may be a fundamental basis for an ethnic group to assert some special privileges in, even ownership of, an area. As a basis for nationalism, the homeland concept can lead to attitudes of closure and boundedness with the associated claim that some people belong in the homeland while others do not. Further, there is a tendency for people to withdraw into the known, their home, and also stereotype the unknown, others in other homes. Even where such stereotypes are positive, they represent a labeling of individuals according to some presumed larger group identity. Group identity, often expressed as ethnicity, can be interpreted as a reaction against cosmopolitanism.

Accounts of human identity are usually based on either Freudian psychoanalysis or symbolic interactionism in sociology. According to Freudian identification theory, the child gradually assimilates external persons and objects, effectively taking over the features of another person, as a means of reducing tension. The child first identifies with parents and subsequently with others who seem to be successful in gratifying his or her needs. In symbolic interaction theory, the self emerges such that the individual is able to reflect on his or her position in the social world, and identification is the process whereby individuals place themselves in socially constructed categories.

But perhaps the most substantive recent work on human identity moved beyond the individual scale to stress the circumstances and motivations for identity construction in the context of the contemporary network society. Thus, Castells (1997:7) stated: 'I propose, as a hypothesis, that, in general terms, who constructs collective identity, and for what, largely determines the symbolic content of this identity, and its meaning for those identifying with it or placing themselves outside of it.' Given this hypothesis, three forms and origins of identity construction were distinguished:

- *legitimizing identity*—this form is introduced by the dominant institutions in society, is used to rationalize their domination of others, and leads to the creation of civil society
- *resistance identity*—this form, often discussed as identity politics, is introduced by those who are in some way excluded and/or disadvantaged, and leads to the formation of communities
- *project identity*—this form is introduced by a group as a new identity that serves to redefine their position in the larger society with the goal of transforming that society (Castells 1997:8–10)

Much of the discussion in the following section incorporates examples of the second of these

three forms of identity construction, especially as this relates to the way in which dominant groups interpret others and the worlds they occupy. In this context, there is an increasing interest in identity politics, although there is also evidence of the fluidity of identity, and both of these issues are outlined in Box 8.3.

Others and Other Worlds

Two principal promptings of the interest in the identities and places of others are as follows.

- First, there is the realization that dominant societies have constructed landscapes that take certain characteristics for granted. For example, urban areas in the Western world have typically been built with the assumption that families are heterosexual and nuclear, that women are dependent on men, and that people are able-bodied. These assumptions have resulted in seeing those who do not conform to societal expectations as different—often as excluded and disadvantaged others. Intrusions of these others into landscapes not constructed for them has proven to be very controversial both culturally and spatially. Certainly, a distinctive feature of the contemporary world is the unsettling effect that expressions of other identities and their crossing of spatial boundaries have on dominant groups. In this context, recall the reference to different humans included in Chapter 2 that noted the usual practice of relegating others, such as different cultural groups, the insane, poor people, children, or women, to a lesser animal status.
- Second, there is the postcolonial theoretical work of Said (1978, 1993) that focuses especially on the way in which, in a colonial situation, cultural identities of both colonizers and colonized were a product of the process of colonization. Known as Orientalism (see Glossary), this idea implies a relational concept of culture that involved colonizers seeing those who were colonized as others, different, and somehow less than those who were doing the colonizing. For Said, the Orient was constructed by Europeans in counterpoint to the West. Shurmer-Smith and Hannam (1994:19) noted: 'It is the function of the Orient to play

"Other" to the West's "Same". An imperial image arose involving the idea that westerners had the right to intervene elsewhere and to use others and their places as they chose. Through this imperial image, places outside of Europe were thus colonized both conceptually—for example, through a renaming process—and practically—with the image manifesting itself in slavery and in such economic activities as resource extraction and plantation agriculture. The concept of otherness may be an important component of the foreign policies of many Western states, especially when a conservative minded leader is in power, such that the essence of a geopolitical strategy of national security may involve the intentional exclusion of others. The basic argument has also been extended by feminists who interpret the colonial process as both implicitly and explicitly gendered. (See Your Opinion 8.2.)

This section considers examples of others and their worlds, as these identities are imposed by dominant groups, along with examples of resistance identities. As Chouinard (1997:379) noted: 'If one were to try to identify a single theme that resonates throughout intellectual and political debates in the late 20th century, it might well be "the difference that difference makes". Although it is still often the case that those who share an identity also share a place, there are many examples of others and of resistance identities that do not share a space. Accordingly, some examples of both place-based and non-place-based others and resistance identities are noted, and it is acknowledged that both the identities and 'the meaning ascribed to the role of place in identity politics are highly contestable' (Agnew 1997:254). The examples noted in this section are similar to the accounts of ethnicity in Chapter 7, although the primary concern in that

Box 8.3: Identity Politics

The tendency for groups of people—who may or may not be associated with a specific place—to involve themselves in a struggle to establish a distinctive identity for themselves that is usually in opposition to some dominant identity is a relatively recent phenomenon. Possibly it is a trend that reflects globalization in general and, more specifically, a weakening of the more established bases of identity such as kinship or religion. Thus, groups identified in the context of environmental concern (as noted in Chapter 5), in the context of gender or ethnicity (as noted in Chapter 7), or in the context of gender and sexuality (as noted later in this chapter) can all be interpreted as instances of new cultural movements, what might be called emergent alternative cultures. These are comparable to resistance identities and may be described as 'instances of cultural and political praxis through which new identities are formed, new ways of life are tested, and new ways of community are prefigured' (Carroll 1992:7).

Substantial and sympathetic conceptual support for these ideas is included in the Gramscian concept of hegemony (see Glossary) and in various accounts of new social movements. But there is, of course, more than one way to view this new identity politics, with some theorists viewing new identities—based on such variables as religion, ethnicity, and sex—as a form of

tribal mania that is tolerable only if they are not treated seriously. There is a clear link with the Marxist model of class consciousness whereby some subordinate group develops an identity and pursues some political path accordingly. However, most emerging identities are based on considerations other than class. What is happening is that there is a dialectic of culture, identity, and politics that acts to initiate social change.

The distinctions between dominant, residual, and emergent forms of culture made by Williams (1977) are also helpful in understanding identity politics. Residual forms represent some earlier institution or tradition, often an ethnic identity, while emergent forms are newer cultural expressions, such as a sexual identity; both are in a process of continuous tension with the dominant culture and may explicitly oppose that culture.

But perhaps the most substantive concern about analyses that tend to focus on these emergent cultures is that—rather like the dominant discourse that equates culture, ethnicity, community, and place to produce stereotypes—they tend to reify culture and to assume a single monolithic cultural identity for a group, and thus may well omit some critical nuances of individual identity. This is a critical point to consider in any discussion of others and their worlds, a topic that is discussed further in Box 8.5.

chapter was with issues of inequality rather than with identity.

Many of the discussions in this section can also be interpreted as examples of transgressions of normative spatial behavior. As noted, landscapes are constructed with certain assumptions concerning who uses those landscapes and how they are used. However, there is increasing evidence that landscapes are being used differently in that the behavior that occurs in landscape is not considered usual. Such behaviors in landscape may be seen as challenges to authority and as expressions of resistance that result in the creation of some contested landscapes.

Gender

Even a casual observer of the visible landscape is able to recognize some expressions of ethnicity, class, language, or religion in what they see around them. But what about gender?

- Is gender reflected in the visible landscape? The short answer is yes because gender is a key element of our human identity. Recall that we are socialized into particular gender roles, and that these roles are reinforced by powerful ideologies, notably by patriarchy.
- But does our casual observer see gender in landscape? The answer is no. A principal reason why gender is relatively invisible is because gender itself is a part of our taken-for-granted world (see Glossary). One of the key contributions of feminism to cultural geographic research is that of uncovering what was always there, but what was not typically seen. Not surprisingly, most work in this area has been accomplished by those with a specifically feminist agenda and has focused on matters concerning women and landscape.

LANDSCAPES OF GENDER

A fundamental argument initiated by feminist geographers is that the cultural landscape is made by men and for men, both as a reflection of patriarchy and as a means of maintaining the privileged position occupied by men. There is little debate concerning the claim that landscapes reflect patriarchy, but the assertion that such building is intentional is clearly more controversial. The following three examples address aspects of these questions.

First, consider the design of houses and also of the suburbs in which they are located. The gendering of space in houses in the Western world means that the kitchen— designed as a separate area for the unpaid work of women—has typically been a relatively small space often located at the rear of the home. Further, the suburban expansion so characteristic of urban areas in the second half of the twentieth century involved several features that limited the ability of women to work away from home, including an absence of child care facilities and limited public transportation. Rose, Kinnaird, Morris, and Nash (1997:149) stated: 'As well as the design of individual living units, then, the design of housing estates also articulates particular assumptions about who will be living in this environment and what they will be doing there.' The argument is that these assumptions are those of a society of nuclear families with the man working outside of the home and the woman working in the home. It is because of these circumstances that place and home can mean very different things to women and to men,

---

Your Opinion 8.2

There is little doubt that our appreciation of others and the worlds that they occupy has been much enhanced by recent developments in social theory, but there is also little doubt that some of the key ideas have long been acknowledged, albeit without the sophisticated conceptual support. For example, in light of the considerable attention given to the ideas discussed above, it is interesting to recall a discussion of geography and development by Spencer (1960:36), who noted that the 'conventional language employed by Occidentals who judge Malaya and Malayans, not from the Malayan point of view but from their own particular biases' and that 'contributed to an impression of Malaya that can be made to fit the Occidental generalization "underdeveloped".' Although necessarily much less sophisticated in conceptual terms, this basic argument anticipates the later work.

a fact that has not typically been acknowledged by humanistic geographers. Further, it has also been shown that local labor markets discriminate against women, as well as against some ethnic groups and the poor, in the way that different groups are limited to different workplaces.

Second, landscapes in urban areas are also gendered in that they explicitly reflect and reinforce the dominant role of men through the building of monuments and other structures that reflect the achievements of men. Most such structures communicate a message of male power and achievement both reflecting and reinforcing patriarchal gender relation. More generally, the typical shopping mall landscape reflects gender in that it is assumed that women are the principal consumers and the mall landscape is structured accordingly.

Third, one way to interpret the gendering of human-made landscapes is to appreciate that women are often used by men as the other, as a backdrop against which men are able to articulate and reinforce their masculine identity. This is an argument that Raglon (1996) used in a discussion of the Canadian wilderness that stressed differences between women and men; that is, between domesticity and the home on the one hand and the outdoors and wilderness on the other. In this instance, men are seen as framed within nature, which is traditionally female. (See Your Opinion 8.3.)

GENDER, LANDSCAPE, AND FEAR

One reason why the lives of men and women—in terms of attitudes, feelings, and behavior—are different is because of a varying likelihood of personal assault. Specifically, feminist geographers have stressed the threat of men's sexual assault of women as one of the ways in which the behavior of women, especially outside of the home, is constrained. This threat can be interpreted as one of the means by which patriarchy is reinforced as women may consider it necessary to become dependent upon a man. Further, it has been argued that the design of urban environments has not taken into account the particular circumstances of women when it comes to a question of personal safety, with public spaces in particular effectively being masculinized spaces. Areas of risk for women are those areas where the behavior of men sharing that space is less easy to predict and control; such areas include any less-frequented areas such as parks or even rural areas in general, and also small relatively enclosed areas such as subway stations and multistory car parks.

- In a study conducted in areas of Edinburgh, Scotland, Pain (1997:235) found that 70 per cent of women worried about sexual assault by a stranger outside of their home and 59 per cent worried about physical assault in similar circumstances. Fears of assault are also much higher during hours of darkness when it is not unusual for young males to dominate many urban public spaces. One result of such concerns is that women develop mental maps of the areas that they need to negotiate, with these maps reflecting their assessments of the degree of danger, and then they behave accordingly.

- With reference to Reading, England, Valentine (1989:386) noted that the 'predominant strategy adopted by women I interviewed is the avoidance

---

Your Opinion 8.3

A similar argument might be appropriate for the way in which agricultural landscapes are sometimes seen as masculinist landscapes, often depicted as examples of the achievements of men in mastering what was previously wild. In some cases, especially in areas that have emerged from relatively recent settlement experiences, it may be possible to read the agricultural landscape as masculine, as marginalizing women, and as a construction of subservient femininity. Rather confusingly, this coding of rural landscapes as masculine and urban landscapes as feminine contradicts the types of landscape gender codings found in the United States, for example, where the rural landscape has typically been associated with nature rather than with culture and has accordingly been coded as feminine with urban landscapes coded as masculine. What are your reactions to these rather divergent ideas? Is it possible that both arguments are correct and that the differences reflect particular settlement and larger cultural differences?

of perceived "dangerous places" at "dangerous times'".

- Such dangerous places can include homes as 'a significant proportion of younger women display certain levels of fear in their own homes at night' (Pawson and Banks 1993:61).

It is not difficult to appreciate that the meaning of many places varies considerably for men and for women. The possible different understandings of home noted earlier are but one component of the way in which the meanings and experiences of places are bound up with gender.

Sexuality

There are two rather different processes at work in the creation of groups that are excluded from some larger legitimized identity.

- Thus, and as most usually discussed, some subordinate populations are excluded by the dominant group.
- But, in addition, many groups that experience exclusion have intentionally reinforced that exclusion through a reinterpretation of their identity as positive, not negative. This reinterpretation effectively inverts the perceived value of their identity by means of placing a high value on an identity that was previously, and perhaps continues to be, disparaged by the dominant group. Castells (1997:9) described this process as *'the exclusion of the excluders by the excluded'.* The principal example of this phenomenon that is discussed in this section concerns identities based on sexuality.

Since about 1990, cultural geographers have shown a great interest in sexuality—especially sexualities other than heterosexuality—as an expression of identity, as a basis for recognizing urban landscape regions, as one way in which the dominant heterosexual landscape can be challenged, and as a basis for helping to understand violence against those seen to be different. It is now generally accepted that sex is not simply some form of instinctual drive designed to ensure species continuity; specifically, human sexuality is not a biological given. Rather, sex is a multifunctional

behavior that can only be fully understood in a larger cultural context. Accordingly, a particular sexual behavior such as **homosexuality** may be unacceptable in one cultural context and yet may be acceptable in some other context; similarly, the attitudes that a culture has toward homosexuality, for example, may change through time. Distinguishing between what is in some sense 'right' or 'wrong' clearly depends upon what criteria are being used to make such judgments; homosexuality may be dysfunctional from the perspective of reproduction, but it may serve to enhance a sense of well-being. Most critically, it is clear that consensus judgments about particular sexual behaviors usually reflect power relations, with some feminists arguing that the critical power relation in this context is patriarchy, which is in turn based on heterosexuality.

It is helpful to appreciate the considerable changes that are taking place in the Western world concerning attitudes toward homosexuality and homosexual relationships. For example, prior to 1973 homosexuality was identified by the American Psychiatric Association as a sexual deviation, but that designation was eliminated on the grounds that homosexuality is just one of the ways in which sexual preference is expressed. Further, same-sex partner relationships are increasingly being legally recognized as legitimate and socially acceptable lifestyle choices. There is, of course, some considerable opposition to such changes, especially from conservative Protestant denominations, from Catholicism, and from Islam. Two important conferences sponsored by the United Nations in the mid-1990s, namely the third International Conference on Population and Development held in Cairo in 1994 and the fourth World Conference on Women held in Beijing in 1995, exposed substantial differences of opinion on the matter of new forms of family structure. But in the Western world especially, it is clear that the status of sexual minorities is changing (Box 8.4).

LANDSCAPES OF SEXUALITY

A principal concern is the identification of both gay and lesbian landscapes within larger, usually urban, areas that are seen as designed for heterosexuals. Some of the earliest such analyses involved studies of prostitution and also indirect attempts to delimit the landscapes occupied and modified by homosexual groups, studies that were often in the urban ecological tradition. But the current emphasis on landscapes of sexuality is rather different as it stresses sexualities as identities and is typically based on ethnographic research. Bell and Valentine (1995:5) noted that the earlier form of research has been criticized for being 'patronising, moralistic and "straight"' in comparison to the latter form, which was seen as 'sex-positive'.

Central to much of this work on geographies of sexuality is the recognition that heterosexuality dominates, defined both in terms of preferred sexual behavior and in terms of power relations in everyday environments. Those who are different tend to feel out of place in most environments that simply assume that all those who use the environment are heterosexual. The most extreme version of this norming of space concerns acts of violence against those who are not heterosexual.

Box 8.4: Resistance Identities—Challenging Heterosexuality

A principal example of emerging resistance identities relates to the questioning of **heterosexuality** as the norm, such that sexual liberation is a new form of self-expression. Homosexual, both gay and lesbian, populations especially—groups that typically have been excluded and disadvantaged—have begun the process of community formation in explicit opposition to the social norm. Probably the most important consequence of these social movements is the increased challenge to patriarchy as a central structure of contemporary societies because patriarchy is, of course, founded in the traditional family structure. It is inevitable that increased social acceptance of other family structures based, for example, on two adult females or two adult males leads to a weakening of patriarchy.

One way to interpret the rise of resistance identities linked to sexuality is to think in terms of the social construction of sexuality. Recall that, as noted in Chapter 7, the conventional feminist and feminist geographic understanding is that gender is rooted in the biological distinction between female and male such that femininity and masculinity are the social constructions imposed on female and male. But other interpretations of both gender and sex are possible, interpretations that assert that it is the two sexes, female and male, that are the primary social constructions. Gregson, Rose, Cream, and Laurie explained this idea as follows:

> The notion that we all fit into either a male or female body is just that, notional. It cannot be sustained, either over time or space. Simply put, there is nothing 'natural' about 'male' and 'female' bodies. There is nothing natural about everyone being forced into one sex or the other. Rather, our belief in the existence of two, and only two, sexes is structured by our ideas about gender. What this means is that our understanding of gender (man and woman) is not

determined by sex (male and female) but that our understandings of sex itself are dictated by an understanding that man and woman should inhabit distinct and separate bodies. So, sex does not make gender; gender makes sex (Gregson, Rose, Cream, and Laurie 1997:195).

Thus, from an essentialist perspective, sexual identity, such as male, female, or homosexual, is a fact that cannot be changed because it is innate. Constructionism, on the other hand, as applied to sexual identity, asserts that sexual identities can only be understood with reference to cultural context, such that our understanding of our sexual identity relates to the world we live in and not to biological considerations. This constructionist view of sex and the **body** involves what may be described as a destabilizing of sexual identities. Much work in this vein proceeds to develop these ideas further by building on Butler (1990) regarding the idea of sexual identity as performance. This is a means of explaining why some people do not conform to the presumed heterosexual norm, choosing instead to engage in different behaviors at different times and in different places. Clearly, such an approach goes beyond that of constructionism with the emphasis on the fluidity of identity.

Although it is the constructionist view and not essentialism that is in accord with much contemporary social thought, it is clear that such a view weakens the political commitment and agenda of some groups. One conceptual resolution to this difficulty involves the claim of strategic essentialism, involving the decision to shelve, perhaps temporarily, the constructionist argument in order to benefit politically from the essentialist argument. Whether or not it is appropriate, intellectually speaking, to reject the constructionist position for pragmatic political reasons is clearly open to debate.

Some work on the diversity of sexual identities has turned to a set of ideas labeled **queer theory**, an uncertain and rather controversial term that refers to a concern with all people who are seen as and/or have been made to feel different, such that there is a common identity on the fringes; this form of theory also emphasizes the fluidity of identities and is concerned about empowering those who lack power. A queer epistemology may be needed in order to challenge the way in which sexual dissidents have been treated within geography. Several studies have considered the ways in which particular sexual identity groups attempt to challenge the mainstream perception of space as heterosexual.

- For example, with reference to the participation of a gay, lesbian, and bisexual group in the annual St Patrick's Day Parade in Boston, Davis (1995:301) noted that: 'Symbolically, the very existence of alternative sexualities was a threat to the locally prevailing notion of what it meant to be Irish.'
- Similarly, L. Johnston (1997) discussed the contested site of a gay pride parade in Auckland, New Zealand, noting that the first two gay pride parades, in 1994 and 1995, took place outside of an area that is generally accepted to be gay, and hence were seen as challenges to the larger heterosexual community, as a queering of the streets. The 1996 parade took place within an area widely perceived to be gay, and proved to be less controversial. This is a good example of the unsettling effect, for those who are not gay, of boundary crossing by those who are gay.

Geographies of the Disadvantaged

The cultural geographic concern with others and their identities and places, linked as it is to a diverse body of postmodern and other theory, has expressed itself primarily in the study of ethnicity, gender, and sexuality. But there are, of course, numerous additional groups who have been regarded as others, such as homeless populations, the unemployed, the disabled, and the elderly. Some of these groups have a visible presence in the landscape, while others may be less visible, and some of these may be controversial to the majority, while others may not be viewed negatively.

The accounts of gendered and sexual landscapes observed that one of the basic premises behind much research is the claim that the landscape in general is made by and for heterosexual men; indeed, much feminist work asserts that this is intentional in order to perpetuate patriarchy. But it is clear that landscapes also 'presume able-bodiedness, and by so doing, construct persons with disabilities as marginalized, oppressed, and largely invisible "others"' (Chouinard 1997:380; see also Golledge 1993). Discrimination can thus take the form of **ableism** or **ageism**.

As noted above, some public spaces in urban areas may be threatening for women, and also for those who display a sexual preference other than heterosexuality. A similar line of argument also applies to the elderly, disabled, and to members of many ethnic groups. The fact is that many public spaces are unsafe for some members of the larger population if they are not seen as conforming to a particular identity that some members of the dominant group judge to be appropriate. It is important to acknowledge, however, that those individuals who are responsible for making public places unsafe for others are not typically representative of the dominant society by virtue of the very fact that they are prepared to resort to harassment, intimidation, or violence. Notwithstanding this fact, it is the case that crime or the threat of crime is one of the means by which the behavior of some people is severely constrained.

There are areas in some cities, usually an inner city area, that are seen as different and are treated accordingly because they lack some of the services that the larger urban area enjoys; in such cases difference can be disadvantage. Indeed, the very labeling of an area as inner city is interpreted in a negative light. Certainly, it is often the case that disadvantage is perceived negatively by others, thus contributing further to the disadvantage. Boundaries may be drawn and redrawn both spatially and culturally as a means to distinguish between those who belong and those who do not belong. Box 8.5 discusses the example of Gypsy communities. Clearly, much contemporary cultural geography is premised on the assumption that women, those who are not heterosexual, the disabled, and numerous others do not fully belong either culturally or spatially such that they are, at least in part, excluded.

Geographies of Resistance

It has been usual in much social science to link resistance and conflict with social class although, not surprisingly, cultural geographers working in the Sauerian tradition showed little interest in class as a basis for delimiting groups and describing landscapes. Rather more surprisingly perhaps, class has not emerged as a key concern in the current postmodern fascination with identity and place. It might be argued that the rise of feminism effectively displaced the traditional Marxist concern with class in many sociological accounts of conflict, and similarly functioned to limit any possible geographic interest to a few conceptual discussions and empirical analyses. When class is employed, it is often linked to the concept of resistance identities as defined earlier in this chapter.

Arguing from a Marxist perspective, Blaut (1980) noted that classes are an appropriate variable for delimiting cultures, and that certain individuals or classes typically exert power over others, a situation drastically affecting the behavior of groups. Most work focusing on class has also often focused on the conflict between classes. Harvey (1993) detailed a powerful example of a geography of resistance that was interpreted in terms of class power relations, especially the way in which industry was prepared to exploit rural poverty and to minimize expenses at any cost. This analysis was of an industrial accident, a fire that occurred in a chicken-processing plant, the exit doors of which were locked, killing twenty-five workers and seriously injuring fifty-six.

The various new social movements are sometimes seen as new precisely because they are not usually class based. Recall that, for Castells (1997:113), such social movements can be viewed in terms of three characteristics, namely, identity, adversary, and goal (see Table 5.1). Certainly, it appears that most types of resistance identities are not directly tied to class, although they are most appropriately interpreted in terms of their opposition to power. The environmental movements

Box 8.5: A Landscape of Exclusion

Throughout much of Europe, Gypsies are widely recognized as a distinctive group living on the margins of a dominant culture; they are often labeled an ethnic or cultural group. Although there are many different versions of Gypsy culture today, they are most easily distinguished by their distaste for waged labor and the lifestyle that this implies.

But Gypsies are not only a minority cultural group in several European countries, they are also a group that do not fit easily into the standard classifications imposed on people by a dominant culture. This uncertain status is related to their characteristic mobility and to the fact that they are usually viewed as different, as outsiders. But this status is also related to the discrepancy between their romantic rural image and their typical lifestyle. For example, in a country such as Britain, the stereotypical romantic and rural perception of Gypsies remains, notwithstanding that the spaces they occupy are as likely to be urban wasteland as they are to be rural. Certainly, the places that Gypsies occupy are, like the people themselves, judged to be different and outside. They are landscapes of exclusion that are the consequence of and that contribute to the outsider status of the group.

Sibley (1995:68) also explained the popular association of Gypsies and dirt: 'Here, the problem is Gypsies' dependence on the residues of the dominant society, scrap metal in particular, and their need to occupy marginal spaces, like derelict land in cities, in order to avoid the control agencies and retain some degree of autonomy, confirm a popular association between Gypsies and dirt.'

Places occupied by Gypsies are avoided by members of the dominant population because they are seen as threatening. As Sibley (1992:112) noted, 'a fear of the "other" becomes a fear of place'. Further, the apparently disorganized character of Gypsy space compared, for example, to a homogeneous suburb results in a devaluing by the dominant culture of that space and of the people occupying that space. In the city of Hull, England, where Gypsies have lived for about 100 years, numerous conflicts eventually prompted the building of two permanent locations for what was widely regarded as a deviant group. One was in a heavily polluted industrial area and the other was in an old quarry that was used for the dumping of garbage. Outsiders relegated to the outside; deviants to residual places.

noted in Chapter 5, the various ethnic separatist and nationalist movements noted in Chapter 7, and various anti-war movements are obvious examples. An analysis of the Baliapal social movement in India, which arose in opposition to the decision to locate a military establishment that would result in evictions of some of the local population, resulted in the area becoming a 'terrain of resistance' (Routledge 1992:588).

Many other subcultures, especially those associated with youthful populations, are characterized by their opposition to mainstream lifestyles and mainstream places. Often such groups choose to display their difference, and thus to express their resistance to and rejection of dominant cultural values, through the adoption of alternative lifestyles including musical and clothing preferences. Certainly, one of the ways in which it is possible to interpret many social trends is in terms of their apparent or sometimes explicit opposition to the conventional.

SITES OF CONTESTATION

Recognizing that there are resistance identities is one component of the larger cultural geographic concern with the fact that different places mean different things to different people; indeed, 'places are the results of tensions between different meanings' and 'are also active players in these tensions' (Cresswell 1996a:59). In some cases, different meanings might comfortably coexist, while in other cases they might result in confrontation. Inevitably then, some places are contested sites because they are valued in different and incompatible ways by more than one group.

- Jerusalem, for example, contains sites that are sacred in Judaism, Christianity, and Islam, and access to these sites at particular times, and even ownership of these sites, is a contributing factor to the ongoing difficulties of the larger region.
- In a rather different context, Morehouse (1996:6) discussed issues of changing power, influence, and control as these related to the question of the boundaries of the Grand Canyon National Park, and thus told 'a story of the contests that have taken place regarding how the area should be shared, inhabited, protected, and used'. Among the many groups that have contested this matter with the Park Service are various Aboriginal populations, other government agencies, transportation companies, specific land use interest groups, and recreational interest groups.
- Cresswell (1996a) employed the term 'heretical geographies' to refer to behaviors in place that are in some way different from those considered normative. For example, with reference to the protest by women outside the Greenham Common military base in England during the early 1980s, women were popularly seen to be out of place both because they deserted their homes and because they occupied an especially unfeminine place. The fact that many of the women present were reported to be indulging in lesbian behavior further added to their otherness.

There is also an interest in considering the implications of the different evaluations that are made of, for example, buildings and particular lifestyles by cultural groups on the one hand and by the state on the other hand. The essential justification for such concerns is the postmodern recognition that societies comprise a plurality of cultural groups such that there are always conflict issues arising over who is responsible for exercising control over both places and the production of culture. An analysis of religious buildings in Singapore focused on 'the oppositional meanings and values invested in religious buildings by individuals on the one hand and the state on the other', but noted that the different meanings and related tensions did not usually result in conflict, although some resistance was involved because individuals adapt the meanings that they invest in religious buildings to avoid conflict (Kong 1993a:342).

Box 8.6 extends these ideas along more clearly humanistic lines in a consideration of the value of using literature and other forms of artistic endeavor as a means of enhancing our understanding of the identities of people who are disadvantaged and whose literature may serve as a principal expression of resistance.

Individual Rights and Group Rights

Recall that the conclusion to Chapter 8 included a brief reference to multiculturalism and to the related question of individual rights and group rights. The material included in the present section provides a substantial context for some reconsideration of this important matter, and Box 8.7 introduces discussion of this sensitive topic through reference to the notion of contested cultures.

It seems clear from the discussions in this section that when groups regard one another as different, then that difference is very often understood as otherness. Whether the groups in question are relatively equal or not, the result of this understanding is one of seeing the groups as mutually exclusive categories. There have been two typical policy responses to this situation.

- Liberal individualism favors an assimilationist society where group membership is voluntary, essentially private, and not enshrined in any political or legal policies or practices.

Box 8.6: 'The Soul of Geography'

Humanistically inspired cultural geographers have studied places and people especially as they are represented in various fields of human endeavor, notably literature, art, and music. Most such studies are based on the premise that impersonal objective research cannot meaningfully expose the experiences of identity and place as can an astute observer or, especially, a participant in the making of that identity and place. Most critically, it is because cultural geographers are unable to experience everything that they might choose to research—especially the lives and places of those who live outside the mainstream, who lack power, or whose identities and places are not valued by the majority. Accordingly, some cultural geographers have considered the way that people express their feelings about identity and place through their artistic activities. Indeed, for Macphail (1997:36), the 'entire spectrum of artistic human endeavour has the potential to be embraced by humanistic geography', not only popular sources such as novels and landscape painting, but also poetry, music, sculpture, and dance.

Responding particularly to the suggestion that some cultural geography was failing to represent many ordinary people, literary and other sources are increasingly being accessed to allow those whose experiences are not typically heard to become participants in research through seeing how they have been represented in literature and other fields of human endeavor, through reading their literature, and through viewing and interpreting their artistic works. For example, in an overview of Canadian prairie literature, Avery (1988:272) stressed that a careful reading facilitated understanding of the geography of the region and especially the contribution made by women to that geography: 'The female voice may not have had an impact on the actual settlement of the prairies, because of the relatively powerless position of its speakers at the time of settlement, but it

can be heard in prairie literature.' Of course, such sources are recognized as being but one representation and as potentially misleading.

One discussion of the need for literature to be acknowledged as a source of understanding for the cultural geographer particularly stressed the value of poetry as a way of listening to those who are oppressed and not normally heard: the need to listen to and respond to voices that are not a part of the mainstream prompted Watson (1983:393) to assert that 'the soul of geography is the geography of the soul'. Poetic expressions of the polluted landscape of the mining settlement of Kirkland Lake, Ontario, and of the oppressed lives of mine workers were described as 'primal knowledge' (Watson 1983:398). These thoughtful claims are to be understood in the context of humanistic interests rather than in terms of any involvement with the cultural turn.

Similarly, a humanistic study of Soweto poetry during the apartheid era—Soweto is the major Black residential area outside of Johannesburg—emphasized the inability of Whites to understand the Black experience, such as the fear experienced on an ongoing basis, the lack of belonging related to some of the apartheid legislation, and most critically the dehumanizing effects of that legislation. Such poetry is a form of resistance, and White academic authors 'cannot capture the same experiential passion which black urban residents have embodied in Soweto poetry' (Macphail 1997:40)

With reference to ethnic groups in large urban areas Watson (1983:397) asserted: 'Their eye may have been full of prejudice, and therefore on a purely impersonal, objective basis, not to be trusted, but that prejudice dominated the places they valued, sought, fought for, were displaced from, or held. This should be an axiom in geography: *People generate prejudice; prejudice governs place.*'

- A range of ideologies—but especially those associated with members of groups that have experienced oppression—favor a separationist policy involving the creation of a society that includes a degree of political and legal autonomy for some groups. It is this second policy that has come to dominate in the Western world.

In this context, recall that, as noted in Chapter 4, it can be argued that in the specific case of the United States most rural ethnic landscapes and ethnic neighborhoods in cities are American rather than some more localized ethnicity, reflecting cosmetic rather than fundamental differences. Zelinsky (1997:161) claimed: 'Yes, it is certainly important to look beyond the dominant culture, to learn how all those many alien peoples have fared as they tried to cope with that huge, absorbent phenomenon we call the American cultural system. What I question is the effectiveness of examining pseudo-ethnic landscapes as a strategy for getting at cultural adjustment or survival.'

However, since about 1970, both conservative and radical perspectives have contributed to the dominant discourse that sees many societies as comprising distinctly different groups of people, usually distinguished according to some ethnic criteria; it is a discourse that shows ideological plasticity. According to this argument, and regardless of ideological perspective, it has been contended that there are ethnic groups (often called racial groups), and that these groups can be defined by reference to a distinctive culture that is shared by all individual members. Further, just as cultures are reified by those who are outside, so they can be reified by those who are inside. (See Your Opinion 8.4.)

If the policy of group opportunity that is central to multiculturalism is indeed found wanting, as the alternative policy of individual opportunity was found wanting because of the legacy of an unequal past, then the challenge to policy makers is that of identifying and implementing some alternative policy. Some favor a society characterized by social equality but also by explicitly differentiated groups, while for others the solution rests in the idea of equality of life chances, an idea that implies the introduction of social policies that target the most disadvantaged members of a society regardless of their cultural identity. Certainly, it might be claimed that such policies are logical given the postmodern circumstances of contested cultural identities as described in Box 8.7, specifically the fact that any policy that identifies groups as different effectively argues that there is a single and almost unchanging identity for that group.

Box 8.7: Contesting Culture

There is much evidence to suggest that many people do not regard themselves as members of one group, but rather as members of many groups, with the specific affiliation at any one time varying according to circumstances. Although it may be that such a situation is particularly characteristic of any plural society (see Glossary), it is also well established in the sociological literature that each person has many selves and plays many roles and, of course, it is also a central idea in much postmodern thought. This simple observation has implications for any cultural geographic analysis that stresses the links between culture, ethnicity, community, and place.

Thus, with reference to multiethnic Southall, London, and on the basis of extensive fieldwork, Baumann (1996:5) observed that 'the vast majority of all adult Southallians saw themselves as members of several *communities*, each with its own *culture*. The same person could speak and act as a member of the Muslim *community* in one context, in another take sides against other Muslims as a member of the Pakistani *community*, and in a third count himself part of the Punjabi *community* that excluded other Muslims but included Hindus, Sikhs, and even Christians.'

This observation is important because it is usual for the dominant discourse to rely on equating culture, ethnicity, and community, and sometimes place, and if this is not possible, as it clearly is not in the case of Southall, then many of these simple associations literally fall apart. To describe the tendency for people to create new communities, or to subdivide existing communities, or to fuse existing communities, Baumann (1996) introduced the term 'demotic discourse' as an alternative yet often overlapping concept to that of dominant discourse.

Symbolic Landscapes

Rethinking Cultural Landscape

Although cultural geographers continue to pursue a wide range of empirical interests, there is a persistent concern with landscape. This interest exists regardless of whether analyses are informed by the Sauerian tradition, by humanism, by postmodernism, or by some other conceptual basis, and, according to Groth (1997), contemporary research into cultural landscapes, although lacking a single coherent paradigm, is characterized by six widely held tenets—these six are introduced in Box 8.8.

Both this and the following section have landscape as their primary concern and, although any classification imposed on the material covered in these sections is necessarily somewhat arbitrary, the current section focuses essentially on the symbolic meaning of landscape, on the way in which humans stamp their identity on landscape through their language and religious beliefs, on the sacredness of some landscapes, on the textual reading of landscape, and on landscapes as used in literature. The following section discusses ordinary landscapes with a distinction drawn between folk culture landscapes and popular culture landscapes.

The general acceptance within twentieth-century North American cultural geography of Sauer's definition of landscape as the cultural transformation of the natural world has weakened since about 1970, especially because of the increasingly diverse scholarly interest in landscape. Use of the term 'explosion' is doubtless something of an exaggeration, but there is certainly an ongoing and seemingly ever-growing concern with the cultural landscape both within and beyond cultural geography, a concern that is at least partly based on a wider acceptance of the idea that landscape can be interpreted as scenery. But, more important, within geography the original Sauerian interest has been both modified and complemented by additional interests. For example:

- the claim that it was cultures that authored landscapes has been challenged (Duncan 1980)
- a humanistic perspective of landscape as a *way of seeing*, typically reflecting only the views of the dominant group' has been employed (Cosgrove 1985:46) and subsequently challenged (Olwig 1996b: 645)
- a dialectical methodology has been proposed (Kobayashi 1989)
- the important role played by social power relations has been acknowledged (Anderson and Gale 1992)
- landscape has been viewed as theater, spectacle, or carnival (Cosgrove 1997a)
- the idea that landscapes may be understood as texts and as outcomes of discourses has become popular (Duncan and Duncan 1988)

Commenting on these developments, Demeritt (1994:167) noted that cultural geographers are setting 'aside the hiking boots preferred by Sauer for

---

Your Opinion 8.4

It might be that a principal reason why many minority groups in larger societies willingly embraced the reification of their culture by the dominant group concerned their seeking some distinctive rights for their group. Often such group rights were justified on the grounds that the group had been treated unfairly in the past, usually in the context of colonialism, and on the grounds that the principle of individual equality of opportunity, the liberal individualism claim, was not adequate to overcome the earlier period of discrimination against a group. Acceptance of this argument—and it is indeed the dominant discourse in many countries, including the United States, Canada, and Britain—resulted in the targeting of specific groups for specific rights that are in addition to those enjoyed by the majority population. In cultural geographic terms, what is happening here is that culture—group identity—is being reified and used to justify some form of political and intellectual separatism. It is possible that the cultural geographic interest in identifying different and distinct cultural landscapes has contributed to this discourse. Critics of multiculturalism, such as Turner (1993:414), have argued that this dominant discourse has come to favor 'cultural nationalists and fetishists of difference. It can also be argued that the most disadvantaged individuals in a society tend to miss out on many of the opportunities available to their group.

the patent leather shoes more appropriate to field-work in the cafés and art museums now of empirical interest'.

Most of these modifications of and additions to the Sauerian focus are associated with humanism or with the cultural turn. More generally, these interests in landscape reflect an expanding concern with identity and symbolism: Cosgrove (1984:13–14) noted both that landscape 'is not merely the world we see, it is a construction, a

composition of that world', and also that 'landscape is a social product, the consequence of a collective human transformation of nature'. Individuals as members of groups construct landscapes intentionally and not so intentionally. Three additional points merit mention:

- First, although the discussions in this chapter focus on work by cultural geographers, it is recognized that the interest in landscape shown out-

Box 8.8: Six Tenets of Cultural Landscape Studies

Observing that—despite the centrality of the *Landscape* magazine tradition initiated by J.B. Jackson—there is no one monolithic approach to studies of the cultural landscape, Groth (1997) identified six widely held tenets of such studies.

- First, ordinary, everyday landscapes are important and worthy of study because, given an interest in cultural meaning, it is our everyday experiences that are essential in the formation of that cultural meaning. Logically, then, the landscapes in which we are interested are not restricted to those that are in some way different, but include those in which we live and work. It is usual to trace this idea back to the introduction of *Landscape* magazine in 1951, but, of course, the traditional Sauerian approach by no means excluded the ordinary, although it certainly emphasized the rural and did not encourage studies of the ordinary in urban areas. As Groth (1997:6) noted: 'For every forty studies of barns and fields, there has been only one about urban factories, workshops, offices, or corner stores as workplaces.'
- Second, contemporary research into cultural landscape is as likely to be urban as it is to be rural and as likely to be concerned with consumption as with production. These are substantive changes to earlier interests that mean more than just a changing scene; the interest in urban landscapes particularly is linked to questions of power and inequality in the landscape.
- Third, cultural landscape studies sometimes choose to stress the uniformity of a given landscape as a reflection of some overarching national considerations, and sometimes stress local identities and differences from place to place as a reflection of ethnic variations. These two emphases are best seen as complementary. Relevant questions here concern whether or not a particular national culture, perhaps interpreted in superorganic terms, proves able to

override many of the more local variations in culture and identity as these are reflected in landscape. Is the landscape best viewed as 'one book' or as 'multiple, coexisting texts' (Groth 1997:7)?
- Fourth, writing about landscape emanates from both scholarly and more popular sources, reflecting both academic and literary styles. Indeed, many of the articles that appeared in *Landscape* magazine exemplified the literary style, even when written by an academic, and the collection of landscape readings edited by Thompson (1995b) continues this tradition of incorporating both the scholarly and the popular. It is not difficult to understand that landscape attracts writers in both scholarly and popular veins, just as landscape also attracts artists, photographers, filmmakers, poets, and novelists.
- Fifth, the fact that there are numerous and quite different approaches to the study of cultural landscape is a consequence of the interdisciplinary character of the enterprise. Particularly significant is that some writers elect to employ a theoretical basis while others favor a more explicit empirical focus. Some studies of landscape stress description while others stress interpretation. Of course, this is an issue that arises frequently within this textbook, especially as it relates to the different perspectives taken by Sauerian and new cultural geographers.
- Sixth, most landscape analyses are based on spatial and visual data, such that the actual landscape is often studied directly. Landscapes are, of course, accessible to us and can be experienced differently from place to place. One of the long-standing complaints about landscape studies as practised by cultural geographers relates to the seeming overemphasis on what can be observed as opposed to what is recorded in written texts. Notwithstanding such concerns, the emphasis on seeing and recording and interpreting what is seen remains fundamental.

side of geography is considerable: 'It is important to understand that *landscape*—as revealed in *place*—is not the province of one, two, or three academic disciplines, but is the concern of at least a score of art forms and academic fields' (Thompson 1995a:xii). Or, as expressed by Lowenthal (1997:180): 'Suddenly, landscape seems to be everywhere—an organizing force, an open sesame, an avant-garde emblem, alike in fiction and music, food and folklore, even for professors and politicians.'

• Second, although both this and the following section focus on some of the more recent approaches to the study of cultural landscape, it must not be forgotten that there remains a substantial interest in the evolution and regionalization of landscape and also an interest in landscapes as expressions of human and land relations (chapters 3, 4, and 5, respectively). Perhaps most significantly, the concern best articulated within the *Landscape* magazine tradition continues to be evident in the humanities and social sciences generally.

• Third, in some cases, different understandings of landscape are explicitly opposed to one another in some substantive sense. Most obviously, there is the relatively clear difference of opinion between the Sauerian emphasis on visible and material landscape and other more recent meanings that focus on the symbolic interpretation of landscape. Thus, Hart (1995:23) indicated satisfaction with the idea of landscape as being what we see, challenged the symbolic approach, and objected to those who 'have endowed the concept of landscape with magical, mystical, or symbolic significance, or have loaded it with metaphysical connotation'.

To summarize this discussion: in addition to the idea that landscape is what we live in (nature transformed) and the idea that landscape is what we look at (the visible world around us), there are also a variety of meanings of landscape that center more generally around the idea of landscape as symbol, and it is with these meanings that this section is concerned.

Writing the World: Identity in Landscape

Regardless of conceptual inspiration, there is agreement that cultures express themselves in the landscapes that they create.

• Recall that for Sauer (1925:30), culture was 'the impress of the works of man upon the area'. Much of the content of this textbook has reflected this basic idea.

• Clearly, however, the idea that cultures are expressed in landscape invites an interpretation of landscape as something more than material and visible, namely as symbolic. With reference to the national scale, Meinig (1979c:164) noted that a 'mature nation has its symbolic landscapes. They are part of the iconography of nationhood, part of the shared set of ideas and memories and feelings which binds a people together.' More generally, all landscapes have a meaning and value for their creators and for those who subsequently occupy and recreate the landscape.

The current concern is with the way that cultural groups may assert their identity on landscape, intentionally and otherwise, such that the landscape has a symbolic identity. A description and interpretation of landscape that focuses on uncovering symbolic meanings, known as **iconography**, is premised on the idea that the identity of a landscape is expressed through symbols. Necessarily, this interest has been evident previously, most obviously in the regional discussions in Chapter 4, especially the accounts of vernacular regions and homelands. The substantial examples of the Hispano and Mormon homelands involved cultural groups adapting to physical environments, changing that environment through the impress of their culture and creating a sense of place—their landscape. In this sense, homelands are landscapes that are laden with meaning, that reflect the attitudes, beliefs, and values of the occupying cultures. Clearly, such landscapes can also be read as texts, an emphasis that is introduced later in this chapter.

The specific interest in the present discussion is with the impacts of selected aspects of language and ideology. Language, through the mechanism of the naming and related mapping of places, and ideology, through the creation of sacred spaces, are two of the components of culture that function to contribute to the symbolic character of landscape. The logic behind these accounts is similar to some of the earlier material on ethnicity, national identity, gender, and sexuality, all of which can contribute to the creation of symbolic landscapes.

NAMING PLACES

It is often claimed that language is central to individual and group cultural identity as it is both a means of communicating and also a symbol or emblem of groupness. Although there is debate about this matter, as noted in Chapter 7, the symbolic function may render language potent for ethnic and nationalist sentiments, and indeed in some instances language may operate primarily as a symbol and serve only a minimal communication function—such is the case with the Irish language. In Wales, the Welsh language and culture have often been seen as inextricably entwined such that 'the language may be considered the matrix which holds together the various cultural elements which compose the Welsh way of life' (Bowen and Carter 1975:2), and in Ireland, survival of the Gaeltacht, the area where Irish is spoken, was argued to be 'synonymous with retention of the distinctive Irish national character' (Kearns 1974:85). Further, the symbolic interactionist perspective, as discussed in Box 5.2, involves the idea that social realities are constructed and maintained by communication, thus acknowledging the centrality of language as a variable affecting both the spatial patterning of society and the social character of space. One interest in language concerns the particular context of place names or toponyms.

- Some cultural geographers, inspired by the Sauerian school, have analyzed place names as examples of the ability of a particular group, usually in their capacity as first effective settlers, to impress their identity on landscape; place names have also been used as a surrogate for culture spread and as a basis for regional delimitation. There are numerous place name studies in the broad Sauerian tradition. In a series of United States analyses, Zelinsky (for example, 1967, 1983) considered various aspects of place name geography, frequently relating specific distributions to the rise of nationalism. With reference to Finnish place names in Minnesota, it is clear that the number of names a group could bestow was related especially to variations in physical landscape and to the arrival date of the group—that is, the first effective settlement concept. This example, detailed in Box 8.9, stressed the important role played by place names as one means of helping to build an environment in which a cultural group felt comfortable and as one means by which ownership was confirmed.

- Other cultural geographers, inspired more by identity politics as discussed in Box 8.3, have analyzed the naming of places in terms of unequal power relations and related identities. In both cases, there is a concern with cultural identity, with place names as symbols—one of the ways by which people attach meanings to places—along with a recognition that the process of naming places is not entirely an innocent one. Place name studies that are informed by identity politics build upon earlier cultural geographic work to emphasize that the naming of places is one of the means by which a group strives to dominate place and, by extension, dominate other groups. Most such work has addressed the naming of places by Europeans in overseas areas and centers on the contestation of place and the way in which European place names assisted both in creating a landscape that is symbolically and materially European and in asserting state ownership and control of place (Box 8.10).

Regardless of conceptual inspiration, it is clear that place names contribute to the social construction of place and that they are able to generate powerful emotions. To name places is to write upon the world.

SACRED SPACES

Ideology, especially in the context of religion, is one of the principal ways in which identity is constructed. In some areas, particularly those experiencing an upsurge in fundamentalism, religion is of increasing importance, while in other areas where there is an increasing secularization of society, religion in the conventional sense of the term may be of lessening importance. Where the latter is the case, as in much of the United States, it may be that another bonding factor, notably a civil religion or nationalism, is replacing religion.

As discussed in Chapter 4, spatial variations in religion are often used by geographers to facilitate the delimitation of both cultural regions and cultural groups. Indeed, for some groups religion is the principal binding element, such that there may be close relationships between religion and the evolution, regionalization, material identity, and symbolic identity of landscape.

The concept of **sacred space** is one aspect of the important idea that all human landscape has a cultural meaning in that sacred space is imbued with some particular meaning for an individual or a group. It can be contrasted with profane or ordinary space, which, although it conveys meaning, lacks particular symbolic quality. Sacred space implies some especial attachment to place, perhaps spiritual devotion, reverence, affection, pride, nostalgia, or a more general sense of belonging. As these possible implications suggest, the degree of sacredness of sacred spaces is highly variable, and a sacred space may be long lasting or quite ephemeral dependent on the type of space, the number of people involved, and the degree of sacredness. Sacred spaces also vary in size and impact (if any) on the visible landscape. Given such a wide variety of circumstances, Jackson and Henrie (1983) proposed a classification that recognized three types of sacred space:

- The *first*, mystico–religious space, includes those spaces associated directly with religious beliefs and experiences. There are many such places. Some are sacred because a specific religious event occurred there or is believed to have occurred there; in some cases, an event is believed to have occurred that is inexplicable without reference to

Box 8.9: Naming Places: Finns in Minnesota

For Kaups (1966:381), the principal factors responsible for the number, distribution, and types of Finnish place names in Minnesota were the 'Old World background of the immigrant group, the occupational structure of the immigrants, the time and place of their arrival, the nature of the physical environment in which they settled, and decisions made by American officials'. Figure 8.1 locates ninety-two such names, primarily in northeastern Minnesota and secondarily in the west-central part of the state; all of these names are in rural areas.

The first Finnish settlers in the state arrived in the mid-1860s and located in areas where other settlers were already present, such that there was no opportunity for the incoming Finns to name places. During the 1870s, other Finns located in the west-central area, but the majority arrived during the 1880s and 1890s and settled in ethnic enclaves in the northeastern area as pioneer agriculturists. Accordingly, it was here that Finns were able to dominate the rural scene as the first effective settlers. But numerical dominance was not sufficient to ensure that the Finns were always able to bestow place names of their choosing; for example, a 1912 petition by Finnish settlers proposed the Finnish name 'Salmi' for a new township, but local county officials decided, without any explanation, to name the township 'Vermilion'.

Kaups employed a classification of both physical and cultural place names that distinguished three types, namely, possessive (further subdivided into personal and ethnic), descriptive, and commemorative (Table 8.1). The majority of names belong in the possessive category, and most of these are personal.

- Possessive names are a way of indicating an association between a Finnish individual or group and a particular place.
- The descriptive names identify the character of a place through incorporation of a Finnish word in the place name; for example, the place name 'Kumpula' describes the location through use of the Finnish word *kumpu*, meaning hill or mound.
- Most of the commemorative names are those of important Finnish literary figures.

The majority of possessive names refer to physical features, the majority of commemorative names refer to cultural features, and the descriptive names are evenly divided between the two. In addition, fourteen abandoned names were identified, thirteen of which referred to cultural features.

Not included in Figure 8.1 or in Table 8.1 is another group of Finnish names, labeled as in-group names. These are names that are used by the Finns, but that have not received official sanction nor appeared on official maps. Fifty-three such names were noted, located in the same areas of the state as the official names. These in-group names arise partly because ethnic groups are not usually free to name places as they choose, and such names are one means of creating a local environment that is supportive of group identity in new and possibly difficult surroundings.

Figure 8.1: Finnish place names in Minnesota. Finnish place names are located in the northeastern and west-central parts of the state and in a few other scattered locations. Of the ninety-two place names identified, fifty-nine are names of physical features, mostly lakes and creeks, and thirty-three are names of cultural features, mostly hamlets. The relative profusion of physical place names in the northeastern area reflects not only the number of Finns and the timing of their arrival but also the varied physical environment that affords numerous place-naming opportunities.

Source: Adapted from M. Kaups, 'Finnish Place Names in Minnesota: A Study in Cultural Transfer', *Geographical Review* 56 (1966):379.

religion. Most religions create sacred spaces through the expression of their identity in landscape; buildings, shrines, and symbolic markers

are examples of this process. 'Indeed, symbols of religious worship are woven into the very fabric of many areas and give them a special and sometimes unique identity', an identity that is sacred to members of the religious group and that is readily recognizable by others (Park 1994:199). Parts of the physical environment are often considered sacred, especially rivers such as the Ganges River in Hinduism and mountains such as Mount Fuji in Shintoism. Members of the religious group visit almost all such places, but some places are so frequently visited that they may be considered pilgrimage sites. Figure 8.2 locates principal pilgrimage places for Hindus in India. For members of the Islamic faith, Mecca and Medina are so sacred that pilgrimage as least once is required of all whose circumstances permit—some of these sites even have their sacredness enhanced by the exclusion of others from those places.

- *Second*, some homelands qualify as sacred space because they represent the roots of each individual and of the larger group. Some individuals may perceive their place of birth and/or their area of upbringing as sacred while others may not. The homeland concept as it relates to identity was discussed in Chapter 4 and also earlier in this chapter.
- *Third*, there are historical sacred spaces. These are places that have been assigned sanctity as a result of some important event either occurring there or being remembered there. Most national and ethnic groups identify several such places. Indeed, landscapes may include places that play an important role in the construction of identity and that have accordingly 'been denoted as essential

Table 8.1: Classification of Finnish Place Names in Minnesota

Number of Place Names	Possessive (personal)	Possessive (ethnic)	Descriptive	Commemorative	Other
Physical	37	8	10	3	1
Cultural	5	2	9	13	4
Total	42	10	19	16	5

Source: M. Kaups, 'Finnish Place Names in Minnesota: A Study in Cultural Transfer', *Geographical Review* 56 (1966):377–9.

codes in the national signifying system, while other sites and landscapes, or elements attached to them, have been consciously or unconsciously excluded from this, being either overlooked or denied' (Raivo 1997:327–8). For many Americans, historical sacred spaces may be associated more with national circumstances than with culture-specific circumstances. Although church membership is common in the United States, a commitment to religion is less common and religious places and religious occasions are evidently becoming less and less sacred. Of course, it is possible to interpret religion very broadly, such that nationalism or civil religion can be thought of in similar terms to the more conventional understanding of religion and can therefore be associated with the creation of sacred space. Examples of such sacred spaces include military cemeteries, monuments, and preserved historical sites. In many instances the sacredness is emphasized through use of national symbols, most usually a flag. Although this circumstance is best documented by cultural geographers in the context of the United States, Zelinsky (1988:175) suggested that 'the workings of the state have set their mark upon the land' in all countries.

Jackson and Henrie (1983) employed this tripartite classification of sacred space in a study of spaces sacred to Mormons and Figure 8.3 locates the principal such sites in the United States. In this research, a first open–ended questionnaire was used to identify all–important places, and a second questionnaire required respondents to rank places. This innovative study concluded that, for Mormons, the most sacred of spaces are those associated with their religion.

In addition to the above classification, and as suggested earlier, vernacular regions can be interpreted as an example of a particular type of space, recognized as different both by those inside and those outside the region. However, as is the case

Box 8.10: Naming and Renaming Places

The question of renaming places in former colonial areas is a source of considerable debate. Accepting the argument that to name is to claim, an attempt to rename can be interpreted as an attempt to reclaim. Berg and Kearns (1996) employed discourse analysis to facilitate understanding of attempts to reinstate Maori place names in favor of European names in one area of New Zealand in the late nineteenth century. An analysis of proposals for and subsequent opposition to renaming highlighted the perceived importance of names to both groups involved. This New Zealand study also stressed the masculinity of the naming of places with many places named for male and often military figures.

But the situation may be more complex than this suggests. 'Returning to a precolonial name can be read as a simple postcolonial strategy of legitimating forms of knowledge and experience demeaned or suppressed under colonialism. Alternatively, this act of reinstating the "original" or "authentic" name can be read as a return to problematic notions of cultural purity and authenticity' (Nash 1998a:1). Regardless of the interpretation favored, place renaming is an example of place contestation, such that in some cases, place names may become a site of resistance. Nash (1998b:74) interpreted the Irish landscape as being 'in deliberate opposition to English landscape aesthetics', noting that 'the cultural significance of the Irish landscape often resides in the stories associated with particular places and in the place names that connect language and locality'.

In addition to asserting authority over and potentially the ownership of a landscape through the naming of places, dominant groups also achieved these goals through the related mapping of landscapes, a circumstance that has been central to the cartographic enterprise for many years as evidenced by the fact that in early China map making was the responsibility of government officials. Naming and mapping and the related production of knowledge can be thought of as exercises of power.

The naming and renaming and related mapping and remapping of places contribute to the making of the world in which we live. Names and maps can direct our attention to things that we might not have seen before and both provide an indication of the circumstances of the various groups occupying a landscape. Further, both place names and mapping are cases of cultural writing on the world and such writing can be read as a text.

The Ganges River is sacred among Hindus, who believe that those who bathe in its waters will attain heaven (*Miles Ertman/ Masterfile*).

with sacred spaces, it is usual for the meaning of a vernacular region to be different for these two groups. Typically, for those living in the vernacular region, that region is home and is valued positively, while for those living outside the region, it is someone else's home. The vernacular name, Bible Belt, which is used to refer to parts of the southeastern and south-central United States, is a regional name in which many residents take great pride—thus, Oklahoma City proclaims itself the 'Buckle on the Bible Belt'. However, for some who live outside the region, the name is used in a derisive manner—thus, there is a 'Bible and Hookworm Belt' or a 'Bible and Lynching Belt'.

Other vernacular regions, such as the French Riviera, comprise what might be called élitist space. The rise of a distinct French Riviera was a response to the requirements of a particular class in French society and to their desire to identify with a given environment. In much the same fashion some urban neighborhoods and communities might be regarded as vernacular regions created by particular groups to help express their identity. Most generally, vernacular regions are named places, and it is the fact of being named that implies that the identity of the place is important, that the area has meaning to the occupants. For the Ozark region of the United States, Miller (1968) used folk materials to delimit a distinct region and to identify the self-sufficient family farm as the principal way of life. The inhabitants of a vernacular region may thus share distinct values and an especially clear link between people and place is implied.

Reading the World: Landscape as Text

Reading the landscape is an important tradition in cultural geography and the metaphor of landscape as text—a manuscript that has been written upon by people through time—is therefore not a new one, although it is certainly an idea that has experienced

Figure 8.2: Hindu pilgrimage sites in India. Visiting pilgrimage sites, often on a regular basis, is one of the means by which individual members of a religion reinforce their identity as a member of a group. Certainly, a shared Hindu identity is generally considered one of the factors that helps promote unity among an otherwise diverse Indian population. This map shows the principal pilgrimage sites, some of which, such as Varanasi on the Ganges, are visited by more than a million people annually. At many of these locations, Hindus seek help from deities in their personal and social lives. Brahma (symbolizing creation), Vishnu (symbolizing preservation), Shiva (symbolizing dissolution), and the Mother Goddess (symbolizing energy) are four of the more important deities.

Source: J.O.M. Broek and J.W. Webb, *A Geography of Mankind*, 2nd edn (New York: McGraw-Hill, 1978):143.

something of a renaissance in connection with the cultural turn. Thus, before about 1970, it was quite usual for historical and cultural geographers to conceive of a landscape as a document, written on by successive groups of people with each writing obliterating some of what was written previously and also adding something new. The approach of sequent occupance, which is discussed in Chapter 3,

implied such an interpretation of landscape—recall particularly the idea of landscape as palimpsest (see Glossary). In this Sauerian tradition, the writing was accomplished by cultural groups and typically resulted in the formation of cultural regions and, accordingly, these ideas are at the core of chapters 3 and 4, but this traditional view of landscape as text has been further informed by:

Figure 8.3: Mormon sacred sites in the United States. Sites that are sacred for Mormons include those associated with Christianity and also a number of specific Mormon places. As noted in Table 8.2, the three highest-ranking sites are all mystico-religious places specific to Mormonism, namely the Salt Lake Temple, the future city of Zion, and the Sacred Grove. The homeland space seen as most sacred was the state of Utah, with spaces such as present and childhood home considered to be significantly less important. Sacred historical sites were those associated with the history of Mormonism, such as Sharon, the birthplace of Joseph Smith, and Carthage Jail, the site of his death. The importance of historical sites reflects the Mormon concern with historical and religious roots. National sites, such as the Lincoln Memorial, proved to be less sacred.

Worldwide, Salt Lake Temple is *the* religious symbol of the Church of Jesus Christ of Latter-day Saints. There are six major towers and finial spires that signify the restoration of priesthood authority and also numerous other symbols. It is a principal tourist attraction for Mormons and others. Jackson County in Missouri is considered a possible location for the future city of Zion—Zion refers to the idea of a gathering place for those who are pure in heart. The Sacred Grove is a treed area on the Smith farmstead where Joseph Smith experienced his first vision, and the Hill Cumorah is the site that Smith was directed to in order to retrieve the gold plates from which the Book of Mormon was translated. Far West, Missouri, was settled by Mormons in 1836 and is important as the place where a temple site was dedicated and seven revelations received. Kirtland, Ohio, and Nauvoo, Illinois, were the church headquarters from 1831 to 1838 and from 1839 to 1846 respectively.

Source: R.H. Jackson and R. Henrie, 'Perception of Sacred Space', *Journal of Cultural Geography* 3, no. 2 (1983):101. Reprinted by permission.

- humanist ideas, with landscapes being studied as 'our unwitting autobiography' (Lewis 1979:12)
- postmodern ideas, with landscapes regarded as systems of communication that may be appropriated by dominant groups

Given the renewed emphasis on the metaphor of landscape as text, studies by cultural geographers are increasingly being informed by work in literary theory that considers how texts are read, this is sometimes described as the linguistic turn. Further, there is an interest in uncovering the various writings through deconstruction (see Glossary). Recall from Chapter 7 that deconstruction seeks to uncover what is not said in a text, such as

the role played by power relations, and what cannot be said because of the constraints imposed by language, especially given the limitations imposed by a particular discourse. For some cultural geographers, reading a landscape is often an exercise in reading social relations and in understanding individual and collective action.

Although this metaphor of landscape as text is a well-established one, it continues to be a source of critical comment and debate, especially concerning the relative emphases placed on the metaphoric and symbolic content of landscape on the one hand and on the 'real' content of landscape on the other hand. Thus, Mitchell (1996:95) noted: 'Because landscape is *partially* a text or a represen-

Table 8.2: Ranking of Selected Mormon Sacred Space

Place	Mean
Salt Lake Temple	1.92
Future City of Zion	2.12
Sacred Grove	2.39
Utah	2.76
Temples other than Salt Lake City	2.95
Bethlehem	3.02
'Holy Land' (Israel)	3.48
Joseph Smith birthplace	3.70
Nauvoo and Kirtland	3.70
Chapels	4.63
Carthage Jail	4.65
Present home	4.74
Present state of residence	4.79
Utah's mountains	4.95
Regions surrounding Utah	4.95
Childhood home	5.41
Present day Jackson county	5.41
Lincoln Memorial	5.86

Note: Values represent how respondents valued each place on a scale of 1 to 7, with 1 as most sacred and 7 indicating no sanctity.

Source: R.H. Jackson and R. Henrie, 'Perceptions of Sacred Space', *Journal of Cultural Geography* 3, no. 2 (1983):102. Reprinted by permission.

tation we have often tried to understand it *entirely* as a text or representation. Certainly, much contemporary cultural geography emphasizes the reading of landscapes 'as though they were texts' (Peet 1996b). But for many contemporary cultural geographers, reading a landscape involves both reading the text and also reading interpretations of the text, not necessarily a straightforward task. Further, understanding landscapes in these terms requires appreciation that they are related to a number of usually competing discourses, defined as the 'social framework of intelligibility within which all practices are communicated, negotiated, or challenged' (Duncan 1990:16).

Several of the landscape discussions that appear earlier in this chapter—especially in the accounts of feminism, sexuality, place names, and sacred spaces—involved reading landscapes as texts and in terms of prevailing discourses, but these ideas are more explicitly utilized in the following examples.

READING THE URBAN LANDSCAPE

Landscapes do communicate in a meaningful way and the language of landscape can be read and comprehended. Building styles, for example, are a visual text and, in some cases, landscape symbolism may seem to be self-evident and therefore relatively easy to read, such as a tall building in an urban area that is the home of a multinational company, while in other cases a much more subtle and detailed reading may be required, such as in small variations in housing styles, coloring, and ornamentation. Domosh (1992:475) noted: 'The city, like a painting, is a representation, an image formed out of the hopes and ideas of the cultural worlds in which we live. As a cultural construct, the meaning of the city can be deciphered by closely examining its complex relationship with the culture of which it is a part.' The most popular approach to the interpretation of landscape symbolism, to the understanding of place, is that of iconography (see Glossary). This approach originated in art history as a means of interpretation and has accordingly often been employed in studies that stress the visual landscape. Three other related interests are as follows:

- *First*, cultural geographers are recognizing that powerful individuals and institutions, including the state, act as agents to shape the landscape, and landscapes are being read accordingly. In a discussion of the circumstances surrounding a planning dispute concerning proposals for urban expansion in Lexington, Kentucky, McCann (1997) identified two interested groups. Opposing expansion were environmentalists, preservationists, and some horse-farm owners; favoring expansion were developers, the construction trade, and the chamber of commerce. These two groups, called textual communities, read the symbolic landscape surrounding Lexington differently—the former seeing 'an interconnected, living entity, the existence of which benefited the population of the entire region', and the latter seeing 'a series of individual, privately owned land parcels, the use of which was primarily a concern

Some of the buildings surrounding Salt Lake City's Temple Square include the Salt Lake Temple, Mormon Tabernacle, and the Gothic Assembly Hall, among others (*William Norton*).

of individual landowners' (McCann 1997:647). The debate between these two groups—involving two very different readings of the landscape—was played out in the context of the planning process and McCann (1997:660) concluded: 'The morphology of landscape in capitalist societies is, therefore, mediated by the contexts and discourses of the planning process which make use of the legitimacy and power associated with institutional sites in order to maintain a liberal ideology of equality and fairness.'

- *Second*, although it is clear that tall buildings in an urban landscape symbolize economic achievement, it may be that landscape meaning, the sense of place, is the consequence of historical processes that are not quite so visible. For example, each of three buildings in Honk Kong represents a different claim about political and economic power. Thus, the Hong Kong and Shanghai Banking Corporation headquarters has remained at the same address since 1864—the *fengshui* of the location is excellent and this bank has functioned as an agent of British political and economic interests. The Bank of China Hong Kong branch, a Chinese national project completed in 1988, symbolizes the uncertainties relating to the 1997 Chinese takeover of Hong Kong. Central Plaza, the tallest building in Hong Kong, has reoriented the skyline having been built in the previously socially marginal area of Wanchai; the building has multiple tenants and, most significantly, was built by a local group and not by the British or by the Chinese. Three buildings with three different interpretations of power and authority.

- *Third*, landscapes and their meanings are both to be understood as parts of larger discourses. Schein (1997) conceptualized the landscape of one neighborhood in Lexington, Kentucky, as the intersection of several discourses, namely those of landscape architecture, insurance mapping, zoning, historic preservation, neighborhood associations, and consumption. Each of these discourses included ideals, guiding principles, and often even explicit and institutionally mandated rules that functioned to discipline the residents of the neighborhood. Of course, different discourses play roles in many other landscapes, and Box 8.11 discusses the representations of landscape that are promoted by the Australian Tourist Commission in order to help maintain a particular national identity.

MONUMENTS IN LANDSCAPE

Most features in the cultural landscape serve a material purpose—houses provide shelter, roads allow for movement, and many fences keep animals in their place. Other features are essentially symbolic in that they are intended to communicate a message—some fences or hedgerows around

properties may serve the purpose of indicating ownership. Perhaps most obviously, monuments are a landscape feature that are built to expressly make a statement—recall that the earlier account of gender in landscape suggested that such statements might be explicitly masculinist and thus serve to reinforce patriarchy—and, as such, they may prove to be especially popular as sites for political and other activities. Monuments are usually key points of hegemonic meaning in landscape and, accordingly, their meaning may be contested.

Harvey (1979) offered a pioneering Marxist historical geography of the powerful political symbolism associated with the Basilica of Sacré Coeur in Paris. Other studies have focused more on textual analysis and on those textual communities that share an interpretation. A particular concern

Box 8.11: Representing Australia

It is often said that image is everything. Certainly, the importance of creating a favorable image of a product through advertising and other means as appropriate is well accepted. Landscapes and identities are no exception, as evidenced by the abundance of literature produced in many areas designed to attract tourists. Understanding such literature requires a knowledge not only of the landscape and identity being represented but also of the authors of the texts and of their specific motivations—the discourses of which they are a part. In this context, recall the discussion in Chapter 6 of images, image makers, and image distortion during the period of European overseas expansion.

In the case of Australia, official representations of identity prior to the 1970s 'embraced mateship, the bush and egalitarianism' with the principal symbolic landscape being the Australian semiarid interior (Waitt 1997:49). This image was reinforced by leading Australian painters, and the principal human component was the White male, with women and other groups excluded or serving as background. More recently, the image of Australian identity conveyed by the Australian Tourist Commission is more in accord with a multicultural focus, although such an identity continues to be defined by the dominant group. The recent image of landscape being conveyed is one aimed at fulfilling the tourists' desire for adventure and paradise.

- Adventure is most clearly suggested in images of the

bush, stressing the isolation and frontier conditions. In these images, the Aboriginal population serves as an important component: 'Aboriginality is socially constructed as the other to serve a particular political context and a deliberate economic strategy, that of selling Australia as an escape from civilisation to a primordial, timeless world, and/or a return to Nature where Aboriginals as the "original conservationists" live in perfect harmony with their environment' (Waitt 1997:50). White males are depicted in positions of power and females in more submissive and accepting roles.

- Paradise is signified through the depiction of parts of the Australian landscape, especially coastal and subtropical areas, as luxuriant and fertile, with emphasis on rainforest, waterfalls, beaches, and coral reefs; it may be that such depictions are designed to stress the femininity of landscape.

Clearly, because these and all other images are texts to be read, they can be interpreted in more than one way, and it is interesting to think about any one interpretation in terms of the positionality of the interpreter. Certainly, as the overall content of this chapter indicates, it is quite usual in much current cultural geography to interpret from a cultural studies perspective, through the eyes of the cultural turn, and thus to stress aspects that relate to issues of power, authority, and otherness.

has been the interpretation of monuments that were constructed as a means of creating or reinforcing some ethnic or national identity. War memorials are often both a readily visible and an easily understood symbolic landscape feature and, in the case of Australia, the building of numerous memorials to those who died during the First World War made a significant contribution to national identity and unity. However, while war memorials are dominantly interpreted as celebrations of bravery and national identity, they can also be seen in terms of a national failure to protect citizens, so their meaning is contested.

There has also been a concern with the contradictory discourses surrounding the construction of deathscapes, such as memorials and cemeteries. For example, in a discussion of roadside memorials in a part of Australia, Hartig and Dunn (1998:5) observed that 'contradictory discourses condemning and condoning youth machismo circulate around these deathscapes'.

Ordinary Landscapes

The cultural geographic and related literature dealing with the **ordinary landscape** is a substantial and varied one.

- Zelinsky (see especially 1994) is a principal contributor.
- Both *Landscape* and the *Journal of Cultural Geography* are important outlets.
- An atlas of North American societies and cultures edited by Rooney, Zelinsky, and Loudon (1982) is

Airlie Beach, Queensland, just off the Great Barrier Reef, offers sailing, fishing, snorkelling, and diving. Most of the islands off the Queensland coast are national parks (*Victor Last*).

The National War Memorial in Ottawa, designed by Vernon March, features twenty-three bronze figures symbolizing those who fought in the First World War and the sacrifices they made for peace. The memorial was unveiled by King George VI in 1939 during his visit to Canada (*Bill Brooks/Masterfile*).

a comprehensive description of one continental region.
- There are important volumes by Jakle (1987), Thompson (1995b), and Groth and Bressi (1997).

This section considers ordinary landscapes through discussions of the personality of place and of the landscapes of folk and popular cultural groups with particular emphases on the topics of music and consumption.

The Personality of Place

As noted in Chapter 3, there is a long tradition in cultural geography that is concerned with the personality of place—recall the idea that landscapes become, 'as it were, a medal struck in the likeness of a people' (Vidal, quoted in Broek and Webb 1978:32). Both Vidal and Sauer and the traditions that they initiated related the emerging personal-

ity of place to the way of life, the cultural occupance, of that place. One version of this intellectual tradition is evident today in popular geographic magazines such as *National Geographic* (United States), *Canadian Geographic* (Canada), and *Geographical Magazine* (Britain), but it is also an integral part of the *Landscape* magazine tradition and, more generally, of the current concern with the ordinary landscape, both folk and popular, that has been evident on several earlier occasions in this textbook.

Perhaps ironically, since about 1960, this traditional and often popularly appreciated form of cultural geographic writing has proven to be rather unappealing to some academic geographers, especially because of the challenges from such approaches as positivism and Marxism to what might be considered mere description uninformed by any conceptual framework. These critical ideas comprise one aspect of the debate surrounding the decline of traditional regional geography as this is discussed in Chapter 4. But, from a different perspective, Barnes and Curry (1983:468) suggested that two important characteristics of the traditional studies are, first, that they 'aim to inform us, to increase our understanding of the world, and that insofar as they succeed they are fulfilling the central function of the social sciences', and, second, that they do not fall into the essentialist trap of deciding what is valid knowledge before initiating a study. Certainly, an interest in the personality of place, ordinary or otherwise, continues to be a vital component of contemporary cultural geography and the ideas and analyses discussed in this section reflect this interest.

The discussion of ordinary landscapes at the heart of the *Landscape* magazine concern is not closely related to the advances in cultural studies initiated by Williams (1958), who built on the idea that culture is ordinary, but the principle is similar, namely that it is inappropriate to include some and exclude other cultural traditions and their built landscapes. For the many writers building on this idea, culture and landscape respectively can be found in a wide variety of texts, including in everyday life and in ordinary landscapes. This important tradition was introduced in Chapter 3.

The idea that ordinary landscapes are places and can be thought of as having character or a personality was a basic premise for Jackson (1951:5), who wrote in the first issue of *Landscape* that 'there is really no such thing as a dull landscape or farm or town. None is without character, no habitat of man is without the appeal of the existence which originally created it.' Such a view is now widely accepted within cultural geography, although it does not necessarily mesh well with the humanistic claim that the historical period of modernity has resulted in the loss of place identity, the creation of placelessness, and perhaps **topocide**. (See Your Opinion 8.5.) Thus, Relph claimed:

> The premodern logic was that place identity grew from the location and its traditions, and this revealed itself in geographical diversity. Customs and styles did move from region to region but these processes of borrowing were relatively subservient to local distinctiveness. The modernist logic was that place was irrelevant and geographies should be determined by international economic forces and fashions; the manifestation of this was placelessness in that locality was subservient and places came to look increasingly alike (Relph 1997:219–20).

Folk Culture and Popular Culture

The personality of places and related human identities are discussed utilizing a convenient, but by no means absolute, division of the ordinary landscape into the two categories of folk and popular, a conventional division in cultural geography.

A **folk culture landscape** is, for most of us, different in that it emphasizes tradition, the oral transmission of songs, local history, an integration of nature and culture, and is often expressed through ritual. The idea of a folk culture refers to those cultures that prevailed in the preindustrial and precapitalist period but that remain evident today, albeit in some limited manner. In some cases, ethnic and national identities are traced to their folk roots such that folk is taken to mean some original unadulterated cultural condition that is evident usually in song, dance, crafts, and dress; this is comparable to the fourth of the six components of ethnicity referred to in Chapter 7. Folk culture landscapes are occupied by groups that intentionally choose to reject at least some of the cultural and technological changes that

---

Your Opinion 8.5

Notwithstanding the basic logic of Relph's claim, it seems difficult to accept the argument that landscapes of modernity—such as residential subdivisions—really are placeless and lacking in identity. Do you agree that this might simply seem to be the case to those who do not live in and experience those landscapes? Certainly, the assertion from Jackson concerning the fact that there is no such thing as a dull landscape does seem compelling.

are typically embraced by most people, and in this sense there is a distinction between folk cultures as local cultures and a single global popular culture.

Several examples of folk cultures and related landscapes have been referred to earlier in this textbook, especially in the discussions of the traditional approach to diffusion in Chapter 3 and of ethnicity and cultural islands in Chapter 4. Cultural geographic analysis of folk culture, especially as reflected in housing styles and building materials, is closely associated with the works of Evans and Kniffen as introduced in Chapter 3. Indeed, there is a voluminous and continuing cultural geographic literature on folk culture landscapes as these are evidenced by preferences in food and drink, music, sports, and material landscape features. Many of these studies belong in the Sauerian tradition.

A **popular culture landscape** usually implies culture for the people, and it is therefore associated with change, with first modernity and later postmodernity, and with everyday life for the masses, including sites of consumption, such as malls and sports stadia and sites of **spectacle** such as celebratory events. As a product of modernity and more recently of postmodernity, popular culture is the culture of the majority. From a critical humanistic viewpoint, Relph claimed that:

> The postmodern logic of places is that they can look like anywhere developers and designers want them to, and in practice this is usually a function of market research about what will attract consumers and what will sell. In postmodernity it is as though the best aspects of distinctive places have been genetically enhanced, then uprooted, and topologically rearranged (Ralph 1997:220).

Studies of these popular culture landscapes reflect both the Sauerian tradition and the cultural turn. Although this contrast between folk and popular culture is helpful, popular culture is also often contrasted with high culture, or with mass culture, and the term has a very different meaning depending upon the contrast implied. When contrasted with mass culture, popular culture refers to a particular consumer niche, for example, the market for a novel or film; when contrasted with high culture, as in the early Marxist–oriented British cultural studies tradition, popular culture implies the interests and activities of the working class.

Music, Identity, Place

Studies of musical styles associated with particular folk and popular culture groups and their places are abundant in the cultural geographic literature, having been initiated in the late 1960s, and it has been suggested that there is a continuing need for cultural geographers 'to open their ears to the auditory components of culture—the sounds of people and places' (Carney 1990:45). An edited volume concerned with American folk and popular music incorporated three basic research themes—namely regional and ethnic studies, questions of cultural hearths and cultural diffusion, and the role of place—and included examples of such musical styles as country music, gospel, jazz, rock, folk, Latin, blues, and zydeco (Carney 1994a).

- Not surprisingly, although analyses of the geography of music are informed by various research traditions, most such work has been conducted within the Sauerian tradition. The most researched musical style is country music, and the least researched are the various forms of military, classical, and religious music. The focus on country music is understandable given the especially close links that this style has to folk ethnic groups and to the particular places that these groups occupy. The diffusion emphasis has considered the spatial spread and temporal growth of such indicators as music festivals, music contest, and radio station programing.
- Increasingly, however, there are studies of the geography of music that are informed by the cultural turn. Although there is rarely any meaningful integration of the two traditions, many of the studies that are informed by the cultural turn do acknowledge the Sauerian tradition. A striking exception to this is the work by Leyshon, Matless, and Revill (1995:423) that explicitly set out to introduce the study of 'The place of music' and to 'discuss previous work on music by geographers' and yet made no reference to studies conducted

in the Sauerian tradition or to the research of leading American writers in the field such as Carney, Ford, Francaviglia, Gritzner, Lehr, and Nash, to cite but a few (see Carney 1990).

Although music is, of course, typically a key aspect of popular culture, most styles necessarily have local or regional origins and therefore may reflect a particular place identity, hence the concern with place and diffusion. Some well-known American examples of place-based musical styles are those of New Orleans jazz, Memphis blues, the Detroit Motown sound, and Nashville country, but there are also cultural geographic analyses of, for example, Pacific Northwest rock from 1958 to 1966, 'indie' music in Manchester, England, a single Singaporean artist, and western North Carolina bluegrass music, all of which stress how regional identity is able to play a key role in the evolution of the specific musical style. Other studies focus on links with national as well as regional identity and more generally on the cultural politics of music. Pursuing a rather different argument, the music of the group, U2, was used by McLeay (1995:1) to 'show how geographic imagery is used for political purposes'.

Of course, recognizing the role that ethnic identity and place can play in the evolution of a particular musical style is not to deny the often overwhelming importance of commercialism and global marketing in the musical industry. Much popular culture is part of what might be considered global culture. Indeed, performers that are generally associated by consumers worldwide with a specific place may not be readily embraced by that place; according to Cohen (1997), this is the situation with the Beatles and Liverpool. In most cases of globally successful artists there is no real association with place, except perhaps for a national identity.

Box 8.12 discusses two related aspects of the cultural geographic study of country music, namely an essentially Sauerian analysis of the origins and diffusion of bluegrass and a study that centers on the rise of Branson, Missouri, as a country music center. These analyses share a concern with the relevance of place and local identity, but also need to be understood with reference to larger commercial strategies. This box discussion leads naturally into a consideration of the geography of tourism and consumption.

Consumption and Spectacle

Some places are being created as cultural resources and as sites of consumption, while other places, and sometimes people, are being recreated or manipulated in order to increase their appeal in these contexts. Further, consumption—of places and other cultures as a part of tourism, of the activities of others through observing sports, festivals, and carnivals, and of food and drink—is increasingly being analyzed by cultural geographers whose work is informed by the conceptual advances associated with the cultural turn. Many sites of consumption—places such as hotels, malls, and theme parks—falsify place and time to create both other places, an elsewhereness, and other times. Certainly, much consumption is increasingly exotic in character, a feature that is evident in tourism, retailing, and the food and drink industry. Consuming can be interpreted as a social practice that has two theoretical implications:

- Consumer culture needs to be understood by reference to the institutions of consumption, such as the shopping malls and tourist sites where consumption occurs, and also the advertising that mediates between our needs and wants. These worlds of consumption are continually being socially created and the pleasures of consumption are social pleasures.
- Consumption is a symbolic practice that may be interpreted.

Much contemporary cultural geography is building on these ideas to generate analyses of consumption and the sites of consumption.

TOURISM AND RECREATION

Tourists are primarily spectators, consumers rather than producers, but what is being consumed? The answer is other cultures and other places. Thus, tourism may serve to reinforce the notion of the cultural other. The distinction between observers and those being observed is also a power difference. Indeed, sometimes even the meaning of a tourist site is contested. In the case of the Alamo historic site in San Antonio, De Oliver (1996) argued that it has been culturally refashioned as a site emphasizing the heroic behavior of Anglos rather than as a site of a Spanish mission, thus reaffirming a dominant Anglo identity and a subordinate Spanish identity. More generally, tourism is increasingly important as a 'means by which modern people assess their world, defining their own sense of identity in the process' (Jakle 1985:xi).

Although most studies of consumption relate to what is best considered popular culture, there are three aspects of folk culture tourism that are of interest:

- First, in some cases folk culture characteristics are being artificially preserved to create a tourist attraction. Folk cultures may be intriguing to dominant cultures precisely because such groups are different, or at least are seen to be different, from the majority. The tourist industry and cultural producers may sometimes manipulate a local cultural identity. In the case of Nova Scotia, for example, folk identity has been emphasized through an erasure of those components of mod-

Box 8.12: Country Music Places—Bluegrass Country and Branson

There is a substantial body of cultural geographic writing concerned with country music. Much of this literature is especially concerned with the origins in particular folk lifestyles, with the mechanics of diffusion, and with the cultural infrastructure that facilitated diffusion. There are also analyses of the links between the music industry and tourism as they relate to urban growth and larger commercial issues. The American cultural geographer, George Carney, is the most prolific writer, and this box summarizes two of his studies of country music that reflect the above interests.

Studies of origins and diffusion of particular musical styles are usually explicitly framed within the Sauerian diffusion tradition as discussed in Chapter 3. In the case of the distinctive bluegrass sound, the geographic origin or culture hearth has long been debated, with suggested source areas including the Pennyroyal Basin of western Kentucky in Appalachia and the Bluegrass region of western Kentucky. However, research conducted by Carney (1996) showed that the source area was the mountain and piedmont regions of western North Carolina during the period from about 1925 to 1965. Groups and individual musicians in this region—Carney (1996:66–74) referred to 'human innovators and place incubators'—pioneered the use of the distinctive combination of instruments, playing techniques, and vocal styles that together make up bluegrass. The period of innovation was followed by growth and spread, locally through such mechanisms as homes, schools, and churches, and regionally through fiddle contests, music festivals, radio, and recordings.

The story of Branson, Missouri, 360 mi. (579 km) west of Nashville, the new 'mecca' of country music, is a fascinating mix of local country music activities and the larger commercial world of tourism and consumption. As a small resort town located in the Ozark country music region, Branson introduced shows for evening entertainment of tourists in 1959, and the town gradually emerged as a center for showcasing local country music talent. There was a close association between this growing musical business and the established tourist attractions in and adjacent to the town, with performances held at the Silver Dollar theme park and a local commercial complex of caverns, and there was also a steady growth of theaters along Highway 76, the main street known as '76 Country Music Boulevard' (Carney 1994b:20). By 1981, when the Ozark Country Jubilee moved from Springfield, Branson was established as a major regional center with related resort and musical tourist attractions.

Transformation of Branson into something more than a regional center began in 1983 when a major national star, Roy Clark, opened a theater on the main street that subsequently served as a venue for other nationally known performers. Branson proved attractive for major stars for several reasons: it offered the opportunity for them to maintain a semipermanent location (in addition to the major center of Nashville) that audiences came to rather than the performers having to visit audiences through a series of one-night stands at various locations; the Branson theaters were seen to be a healthier and more secure venue than many of those encountered during tours; the Ozark

ern identity related to urbanization, industrialization, and also those identity differences based on ethnicity, class, or gender.

- Second, in other cases, there is clearly tension within a folk community as some members display a folk culture, putting on a performance for tourists, while other members reflect a more contemporary lifestyle. This point may apply especially to those expressions of cultural identity expressed through festivals.
- Third and most generally, it is often the case that tourism is capable of destroying the very thing being sought, namely folk cultural authenticity.

The result of these three circumstances is that tourists only rarely have access to authentic cultural experiences. Principal exceptions to this generalization are those tourist sites and activities that do have meaning in that they reflect values, traditions, and a sense of place. For example, some Amish groups in North America stage events for tourists that reflect their lives but do not intrude on their privacy.

The most evident sites for contemporary tourism and recreation are those landscapes of consumption and spectacle associated with post-

Roy Clark and Boxcar Willie were two of the first entertainers to come to Branson, Missouri, which has since blossomed as one of the best-known music towns in North America (*Branson Chamber of Commerce and Visitors Bureau*).

region was an attractive place both physically and culturally; and, for many performers, Branson proved an excellent investment in land, facilities, and media. During the 1990s, the style of Branson country entertainment has broadened as many younger artists are drawn to the new large capacity theaters.

Remarkably, Branson is more successful in the group tour market than is Nashville as a result of the small-town atmosphere that appeals to many fans, the large number of theaters and shows that are in close proximity, the low cost of food, lodging, and entertainment, and, of course, as a result of successful marketing.

The Alamo, an old Spanish mission in San Antonio, Texas, was the scene of a defense in the war between Texas and Mexico. In the war for Texan independence, a band of defenders, who held out at the Alamo for thirteen days, were attacked by 4,000 Mexicans led by General Antonio López de Santa Anna. The siege ended when the Mexicans overpowered and slaughtered the Texans on 6 March 1836 (*Victor Last*).

modernity and with the rise of a global culture as these ideas are discussed in Chapter 7. Among the many such places are culturally enhanced physical attractions such as Niagara Falls, culturally enhanced historical sites such as Gettysburg, and cultural constructions including world fairs and theme parks such as Disneyland. In the case of Dragon World, the first cultural theme park in Singapore, there is clear removal of the everyday along with an emphasis on the faraway, remote, and exotic.

Similarly, major sporting occasions, such as the Summer Olympics, the World Cup of soccer, and the Super Bowl, are spectacles or theater that, at least temporarily, unite diverse cultures in a shared activity. Notably, however, much work on the geography of sport has maintained a relatively traditional focus with an emphasis on regional, place-based, sports and sporting occasions. Examples of studies include stock car racing in the

American South and a focus on golf and football in the American context.

RETAILING AND FOOD AND DRINK

Retailing sites in the contemporary developed world offer much more than the opportunity to purchase goods and services; increasingly they are offering the opportunity for leisure, recreation, and other activities. The much-discussed example of West Edmonton Mall in Edmonton, the first megamall in the world, is detailed in Box 8.13 (see especially the Special Issue of the *Canadian Geographer* on the subject of West Edmonton Mall and megamalls generally, volume 35, no. 3. 1991).

Related to the cultural geographic interest in retailing is a specific concern with food and drink consumption. In addition to studies on food and place and on food and ethnicity, a principal focus is the role played by franchising, initially in the context of the mobility of American society and

Box 8.13: The Mall as a Postmodern Consumption Site?

The West Edmonton Mall Map and Directory (West Edmonton Mall n.d.) leave little doubt that this place is intended to be more than a collection of retail establishments: 'How to shop West Edmonton Mall. It's a lot easier than you might expect . . . and even more fun. Whether you're shopping in a hurry or in the mood to explore, it's all here under one roof. With over 800 stores, there's always one nearby that has what you're looking for. And there's always something exciting to see or do.' Certainly, the inclusion of an hotel, an ice rink, a deep-sea adventure setting that includes submarine rides and dolphin presentations, a waterpark that has artificial waves, a bungee jump, a large amusement park, theaters, clubs, a casino, and bingo, numerous eating places, and financial and other services, in addition to the over 800 stores, leaves little doubt that this is indeed much more than a shopping mall. It is also a tourist attraction, an entertainment center, and a business center and is therefore concerned with more than one aspect of consumption.

How are cultural geographers whose research is inspired by the cultural turn studying the mall? They are reading it as a communicative text. According to Goss (1993:19), developers are 'designing into the retail built environment the means for a fantasized dissociation from the act of shopping' as one means of manufacturing 'the illusion that something else other than mere shopping is going on'. The mall is being interpreted as a postmodern consumption site, 'a landscape of myths and elsewhereness' (Hopkins 1990:2). As with theme parks such as Disneyland, the mundane is seen as unacceptable in the megamall.

Gregson (1995:137) expressed concerns about the direction of such research, specifically questioning 'the omission of the fundamental and the mundane (notably shopping) from these readings?' Noting that most shopping is accomplished by women and that women comprise the majority of retail workers, Gregson (1995:137) contended: 'When we look at geographers' readings of the megamall what we find then are masculine and masculinist representations masquerading as universal and homogeneous tendencies in the world of consumption.'

Tourists from all over the world visit West Edmonton Mall, which features a man-made lake, a replica of the *Santa Maria*, a waterpark, and a miniature golf course in addition to hundreds of stores (*Victor Last*).

more recently in the context of the emergence of a global culture: 'Today, McDonald's golden arches are one of the most recognizable symbols of American popular culture throughout the world, from Tokyo to Moscow' (Carney 1995b:95).

There are also studies that center on food and drink consumption as a process imbued with symbolic meaning, especially when it is seen as an increasingly exotic activity. Building on the work of Said (1978) concerning Western fascination with the Orient as other, May (1996) discussed the tendency of some westerners to consume exotic foods and drinks, and indeed other exotic goods, as one means of saying who we are, both to ourselves and to those around us. Viewed in this way, exotic food and drink consumption is yet another means of creating self-identity, not dissimilar to such considerations as ethnicity and sexuality as discussed previously; it is a part of our cultural capital. Of course, such ideas are well established in the social psychological literature where there has been much discussion of how people present themselves, especially through their choice of clothing.

Given the increasing interest in consumption and related presentation of self, it is not surprising that there is also an emerging concern with advertising. Jackson and Taylor (1996:356) noted that advertising 'is an inherently spatial practice, playing a crucial role in an increasingly mediated world as part of the national and international expansion of markets; creating uneven patterns of demand across space; and striving for universality but constantly subject to local variations in meaning and interpretation'.

Concluding Comments

As discussed at the beginning of Chapter 7, much of the content of chapters 7 and 8 reflects the cultural turn as it has been applied in cultural geography. The traditional Sauerian view has been modified and complemented—some cultural geographers might even say replaced—by new conceptual inspirations. Further, prior to the cultural turn, human geography generally placed emphasis on economic and political considerations and on a positivist, traditional Marxist, or humanist approach. Key components of the cultural turn

include a rejection of all theories that purport to offer some single correct answer, such as Marxism, and a celebratory concern with human differences. All areas of human geography, then, not only cultural geography, have been affected by this cultural turn.

Questions about human identity have surfaced regularly in this chapter, with the basic questions being:

- how people identify with each other
- how they construct and reconstruct these identities in opposition to others

The answers given are based on the idea that identities are not given to people; rather, they are created by people, with the result that identities can change and also have a meaning that is open to dispute. It is increasingly popular for cultural geographers to study human identities as the people involved understand them rather than as they are assumed and imposed by the researcher. According to Hall (1996:5), the process of identity creation is a strategic and positional activity such that identities are 'the products of the marking of differences and exclusion'. It is this understanding of identity that has been at the core of much of the material in this chapter and the preceding chapter. The distinction with the prevailing understanding of culture in earlier chapters of this book is quite substantial.

Related to the issue of identity, and also central to much of the content of this chapter, is the concept of place. The humanistic interpretation of place stresses the idea that place is not so much a location, a thing, but rather a setting for human behavior, a relationship. From a humanistic perspective, the meaning of a place cannot be understood without an awareness of the identity or identities of those who occupy the place, but there are also more recent interpretations of place that are conceived in terms of the way places are controlled by those in authority and also challenged by others. Declaring a place to be a 'no go' area or to be 'out of bounds' results in that place having an increased symbolic significance. Both of these interpretations of place recognize that, because we segment the world, we believe that there are proper and therefore also improper places for some things and some behaviors. We and our behaviors can be in place or out of place.

Much of the material in this chapter, especially that included in the sections on others and other worlds and on symbolic landscapes, has been concerned with the boundaries and the breaking of those boundaries that these circumstances imply. As Cresswell (1996a:149) stated: 'The geographical ordering of society is founded on a multitude of acts of boundary making—of territorialization—whose ambiguity is to simultaneously open up the possibilities for transgression'; further, the 'geographical classification of society and culture is constantly structured in relation to the unacceptable, the other, the dirty'.

Building upon questions of human and place identity, this chapter has stressed the symbolic character of landscape—landscapes as representations of cultural values and ideas. Whereas much, but by no means all, of the content of earlier chapters in this book considered the visible and material landscape in an essentially Sauerian tradition, this chapter has considered landscapes as expressions of human identity and has attempted to read landscapes as outcomes of competing discourses.

Further Reading

The following are useful sources for further reading on specific issues.

Duncan (1994b:362) on the fragmentation of the new cultural geography.

Barnes and Curry (1983), Duncan and Ley (1993), and Duncan and Sharp (1993) on modes of representation in cultural geography.

Ley and Samuels (1978) on a humanistic tradition in geography.

Berdoulay (1989) on Vidalian cultural geography.

Tuan (1977, 1989) on the humanistic concept of place.

Dowling and Pratt (1993) and Wagner (1996:80) on home as a site of oppression for women.

Social scientists have discovered place often as a consequence of particular readings of postmodern theorists, and Yaeger (1996:18) was prompted to

ask: 'Why are scholars from a range of disciplines suddenly reinvested in the energetic pursuit of geography?' A postmodern inspiration for what was called the new regionalism in history was identified by Wilson (1998:xvi), with the claim that cultural identity 'hardly exists apart from social relations in specific places and contexts'. In sociology, Cuba and Hummon (1993:112) observed that 'place identities are thought to arise because places, as bounded locales, imbued with personal, social, and cultural meanings, provide a significant framework in which identity is constructed, maintained, and transformed'.

Keith and Pile (1993), Pile and Thrift (1995:13–51), and Longhurst (1997) on some recent advances in psychoanalytic theory that offer insights into the relations between the social and the individual.

Touraine (1981) and Geyer (1996:xiv) on new social movements.

Dalby (1991) on the concept of otherness as it applies in a political context and McEwan (1996) on the gendered colonial process.

Hanson and Pratt (1995) on the way local labor markets discriminate against women.

Monk (1992) and Winchester (1992) on gendered landscapes in urban areas.

Berg and Kearns (1996) and Liepens (1996) on the gendering of the New Zealand agricultural landscape.

Valentine (1989) and Pain (1999) on gender and landscapes of fear.

Bell and Valentine (1995:23) and Gregson et al. (1997:74) on strategic essentialism.

Callard (1998) on cultural geography and the body.

Symanski (1974, 1981) on the geography of prostitution.

Lyod and Rowntree (1978) was an early study of landscapes of homosexuality; more recent studies include those by Knopp (1995), Myslik (1996), and Valentine (1993, 1996).

Bell and Valentine (1999) on geographies of sexuality.

Boone (1996) and Binnie (1997) on geography and queer theory.

Ford (1998) on a British survey that determined that 3 per cent of the white-skinned population, 4 per cent of the Indian population, and 10 per cent of the black-skinned population avoided soccer games because of the fear of violence, a fear that was also reflected in activities such as movie theater attendance.

Hodge (1996) identified the western part of Sydney, Australia, as an area of neglect, disadvantage, unemployment, lack of services, and high levels of criminal activity, while Dear and Wolch (1987:8–27) discussed the 'social construction of the service-dependent ghetto'.

Bastian (1975) on class and traditional cultural geography.

Skelton and Valentine (1998) on geographies of youth cultures and Matthews and Limb (1999) on the need for a geography of children.

Pile and Keith (1997) on examples of resistance movements, especially as they are tied to place, and Routledge (1997) on the subculture of resistance that has emerged surrounding the construction of a freeway in Glasgow, Scotland.

Cresswell (1993) on mobility and resistance through a reading of the book, *On the Road* (see also Cresswell 1996b; McDowell 1996).

Kong (1993b) on religious buildings in Honk Kong.

Kong (1995) on the production of popular music and the way in which it can be a form of cultural resistance that can be used by the state to maintain and reinforce particular ideologies.

Pocock (1981), Barrell (1982), Rees (1984), and Daniels (1994) on geography and artistic endeavor.

Baumann (1996) on the idea that societies comprise distinctly different groups of people.

Howe (1998) on Afrocentrism.

Young (1993) and Wilson (1987) on some alternatives to multiculturalism.

Zelinsky (1973, 1988, 1997) on the impacts of a national culture on landscape; much recent work, sometimes informed by postmodernism, identifies the impacts of local or regional identities (Hayden 1995).

Muir (1999) on the current diverse yet incomplete geographic approaches to landscape; Granö (1997) on a traditional European interpretation of landscape.

Baker (1992) on ideology and landscape.

Hirsch and O'Hanlon (1995:22), in an apparent desire to establish an anthropological interest in landscape seen to be different from the interests of cultural geographers, proposed the idea of landscape as process—'one which relates a "foreground" everyday social life (us the way we are) to a "background" social existence (us the way we might be)' (see also Bender 1993).

The wealth of interest in ordinary and sometimes not quite so ordinary landscapes is reflected in the introductions to and some of the contributions included in the edited volumes, *Landscape in America* (Thompson 1995b) and *Understanding Ordinary Landscapes* (Groth and Bressi 1997), and also in Hart (1998).

Francaviglia (1994) on the understanding of the American Southwest as it has been continuously redefined through a series of cultural occupances.

Carter (1987) on the exploration and mapping of Australia as this related to the production of knowledge, and Ryan (1996) on the exploration and description of social space by European explorers.

Tuan (1991b) on the naming of places.

Eliade (1959) on sacred and profane space; Carmichael, Hubert, Reeves, and Schanche (1994) on global issues of sacred spaces and heritage issues.

Biswas (1984) on sacred buildings.

Gade (1982) on the French Riviera as élitist space.

Duncan (1990) on the early nineteenth-century landscape in Sri Lanka; this is an influential landscape study in the newer tradition of landscape arguing that accounts of the world and the world itself are both intertextual.

Walton (1995) on the idea of landscapes as texts.

Cosgrove and Daniels (1988) on iconographic analyses. Also Cosgrove (for example, 1984) on the landscapes of Renaissance Venice and Vicenza within their larger cultural framework; Domosh (1989) on a single newspaper building, the New York World Building constructed in 1890 in New York City, in terms of both personal egos and industrial capitalism; Eyles and Peace (1990) on iconographic analysis to help understand two prevailing but contradictory images of Hamilton, Ontario, namely those of smokestack city and cultured city.

Cartier (1997) on the symbolism of bank buildings in Hong Kong.

Osborne (1988) on the iconography of national identity that was forged by painters, stressing the role played by the early twentieth-century 'Group of Seven' and their concern with the Canadian North (see also Osborne 1992).

Zelinsky (1988), Jeans (1988), and Johnson (1995) on monuments as a reflection of ethnic or national identity; Bell (1999) on the redefinition of national identity in Uzbekistan.

Teather (1998) on the symbolic significance of Chinese cemeteries in Hong Kong; Auster (1997) on several different interpretations of a memorial column on a hillside outside Armidale, Australia; Kong (1999) on deathscapes and the new cultural geography.

Lutz and Collins (1993) on a provocative 'reading' of *National Geographic* that challenges the claim that this magazine serves as a mirror of the world.

Featherstone (1995) on local folk cultures and a global popular culture.

Carney (1998) and Francaviglia (1996) on folk culture landscapes.

Carney (1995a) on popular culture as the culture of the majority.

Gill (1993) on Pacific Northwest rock from 1958 to 1966; Halfacree and Kitchin (1996) on 'indie' music in Manchester, England; Kong (1996) on a single Singaporean artist; Carney (1996) on western North Carolina bluegrass; McLeay (1997b) on regional identity and the development of a particular musical style.

McKay (1994) on the way in which the tourist industry and cultural producers have manipulated the cultural identity of Nova Scotia; Kneafsey (1998) on tourism and place identity; Ringer (1998) on the cultural landscapes of tourism.

Jackson (1992), Getz (1995), Abram (1997), and Waterman (1998) on the tensions that may be evident in a community when some but not all members perform for tourists.

Ley and Olds (1988) and Squire (1994) on landscapes of consumption and spectacle.

Yeoh and Teo (1996) on Dragon World in Singapore.

Pillsbury (1974, 1989) on stock car racing in the American South; Rooney (1974, 1993) and Rooney and Pillsbury (1992) on golf and football.

Hopkins (1991) and Goss (1992, 1999) on malls as sites of consumption.

Rooney and Butt (1978), and Shortridge and Shortridge (1995, 1998) on food as it relates to place and ethnicity.

Bell and Valentine (1997) on food and drink consumption as part of our cultural capital.

Tuan (1982) on behavior and place.

The Status of Cultural Geography

This book is a journey through a body of changing ideas and changing practices, specifically through the changing concepts that inform the work of cultural geographers and the changing analyses that cultural geographers perform. This final chapter is a brief assessment of where cultural geography stands today, and of where it may be heading. Both these tasks are important but difficult.

Describing where we stand is not as easy as it might sound.

- There continue to be questions about the integration of physical and human geography in general, questions that have real implications for cultural geographers, especially those concerned with ecological issues.
- There continue to be debates about the relative merits of traditional Sauerian approaches and the more socially and theoretically informed approaches, especially those associated with the cultural turn.
- There continue to be discussions of philosophical questions, especially concerning the legitimacy of a naturalistic focus for some analyses—about whether to search for cause or to seek understanding.
- There continue to be questions about the interweaving of culture, politics, and economy.
- There continue to be related debates about the appropriateness of particular writing styles, especially concerning whether or not it is possible for writing to in some sense mirror a real world.

In short, cultural geography is contested terrain, with different scholars favoring often quite different interpretations. In this book, an attempt has been made to reflect, incorporate, and, on a few occasions, reconcile these differences. Given this context of changing ideas, analyses, and opinions, it is noted that 'it would be a moribund discipline indeed, in which internal dialogue was not bubbling with the constant ferment of new and conflicting ideas' (Clark 1977:4). An appropriate analogy for the development of cultural geographic research may be that of a bush with many branches and twigs, some of which generate new growth and some of which do not.

Informing Cultural Geographic Analyses

Consider the first of the two important claims that were referred to at the beginning of Chapter 1, namely the assertion by Zelinsky (1996:750) to the effect that cultural geography is a 'scholarly discourse that has shifted from the comfort of placid marginality toward the overheated vortex of ferment and creativity in today's human geography'. The evidence here seems undeniable and primarily relates to the activities of those who consider themselves to be cultural geographers, to the activities of practitioners from other geographic subdisciplines who have discovered the cultural, and also to work being conducted in other academic disciplines that involves increasing awareness of the significance of space and place. Within cultural geography there are two principal bodies of work.

- On the one hand, the work of Sauer and the Sauerian school continues to be debated and the basic approaches and concepts first introduced in the 1920s continue to be implemented.

On the other hand, the cultural turn and all that this implies is an integral part of contemporary cultural geography, and indeed is quite evident throughout much of contemporary human geography.

Thus, the traditional and the new cultural geography are both with us today, and both are changing. The new—much of which is no longer quite so new, of course—has certainly proven to be an enrichment, and most of the recent statements applauding the vitality of cultural geography are reflective of what can be seen as a 'transformation' (McDowell 1994:146) or possibly merely a 'reinvention' (Price and Lewis 1993a) of the earlier tradition. As is evident from the contents of this volume, the use of both terms is understandable. Thus, the new certainly reflects significantly different conceptual content and includes a different set of empirical issues. In this sense, the new was not something that evolved easily from the traditional; rather, it was a significantly different set of ideas and practices often either being added to or perhaps even replacing previous ideas and practices. However, some aspects of the new may still contain some of the limitations of the traditional, especially concerning the possible reification of culture. Regardless, many links can be identified between the traditional and the new, particularly a continuing interest in the cultural landscape, but also a mutual mistrust of scientific method as contained within positivism, an interest in human and land relationships, a predilection for fieldwork, a concern for regional identity, and a willingness to conduct research at various spatial, temporal, and social scales.

An exchange of views concerning the categories of nature and culture has highlighted some of the issues involved in this ongoing debate concerning an appropriate conceptual basis for cultural geography.

- Focusing on the West Coast of Canada, Willems-Braun (1997a) explored ways that colonial power relations were established through the construction of nature as separate from culture and how such relations are reproduced in the present. Thus, discussions of nature tend to stress the dif-

ferences of opinion between those who see nature as a resource to be exploited (such as logging companies) and those who see nature as a wilderness to be preserved (such as conservationist groups). But more important, it can be argued that neither of these options allows for meaningful appreciation of Native understandings of nature and the approach taken was to discuss some current conflicts concerning the temperate rainforest ecosystem as these reflect colonial relations established at an earlier time and also to seek an understanding of these through a cautious use of selected postcolonial theory.

- Commenting on this work, Sluyter (1997:700) noted the absence of any reference to the Sauerian tradition that typically criticized the idea that North America was a primordial wilderness prior to European involvement: 'in his avid digging-up of the colonizers' taken-for-granted epistemologies, Willems-Braun indiscriminately buries other epistemologies. In concert with many smitten by the uptown cooliality of postmodernism, he negates the work of other geographers who have striven to understand the ways in which colonizers rhetorically and materially invented colonized peoples and natures.'
- Responding to these comments, Willems-Braun (1997b:706) expressed concern at the perceived suggestion that the Sauerian tradition was 'a benchmark against which all others must situate their work', and justified excluding the Sauerian tradition on the grounds that it was not an appropriate conceptual basis for work that emphasized questions relating to power and politics, and also on the grounds that the tradition incorporated an imperialist nostalgia.

However, to suggest that work of interest to contemporary cultural geographers is comprised solely of activity that can be conveniently labeled as traditional (meaning essentially derived from the pioneering statements of Sauer) or, alternatively, as new (meaning largely informed by a variety of recently favored social theories) is itself misleading. Neither Huntington, author of *The Clash of Civilizations and the Remaking of World Order*, a book that was based on a cultural world regionalization, nor

current state of excitement in cultural geography but that are neither traditional nor appropriately seen as part of the cultural turn. As one compositional subdivision of human geography, cultural geography continues to embrace a diverse set of ideas and to address issues with multiple layers of relevance.

The Cultural Landscape

As is evident throughout this textbook, the study of cultural landscape has been a persistent concern of cultural geographers from the late nineteenth century to the present. Since it was first explicitly formulated by Ratzel, further developed by Vidal and Schlüter, and then decisively introduced into English language geography by Sauer, the study of cultural landscape has proven to be an almost endless source of conceptual inspiration and a rich vein for local and regional understanding. The impact of Sauer's methodological statements and the example set by the empirical studies conducted both by Sauer and his students is difficult to exaggerate. Most such studies focused on evolution, regionalization, and on human and land relationships. A principal concern in this tradition is with understanding the impact of humans, typically seen as

Temperate rainforests on the West Coast have the most productive growth conditions in the country. Trees in these forests include the giant Sitka spruce, Western hemlock, and red cedar, which are harvested by British Columbia's forestry industry (Philip Dearden).

Kaplan, author of *The Ends of the Earth: A Journey at the Dawn of the 21st Century*, a book that offered global overviews of ecology and human quality of life, are cultural geographers, but their work certainly contributes to established traditions in cultural geography. Further, within cultural geography both the new cultural ecology and the political ecological approach are examples of recent bodies of work that are contributing to the

members of cultural groups, a concern that broadened in the 1970s through studies that stressed the need to read that impact. The impact of Sauer is of course well known within the discipline and in related areas of scholarship, but it is interesting to note that there was also considerable interest shown in his work after the Second World War by a group of avant-garde poets known as the Black Mountain School.

Of course, reading a landscape intelligently requires an awareness of authorship, and one of the earliest and most critical contributions from the new cultural geography was that of questioning the emphasis placed on cultures as groups and the related apparent rejection of human agency. Since about 1980, approaches to cultural landscape study have diversified as one reflection of the proliferation of theoretical bases, such that there are concerns with symbolic landscapes, landscapes linked to prevailing social power relations, and landscapes of both production and consumption. Further, there are explicit analyses of landscapes as texts. For Sauer, at least in principle, there was a universal model that could be used to inform analyses of cultural landscapes, whereas the contemporary interest in cultural landscape incorporates a wide variety of approaches. It is possible to interpret this change as being one part of the modern to postmodern transition. *Indeed, perhaps the most meaningful difference between the Sauerian approach to landscape and the more recently developed approaches to landscape lies in this transition from a concern with Culture to a concern with cultures.*

Not all cultural geographers have welcomed this conceptual broadening of interest in cultural landscape. Thus, Hart (1995:23) wrote: 'I am quite content with the simple vernacular definition of landscape as "the things we see," and I am saddened by the way the meaning of the word has been transmogrified. Some people have endowed the concept of landscape with magical, mystical, or symbolic significance, or have loaded it with metaphysical significance.' In turn, Henderson (1997:95–6) criticized Hart's point of view as one part of a more general criticism of those landscape studies that stressed the richness and diversity of landscape at the expense of such considerations as human inequalities and related social struggles.

Cultural landscapes are perhaps best seen as being one aspect of the interaction between people and place; they are constructed by people and interpreted by people. Ordinary landscapes are a consequence of the actions of ordinary people. They can be read partially through observation, although a more complete reading requires consideration of other textual sources such as historical records placed into appropriate contexts, discourses, to facilitate understanding.

Regardless of perspective, one key point emerges from this ongoing and diversifying interest in cultural landscape, namely that landscapes as the places we create are important in any attempt to understand who we are—our human identity—and also how we live—our social, political, and economic worlds. Inevitably, then, such studies also facilitate appreciation of our place in the larger world. The interest being shown by other disciplines in this traditional geographic concern speaks clearly to the increasing recognition of this point. The importance of landscape is reflected in the important idea that each of us can learn through landscape.

A Practical Discipline

Now consider the second of the two important claims referred to at the beginning of Chapter 1, namely the assertion by Wagner (1990:41) to the effect that 'a theoretically well-grounded, intellectually vigorous, and practically effective social and cultural geography might well assume, in time, a major role in guiding and guarding the evolution of humanity's environments'.

Given the concern with understanding people and place, cultural geography is well positioned to contribute in various applied areas. The ambitious claim by Wagner has much merit as the discussions of ecology and of cultural and regional inequalities have demonstrated, and there seems little doubt that cultural geographers will continue to frame analyses within the context of humans and nature, sometimes separating these two and sometimes integrating the two as one. Certainly, the larger discipline of geography is especially well positioned to make meaningful contributions to discussions of the human use of nature and the social construction of nature, including questions of landscape perception and issues of preservation. Two important related research directions at the global scale are now noted.

Global Cultural Geography

There is one critical aspect of human and land relationships that cultural geographers are not emphasizing at present, notwithstanding the pioneering work of Sauer—in short, little is being said about questions of global environmental and human concern. Some studies within cultural ecology offer analyses of particular regional circumstances, but the global seems quite another matter. Not much is being written concerning the all too evident gross inequalities of the human condition. For the early twenty-first century, it is estimated that about 650 million people live on the very edge of existence, lack a reliable food supply, access to clean water, and sustained medical care. Facts such as these identify urgent global issues and are but suggestive of a long and depressing list of problems. While all scholarly work can be of importance, it might be suggested that studying the consumption behavior of the undernourished and the malnourished in the less developed world with the aim of identifying problems and proposing solutions is one aspect of the geography of consumption that merits additional effort from cultural geographers. Certainly, the major global differences of wealth and power are being largely neglected. Recall from Chapter 7 the following quote from a political geographer: 'We need to construct a new global geography that focuses on the world map of global inequalities. We should very consciously proclaim this to be *the most important map in the world* and focus our Geography accordingly' (Taylor 1992:20). Cultural geographers might profitably pursue this line of argument.

Closely related to the above is the issue of cultural futures as these involve globalization, regionalization, and localization. As noted especially in Chapter 7, there is much discussion about the possible emergence of a global culture and of the related reassertion of some regional and local cultural identities in explicit opposition to globalization. In this context, cultural geographers might usefully evaluate, for example, the six possible global cultural scenarios and the various possible regional scenarios outlined by Masini (1994:20–5). The six global scenarios are:

- Cultures, particularly living community cultures, are reduced to just a relict, museum, or tourist role and are unable to challenge the dominant

global identity that reflects Western technology and values.
- There is a combination of culture and change that allows the core elements of cultures to remain while also involving some acceptance of global trends.
- Many cultures resist the dominant global trend with most of the cultural change being internally generated.
- The Gaia view involving a widespread acceptance by most cultures that no one culture is complete in itself.
- All cultures are influenced by information from other cultures, with Western culture as the most influential because of advances in information technologies.
- A non-Western culture, most likely an Asian culture, becomes globally dominant.

Further, as noted in Chapter 4, cultural geographers have had relatively little to say about the formation of cultural regions at the global scale, despite the suggestion that this is 'the greatest topic in historical geography' (Meinig 1976:35). It is possible that cultural geographic interest in this topic will increase significantly.

Cultural Geographies of Difference

The material included in the sections on ethnicity in Chapter 7, and on gender, sexuality, and the disadvantaged in Chapter 8, reflect a number of concerns that are central to much contemporary research in cultural geography, notably concerns about identity, place, and the relations between identity and place. Although it is common to note that these interests reflect a concern with difference, they are primarily a concern with others— that is, those who are excluded in some way from a larger society. It seems clear that ideas and analyses in these areas are changing rapidly.

Thus, the body of theory used to inform such work, which can be loosely described as postmodernism, is both changing and contested. There is, for example, more than one way to interpret claims about the emergence of alternative cultures, such as those based on a particular sexual preference. In the context of a discussion of the globalization of culture, Albrow (1997:150) stated: 'The multiplication of worlds means that individuals can inhabit several simultaneously, but secondly that each severally can only make a small selection from the many which coexist. The result of a plurality of individuals making their own selections is that each builds a different repertoire, and its total scope is obscure to everyone else.' This is a challenging claim implying that the membership of individuals in particular groups is both far from fixed and, more critically, not necessarily a meaningful marker of their identity (see Box 8.4).

Further, there is a complex intermixing of the academic and the political in much current work, as evidenced by the references to strategic essentialism and to queer theory in Chapter 8. Of course, such intermixing has a healthy pedigree, with Marxism and feminism both having political agendas, but it might contribute to some academic failings. Further, it is not unusual for researchers, especially those studying sexualities other than heterosexuality, to explicitly state their insider status. Because of this, it might be that cultural geography along with much other social science—in the desire and willingness to focus on particular disadvantaged others—tends to neglect those others who do not have a political agenda, or whose political agenda is less palatable to researchers. It is possible, for example, that more conservative groups, including some religious cults and North American militia groups, are less likely to be studied because their different identities are not in sympathy with the identities of cultural geographers attracted to this type of work and also because such work typically involves insider research.

Moreover, although there are analyses of a diverse range of sexual identities and related behaviors, there is limited evidence of a cultural geographic concern with some other identities and behaviors, such as introvert and extravert, or authoritarian and submissive, or of those who suffer from phobias. The most common phobia is **agoraphobia**—the fear of leaving home to go into public space—and this has obvious geographic implications in terms of the constraints imposed on movement. Psychologists also recognize a wide range of anxiety states, stresses, mood disorders,

schizophrenic states, and personality disorders, all of which might be interpreted as differences of interest to cultural geographers. Certainly, it can be argued that, largely for ideological reasons, cultural geographers have chosen to focus on those different identities that can be classed as others, usually disadvantaged groups. Thus, the interest shown in difference is selective rather than all-encompassing, and it seems reasonable to suggest that there will be a broadening interest in difference. There are three other closely related matters that might be of concern regarding the interest in difference.

- First, the current preoccupation with those who are different might be a case of overcompensating for the earlier emphasis placed on the dominant culture.
- Second, as noted in Chapter 7, an emphasis on human difference rather than human unity might have the unfortunate consequence of implying not just difference but also inequality. Europeans saw numerous others as different, especially during the period of colonial expansion, and inequality accompanied or quickly followed this recognition. Similarly, any contemporary expression of pride in a particular identity, a particular difference, can easily, although not necessarily, lead to ideas of superiority. This simple point is one of the lessons of object relations theory. The need to listen to others seems clear, but the need to stress difference might be less clear.
- Third, it is increasingly acknowledged that cultural geographers, especially when discussing questions of power and authority as these relate to human differences, need to be more explicit concerning the values that underlie critical analyses of, for example, racist and sexist attitudes and practices.

Some Final Questions

Is There a Distinct Subdiscipline of Cultural Geography?

A key question to ask about cultural geography—really a question about the discipline of geography or indeed about the way in which knowledge is structured—concerns the viability of identifying a subdiscipline labeled cultural geography. Described in Chapter 1 as a compositional subdiscipline—one concerned with some particular subject matter—cultural geography is always being questioned as are all other compositional subdisciplines. Primarily because of the cultural turn, the links with social, political, and economic geographic subdisciplines are increasingly close. For example, in a discussion of economic geography, Crang (1997:4) considered five options for relating the economic and the cultural:

- a continued opposition of the economic and the cultural, effectively resisting the cultural turn
- an export of the economic to the cultural
- the economic seen as embedded in a culturally constructed context
- the economic seen as represented through the cultural, acknowledging that places are cultural constructions
- the cultural understood as materialized in the economic

Similarly, one study of the discipline of geography was structured around themes rather than subject matter using four principal headings (Mabogunje 1997):

- conceptual fundamentals relating to space, time, theory, and methodology
- environment and society, including human transformation of earth, climate change, and human response to hazards
- spatial structures and social processes, including market and cultural forces, particular modes of production, transportation, and trade
- spatial organization and globalization processes, including the role played by the state, transnational corporations, and megacities

Such a framework for writing about geography effectively incorporates the cultural into a larger geographic context.

Given these comments, what might be the future of cultural geography? As this textbook has often had occasion to stress, cultural geography can be viewed as contested intellectual terrain and if this is the case, then describing where we are heading is of course well nigh impossible. In most cases, textbook authors outline the way ahead as a continuation of current trends—at best, such predictions are but a partial truth. Certainly, it is reasonable to assume that the principal current trends will continue, but it is also reasonable to assume that new trends will appear. In this context, cultural geography might be described as a heterotopia (a site of incompatible discourses) and with this view in mind, Duncan (1994a:402) wrote approvingly of the fact that cultural geography 'is no longer as much an intellectual site in the sense of sharing a common intellectual project as it is an institutional site containing significant epistemological differences'.

With reference to the cultural geographic concept of culture, it is fair to state that a particular understanding of culture is central to most of the work in both traditions, but two qualifications of this statement are worth noting.

- First, with reference to landscape school geographers through to the 1970s, Mikesell (1977:460) observed that 'most cultural geographers have adopted a laissez-faire attitude towards the meaning of culture'.
- Second, some cultural geographers have questioned the intent of the new cultural geography and sought a major reorientation. Thus, it has been argued that there is no such thing as culture, but rather only the idea of culture (Mitchell 1995), and that there is no 'need of a concept of culture as such', rather a need for 'a much deeper, clearer, more operational conception of human behavior and development' (Wagner 1994:5).

Does the Cultural Turn Challenge the Legitimacy of Cultural Geography?

Although the impact of the cultural turn on cultural geography is considerable, there are many questions that can be asked. In the opening section of Chapter 7 reference was made to some critical evaluations of the cultural turn and there have also been issues raised in Chapter 8 and earlier in this concluding chapter. Now is an appropriate time to venture what may be an especially troubling concern.

Contemporary cultural geography clearly welcomes the importation into the subdiscipline of ideas from elsewhere, especially from the cultural studies tradition, but there may be a real difficulty in such movement because cultural studies is seen by many of its practitioners to be a postdisciplinary and critical practice. Thus, from a cultural studies perspective, it does not necessarily make sense that ideas be exported to disciplines and subdisciplines and then be employed within precisely those disciplinary frameworks that cultural studies challenges. It might be suggested that cultural geographers seem to want the best of both worlds—both the critical theories developed in cultural studies and the traditional subdiscipline of cultural geography in which to apply these critical theories. But, for some practitioners of cultural studies, these two are irreconcilable, not least because disciplines and subdisciplines are seen as something other than neutral when it comes to discussions of such topics as racism and sexism. Cultural studies does not take the existence of disciplines for granted whereas most new cultural geography implicitly accepts the existence of a body of knowledge labeled cultural geography. Simply put, there is a very real challenge involved in attempting to employ theories derived from an avowedly postdisciplinary tradition in a disciplinary context. Indeed, for many in cultural studies it is a challenge that should not be taken up. This contradiction seems implicit in any discussion of an academic discipline in postmodern terms and was noted in the discussion of postmodernism in Chapter 7.

Is It Possible to Integrate Traditional and New Cultural Geography?

Finally, it is worth emphasizing that this book has an ambitious and, perhaps for some, a questionable goal, namely that of integrating, or at least incorporating under some common headings, work that might be variously labeled traditional or new cultural geography.

- Few would argue with the observation that the goal of integration is an ambitious one. There is one principal reason—in addition, of course, to authorial limitations—why this goal is at best only partially attained. Thus, cultural geography is all too clearly a moving target, with new, often imported, concepts informing increasingly diverse analyses, and also with ever closer connections to such other traditional compositional subdivisions of human geography as social, political, and economic geography. These are certainly healthy signs but, equally certainly, it makes accommodating the contents of cultural geography within a few key themes—chapters in the context of this book—an unenviable task. Indeed, there is a tendency for reviewers of cultural geographic literature to focus on selections of material rather than on addressing some perceived totality.

Your Opinion 9.1

Does it seem reasonable to suggest that the traditional and the new will merge in some way to become one or might the two continue as related but essentially different approaches? Certainly, the more traditional thematic approaches employed by cultural geographers are likely to persist for some time, both for reasons of academic conservatism and because different approaches all have their merits. According to Newson (1996:279), 'in the USA traditional cultural geography remains an active, and largely distinct, research field that is unlikely to be totally overtaken by new approaches'. Further, there is little doubt that the various stylistic tactics employed by some of the cultural geographers who are deeply involved in postmodernism are not always well received. Walker (1997:172–3) expressed it this way: 'I become easily irritated with the posturing of the postmodernist and the mannered style of discourse that glibly condemns the linear, logical, and evidentiary essay in favor of fragments of literary allusion and freely tossed Lacanian word salads, which leave a faint and convoluted trail of simulacrumbs for the poor reader to follow.' What are your reactions to these final comments and, more generally, to the contents of this textbook?

- The goal of integration, as it might appear to imply a unity, may be seen as questionable by some cultural geographers because it 'reflects the modernist will to order and discipline, to be governed under a master narrative' (Duncan 1994a:402).
- Whatever the merits of the claim that integration is a questionable goal, this text accepts that there is value in discussing cultural geography, all cultural geography as broadly conceived, within a single organizational context, a value that is especially evident when the needs of student geographers are being considered. Organizational frameworks such as that adopted in this book facilitate comprehension of diverse ideas and analyses without in any sense imposing unwavering identities on ideas and analyses or suggesting that the framework used is in some sense correct and uncontestable, a point that was made in the opening chapter of this textbook.
- But, most important, see Your Opinion 9.1.

Further Reading

The following are useful sources for further reading on specific issues.

Parsons (1996) on the Black Mountain School.

Salter (1977) on learning through landscape.

Dowling (1997) on some issues relating to cultural planning in Australia in the context of contemporary cultural geography.

Philo (1989), Parr and Philo (1995), and O'Dwyer (1997) on particular identities and behaviors.

Sayer and Storper (1997) on the need to understand the values that inform critical cultural geographic analyses.

Harris (1999) on the changing disciplinary landscape.

Barnes (1996) on the cultural understood as materialized in the economic.

Chambers (1994:121–2) on cultural studies and disciplinary identities.

Glossary

ableism
Ideas and practices that assume that all individuals are able–bodied.

acculturation
The process by which a minority or politically weaker cultural group gradually experiences cultural change to accord more closely with the character of a majority or dominant cultural group. Does not involve a process of complete cultural change.

acid rain
An environmental problem. The burning of coal, oil, and natural gas releases the oxides of sulfur and nitrogen into the atmosphere, resulting in acidified rain that causes damage to plant and animal life.

adaptation
Human adjustment to environmental conditions, both physical and cultural, in order to ensure provision of the needs of the group and to reduce conflict. It is usual to refer to an adaptive strategy as the particular way in which a group meets those needs. The term 'cultural adaptation' refers to the idea that culture is an adaptive system.

ageism
Discrimination or prejudice on the basis of age.

agoraphobia
A fear of public places and of people in those places.

agricultural revolution
A series of gradual changes that commenced about 12,000 years ago near the end of the Pleistocene involving a change from hunting, fishing, and gathering to food production through a process of animal and plant domestication. Marks the end of the cultural period known as the Mesolithic and the beginning of the Neolithic.

alienation
The circumstance in which a person is indifferent to or estranged from nature or from the means of production. The sense that our human abilities are taken over by other entities.

anthropocentric
Regarding humans as the central fact of the world.

apartheid
An Afrikaans term that translates literally as 'apartness'. Refers to policies of spatial separation of groups of people distinguished on the basis of a racial classification, and to the social distinctions made between those groups. These policies and practices were applied in South Africa between 1948 and 1994.

artificial selection
As distinguished from natural selection, the process by which humans choose specific members of an animal or plant species to live and reproduce.

assimilation
The loss by a cultural group of all their previous cultural traits through the acceptance of the traits of some other dominant group with which they are in contact. Involves a loss of identity.

authority
The right, usually by mutual recognition, to require and receive submission by others.

body
Increasingly, the body is being interpreted as one of the concerns of cultural geographers, being variously interpreted as a site of resistance, as in some feminist and postcolonial theory, as a producer of meanings, and as a declaration of subjectivity.

capitalism
A social and economic system for the production of goods and services that is based on private enterprise, that involves a separation of the

producer from the means of production, and that allows relatively few individuals to have access to resources.

carrying capacity
The number of people that a given region is able to support given specific technological and other cultural characteristics.

Cartesianism
Descartes was a key figure in the emergence of modern philosophical and scientific thinking. Cartesianism refers to the need for mechanical and mathematical explanations in physical science. The Cartesian theory of the mind, dualism, maintained that the mind is entirely separate from the body; this view of the mind was famously referred to by Gilbert Ryle as a 'ghost in the machine'.

catastrophism
In contrast to uniformitarianism, the idea that the world has experienced and results from a series of catastrophic events, such as the Flood.

civilizations
A term with various contested meanings. Often used to refer to cultures that had an agricultural surplus, a stratified social system, some specialization of labor, a form of central authority, and a system of keeping records. More generally used to refer to cultures of global significance.

class
Frequently used but not agreed upon term. Generally, a large group comprising individuals of similar social status, income, and culture. The presence of classes in a society explicitly reflects divisions usually involving an unequal distribution of economic goods, political power, and cultural status.

cognition
Human thought processes involving perception, reasoning, and remembering.

cognitive map
For each individual, this is the model of the world in which they live.

colonialism
The policy of a state or people seeking to establish and maintain their authority over another state or people.

conservation
A general term referring to any form of environmental protection including preservation.

constructionism
This is a position that claims that all of our conceptual underpinnings, such as ideas about identity, are necessarily contingent and dynamic, not given or absolute (compare essentialism). Also known as constructivism. See social construction.

consumer culture
As discussed by Baudrillard, consumer culture is understood with reference to institutions of consumption, such as shopping malls and advertisements, and is a symbolic practice that needs to be interpreted.

contextualism
Broadly, the claim that it is necessary to take into account the specific discourse or context within which any account of ideas and facts takes place.

cornucopian thesis
The argument that advances in science and technology will continue to create resources sufficient to support the growing world population.

counterfactual
A statement that asks what might have been the consequences if some detail of history had been different. They may be used to analyze presumed cause and effect relationships through a comparison of the observed world and some hypothetical world derived from contemplating a different history.

creation myths
Stories, often invoking the supernatural, that explain both the origin of the world and the origin and early history of people in the world.

cultural ecology
Sometimes used to refer to the discipline of cultural anthropology. Study of the interactions between humans and nature. More specifically, the analysis of culture as an adaptive system.

cultural hearth
Usually also cultural core. The place of origin, the heartland, of a culture from which it diffuses outwards.

cultural realm

An extensive, often continentally based, area within which some uniformity of cultural practice is evident. May be a cluster of related cultural regions. Necessarily, a realm is less internally homogeneous than is a region.

cultural region

An area that is occupied by a cultural group and that reflects that occupation both in the visible material and symbolic landscapes.

cultural studies

An area of academic interest concerned with various types of texts as these aid understanding of the production of values, meanings, and identities. Focuses primarily on the political meaning of culture—on gender, sexuality, ethnicity, nationality, and class.

cultural trait

One element in the normal practice of a cultural group. A cultural complex is some related set of traits.

cultural turn

Changes in social science prompted by developments in philosophy and social theory, recognizing that culture is too important to be reduced to economics and politics.

culture

There are numerous specific interpretations of this word. Refers generally to the way of life of the members of a society as evident in their values, norms, and material goods.

Darwinism

The view that the development of a species results from competition among and within species, gradually eliminating the least fit and permitting the fittest to survive. The fittest are those having genetically based features that give them some competitive advantage over fellow members of their species. Darwinism is non-teleological and materialist.

deconstruction

The principal method of studying texts employed in the postmodernist tradition. Based on the assumption that a text is not a self-contained entity, the meaning of which is determined by the intent of the author, but rather that a text exists in a context with both the production (writing) and receiving (reading) being affected by other texts. The unquestioned, taken-for-granted, philosophical assumptions that underpin a text, assumptions that are inherited and typically contain uncritically accepted dualisms, are exposed.

dependency theory

The idea that European overseas expansion resulted in a symbiotic relationship between the development of dependence and the development of underdevelopment.

determinism

The general idea that everything that happens is an effect; events are necessitated by earlier events. Philosophers such as Hobbes and Hume were determinists. There are many versions of determinism, mostly in physical science, with the principal example being the mechanistic clockwork universe of Newtonian physics. A dominant view in the nineteenth century, determinism was liberally applied in the social sciences in a series of varyingly effective versions. Philosophically, a principal concern with determinism is the application to the human world, with the implication that humans are unable to exercise free will and with related concerns about moral responsibility.

development

A term to be used cautiously because it often involves an implicit general acceptance of a particular perspective (ethnocentrism); typically interpreted to mean a process of becoming larger, more mature, and better organized; often measured by economic criteria.

developmentalism

Analysis of cultural and economic change that treats each country or region of the world separately in an evolutionary manner; assumes that all areas follow the same stages and that they are autonomous.

dialectic

The resolution of contradictions in the pursuit of truth. A method of reasoning, with thought proceeding from thesis to antithesis to synthesis.

diffusion
The process of spread over space and growth through time of cultural traits, ideas, disease, or people, from a center or centers of origin.

discourse
Language in which word meanings are specific to a community of users; for example, the technical vocabulary that distinguishes academic disciplines from each other. More generally, the various social practices that enable the world to be made intelligible. Now a widely used and often confusing term. Although sometimes used interchangeably with the term 'text', 'discourse' is more appropriately seen as a way of connecting texts that share a common point of view, as in sexist discourse.

disorganized capitalism
In contrast to organized capitalism, refers to a new form characterized by a process of disorganization and industrial restructuring.

dualism
In the classic Cartesian formulation, based on earlier Greek ideas, the idea that mind and matter are separate. More generally refers to the separation of two things that might alternatively be seen as one—humans and nature, for example. Often contrasted with holism.

ecocentric
Emphasizing the value of all parts of an ecosystem rather than, for example, placing humans at the center, as in an anthropocentric emphasis.

ecosystem
An ecological system; comprises a set of interacting and interdependent organisms and their physical, chemical, and biological environment.

egalitarian
A society in which all people are essentially similar in terms of wealth and power, with the only major distinctions being those based on age and sex.

empiricism
An empiricist epistemology asserts that all factual knowledge is based on experience, with the human mind being a blank tablet (*tabula rasa*) before encountering the world. Empiricism is a fundamental component of the philosophy of positivism. It is also related to pragmatism.

energy
The capacity of a physical system for doing work.

Enlightenment
This European intellectual movement, the Age of Reason, of the seventeenth and eighteenth centuries is contrasted with the earlier period of irrationality and superstition. It is not easy to summarize but, generally, two assumptions reflect the philosophical ideas of the period, an empiricist epistemology and the idea that the aim of social science was to enhance social progress through revealing truths about ourselves.

environmental determinism
The argument that the physical environment is the principal cause of human behavior and human landscape creation.

episteme
The world views, structures of thought, that a society holds at a particular time and that impose the same standards on all branches of knowledge; associated with Foucault.

epistemology
The branch of philosophy concerned with the nature, sources, and justification of knowledge. Epistemological questions are about how we know knowledge is knowledge.

essentialism
A belief in the existence of fixed unchanging properties; attributing essential characteristics to a group rather than seeing such characteristics as being constructed socially (compare with constructivism).

ethnic
A group whose members perceive themselves as different from others because of a common ancestry and shared culture.

ethnic cleansing
The forced removal of an ethnic group by another group. Often involves extermination.

ethnic religion
A religion associated with a particular group of people and that accordingly does not seek to convert others.

ethnocentric

Making judgments about a culture based on the values of one's own culture and, as a consequence, misrepresenting that culture. For example, Eurocentric refers to ideas that fail to acknowledge that the priorities of European discourse are not necessarily shared by other ethnic groups.

ethnography

A general research approach requiring researcher involvement in the subjects studied; refers to any approach that is based on first-hand observation in the field; qualitative methods such as participant observation are a part of ethnography; typically an intensive study of a group defined in cultural terms.

evolution

The idea that organisms have developed from initially primitive forms, through natural processes, to more complex forms. In Europe, a product of Enlightenment thinking; prior to this the prevailing view was that of creation. Lamarck was the first important evolutionist, but the critical contribution was from Darwin. There are many applications of an evolutionary epistemology to cultures, with Spencer as the major figure.

existentialism

Humanistic philosophy; concerned with human existence; often emphasizes human estrangement from larger world with the aim of reunification.

feminism

There are numerous versions of feminism. Fundamentally, it refers to the advocacy of the rights of women to equality with men, recognizing that sexism prevails, is wrong, and needs to be eliminated. Patriarchy is usually seen as the fundamental issue. Increasingly, feminism is concerned not only with the oppression or subordination of women, but also with other oppressions, such as those based on social class, skin color, income, religion, age, culture, and geographic location.

fengshui

A Chinese folk-belief system that ascribes auspicious qualities to both the physical and human environments.

feudalism

A social and economic system prevalent in Europe prior to the Industrial Revolution that involved two principal groups; the land was controlled by lords while the peasant was bound to the land and subject to the authority of the lord. The clergy were also a distinct group.

first effective settlement

Phrase popularized by Zelinsky that refers to the group that first establishes a viable, self-perpetuating community in an area undergoing settlement from outside. Similar to the initial occupance concept noted by Kniffen.

flexible accumulation

Industrial technologies, labor practices, relations between firms and consumption patterns that are increasingly flexible.

folk culture landscape

Often compared to popular culture landscape. The landscape associated with a group that is relatively unchanging and usually small in number; often characterized by traditional cultural traits associated with clothing, food, architecture, and religion.

forces of production

A Marxist term that refers to the raw materials, tools, and workers that actually produce goods.

Fordism

A group of industrial and broader social practices initiated by Henry Ford and dominant until recently in most industrial countries. Characterized by standardized products of mass production in large factories.

formal region

A formal cultural region is occupied by a relatively distinct cultural group and displays a relatively uniform landscape.

frontier

A zone of advance penetration by an incoming group; often an area of conflict between an existing culture and an incoming culture.

frontier thesis

As introduced by the American historian, Turner, an environmental determinist view that sees the American frontier as the place where civilization encountered savagery and where civilization was continually, over a 300-year period, conditioned. The result of this extended experience was a new American, not European, culture.

functional region
A functional cultural region functions politically or economically as an integrated unit. May be related to some homogeneity in people and/or landscape.

functionalism
A concern with the analysis of functions. A form of teleological philosophy that explains social situations through an account of roles.

Gemeinschaft
A term introduced by Tönnies; a form of human association based on loyalty, informality, and personal contact; assumed to be characteristic in traditional village communities.

gender
The social aspect of the relations between the sexes. Does not refer to physical attributes, but rather to the socially formed traits associated with masculine and feminine categories. Masculinity and femininity are not, therefore, naturally occurring; rather they are the consequences of human history.

gender relations
The idea that gender roles are explained in terms of power relations between women and men, especially that of patriarchy.

gender roles
A set of behaviors traditionally assigned to women characterized by passivity and relationship behavior (actions that facilitate human interaction)—femininity; and a set of behaviors traditionally associated with men characterized by activity and instrumental behavior (goal-oriented actions)—masculinity.

Gesellschaft
A term introduced by Tönnies; a form of human association based on rationality and depersonalization; assumed to be characteristic in urban areas.

ghetto
A residential district in an urban area with a concentration of a particular ethnic group.

globalization
A process whereby the population of the world is increasingly bonding into a single culture and economy. Often related to the emergence of a world system.

green revolution
A series of agricultural innovations applied in the less developed world context. The development of higher-yielding and faster-growing crops, especially cereals.

greenhouse effect
Usually used to refer to the human addition to the naturally occurring greenhouse effect. The atmospheric consequences of human activities, especially fossil fuel burning, that add carbon dioxide and other gases to the atmosphere that result in the earth retaining more of the warmth that comes from the sun than would otherwise be the case.

group
Perhaps the most general term used to refer to some collection of human beings. Usually implies that there is set of relations existing between individual members of the group, and also that each member is conscious of the existence of the group.

hegemony
The ability of a group to exercise control over others without needing to rely on laws or the use of force. Involves an acceptance by others of fundamentally unequal circumstances.

hermeneutics
The study and interpretation of meaning. The interpretation of texts. The uncovering of cultural meaning in everyday life by understanding the signs and symbols of your own group in context with those of other groups. There are many versions; all give a key role to the mental quality of humans. Compare with the term 'naturalism'.

heterosexuality
The orientation of emotions and/or sexual activity toward those of the opposite sex.

historical materialism
A method, associated with Marxism, that centers on the material basis of society and that attempts

to understand social change by reference to historical changes in social relations.

holism

Refers to the idea that the properties of individual elements in some complex are affected by the relations with other elements. Implies the value of studying groups of things together rather than apart; ecology is holistic. Often contrasted with dualism.

Holocene

The modern geological epoch, which began some 10,000 years ago.

homeland

A type of culture region that involves interaction with the physical environment to evoke emotional attachment and bonding. Usually associated with a particular group defined in ethnic terms.

homosexuality

The orientation of emotions and/or sexual activity toward those of the same sex.

humanism

An approach to the study of humans and human behavior that gives priority to the fact of being human. Although the term has many connotations, modern humanism flowered in the nineteenth century as one component of the conflict between science, as exemplified by Darwinism, and fundamentalist interpretations of religion. Humanists emphasize our ability to make choices and our responsibility for the actions we take. The humanistic philosophical tradition includes such figures as Giambattista Vico (1668–1774) and George Berkeley (1685–1753), both of whom questioned the idea that mathematics and mechanics are keys to understanding the world, and Hegel, whose work was an inspiration for various idealist perspectives.

iconography

The description and interpretation of visual images, including landscape, in order to uncover their symbolic meanings. The identity of a region as expressed through symbols.

idealist

A group of philosophies that suggest that what is real is related to and created by the contents of the human mind; therefore, opposed to both materialism and realism. According to Berkeley, nothing exists outside the mind. According to the transcendental idealism of Immanuel Kant (1724–1804), the categories of terms that are employed to describe the world are not objective characteristics of the world but are, rather, structures imposed by the mind. According to Hegel, history comprises a progressive realization of a single spirit (*Geist*). Generally, a metaphysical view that only minds and ideas exist.

idealism

As introduced into cultural geography, a specific version of the larger idealist perspective, derived from the historical idealism proposed by the historian, Collingwood; explicitly concerned with rethinking the thoughts behind human actions.

identity

Refers to sameness and continuity; a term that became popular with the rise of mass society and the related quest for identity. Accounts of identity derive from both sociological and psychoanalytic inspirations.

ideology

A socially ordered system of cultural symbols. A body of ideas or a way of thinking. Closely related to the concept of power as ideological systems serve to legitimize the differential power that groups possess.

idiographic

A method stressing the individuality and uniqueness of phenomena rather than the similarities between phenomena (compare with nomothetic).

image

The perception of reality held by an individual or group.

imperialism

A relationship between states with one being dominant over the other. The process of empire establishment that took place during the period of political colonialism.

Industrial Revolution

A process that converted a fundamentally rural society into an industrial society beginning in mid-eighteenth century England; primarily a technological revolution associated with new energy sources.

infrastructure (base)
A Marxist term that refers to the economic structure of a society especially as these give rise to political, legal, and social systems.

innovation
An idea that leads to change, often increasing individual and/or group productivity, or a cultural trait that is new to a group. An innovator is a person who leads change.

intertextuality
The idea that meaning is produced from one text to another rather than being produced between the world, including the author(s) and the historical context, and any given text.

Lamarckianism
An evolutionary process proposed by Lamarck, essentially abandoned after the acceptance of natural selection, to the effect that characteristics acquired by individual members of a species through their experiences can be inherited by their offspring. For some, it is appropriate to view cultural evolution, as opposed to biological evolution, in Lamarckian terms.

language family
A group of related languages derived from a single common ancestral language.

late (taken–for–granted) capitalism
Characterized by multinational capitalism and by the commodification of culture; associated with postmodernity.

lebensraum
Literally, living space. As employed by Ratzel, the argument that a political state was similar to a living organism in that it might require space to grow.

limits to growth
The argument that, in the future, both world population and world economy may collapse because available world resources are inadequate.

locales
The settings or contexts for social interaction; a term introduced in structuration theory that has become popular in human geography as an

alternative to place. The term may be employed at a range of scales.

logical positivism
A philosophical approach initiated by the Vienna Circle group of scholars in the 1920s that rejected all metaphysics in favor of a scientific approach.

maladaptation
An adaptive strategy that either fails to achieve the desired goals of the group or that involves damage to the environment.

material culture
The visible physical objects that are made by or used by a group, including clothing, buildings, and tools.

materialism
The idea that humans are dependent on the natural world. Forms of materialism include the mechanistic materialism of early science and the dialectical materialism of Marxism. Sometimes used as a synonym for naturalism; contrasted with idealism.

mechanistic
For some, a mechanistic world is one that functions like clockwork; it exhibits regularity and predictability and is subject to the operation of laws. A term that is closely related to Cartesianism and determinism and more generally to the scientific approach emphasizing cause and effect.

mental map
The already constructed images that humans have in their minds and that affect behavior.

mercantilism
A school of economic thought, dominant in Europe in the seventeenth and early eighteenth centuries, that argued for the involvement of the state in economic life so as to increase national wealth and power. Acquisition of precious metals and a favorable trade balance were desirable.

Mesolithic
Cultures in early Holocene Europe between the end of the Upper Paleolithic and prior to the agricultural revolution of the Neolithic. Characterized by a number of cultural adaptations related especially to changes in plant and animal communities as ice sheets retreated. The

term literally means 'middle stone', referring to a transitional stone tool technology between the Paleolithic ('old stone'), and Neolithic ('new stone') periods.

metaphysics
Refers to attempts to explore the world of the suprasensible, that is, the world beyond experience. Claims to deal with issues that science cannot comprehend.

minority language
A language spoken by a minority group in a state in which the majority of the population speaks some other language; may or may not be an official language.

mode of production
A Marxist term that refers to the organized social relations through which a human society organizes productive activity; human societies are seen as passing through a series of such modes.

modernism
A view that assumes the existence of a reality characterized by structure, order, pattern, and causality.

multiculturalism
A policy that endorses the right of ethnic groups to remain distinct rather than being assimilated into a dominant society.

multilinear evolution
A view of evolution claiming that there are various paths of development that cultures may follow.

multinational corporations
Large business organizations that operates in two or more countries; sometimes called transnational companies.

nation
A group of people sharing a common culture and an attachment to some territory; a difficult term to define objectively.

nation state
A political unit that contains one principal national group and which identifies itself and its territory with that group.

nationalism
The political expression of nationhood or aspiring nationhood; reflects a consciousness of belonging to a nation.

natural selection
The principal contribution of Darwin to the question of evolution. Refers to the 'survival of the fittest' (a term coined by Herbert Spencer in 1852). It is a materialist explanation for evolutionary change.

naturalism
The view that all things are natural. Thus, the idea that human behavior is explicable by reference to mechanistic laws similar to those that apply in physical science. A fundamental decision that any social scientist must make is whether or not to adopt a naturalist philosophy. Compare with the term 'hermeneutics'. Also the belief that what is studied by the physical and human sciences is all there is. Includes acceptance of natural selection.

nature
It is not helpful to suggest that there is a neat and clear definition of this term that is generally agreed to by cultural geographers and others. There are numerous specific meanings dependent on the particular context in which the term is employed. To suggest a contrast of nature with culture is a reasonable general understanding given the cultural geographic tradition.

neocolonialism
Colonial economic and cultural circumstances under a different guise from the original political colonialism.

Neolithic
A cultural phase characterized by the disappearance of foraging and the appearance of animal and plant domestication as the principal means of subsistence. Such domestication may have begun as early as about 12,000 years ago. The name is inappropriate in that Neolithic literally means 'new stone', referring to the production of new types of stone tools, although the key distinguishing characteristic is domestication of animals and plants.

new cultural geography
First proposed in the 1980s, an agenda for cultural geography that stressed the need to understand rather than merely describe cultures and places. The Sauerian concept of culture as

cause was rejected and emphasis was placed on the diversity of cultures, on the critical role played by power relations, and on such markers of human identity as 'race', gender, and sexuality. As one part of the modern to postmodern transition, this approach to cultural geography stressed the constructive power of language.

niche
An ecological address; the space occupied by an organism and the activities that allow it to survive.

nomothetic
A method stressing the similarities between phenomena, seeking laws, associated with positivism (compare with idiographic).

non–material culture
The oral components of culture, such as beliefs, customs, songs, paintings, and poetry.

norms
Rules of conduct that identify appropriate and inappropriate behavior for members of a group. Norms are reinforced by sanctions of some form.

ontology
The assumptions about existence that underlie any particular system of ideas. It is usual to distinguish between the ontology and epistemology of knowledge.

ordinary landscape
The landscapes of everyday urban space, rooms, buildings, backyards, streets, and neighborhoods, and of everyday rural space, fields, fences, barns, and farmhouses.

organized capitalism
A form of capitalism that developed after the Second World War characterized by increased growth of major (often multinational) corporations and increased involvement by the state (often in the form of public ownership) in the economy.

Orientalism
Refers to the particular perspective that Western scholars of the Orient have of that area; implies a view of the periphery from the center. Although now associated with postcolonial theory, primarily with the work of Said, it is clear that the idea of the Orient as a foil for the West dates back to classical Greece.

other
Philosophically, an ambiguous term that derives from the writings of Hegel and that has been used especially by Lacan. Prominent concept in feminist and postcolonial theory. Usually defined in opposition to the same or to self. Based on the assumption that identities are defined not autonomously, but by reference to something that can be either excluded or contradicted. The term may be capitalized when used in this specific context. In cultural geography, it is often used to refer to subordinate groups as these are viewed by and contrasted to dominant groups; implies both difference and inferiority. For example, masculine identity is defined in terms of the exclusion of the feminine other.

palimpsest
A writing material on which writing has been removed or partly removed to allow for subsequent writings. A term adopted by historical geographers to refer to cultural landscapes that contain features from a series of occupations. There is a suggestive association with the idea of a landscape as a text than can be read.

paradigm
A term that describes the stable pattern of scientific activity prevailing within a discipline.

patriarchy
Literally, the rule of the fathers. A social system in which men dominate, oppress, and exploit women. Often seen as constituted by six factors—the family household, employment, sexuality, violence, culture, and the state. Most feminist movements seek to combat patriarchy.

perception
The process by which humans acquire information about physical and social environments.

phenomenology
A variety of philosophies that provide non-empirical descriptions of phenomena. The modern version, based on Husserl, tries to reveal phenomena as intuited essences through direct awareness.

place
Location; in humanistic geography this term has acquired a particular meaning as a context for

human action that is rich in human significance and meaning. Use of this term usually implies a rejection of various scientific approaches, including positivism.

placelessness
Homogeneous and standardized landscapes that lack local variety and character. Sometimes the result of the spread of popular culture at the expense of local cultures. Globalization is an extreme version.

Pleistocene
A geological epoch characterized by a series of about eighteen glacial periods. Began about 1.6 million years ago, ended about 10,000 years ago, and was followed by the Holocene.

plural society
A culture comprising several ethnic groups, each living in a community largely separate from the others.

popular culture landscape
Often compared to folk culture landscape. The landscape associated with a dynamic culture; often an urban area, and possessing traits that reflect recent developments in ideas, values, and preferences.

positivism
A movement introduced by Comte, related to both empiricism and naturalism, which organized knowledge and technology into a consistent whole. Claimed that the history of human thought evolved through three stages—religious, metaphysical, and scientific—and that the sciences form a natural hierarchy ranging from mathematics through to the human science of sociology. The philosophy of logical positivism is often equated with a scientific approach.

post–Fordism
A group of industrial and broader social practices evident in industrial countries since about 1970; involves more flexible production methods than those associated with Fordism. Involves decentralized use of information technologies.

postcolonial studies
Part of the cultural studies perspective, more specifically of postmodernism, that explicitly opposes the ethnocentrism seen as a fundamental component of the European cultural tradition. Sees national cultures of previously colonial areas as defined by the tensions related to the history of colonial domination.

postmodernism
A movement in philosophy, social science, and the arts, arguing that reality cannot be studied objectively, and stressing that multiple interpretations are both possible and legitimate.

poststructuralism
A complex body of philosophical thought that is sometimes loosely equated with postmodernism or even with deconstruction. It is, however, better regarded as a set of ideas that expand upon the logic of structuralism to insist that meaning is produced within language and that human actions are constrained by structures.

power
In general, the capacity to affect outcomes; more specifically, to dominate others by means of violence, force, manipulation, or authority. Characteristic of an oligarchic society in which power is in the hands of an élite group.

pragmatism
The one original American philosophy, assumes that truth is to be determined by reference to practical outcomes. Founded by Charles Saunders Peirce (1859-1914), William James (1842-1910), and John Dewey (1859-1952).

preadaptation
A preadapted culture already possesses the necessary cultural traits to allow successful occupation of a new environment prior to movement to that environment; groups with these characteristics have a competitive advantage.

prejudice
Holding preconceived ideas about an individual or group and being resistant to change even in the light of contrary evidence. Literally, prejudging.

queer theory
Developed in gay and lesbian studies and concerned with oppressed sexualities in terms of both social rights and cultural politics.

race
Subspecies; a physically distinguishable population within a species.

racism
A particular form of prejudice. Attributing char-acteristics of superiority or inferiority to a group of people who share some physically inherited characteristics.

realism
Philosophical view holding that material objects exist independently of sense experiences; contrasted with idealism. Aims to reveal the causal mechanisms through which events are situated within underlying structures.

reductionism
Any doctrine that claims to be able to make the seemingly complex comprehensible in more simple and limited terms.

relations of production
A Marxist term that refers to the ways in which the production process is organized, specifically to the relationships of ownership and control.

renewable resources
Resources that naturally regenerate to provide a new supply within a human life span.

resource
Something material or abstract that can be used by humans to satisfy a need or perceived deficiency.

restructuring
In a capitalist economy, changes in or between the various component parts of an economic system resulting from economic change.

sacred space
A landscape that is particularly esteemed by an individual or a group, usually but not necessarily for religious reasons.

satisficing behavior
A model of human behavior that rejects the rationality assumptions of the economic operator, assuming instead that the objective is to reach a level of satisfaction that is acceptable.

sense of place
Phrase that refers to the deep attachments that humans have to specific locations such as home and also to particularly distinctive locations.

sequent occupance
An approach to evolutionary landscape analysis that recognizes a series of stages during which the cultural landscape is essentially unchanging, with periods of rapid and profound change occurring between stages.

sexism
Attitudes or beliefs that serve to justify sexual inequalities by incorrectly attributing or denying certain capacities either to women or to men.

sexuality
In some feminist and psychoanalytic theory, interpreted as a cultural construct rather than as a biological given. Aligned with power and control.

simplification
A process of cultural change involving the creation of a less complex culture that is often experienced by groups moving to a different environment.

simulation
A method of representing a real process in an abstract form for the purposes of experimentation.

slavery
Labor that is controlled through compulsion and not involving remuneration; in Marxist terminol-ogy it is one example of a mode of production.

socialism
A social and economic system that involves common ownership of the means of production and distribution.

social construction
The recognition that all knowledge reflects the fact that we are all born into an existing society that precedes any individual develop-ment, with that social knowledge becoming a part of our world view and ideology. See constructionism.

Social Darwinism
An interpretation of Darwinian concepts and application to human societies. An evolutionary and naturalistic conception of society; initially proposed by Herbert Spencer.

society
Refers to the system of interrelationships that connect individuals together as members of a culture.

sociobiology

A modern growth area within what some call 'neo-Darwinism'. A concern with the evolutionary interpretation of species behavior, especially concerning social interaction. Behavior is interpreted in terms of strategies that have selective advantage in that they increase the chances of survival. Applied to humans, sociobiology has proven controversial because of the implication that human behavior is genetically determined.

spatial turn

Beginning in the 1950s, changes in the practice of human geography that involved a rejection of the essentially descriptive regional approach in favor of analyses informed by a positivist philosophy; intended to offer explanations of locations through the development of theories and use of quantitative methods.

spectacle

Places and events that are examples of carefully created mass leisure and consumption.

stereotype

A collection of expectations and beliefs about a particular group of people that effectively serves as a blueprint affecting how they are perceived and understood by others.

stock resources

Resources that have evolved over a geological time span and that cannot therefore be used by humans without depleting the total available.

Stoicism

A Greek philosophical tradition; materialist; viewed the earth as designed and fit for human life.

stratification

A stratified society is divided into social classes that comprise a hierarchy. In capitalism, a powerful élite possess capital and control the means of production while the majority of the population are engaged in the productive process.

structuralism

Range of philosophies; all share the view that the empirical world of observable phenomena results from underlying structures.

structuration theory

Social theory developed by Anthony Giddens that aims to integrate knowledgeable human agents and the social structures of which they are a part.

subaltern

A term employed in the cultural studies perspective to refer to groups that are considered, on the basis of class, caste, gender, race, or culture, to be socially inferior to other groups. Closely linked to the concept of hegemony with a focus on relationships between dominant and subordinate groups.

superorganic

An interpretation of culture that sees it as above both nature and individuals and therefore as the principal cause of the human world; a form of cultural determinism.

superstructure

A Marxist term that refers to the political, legal, and social systems of a society.

surrogate

Substitute data used to represent a variable when precise data pertaining to the desired variable is not available.

sustainability

An adaptive strategy that ensures that the environment continues to provide for future generations of the population.

taken-for-granted world

The world of everyday living and thinking, sometimes called lifeworld, and most closely associated with phenomenology. The intersubjective world of lived experience and shared meanings.

technology

The ability to convert energy into forms useful to humans. Also refers to the tools and procedures used by humans to meet their needs.

teleology

The doctrine that everything in the world has been designed by God. Also refers to the study of purposiveness in the world, the idea that some phenomena are best explained in terms of ends (what they have become or what they achieve) rather than in terms of causes. Sometimes used to refer to a recurring theme in history, such as progress or class conflict.

text
Term that originally referred to the written page, but that has broadened to include all human activities, products, and representations that can be read, for example, maps and landscape. Texts can be regarded as indicators of deeper cultural realities. Postmodernists recognize that there are any number of realities depending on how a text is read. Sometimes seen as synonymous with discourse.

topocide
The death of a place, often resulting from industrial or commercial expansion.

topophilia
The affective ties that humans have with particular places; literally, love of place.

transculturation
Cultural borrowing related to the meeting of two cultures that have similar levels of technology and complexity.

uniformitarianism
In contrast to catastrophism, the idea that those physical processes that affected the earth in the past continue to operate today, and vice versa.

unilinear evolution
A view of cultural evolution that claims that all cultures pass through the same sequence of stages.

universalizing religion
A religion that does not have a restricted domain because of the claim that its beliefs are appropriate for all people.

Upper Paleolithic
Cultural phase, describing the stone tool activities of modern humans from about 40,000 years ago to about 10,000 years ago. Paleolithic literally means 'old stone', referring to a stone tool technology that is contrasted with the subsequent Neolithic ('new stone'), phase.

values
Ideas held by an individual or a group concerning what is good, bad, appropriate, and inappropriate. Different cultures often possess different values.

vernacular region
A vernacular cultural region is identified as such on the basis of the perceptions held both by those inside and outside the region. Usually has a generally accepted name.

verstehen
A research method, associated primarily with phenomenology, that involves the researcher adopting the perspective of the individual or group under investigation; German term that is best translated as sympathetic or empathetic understanding.

world system
A cultural system of global dimensions linking different cultures in some key respects.

References

Abler, R., J.S. Adams, and P. Gould. 1971. *Spatial Organization: A Geographer's View of the World.* Englewood Cliffs, NJ: Prentice Hall.

Abram, S. 1997. 'Performing for Tourists in Rural France'. In *Tourists and Tourism: Identifying with People and Places*, edited by S. Abram, J. Waldren, and D.V.L. Macleod, 29–49. New York: Berg.

Adams, W.M. 1997. 'Rationalization and Conservation: Ecology and the Management of Nature in the United Kingdom'. *Transactions of the Institute of British Geographers* NS 22:277–91.

Agnew, J.A. 1989. 'The Devaluation of Place in Social Science'. In *The Power of Place: Bringing Together Sociological and Geographical Imaginations*, edited by J.A. Agnew and J.S. Duncan, 9–29. Boston: Unwin Hyman.

_____. 1997. 'Places and the Politics of Identities'. In *Political Geography: A Reader*, edited by J. Agnew, 249–55. New York: Arnold.

_____, and J.S. Duncan. 1989. 'Introduction'. In *The Power of Place: Bringing Together Sociological and Geographical Imaginations*, edited by J.A. Agnew and J.S. Duncan, 1–8. Boston: Unwin Hyman.

Aitchison, J., and H. Carter. 1990. 'Battle for a Language'. *Geographical Magazine* 42, no. 3:44–6.

Albrow, M. 1997 *The Global Age: State and Society Beyond Modernity.* Stanford: Stanford University Press.

Althusser, L. 1969. *For Marx*, translated by B. Brewster. Harmondsworth: Penguin.

Amedeo, D., and R.G. Golledge. 1975. *An Introduction to Scientific Reasoning in Geography.* New York: Wiley.

Anderson, B. 1983. *Imagined Communities: Reflections on the Origins and Spread of Nationalism.* London: Verso.

Anderson, K. 1995. 'Culture and Nature at the Adelaide Zoo: At the Frontiers of "Human" Geography'. *Transactions of the Institute of British Geographers* NS 20:275–94.

_____. 1997. 'A Walk on the Wild Side: A Critical Geography of Domestication'. *Progress in Human Geography* 21:463–85.

_____, and F. Gale. 1992. *Inventing Places: Studies in Cultural Geography.* Melbourne: Longman Cheshire.

Andrews, H.F. 1984. 'The Durkheimians and Human Geography: Some Contextural Problems in the Sociology of Knowledge'. *Transactions of the Institute of British Geographers* NS 9:315–36.

Appleton, J. 1975a. *The Experience of Landscape.* New York: Wiley.

_____. 1975b. 'Prospect and Refuge in the Landscapes of Britain and Australia'. In *Geographical Essays in Honour of Gilbert J. Butland*, edited by I. Douglas, J.E. Hobbs, and J.J. Pigram, 1–20. Armidale: University of New England, Department of Geography.

_____. 1990. *The Symbolism of Habitat: An Interpretation of Landscape in the Arts.* Seattle: University of Washington Press.

_____. 1996. *The Experience of Landscape*, rev. edn. New York: Wiley.

Arendt, H. 1951. *The Origins of Totalitarianism.* New York: Harcourt Brace.

Arkell, T. 1991. 'Geography on Record'. *Geographical Magazine* 63, no. 7:30–4.

Arnold, D. 1996. *The Problem of Nature: Environment, Culture and European Expansion.* Cambridge, MA: Blackwell.

Aschmann, H. 1965. 'Athapaskan Expansion in the Southwest'. *Association of Pacific Coast Geographers Yearbook* 32:79–97.

_____. 1987. 'Carl Sauer, A Self Directed Career'. In *Carl O. Sauer: A Tribute*, edited by M.S. Kenzer, 137–43. Corvallis: Oregon State University Press.

Atkins, P., M. Simmons, and B. Roberts. 1998. *People, Land and Time: An Historical Introduction to the Relations Between Landscape, Culture and Environment.* New York: Arnold.

Augelli, J.P. 1958. 'Cultural and Economic Changes of Bastos, a Japanese Colony on Brazil's Paulista Frontier'. *Annals of the Association of American Geographers* 48:3–19.

Auster, M. 1997. 'Monument in a Landscape: The Question of "Meaning"'. *Australian Geographer* 28:219–27.

Avery, H. 1988. 'Theories of Prairie Literature and the Woman's Voice'. *Canadian Geographer* 32:270–2.

Axford, B. 1995. *The Global System: Economics, Politics and Culture.* New York: St Martin's Press.

Badcock, B. 1996. '"Looking-glass" Views of the City'. *Progress in Human Geography* 20:91–9.

Baker, A.R.H. 1992. 'Introduction: On Ideology and Landscape'. In *Ideology and Landscape in Historical Perspective*, edited by A.R.H. Baker and G. Biger, 1–14. New York: Cambridge University Press.

_____, and R.A. Butlin, eds. 1973. *Studies of Field Systems in the British Isles*. New York: Cambridge University Press.

Baker, R. 1997. 'Landcare: Policy, Practice and Partnerships'. *Australian Geographical Studies* 35: 61–73.

Barber, B.R. 1995. *Jihad vs. McWorld: How Globalism and Tribalism Are Reshaping the Modern World*. New York: Ballantine Books.

Barker, R.G. 1968. *Ecological Psychology: Concepts and Methods for Studying the Environment and Behavior*. Stanford: Stanford University Press.

Barnes, T.J. 1996. 'Political Economy II: Compliments of the Year'. *Progress in Human Geography* 20:521–8.

_____, and M. Curry. 1983. 'Towards a Contextualist Approach to Geographical Knowledge'. *Transactions of the Institute of British Geographers* NS 8:467–82.

_____, and J.S. Duncan. 1992. 'Introduction: Writing Worlds'. In *Writing Worlds: Discourse, Text and Metaphor in the Representation of Landscape*, edited by T.J. Barnes and J.S. Duncan, 1–17. New York: Routledge.

Barnett, C. 1998a. 'The Cultural Turn: Fashion or Progress in Human Geography?' *Antipode* 30:379–94.

_____. 1998b. 'Guest Editorial: Cultural Twists and Turns'. *Environment and Planning D: Society and Space* 16: 631–4.

Barrell, J. 1982. 'Geographies of Hardy's Wessex'. *Journal of Historical Geography* 8:347–61.

Barrows, H.H. 1923. 'Geography as Human Ecology'. *Annals of the Association of American Geographers* 13:1–14.

Barth, F. 1956. 'Ecologic Relationships of Ethnic Groups in Swat, North Pakistan'. *American Anthropologist* 58:1079–89.

_____, ed. 1969. *Ethnic Groups and Boundaries*. Boston: Little Brown.

Bassett, T.J. 1988. 'The Political Ecology of Peasant–Herder Conflicts in the Northern Ivory Coast'. *Annals of the Association of American Geographers* 78:453–72.

Bastian, R.W. 1975. 'Architecture and Class Segregation in Late Nineteenth-Century Terre Haute, Indiana'. *Geographical Review* 65:166–79.

Baumann, G. 1996. *Contesting Culture: Discourses of Identity in Multi-ethnic London*. New York: Cambridge University Press.

Bell, D., and G. Valentine. 1995. 'Introduction: Orientations'. In *Mapping Desire: Geographies of Sexualities*, edited by D. Bell and G. Valentine, 1–27. New York: Routledge.

_____, and G. Valentine. 1997. *Consuming Geographies: We Are Where We Eat*. New York: Routledge.

_____, and G. Valentine. 1999. 'Geographies of Sexuality—a Review of Progress'. *Progress in Human Geography* 23:175–87.

Bell, J. 1999. 'Redefining National Identity in Uzbekistan: Symbolic Tensions in Tashkent's Official Public Landscape'. *Ecumene* 6:183–211.

Bender, B. 1993. *Landscape: Politics and Perspectives*. Oxford: Berg.

Benko, G., and U. Strohmayer. 1997. 'Preface'. In *Space and Social Theory: Interpreting Modernity and Postmodernity*, edited by G. Benko and U. Strohmayer, xiii–xvi. Malden, MA: Blackwell.

Bennett, J.W. 1976. *The Ecological Transition: Cultural Anthropology and Human Adaptation*. New York: Pergamon.

_____. 1993. *Human Ecology as Human Behavior: Essays in Environmental and Developmental Anthropology*. New Brunswick, NJ: Transaction Publishers.

Berdoulay, V. 1978. 'The Vidal-Durkheim Debate'. In *Humanistic Geography*, edited by D. Ley and M. Samuels, 77–90. Chicago: Maaroufa Press.

_____. 1989. 'Place, Meaning, and Discourse in French Language Geography'. In *The Power of Place: Bringing Together Sociological and Geographical Imaginations*, edited by J.A. Agnew and J.S. Duncan, 124–39. Boston: Unwin and Hyman.

Beresford, M.W. 1957. *History on the Ground: Six Studies in Maps and Landscapes*. London: Lutterworth.

Berg, L.D. 1993. 'Between Modernism and Postmodernism'. *Progress in Human Geography* 17:490–507.

_____, and R.A. Kearns. 1996. 'Naming as Norming: "Race", Gender, and the Identity Politics of Naming Places in Aotearoa/New Zealand'. *Environment and Planning D: Society and Space* 14:99–122.

_____, and R.A. Kearns. 1997. 'Constructing Cultural Geographies of Aotearoa'. *New Zealand Geographer* 53, no. 2:1–2.

Berger, P., and T. Luckman. 1966. *The Social Construction of Reality: A Treatise in the Sociology of Knowledge*. Garden City, NY: Doubleday.

Bernard, F., and D. Thom. 1981. 'Population Pressure and Human Carrying Capacity in Selected Locations in Machakos and Kitui Districts'. *Journal of Developing Areas* 5:381–406.

Berry, B.J.L. 1995. 'Editorial: The Postmodernist Pursuit of *Pragna*'. *Urban Geography* 16:95–7.

_____, E.C. Conkling, and D.M. Ray. 1997. *The Global Economy in Transition*, 2nd edn. Upper Saddle River, NJ: Prentice Hall.

Berry, J.W. 1984. 'Cultural Ecology and Individual Behavior'. In *Human Behavior and Environment*,

Advances in Theory and Research, Volume 4, Environment and Culture, edited by I. Altman, A. Rapaport, and J.F. Wohlwill, 83–106. New York: Plenum Press.

_____. 1997. 'Immigration, Acculturation, and Adaptation'. *Applied Psychology: An International Review* 46:5–68.

Bhaskar, R. 1989. *The Possibility of Naturalism: A Philosophical Critique of the Contemporary Human Sciences*, 2nd edn. Brighton, England: Harvester Press.

Binnie, J. 1997. 'Coming out of Geography: Towards a Queer Epistemology'. *Environment and Planning D: Society and Space* 15:223–37.

Birks, H.H., et al. 1988. *The Cultural Landscape: Past, Present and Future*. New York: Cambridge University Press.

Biswas, L. 1984. 'Evolution of Hindu Temples in Calcutta'. *Journal of Cultural Geography* 4, no. 2:73–84.

Bjorklund, E.M. 1964. 'Ideology and Culture Exemplified in Southwestern Michigan'. *Annals of the Association of American Geographers* 54:227–41.

Blaikie, P. 1978. 'The Theory of Spatial Diffusion of Innovations: A Spacious Cul-de-Sac'. *Progress in Human Geography* 2:270–95.

_____. 1985. *The Political Economy of Soil Erosion in Developing Countries*. London: Longman.

_____, and H. Brookfield. 1987. *Land Degradation and Society*. London: Methuen.

Blainey, G. 1983. *A Land Half Won*. Melbourne: Sun.

Blau, P.M., and J.W. Moore. 1970. 'Sociology'. In *A Readers's Guide to the Social Sciences*, 2nd edn, edited by B.F. Hoselitz, 1–40. New York: Free Press.

Blaut, J.M. 1977. 'Two Views of Diffusion'. *Annals of the Association of American Geographers* 67:343–9.

_____. 1980. 'A Radical Critique of Cultural Geography'. *Antipode* 12:25–9.

_____. 1984. 'Commentary on Nostrand's "Hispanos" and Their "Homeland"'. *Annals of the Association of American Geographers* 74:157–64.

_____. 1993a. 'Mind and Matter in Cultural Geography'. In *Culture, Form, and Place: Essays in Cultural and Historical Geography*, edited by K. Mathewson. *Geoscience and Man* 32:345–56. Baton Rouge: Louisiana State University, Department of Geography and Anthropology, Geoscience Publications.

_____. 1993b. *The Colonizer's Model of the World: Geographical Diffusionism and Eurocentric History*. New York: Guilford.

Blumler, M.A. 1996. 'Ecology, Evolutionary Theory and Agricultural Origins'. In *The Origins and Spread of Agriculture and Pastoralism in Eurasia*, edited by D.R.

Harris, 25–50. Washington, DC: Smithsonian Institution Press.

Boas, F. 1928. *Anthropology and Modern Life*. New York: Norton.

Bondi, L. 1997. 'In Whose Words? On Gender Identities, Knowledge and Writing Practices'. *Transactions of the Institute of British Geographers* NS 22:245–58.

_____, and M. Domosh. 1992. 'Other Figures in Other Landscapes: On Feminism, Postmodernism and Geography'. *Environment and Planning D: Society and Space* 10:199–213.

Bonnett, A. 1997. 'Geography, "Race" and Whiteness: Invisible Traditions and Current Challenges'. *Area* 29:193–9.

Boone, J.A. 1996. 'Queer Sites in Modernism: Harlem/The Left Bank/Greenwich Village'. In *The Geography of Identity*, edited by P. Yaeger, 243–72. Ann Arbor: University of Michigan Press.

Boserup, E. 1965. *The Conditions of Agricultural Growth: The Economics of Agrarian Change Under Population Pressure*. London: Allen and Unwin.

Botkin, D. 1990. *Discordant Harmonies: A New Ecology for the Twenty-First Century*. New York: Oxford University Press.

Boulding, K.E. 1950. *A Reconstruction of Economics*. New York: Wiley.

_____. 1956. *The Image*. Ann Arbor: University of Michigan Press.

Bourdieu, P. 1977. *Outline of a Theory of Practice*, translated by R. Nice. New York: Cambridge University Press.

Bowden, M.J. 1976. 'The Great American Desert in the American Mind: The Historiography of a Geographical Notion'. In *Geographies of the Mind: Essays in Historical Geosophy in Honor of John Kirtland Wright*, edited by D. Lowenthal and M.J. Bowden, 119–47. New York: Oxford University Press.

Bowen, D.S. 1996. 'Carl Sauer, Field Exploration, and the Development of American Geographical Thought'. *Southeastern Geographer* 36:176–91.

Bowen, E. 1981. *Empiricism and Geographical Thought*. Cambridge: Cambridge University Press.

Bowen, E.H., and H. Carter. 1975. 'The Distribution of the Welsh Language in 1971'. *Geography* 60:1–15.

Bowler, P.J. 1992. *The Fontana History of the Environmental Sciences*. London: Fontana Press.

Bowman, I. 1934. *Geography in Relation to the Social Sciences*. New York: Charles Scribner.

Breitbart, M.M. 1981. 'Peter Kropotkin: The Anarchist Geographer'. In *Geography, Ideology and Social Concern*, edited by D.R. Stoddart, 134–53. Totowa, NJ: Barnes and Noble.

Broek, J.O.M. 1932. *The Santa Clara Valley, California: A Study in Landscape Change*. Utrecht: Oosthoek.

_____. 1965. *Compass of Geography*. Columbus: Merrill.

_____, and J.W. Webb. 1978. *A Geography of Mankind*, 2nd edn. New York: McGraw-Hill.

Brookfield, H.C. 1964. 'Questions on the Human Frontiers of Geography'. *Economic Geography* 40:283–303.

_____. 1969. 'On the Environment as Perceived'. In *Progress in Geography, Volume 1*, edited by C. Board et al., 51–80. New York: Arnold.

Browett, J. 1980. 'Development, the Diffusionist Paradigm, and Geography'. *Progress in Human Geography* 4:57–79.

Brunn, S.D. 1963. 'A Cultural Plant Geography of the Quince'. *Professional Geographer* 15:16–18.

Bryant, R.L. 1997. 'Beyond the Impasse: The Power of Political Ecology in Third World Environmental Research'. *Area* 29:5–19.

Buchanan, R.H., E. Jones, and D. McCourt, eds. 1971. *Man and His Habitat: Essays Presented to Emyr Estyn Evans*. London: Routledge and Kegan Paul.

Bunkse, E.V. 1981. 'Humboldt and an Aesthetic Tradition in Geography'. *Geographical Review* 71:127–46.

_____. 1996. 'Humanism: Wisdom of the Heart and Mind'. In *Concepts in Human Geography*, edited by C. Earle, K. Mathewson, and M.S. Kenzer, 355–81. Lanham, MD: Rowman and Littlefield.

Bunting, T.E., and L. Guelke. 1979. 'Behavioral and Perception Geography: A Critical Appraisal'. *Annals of the Association of American Geographers* 69:448–62.

Burton, I. 1963. 'The Quantitative Revolution and Theoretical Geography'. *Canadian Geographer* 7:151–62.

Butler, J. 1990. *Gender Trouble: Feminism and the Subversion of Identity*. New York: Routledge.

Butler, W.F. 1872. *The Great Lone Land*. London: Sampson Low.

Butlin, R.A. 1993. *Historical Geography: Through the Gates of Space and Time*. New York: Arnold.

_____, and N. Roberts. 1995. 'Ecological Relations in Historical Times: An Introduction'. In *Ecological Relations in Historical Times: Human Impact and Adaptation*, edited by R.A. Butlin and N. Roberts, 1–14. Cambridge, MA: Blackwell.

Butzer, K.W. 1980. 'Civilizations: Organisms or Systems?' *American Scientist* 68:517–23.

_____. 1989a. 'Hartshorne, Hettner, and *The Nature of Geography*'. In *Reflections on Richard Hartshorne's The Nature of Geography*, edited by J.N. Entrikin and S.D. Brunn, 35–52. Washington, DC: Association of American Geographers, Occasional Publication.

_____. 1989b. 'Cultural Ecology'. In *Geography in America*, edited by G.L. Gaile and C.J. Willmott, 192–208. Columbus, OH: Merrill.

Buxton, G.L. 1967. *The Riverina: 1861–1891: An Australian Regional Study*. Melbourne: Melbourne University Press.

Callard, F. 1998. 'The Body in Theory'. *Environment and Planning D: Society and Space* 16:387–400.

Cameron, J.M.R. 1974. 'Information Distortion in Colonial Promotion: The Case of Swan River Colony'. *Australian Geographical Studies* 12:57–76.

_____. 1977. *Coming to Terms: The Development of Agriculture in Pre-Convict Western Australia*. Perth: University of Western Australia, Department of Geography, Geowest 11.

Carlson, A.W. 1990. *The Spanish-American Homeland: Four Centuries of Change in New Mexico's Río Arriba*. Baltimore: The Johns Hopkins University Press.

Carlstein, T. 1982. *Time, Resources, Society and Ecology: On the Capacity for Human Interaction in Space and Time. Vol. 1: Preindustrial Societies*. London: Allen and Unwin.

Carmichael, D.L., J. Hubert, B. Reeves, and A. Schanche, eds. 1994. *Sacred Sites, Sacred Places*. New York: Routledge.

Carneiro, R. 1960. 'Slash and Burn Agriculture: A Closer Look at its Implications for Settlement Patterns'. In *Men and Cultures*, edited by A.F.C. Wallace, 229–34. Philadelphia: University of Pennsylvania Press.

Carney, G.O. 1990. 'Geography of Music: Inventory and Prospect'. *Journal of Cultural Geography* 10, no. 2:35–48.

_____, ed. 1994a. *The Sounds of People and Places: A Geography of American Folk and Popular Music*, 3rd edn. Lanham, MD: Rowman and Littlefield.

_____. 1994b. 'Branson: The New Mecca of Country Music'. *Journal of Cultural Geography*. 14, no. 2:17–32.

_____. 1995a. 'Introduction: Culture: A Workable Definition'. In *Fast Food, Stock Cars, and Rock 'n' Roll*, edited by G.O. Carney, 1–14. Lanham, MD: Rowman and Littlefield.

_____. 1995b. 'Part III: Food'. In *Fast Food, Stock Cars, and Rock 'n' Roll*, edited by G.O. Carney, 95. Lanham, MD: Rowman and Littlefield.

_____. 1996. 'Western North Carolina: Culture Hearth of Bluegrass Music'. *Journal of Cultural Geography* 16, no. 1:65–87.

_____, ed. 1998. *Baseball, Barns, and Bluegrass: A Geography of American Folklife*. Lanham, MD: Rowman and Littlefield.

Carroll, W.K. 1992. 'Introduction: Social Movements and Counter-Hegemony in a Canadian Context'. In *Organizing Dissent: Contemporary Social Movements in Theory and Practice*, edited by W.K. Carroll, 1–19. Toronto: Garamond Press.

Carson, R. 1962. *Silent Spring*. Boston: Houghton Mifflin.

Carter, E., J. Donald, and J. Squires, eds. 1995. *Cultural Remix: Theories of Politics and the Popular*. London: Lawrence and Wishart.

Carter, G.F. 1948. 'Clark Wissler: 1870–1947'. *Annals of the Association of American Geographers* 38:145–6.

_____. 1968. *Man and the Land: A Cultural Geography*, 2nd edn. New York: Holt, Rinehart and Winston.

_____. 1977. 'A Hypothesis Suggesting a Single Origin of Agriculture'. In *Origins of Agriculture*, edited by C.A. Reed, 99–109. The Hague: Mouton.

_____. 1978. 'Context as Methodology'. In *Diffusion and Migration: Their Roles in Cultural Development*, edited by P.G. Duke et al., 55–64. Calgary: University of Calgary, Archaeological Association.

_____. 1980. *Earlier Than You Think: A Personal View of Man in America*. College Station: Texas A&M University Press.

_____. 1988. 'Cultural Historical Diffusion'. In *The Transfer and Transformation of Ideas and Material Culture*, edited by P.J. Hugill and D.B. Dickson, 3–22. College Station: Texas A&M University Press.

Carter, H., and J.G. Thomas. 1969. 'The Referendum on the Sunday Opening of Licensed Premises in Wales as a Criterion of a Cultural Region'. *Regional Studies* 3:61–71.

Carter, P. 1987. *The Road to Botany Bay: An Essay in Spatial History*. London: Faber and Faber.

Cartier, C. 1997. 'Symbolic Landscape in High Rise Hong Kong'. *Focus* 44, no. 3:13–21.

Carvel, J. 1997. 'Global Study Finds the World Speaking in 10,000 Languages'. *The Guardian* (22 July).

Castells, M. 1997. *The Power of Identity*. Malden, MA: Blackwell.

Catania, A.C. 1984. 'The Operant Behaviorism of B.F. Skinner'. *Behavioral and Brain Sciences* 7:473–5.

Cater, J., and T. Jones. 1989. *Social Geography: An Introduction to Contemporary Issues*. New York: Arnold.

Cavalli-Sforza, L.L., and M.W. Feldman. 1981. *Cultural Transmission and Evolution: A Quantitative Approach*. Princeton: Princeton University Press.

Chambers, I. 1994. *Migrancy, Culture, Identity*. New York: Routledge.

Chapman, G.P. 1977. *Human and Environmental Systems: A Geographer's Appraisal*. New York: Academic Press.

Charlesworth, A. 1992. 'Towards a Geography of the Shoah'. *Journal of Historical Geography* 18:464–9.

Childe, V.G. 1936. *Man Makes Himself*. London: Watts and Co.

Chisholm, M. 1975. *Human Geography: Evolution or Revolution?* Harmondsworth: Penguin.

Chorley, R.J. 1973. 'Geography as Human Ecology'. In *Directions in Geography*, edited by R.J. Chorley, 155–69. London: Methuen.

Chouinard, V. 1997. 'Guest Editorial. Making Space for Disabling Differences: Challenging Ableist Geographies'. *Environment and Planning D: Society and Space* 15:379–87.

Christensen, K. 1982. 'Geography as a Human Science: A Philosophic Critique of the Positivist-Humanist Split'. In *A Search for Common Ground*, edited by P. Gould and G. Olsson, 37–57. London: Pion.

Christopher, A.J. 1973. 'Environmental Perception in Southern Africa'. *South African Geographical Journal* 55:14–22.

_____. 1982. 'Towards a Definition of the Nineteenth Century South African Frontier'. *South African Geographical Journal* 64:97–113.

_____. 1984. *Colonial Africa*. Totowa: Barnes and Noble.

_____. 1994. *The Atlas of Apartheid*. New York: Routledge.

Church News. 1979. *Special Edition: The Era of Mormon Colonization*. Salt Lake City: Deseret News (26 May).

Clark, A.H. 1954. 'Historical Geography'. In *American Geography: Inventory and Prospect*, edited by P.E. James and C.F. Jones, 70–105. Syracuse: Syracuse University Press.

_____. 1959. *Three Centuries and the Island*. Toronto: University of Toronto Press.

_____. 1975. 'The Conceptions of 'Empires' of the St. Lawrence and the Mississippi: An Historico-Geographical View With Some Quizzical Comments on Environmental Determinism'. *American Review of Canadian Studies* 5:4–27.

_____. 1977. 'The Whole Is Greater Than the Sum of Its Parts: A Humanistic Element in Human Geography'. In *Geographic Humanism: Analysis and Social Action*, edited by D.R. Deskins, Jr et al., 3–26. Ann Arbor: University of Michigan, Department of Geography, Geographical Publications, No. 17.

Clarkson, J.D. 1970. 'Ecology and Spatial Analysis'. *Annals of the Association of American Geographers* 60:700–16.

Clay, G. 1980. *Close Up: How to Read the American City*. Chicago: University of Chicago Press.

_____. 1987. *Right Before Your Eyes*. Washington, DC: APA Planners Press.

_____. 1994. *Real Places: An Unconventional Guide to America's Generic Landscape*. Chicago: University of Chicago Press.

Clements, F.E. 1905. *Research Methods in Ecology*. Lincoln: University of Nebraska Publishing Company.

Cloke, P., C. Philo, and D. Sadler. 1991. *Approaching Human Geography: An Introduction to Contemporary Theoretical Debates*. New York: Guilford Press.

Coates, P. 1998. *Nature: Western Attitudes Since Ancient Times*. Berkeley: University of California Press.

Cobb, J.C. 1992. *The Most Southern Place on Earth: The Mississippi Delta and the Roots of Regional Identity*. New York: Oxford University Press.

Cohen, S. 1997. 'More Than The Beatles: Popular Music, Tourism, and Urban Regeneration'. In *Tourists and Tourism: Identifying With People and Places*, edited by S. Abram, J. Waldren, and D.V.L. Macleod, 71–90. New York: Berg.

Cole, T., and G. Smith. 1995. 'Ghettoization and the Holocaust: Budapest 1944'. *Journal of Historical Geography* 21:300–16.

Conacher, A. 1992. 'Review Essay: Changing Environmentalism'. *Australian Geographer* 23:177–83.

Conzen, M.P., ed. 1990a. *The Making of the American Landscape*. Boston: Unwin Hyman.

_____. 1990b. 'Ethnicity on the Land'. In *The Making of the American Landscape*, edited by M.P. Conzen, 221–48. Boston: Unwin Hyman.

_____. 1993. 'The Historical Impulse in Geographical Writing about the United States, 1850–1990'. In *A Scholar's Guide to Geographical Writing on the American and Canadian Past*, edited by M.P. Conzen, T.A. Rumney, and G. Wynn, 3–90. Chicago: University of Chicago Press.

_____, T.A. Rumney, and G. Wynn. 1993. *A Scholar's Guide to Geographical Writing on the American and Canadian Past*. Chicago: University of Chicago Press.

Cook, N.D., and W.G. Lovell, eds. 1992. *'Secret Judgments of God': Old World Disease in Colonial Spanish America*. Norman: University of Oklahoma Press.

Coones, P., and J. Patten. 1986. *The Penguin Guide to the Landscape of England and Wales*. Harmondsworth: Penguin.

Cosgrove, D. 1978. 'Place, Landscape and the Dialectics of Cultural Geography'. *Canadian Geographer* 22:66–72.

_____. 1983. 'Towards a Radical Cultural Geography: Problems of Theory'. *Antipode* 15:1–11.

_____. 1984. *Social Formation and Symbolic Landscape*. London: Croom Helm.

_____. 1985. 'Prospect, Perspective and the Evolution of the Landscape Idea'. *Transactions of the Institute of British Geographers* NS 10:45–62.

_____. 1997a. 'Spectacle and Society: Landscape as Theater in Premodern and Postmodern Cities'. In *Understanding Ordinary Landscapes*, edited by P. Groth and T.W. Bressi, 99–110. New Haven: Yale University Press.

_____. 1997b. *Social Formation and Symbolic Landscape*, 2nd edn. Madison: University of Wisconsin Press.

_____. 1998. 'Cultural Landscapes'. In *A European Geography*, edited by T. Unwin, 65–81. New York: Addison Wesley Longman.

_____, and S. Daniels, eds. 1988. *The Iconography of Landscape*. New York: Cambridge University Press.

_____, and P. Jackson. 1987. 'New Directions in Cultural Geography'. *Area* 19:95–101.

Crang, M. 1998. *Cultural Geography*. New York: Routledge.

Crang. P. 1997. 'Cultural Turns and the (Re)constitution of Economic Geography: Introduction to Section One'. In *Geographies of Economies*, edited by R. Lee and J. Wills, 3–15. New York: Arnold.

Cresswell, T. 1993. 'Mobility as Resistance: A Geographical Reading of Kerouac's "On the Road"'. *Transactions of the Institute of British Geographers* NS 18:249–62.

_____. 1996a. *In Place/Out of Place: Geography, Ideology, and Transgression*. Minneapolis: University of Minnesota Press.

_____. 1996b. 'Writing, Reading, and the Problem of Resistance: A Reply to McDowell'. *Transactions of the Institute of British Geographers* NS 21:420–4.

Crick, B. 1996. 'Foreword'. In *Race: The History of an Idea in the West* by I. Hannaford, xi–xvi. Baltimore: The Johns Hopkins University Press.

Cronon, W. 1983. *Changes in the Land: Indians, Colonists, and the Ecology of New England*. New York: Hill and Wang.

_____. 1992. A Place for Stories: Nature, History, and Narrative'. *Journal of American History* 78:1347–76.

_____. 1995. 'Introduction: In Search of Nature'. In *Uncommon Ground: Toward Reinventing Nature*, edited by W. Cronon, 23–56. New York: W.W. Norton.

Crosby, A.W. 1978. 'Ecological Imperialism: The Overseas Migration of Western Europeans as Biological Phenomenon'. *Texas Quarterly* 21:10–22.

_____. 1986. *Ecological Imperialism: The Biological Expansion of Europe, 900–1900*. New York: Cambridge University Press.

_____. 1995. 'The Past and Present of Environmental History'. *American Historical Review* 100:1177–89.

Crowley, W.K. 1978. 'Old Order Amish Settlement: Diffusion and Growth'. *Annals of the Association of American Geographers* 68:249–64.

Crumley, C.L. 1994. 'Historical Ecology: A Multidimensional Ecological Orientation'. In *Historical Ecology: Cultural Knowledge and Changing Landscapes*, edited by C.L. Crumley, 1–16. Santa Fe: School of American Research Press.

Crystal, D. 1997. *English as a Global Language*. New York: Cambridge University Press.

Cuba, L., and D. Hummon. 1993. 'A Place to Call Home: Identification with Dwelling, Community, and Region'. *Sociological Quarterly* 34:111–31.

Cullen, I.G. 1976. 'Human Geography, Regional Science and the Study of Individual Behavior'. *Environment and Planning A* 8:397–409.

Cumberland, K.B. 1949. 'Aotearoa Maori: New Zealand About 1780'. *Geographical Review* 39:401–24.

Dalby, S. 1991. 'Critical Geopolitics: Discourse, Difference, and Dissent'. *Environment and Planning D: Society and Space* 9:261–83.

_____. 1996. 'Reading Robert Kaplan's "Coming Anarchy"'. *Ecumene* 3:472–96.

Daniels, S. 1985. 'Arguments for a Humanistic Geography'. In *The Future of Geography*, edited by R.J. Johnston, 143–58. New York: Methuen.

_____. 1994. *Fields of Vision: Landscape Imagery and National Identity in England and the United States*. Cambridge: Polity Press.

Darby, H.C. 1940. *The Draining of the Fens*. New York: Cambridge University Press.

_____. 1956. 'The Clearing of the Woodland in Europe'. In *Man's Role in Changing the Face of the Earth*, edited by W.L. Thomas Jr, C.O. Sauer, M. Bates, and L. Mumford, 183–216. Chicago: University of Chicago Press.

_____. 1973. *The New Historical Geography of England*. New York: Cambridge University Press.

Davidson, W.V. 1974. *Historical Geography of the Bay Islands, Honduras*. Birmingham, AL: Southern University Press.

Davis, T. 1995. 'The Diversity of Queer Politics and the Redefinition of Sexual Identity and Community in Urban Spaces'. In *Mapping Desire: Geographies of Sexualities*, edited by D. Bell and G. Valentine, 284–303. New York: Routledge.

Dear, M. 1988. 'The Postmodern Challenge: Reconstructing Human Geography'. *Transactions of the Institute of British Geographers* NS 13:262–74.

_____, and J.R. Wolch. 1987. *Landscapes of Despair*. Princeton: Princeton University Press.

Demeritt, D. 1994. 'The Nature of Metaphors in Cultural Geography and Environmental History'. *Progress in Human Geography* 18:163–85.

Denevan, W.M. 1983. 'Adaptation, Variation and Cultural Geography'. *Professional Geographer* 35:399–406.

_____, ed. 1992. *The Native Population of the Americas in 1492*. Madison: University of Wisconsin Press.

Dennis, R.J., and H. Clout. 1980. *A Social Geography of England and Wales*. New York: Pergamon.

De Oliver, M. 1996. 'Historical Preservation and Identity: The Alamo and the Production of a Consumer Landscape'. *Antipode* 28:1–23.

de Steiguer, J.E. 1997. *The Age of Environmentalism*. New York: McGraw-Hill

Diamond, J. 1997. *Guns, Germs, and Steel: A Short History of Everybody for the Last 13,000 Years*. New York: W.W. Norton.

Dicken, P. 1992. *Global Shift: The Internationalization of Economic Activity*, 2nd edn. New York: Guilford Press.

Dicken, S., and E. Dicken. 1979. *The Making of Oregon: A Study in Historical Geography*. Portland: Oregon Historical Society.

Dickens, P. 1996. *Reconstructing Nature: Alienation, Emancipation and the Division of Labour*. New York: Routledge.

Dickinson, R.E. 1969. *The Makers of Modern Geography*. London: Routledge and Kegan Paul.

Diesendorf, M., and C. Hamilton, eds. 1997. *Human Ecology, Human Economy: Ideas for an Ecologically Sustainable Future*. St Leonard's, Australia: Allen and Unwin.

Dohrs, F.E., and L.M. Sommers, eds. 1967. *Cultural Geography: Selected Readings*. New York: Crowell.

Domosh, M. 1989. 'A Method for Interpreting Landscape: A Case Study of the New York World Building'. *Area* 21:347–55.

_____. 1992. 'Urban Imagery'. *Urban Geography* 13:475–80.

_____ 1996. 'Feminism and Human Geography'. In *Concepts in Human Geography*, edited by C. Earle, K. Mathewson, and M.S. Kenzer, 411–27. Lanham, MD: Rowman and Littlefield.

Donkin, R.A. 1997. 'A "Servant of Two Masters"'. *Journal of Historical Geography* 23:247–66.

Douglas, I., R. Huggett, and M. Robinson. 1996. 'Preface'. In *Companion Encyclopedia of Geography: The Environment and Humankind*, edited by I. Douglas, R. Huggett, and M. Robinson, ix–xi. New York: Routledge.

Dowling, R. 1997. 'Planning for Culture in Urban Australia'. *Australian Geographical Studies* 35:23–31.

_____, and P.M. McGuirk. 1998. 'Gendered Geographies in Australia, Aotearoa/New Zealand and the Asia–Pacific'. *Australian Geographer* 29:279–91.

_____, and G. Pratt. 1993. 'Home Truths: Recent Feminist Constructions'. *Urban Geography* 14:464–75.

Downs, R.M. 1979. 'Critical Appraisal or Determined Philosophical Skepticism?' *Annals of the Association of American Geographers* 69:468–71.

_____, and D. Stea. 1973. 'Cognitive Maps and Spatial Behavior: Process and Products'. In *Image and Environment: Essays on Cognitive Mapping*, edited by R.M. Downs and D. Stea, 8–26. New York: Arnold.

Dunbar, G.S. 1974. 'Geographic Personality'. In *Man and Cultural Heritage: Papers in Honor of Fred B. Kniffen*, edited by H.J. Walker and W.G. Haag. *Geoscience and Man* 5:25–33. Baton Rouge: Louisiana State University, Department of Geography and Anthropology, Geoscience Publications.

_____. 1977. 'Some Early Occurrences of the Term "Social Geography"'. *Scottish Geographical Magazine* 93:15–20.

Duncan, J.S. 1978. 'The Social Construction of Unreality: An Interactionist Approach to the Tourist's Cognition of Environment'. In *Humanistic Geography*, edited by D. Ley and M.S. Samuels, 269–82. Chicago: Maaroufa Press.

_____. 1980. 'The Superorganic in American Cultural Geography'. *Annals of the Association of American Geographers* 7:181–98.

_____. 1990. *The City as Text: The Politics of Interpretation in the Kandyan Kingdom*. New York: Cambridge University Press.

_____. 1993. 'Commentary on "The Reinvention of Cultural Geography"'. *Annals of the Association of American Geographers* 83:517–19.

_____. 1994a. 'After the Civil War: Reconstructing Cultural Geography as Heterotopia'. In *Re-reading Cultural Geography*, edited by K.E. Foote, P.J. Hugill, K. Mathewson, and J.M. Smith, 401–8. Austin: University of Texas Press.

_____. 1994b. 'The Politics of Landscape and Nature, 1992–93'. *Progress in Human Geography* 18:361–70.

_____. 1998. 'Author's Response'. *Progress in Human Geography* 22:571–3.

_____, and N. Duncan. 1988. '(Re)reading the Landscape'. *Environment and Planning D: Society and Space* 6:117–26.

_____, and D. Ley. 1993. 'Introduction: Representing the Place of Culture'. In *Place/Culture/Representation*, edited by J.S. Duncan and D. Ley, 1–21. New York: Routledge.

Duncan, N. 1996. 'Postmodernism in Human Geography'. In *Concepts in Human Geography*, edited by C. Earle, K. Mathewson, and M.S. Kenzer, 429–58. Lanham, MD: Rowman and Littlefield.

_____, and J.P. Sharp. 1993. 'Confronting Representation(s)'. *Environment and Planning D: Society and Space* 11:473–86.

Dunn, K.M. 1997. 'Cultural Geography and Cultural Policy'. *Australian Geographical Studies* 35:1–11.

Durham, W.H. 1976. 'The Adaptive Significance of Cultural Behavior'. *Human Ecology* 4:89–121.

Dwivedi, O.P. 1990. '*Satyagraha* for Conservation: Awakening the Spirit of Hinduism'. In *Ethics of Environment and Development: Global Challenge, International Response*, edited by J.R. Engel and J.G. Engel, 201–12. Tucson: University of Arizona Press.

Earle, C. 1995. 'Review Article: Historical Geography in Extremis? Splitting Personalities on the Postmodern Turn'. *Journal of Historical Geography* 21:455–9.

Earley, J. 1997. *Transforming Human Culture: Social Evolution and the Planetary Crisis*. Albany: State University of New York Press.

Eder, K. 1996. *The Social Construction of Nature*. London: Sage.

Edwards, C.H., E.W. Brabble, Q.J. Cole, and O.E Westney. 1991. *Human Ecology: Interactions of Man with His Environments*. Dubuque, IA: Kendall/Hunt.

Edwards, G. 1996. 'Alternative Speculations on Geographical Futures: Towards a Postmodern Perspective'. *Geography* 81:217–24.

Eichengreen, B. 1998. 'Geography as Destiny: A Brief History of Economic Growth'. *Foreign Affairs* 77, no. 2:128–33.

Eigenheer, R.A. 1973–4. 'The Frontier Hypothesis and Related Spatial Concepts'. *California Geographer* 14:55–69.

Elazar, D.J. 1984. *American Federalism: A View from the States*. New York: Harper and Row.

_____. 1994. *The American Mosaic: The Impact of Space, Time and Culture on American Politics*. Boulder, CO: Westview Press.

Eliade, M. 1959. *The Sacred and the Profane: The Nature of Religion*. New York: Harcourt, Brace and World.

Eliades, D.K. 1987. 'Two Worlds Collide: The European Advance into North America'. In *A Cultural Geography of North American Indians*, edited by T.E. Ross and T.G. Moore, 33–44. Boulder, CO: Westview.

Ellen, R. 1982. *Environment, Subsistence, and System: The Ecology of Small-Scale Social Formations*. New York: Cambridge University Press.

_____. 1988. 'Persistence and Change in the Relationship Between Anthropology and Human Geography'. *Progress in Human Geography* 12:229–61.

Eller, J.D. 1999. *From Culture to Ethnicity to Conflict: An Anthropological Perspective on International Ethnic Conflict*. Ann Arbor: University of Michigan Press.

Emanuelsson, U. 1988. 'A Model for Describing the Development of the Cultural Landscape'. In *The Cultural Landscape: Past, Present and Future*, edited by H.H. Birks, H.J.B. Birks, P.E. Kaland, and D. Moe, 111–21. New York: Cambridge University Press.

England, K.V.L. 1994. 'Getting Personal: Reflexivity, Positionality, and Feminist Research'. *Professional Geographer* 46:80–9.

English, P.W., and R.C. Mayfield. 1972. 'Ecological Perspectives'. In *Man, Space and Environment*, edited by P.W. English and R.C. Mayfield, 115–20. New York: Oxford University Press.

Ennals, P., and D. Holdsworth. 1981. 'Vernacular Architecture and the Cultural Landscape of the Maritime Provinces: A Reconnaissance'. *Acadiensis* 10:86–105.

Entrikin, J.N. 1976. 'Contemporary Humanism in Geography'. *Annals of the Association of American Geographers* 66:615–32.

_____. 1980. 'Robert Park's Human Ecology and Human Geography'. *Annals of the Association of American Geographers* 70:43–58.

_____. 1984. 'Carl Sauer: Philosopher in Spite of Himself'. *Geographical Review* 74:387–407.

_____. 1988. 'Diffusion Research in the Context of the Naturalism Debate in Twentieth-Century Social Thought'. In *The Transfer and Transformation of Ideas and Material Culture*, edited by P.J. Hugill and D.B. Dickson, 165–78. College Station: Texas A&M University Press.

_____. 1991. *The Betweenness of Place*. Baltimore: The Johns Hopkins University Press.

Ericksen, E.G. 1980. *The Territorial Experience: Human Ecology as Symbolic Interaction*. Austin: University of Texas Press.

Estaville, L.E., Jr. 1993. 'The Louisiana-French Homeland'. *Journal of Cultural Geography* 13, 2:31–45.

Evans, E.E. 1973. *The Personality of Ireland: Habitat, Heritage, and History*. New York: Cambridge University Press.

_____. 1996. *Ireland and the Atlantic Heritage: Selected Writings*. Dublin: Lilliput Press.

Evernden, N. 1992. *The Social Creation of Nature*. Baltimore: The Johns Hopkins University Press.

Eyles, J., and W. Peace. 1990. 'Signs and Symbols in Hamilton: An Iconology of Steeltown'. *Geografiska Annaler* 72B:73–88.

_____, and D.M. Smith. 1978. 'Social Geography'. *American Behavioral Scientist* 22:41–58.

Fairhead, J., and M. Peach. 1996. *Misreading the African Landscape: Society and Ecology in a Forest-Savanna Mosaic*. New York: Cambridge University Press.

Featherstone, M. 1995. *Undoing Culture: Globalization, Postmodernism and Identity*. London: Sage.

Febvre, L. 1925. *A Geographical Introduction to History*. New York: Knopf.

Feder, K.L., and M.A. Park. 1997. *Human Antiquity: An Introduction to Physical Anthropology and Archaeology*, 3rd edn. Mountain View, CA: Mayfield.

Ferguson, N. 1997. *Virtual History: Alternatives and Counterfactuals*. London: Picador.

Firey, W. 1945. 'Sentiment and Symbolism as Ecological Variables'. *American Sociological Review* 10:140–8.

_____. 1947. *Land Use in Central Boston*. Cambridge, MA: MIT Press.

Fitzsimmons, M. 1989. 'The Matter of Nature'. *Antipode* 21:106–20.

Fleure, H.J. 1919. 'Human Regions'. *Scottish Geographical Magazine* 35:94–105.

_____. 1951. *Natural History of Man in Britain*. London: Collins.

Flew, A. 1978. *A Rational Animal and Other Philosophical Essays on the Nature of Man*. Oxford: Clarendon Press.

Fogel, R.W. 1964. *Railroads and American Economic Growth: Essays in Econometric History*. Baltimore: The Johns Hopkins University Press.

Foley, M.A. 1976. 'Culture Area'. In *Encyclopedia of Anthropology*, edited by D.E. Hunter and P. Whitten, 104. New York: Harper and Row.

Foote, D.C., and B. Greer-Wootten. 1968. 'An Approach to Systems Analysis in Cultural Geography'. *Professional Geographer* 20:86–91.

Ford, R. 1998. 'Blacks and Asians "Imprisoned" by Fear of Violence'. *London Times* (13 April).

Foucault, M. 1970. *The Order of Things: An Archaeology of the Human Sciences*, translated by A. Sheridan-Smith. New York: Random House.

_____. 1980. *Power/Knowledge*. New York: Pantheon.

Fox, E.W. 1991. *The Emergence of the Modern European World: From the Seventeenth to the Twentieth Century*. Cambridge, MA: Blackwell.

Francaviglia, R.V. 1978. *The Mormon Landscape*. New York: AMS Press.

_____. 1991. *Hard Places: Reading the Landscape of America's Historic Mining Districts*. Iowa City: University of Iowa Press.

_____. 1994. 'Elusive Land: Changing Geographic Images of the Southwest'. In *Essays on the Changing Images of the Southwest*, edited by R.V. Francaviglia and D. Narrett, 8–39. College Station: Texas A&M University Press.

_____. 1996. *Main Street Revisited: Time, Space, and Image Building in Small-Town America*. Iowa City: University of Iowa Press.

Francis, R.D. 1988. 'The Ideal and the Real: The Image of the Canadian West in the Settlement Period'. In: *Rupert's Land: A Cultural Tapestry*, edited by R.C. Davis, 253–73. Waterloo, Ont: Wilfrid Laurier Press.

Fredrickson, G.M. 1981. *White Supremacy: A Comparative Study in American and South African History*. New York: Oxford University Press.

Freeman, D.B. 1985. 'The Importance of Being First: Preemption by Early Adopters of Farming Innovations in Kenya'. *Annals of the Association of American Geographers* 75:17–28.

Friesen, G. 1987. *The Canadian Prairies: A History*. Toronto: University of Toronto Press.

Gade, D.W. 1982. 'The French Riviera as Elitist Space'. *Journal of Cultural Geography* 3:19–28.

_____. 1997. 'Germanic Towns in Southern Brazil'. *Focus* 44, 1:1–6.

Garrison, J.R. 1990. Review of *The American Backwoods Frontier: An Ethnic and Ecological Interpretation* by T.G. Jordan and M. Kaups. *Annals of the Association of American Geographers* 80:639–41.

Garst, R.D. 1974. 'Innovation Diffusion Among the Gusii of Kenya'. *Economic Geography* 50:300–12.

Gastil, R.D. 1975. *Cultural Regions of the United States*. Seattle: University of Washington Press.

Gay, J. 1971. *Geography of Religion in England*. London: Duckworth.

Geertz, C. 1963. *Agricultural Involution: The Processes of Ecological Change in Indonesia*. Berkeley: University of California Press.

_____. 1973. *The Interpretation of Cultures*. New York: Basic Books.

_____. 1996. 'Afterword'. In *Senses of Place*, edited by S. Feld and K.H. Basso, 259–62. Santa Fe: School of American Research Press.

Gellner, E. 1997. 'Knowledge of Nature and Society'. In *Nature and Society in Historical Context*, edited by M. Teich, R. Porter, and B. Gustafsson, 9–17. New York: Cambridge University Press.

Gerber, J. 1997. 'Beyond Dualism—The Social Construction of Nature and the Natural and Social Construction of Human Beings'. *Progress in Human Geography* 21:1–17.

Gerlach, R.L. 1976. *Immigrants in the Ozarks: A Study in Ethnic Geography*. Columbia: University of Missouri Press.

Getz, D. 1995. 'Event Tourism and the Authenticity Dilemma'. In *Global Tourism: The Next Decade*, edited by W.F. Theobald, 313–29. Boston: Butterworth Heinemann.

Geyer, F. 1996. 'Introduction: Alienation, Ethnicity, and Postmodernism'. In *Alienation, Ethnicity, and Postmodernism*, edited by F. Geyer, ix–xxviii. Westport, Conn.: Greenwood.

Giddens, A. 1984. *The Constitution of Society: Outline of the Theory of Structuration*. Cambridge: Polity Press.

_____. 1987. *Sociology: A Brief But Critical Introduction*, 2nd edn. New York: Harcourt Brace Jovanovich.

_____. 1991. *Introduction to Sociology*. New York: W.W. Norton.

_____, and J.H. Turner. 1987. 'Introduction'. In *Social Theory Today*, edited by A. Giddens and J.H. Turner, 1–10. Stanford: Stanford University Press.

Gilbert, A. 1988. 'The New Regional Geography in English and French-Speaking Countries'. *Progress in Human Geography* 12:208–28.

Gill, W.G. 1993. 'Region, Agency, and Popular Music: The Northwest Sound'. *Canadian Geographer* 37:120–31.

Gills, B.K. 1995. 'Capital and Power in the Processes of World History'. In *Civilizations and World Systems: Studying World-Historical Change*, edited by S.K. Sanderson, 136–62. Walnut Creek, CA: Altimira Press.

Ginsburg, N. 1970. Geography. In *A Reader's Guide to the Social Sciences*, 2nd edn, edited by B.F. Hoselitz, 293–318. New York: Free Press.

Glacken, C. 1967. *Traces on the Rhodian Shore: Nature and Culture in Western Thought from Ancient Times to the End of the Eighteenth Century*. Berkeley: University of California Press.

_____. 1985. 'Culture and Environment in Western Civilization During the Nineteenth Century'. In *Environmental History*, edited by K.E. Bailes, 46–57. New York: University Press of America.

Glasscock, R.E. 1991. 'Obituary: E. Estyn Evans, 1905–1989'. *Journal of Historical Geography* 17:87–91.

Gold, J.R. 1980. *An Introduction to Behavioral Geography*. New York: Oxford University Press.

_____, and B. Goodey. 1984. 'Behavioral and Perceptual Geography: Criticisms and Responses'. *Progress in Human Geography* 8:544–50.

Goldhagen, D.J. 1996. *Hitler's Willing Executioners: Ordinary Germans and the Holocaust*. New York: Knopf.

Goldschmidt, W. 1965. 'Theory and Strategy in the Theory of Cultural Adaptability'. *American Anthropologist* 67:402–8.

Golledge, R.G. 1969. 'The Geographical Relevance of Some Learning Theories'. In *Behavioral Problems in Geography: A Symposium*, edited by K.R. Cox and R.G. Golledge, 101–45. Evanston, Ill.: Northwestern University Press, Studies in Geography, No. 17.

_____. 1987. 'Environmental Cognition'. In *Handbook of Environmental Psychology, Volume 1*, edited by D. Stokols and I. Altman, 131–74. New York: Wiley.

_____. 1993. 'Geography and the Disabled: A Survey with Special Reference to Vision Impaired and Blind Populations'. *Transactions of the Institute of British Geographers* NS 18:63–85.

_____, and R.J. Stimson. 1997. *Spatial Behavior: A Geographic Perspective*. New York: Guilford.

Goode, P. 1926. *The Geographic Background of Chicago*. Chicago: University of Chicago Press.

Gore, A. 1992. *Earth in the Balance: Ecology and the Human Spirit*. Boston: Houghton Mifflin.

Goss, J. 1992. 'Modernity and Post-modernity in the Retail Landscape'. In *Inventing Places: Studies in Cultural Geography*, edited by K. Anderson and F. Gale, 159–77. Melbourne: Longman Cheshire.

_____. 1993. 'The "Magic of the Mall": An Analysis of Form, Function, and Meaning in the Contemporary Built Environment'. *Annals of the Association of American Geographers* 83:18–47.

_____. 1999. 'Once-upon-a-Time in the Commodity World: An Unofficial Guide to the Mall of America'. *Annals of the Association of American Geographers* 89:45–75.

Gould, P. 1963. 'Man Against His Environment: A Game Theoretic Framework'. *Annals of the Association of American Geographers* 53:290–7.

_____. 1969. *Spatial Diffusion*. Washington, DC: Association of American Geographers, Resource Paper, No. 4.

_____, and R. White. 1986. *Mental Maps*, 2nd edn. Boston: Allen and Unwin.

Gould, S.J. 1990. 'The Golden Rule—a Proper Scale for Our Environmental Crisis'. *Natural History* 99, no. 9:24–30.

Graham, B.J., ed. 1997. *In Search of Ireland: A Cultural Geography*. New York: Routledge.

Graham, B.J., and L.J. Proudfoot. 1993. 'A Perspective on the Nature of Irish Historical Geography'. In *An Historical Geography of Ireland*, edited by B.J. Graham and L.J. Proudfoot, 1–18. New York: Academic Press.

Graham, E. 1997. 'Philosophies Underlying Human Geographic Research'. In *Methods in Human Geography: A Guide for Students Doing a Research Project*, edited by R. Flowerdew and D. Martin, 6–30. Harlow: Longman.

Granö, J.G. [1929] 1997. *Pure Geography*, edited by O. Granö and A. Paasi, and translated by M. Hicks. Baltimore: Johns Hopkins University Press.

Gregory, D. 1981. 'Human Agency and Human Geography'. *Transactions of the Institute of British Geographers* NS 6:1–18.

_____. 1994. *Geographical Imaginations*. Cambridge, MA: Blackwell.

Gregson, N. 1987. 'Structuration Theory: Some Thoughts on the Possibilities for Empirical Research'. *Environment and Planning D: Society and Space* 5:73–91.

_____. 1993. '"The Initiative": Delimiting or Deconstructing Social Geography'. *Progress in Human Geography* 17:525–30.

_____. 1995. 'And Now It's All Consumption?'. *Progress in Human Geography* 19:135–41.

_____, G. Rose, J. Cream, and N. Laurie. 1997. 'Conclusions'. In *Feminist Geographies: Explorations in Diversity and Difference*, edited by Women and Geography Study Group of the Royal Geographical Society with the Institute of British Geographers, 191–200. Harlow: Longman.

_____, et al. 1997. 'Gender in Feminist Geography'. In *Feminist Geographies: Explorations in Diversity and Difference*, edited by Women and Geography Study Group of the Royal Geographical Society with the Institute of British Geographers, 49–85. Harlow, England: Longman.

Griffiths, T. 1997. 'Introduction. Ecology and Empire: Towards an Australian History of the World'. In *Ecology and Empire: Environmental History of Settler Societies*, edited by T. Griffiths and L. Robin, 1–16. Seattle: University of Washington Press.

Gritzner, C.F. 1966. 'The Scope of Cultural Geography'. *Journal of Geography* 65:4–11.

Grossman, L. 1977. 'Man-Environment Relationships in Anthropology and Geography'. *Annals of the Association of American Geographers* 67:126–44.

_____. 1984. *Peasants, Subsistence, Ecology, and Development in the Highlands of Papua New Guinea*. Princeton: Princeton University Press.

_____. 1993. 'The Political Ecology of Banana Exports and Local Food Production in St Vincent, Eastern Caribbean'. *Annals of the Association of American Geographers* 83:347–67.

Groth, P. 1997. 'Frameworks for Cultural Landscape Study'. In *Understanding Ordinary Landscapes*, edited by P. Groth and T.W. Bressi, 1–21. New Haven: Yale University Press.

_____, and T.W. Bressi, eds. 1997. *Understanding Ordinary Landscapes*. New Haven: Yale University Press.

Guelke, L. 1971. 'Problems of Scientific Explanation in Geography'. *Canadian Geographer* 15:38–53.

_____. 1974. 'An Idealist Alternative in Human Geography'. *Annals of the Association of American Geographers* 64:193–202.

_____. 1975. 'On Rethinking Historical Geography'. *Area* 7:135–8.

_____. 1982. *Historical Understanding in Geography*. New York: Cambridge University Press.

_____. 1987. 'Frontier Settlement and Human Values: A Comparative Look at North America and South Africa'. In *Abstract Thoughts: Concrete Solutions. Essays in Honour of Peter Nash*, edited by L. Guelke and R. Preston, 181–99. Waterloo: University of Waterloo, Department of Geography, Publication Series, No. 29.

Habermas, J. 1972. *Knowledge and Human Interests*. London: Heinemann.

Hage, G. 1996. 'The Spatial Imagery of National Practices: Dwelling-Domesticating/Being-Exterminating'. *Environment and Planning D: Society and Space* 14:463–85.

Hägerstrand, T. 1951. *Migration and the Growth of Culture Regions*. Lund: Gleerup, Lund Studies in Geography, Series B, No. 3.

_____. 1952. *The Propagation of Innovation Waves*. Lund: Gleerup, Lund Studies in Geography, Series B, No. 4.

_____. 1967. *Innovation Diffusion as a Spatial Process*. Translated by A. Pred. Chicago: University of Chicago Press.

Haggett, P. 1979. *Geography: A Modern Synthesis*, 3rd edn. New York: Harper and Row.

Hale, R.F. 1971. "A Map of Vernacular Regions in America". Ph.D. dissertation. Minneapolis: University of Minnesota.

_____. 1984. 'Vernacular Regions of America'. *Journal of Cultural Geography* 5, no. 1:131–40.

Halfacree, K.H., and R.M. Kitchin. 1996. '"Madchester Rave On": Placing the Fragments of Popular Music'. *Area* 28:47–55.

Hall, C.S., and G. Lindzey. 1978. *Theories of Personality*, 3rd edn. New York: Wiley.

Hall, S. 1996. 'Introduction: Who Needs Identity?'. In *Questions of Cultural Identity*, edited by S. Hall and P. duGay, 1–17. London: Sage.

Hamnett, C., ed. 1996. *Social Geography: A Reader*. New York: Arnold.

Hannaford, I. 1996. *Race: The History of an Idea in the West*. Baltimore: The Johns Hopkins University Press.

Hanson, S. 1992. 'Geography and Feminism: Worlds in Collision'. *Annals of the Association of American Geographers* 82:569–86.

_____, and G. Pratt. 1995. *Gender, Work, and Space*. New York: Routledge.

Haraway, D. 1991. *Simians, Cyborgs, and Women: The Reinvention of Nature*. New York: Routledge.

Hardesty, D.L. 1977. *Ecological Anthropology*. New York: Wiley.

_____. 1986. 'Rethinking Cultural Adaptation'. *Professional Geographer* 38:11–18.

Hardin, G. 1968. 'The Tragedy of the Commons'. *Science* 162:1243–8.

Harris, D.R. 1996. 'Introduction: Themes and Concepts in the Study of Early Agriculture'. In *The Origins and Spread of Agriculture and Pastoralism in Eurasia*, edited by D.R. Harris, 1–9. Washington, DC: Smithsonian Institution Press.

Harris, M. 1968. *The Rise of Anthropological Theory*. London: Routledge and Kegan Paul.

_____. 1979. *Cultural Materialism: The Struggle for a Science of Culture*. New York: Random House.

Harris, R.C. 1977. The Simplification of Europe Overseas'. *Annals of the Association of American Geographers* 67:469–83.

_____. 1978. 'The Historical Geography of North American Regions'. *American Behavioral Scientist* 22:115–30.

_____. 1999. 'Comments on "The Shaping of America, 1850–1915"'. *Journal of Historical Geography* 25:9–11.

Harrison, R., and D.N. Livingstone. 1979. 'There and Back Again: Towards a Critique of Idealist Human Geography'. *Area* 11:75–9.

Hart, J.F. 1975. *The Look of the Land*. Englewood Cliffs, NJ: Prentice Hall.

_____. 1982. 'The Highest Form of the Geographer's Art'. *Annals of the Association of American Geographers* 72:1–29.

_____. 1995. 'Reading the Landscape'. In *Landscape in America*, edited by G.F. Thompson, 23–42. Austin: University of Texas Press.

_____. 1998. *The Rural Landscape*. Baltimore: The Johns Hopkins University Press.

Hartig, K.V., and K.M. Dunn. 1998. 'Roadside Memorials: Interpreting New Deathscapes in Newcastle, New South Wales'. *Australian Geographical Studies* 36:5–20.

Hartshorne, R. 1939. *The Nature of Geography*. Lancaster, PA: Association of American Geographers.

_____. 1959. *Perspective on the Nature of Geography*. Chicago: Rand McNally.

Harvey, D.W. 1969. *Explanation in Geography*. New York: Arnold.

_____. 1973. *Social Justice and the City*. Baltimore: The Johns Hopkins University Press.

_____. 1979. 'Monument and Myth'. *Annals of the Association of American Geographers* 69:362–81.

_____. 1989. *The Condition of Postmodernity: An Enquiry into the Origins of Cultural Change*. Cambridge, MA: Blackwell.

_____. 1993. 'Class Relations, Social Justice and Politics of Difference'. In *Place and the Politics of Identity*, edited by M. Keith and S. Pile, 41–66. New York: Routledge.

Hauptman, L.M., and R.G. Knapp. 1977. 'Dutch–Aboriginal Interaction in New Netherland and Formosa: An Historical Geography of Empire'. *Proceedings of the American Philosophical Society* 121:166–82.

Hawley, A.H. 1950. *Human Ecology: A Theory of Community Structure*. New York: Ronald Press Co.

_____. 1968. 'Human Ecology'. In *International Encyclopedia of the Social Sciences*, vol. 4, edited by D.L. Sills, 328–37. New York: Free Press.

_____. 1998. 'Human Ecology, Population, and Development'. In *Continuities in Sociological Human Ecology*, edited by M. Micklin and D.L. Poston, Jr, 11–25. New York: Plenum Press.

Hayden, D. 1995. *The Power of Place: Urban Landscapes as Public History*. Cambridge, MA: MIT Press.

Haynes, R.M. 1980. *Geographical Images and Mental Maps*. New York: MacMillan.

Head, L. 1993. 'Unearthing Prehistoric Cultural Landscapes: A View From Australia'. *Transactions of the Institute of British Geographers* NS 18:481–99.

Headland, T. 1997. 'Revisionism in Ecological Anthropology'. *Current Anthropology* 38:605–9.

Heathcote, R.L. 1972. 'The Visions of Australia, 1770–1970'. In *Australia as Human Setting*, edited by A. Rapoport, 77–98. Sydney: Angus and Robertson.

Henderson, G. 1997. '"Landscape is Dead, Long Live Landscape": A Handbook for Sceptics'. *Journal of Historical Geography* 24:94–100.

Henderson, J.R. 1978. 'Spatial Reorganization: A Geographical Dimension in Acculturation'. *Canadian Geographer* 12:1–21.

Henderson, M.L. 1996. 'Geography, First Peoples, and Social Justice'. *Geographical Review* 86:278–83.

Heyer, P. 1982. *Nature, Human Nature, and Society: Marx, Darwin, Biology and the Human Sciences*. Westport, CT: Greenwood.

Hilliard, S.B. 1972. *Hog Meat and Hoecake: Food Supply in the Old South, 1840–1860*. Carbondale: Southern Illinois University Press.

Hirsch, E., and M. O'Hanlon, eds. 1995. *The Anthropology of Landscape: Perspectives on Space and Place*. Oxford: Clarendon.

Hobsbawm, E. 1993. 'The New Threat to History'. *The New York Review of Books* (16 December):62–4.

Hodge, S. 1996. 'Disadvantage and "Otherness" in Western Sydney'. *Australian Geographical Studies* 34:32–44.

Hoggart, R. 1957. *The Uses of Literacy: Changing Patterns in English Mass Culture*. Fair Lawn, NJ: Essential Books.

Hoke, G.W. 1907. 'The Study of Social Geography'. *Geographical Journal* 29:64–7.

Homans, G. 1987. 'Behaviorism and After'. In *Social Theory Today*, edited by A. Giddens and J.H. Turner, 58–81. Stanford: Stanford University Press.

Honderich, T. 1995. *The Oxford Companion to Philosophy*. New York: Oxford University Press.

Hooson, D.J.M. 1981. 'Carl O. Sauer'. In *The Origins of Academic Geography in the United States*, edited by B.W. Blouet, 165–74. Hamden, CT: Archon Books.

Hopkins, J.S.P. 1990. 'West Edmonton Mall: Landscape of Myths and Elsewhereness'. *Canadian Geographer* 34:2–17.

_____. 1991. 'West Edmonton Mall as a Centre for Social Interaction' *Canadian Geographer* 35:261–7.

Hopkins, T.K., and I. Wallerstein. 1996. 'The World-System: Is There a Crisis?'. In *The Age of Transition: Trajectory of the World System, 1945–2025*, coordinated by T.K. Hopkins and I. Wallerstein, 1–10. Atlantic Highlands, NJ: Zed Books.

Hornbeck, D., C. Earle, and C.M. Rodrigue. 1996. 'The Way We Were: Deployments (and Redeployments) of Time in Human Geography'. In *Concepts in Human Geography*, edited by C. Earle, K. Mathewson, and M.S. Kenzer, 33–61. Lanham, MD: Rowman and Littlefield.

Hoskins, W.G. 1955. *The Making of the English Landscape*. London: Hodder and Stoughton.

Hough, M. 1990. *Out of Place: Restoring Identity to the Regional Landscape*. New Haven: Yale University Press.

Howe, S. 1998. *Afrocentrism: Mythical Pasts and Imagined Homes*. New York: Verso.

Hsiung, D.C. 1996. *Two Worlds in the Tennessee Mountains: Exploring the Origins of Appalachian Stereotypes*. Lexington: University of Kentucky Press.

Hudson, J.C. 1977. 'Theory and Methodology in Comparative Frontier Studies'. In *The Frontier*, edited by D.H. Miller and J.O. Steffen, 11–32. Norman: University of Oklahoma Press.

_____. 1994. *Making the Corn Belt: A Geographical History of Middle-Western Agriculture*. Bloomington: Indiana University Press.

Hudson, W., and G. Bolton. 1997. 'Creating Australia'. In *Creating Australia: Changing Australian History*, edited by W. Hudson and G. Bolton, 1–11. St Leonards, NSW: Allen and Unwin.

Hufferd, J. 1980. 'Toward a Transcendental Human Geography of Places'. *Antipode* 12, no. 3: 18–23.

Huggett, R. 1980. *Systems Analysis in Geography*. Oxford: Clarendon Press.

Hugill, P.J. 1997. 'Review Article: World–System Theory: Where's the Theory?'. *Journal of Historical Geography* 23:344–9.

_____, and D.B. Dickson, eds. 1988. *The Transfer and Transformation of Ideas and Material Culture*. College Station: Texas A&M University Press.

Huntington, E. 1915. *Civilization and Climate*. New Haven: Yale University Press.

_____. 1945. *Mainsprings of Civilization*. New York: Wiley.

Huntington, S.P. 1993. 'The Clash of Civilizations'. *Foreign Affairs* 73, no. 2:22–49.

_____. 1996. *The Clash of Civilizations and the Remaking of World Order*. New York: Simon and Schuster.

Hutcheon, P.D. 1994. 'Is There a Dark Side to Multiculturalism?'. *Humanist in Canada* 27, no. 2:26–9.

_____. 1995. 'Defining Modern Humanism'. *Humanist in Canada* 28, no. 1:30–3.

_____. 1996. *Leaving the Cave: Evolutionary Naturalism in Social-Scientific Thought*. Waterloo, Ont: Wilfrid Laurier University Press.

Huxley, J.S., and A.C. Haddon. 1936. *We Europeans*. London: Jonathan Cape.

Inglehart, R. 1990. *Culture Shift in Advanced Industrial Society*. Princeton: Princeton University Press.

Innis, H.A. 1930. *The Fur Trade in Canada*. New Haven: Yale University Press.

Isajiw, W. 1974. 'Definitions of Ethnicity'. *Ethnicity* 1:111–24.

Ittelson, W.H., H.M. Proshansky, L.G. Rivlin, and G.H. Winkel. 1974. *An Introduction to Environmental Psychology*. New York: Holt, Rinehart and Winston.

Jackson, J.B. 1951. 'The Need of Being Versed in Country Things'. *Landscape* 1, no. 1:1–5.

_____. 1974. *A Sense of Place: A Sense of Time*. New Haven: Yale University Press.

_____. 1984. *Discovering the Vernacular Landscape*. New Haven: Yale University Press.

_____. 1997a. *Landscape in Sight: Looking at America*, edited by H.L. Horowitz. New Haven: Yale University Press.

_____. 1997b. 'Foreword'. In *The Evolving Landscape: Homer Aschmann's Geography*, edited by M.J. Pasqualetti, vii–viii. Baltimore: The Johns Hopkins University Press.

Jackson, P. 1980. 'A Plea for Cultural Geography'. *Area* 12:110–13.

_____. 1987. 'The Idea of "Race" and the Geography of Racism'. In *Race and Racism: Essays in Social Geography*, edited by P. Jackson, 3–21. Boston: Allen and Unwin.

_____. 1989. *Maps of Meaning: An Introduction to Cultural Geography*. Boston: Unwin Hyman.

_____. 1992. 'The Politics of the Streets: A Geography of Caribana'. *Political Geography* 11:130–51.

_____, and J. Penrose, eds. 1994. *Constructions of Race, Place, and Nation*. Minneapolis: University of Minnesota Press.

_____, and S.J. Smith. 1984. *Exploring Social Geography*. London: Allen and Unwin.

_____, and J. Taylor. 1996. 'Geography and the Cultural Politics of Advertising'. *Progress in Human Geography* 20:356–71.

Jackson, R.H. 1978. 'Mormon Perception and Settlement'. *Annals of the Association of American Geographers* 68:317–34.

_____, and R. Henrie. 1983. 'Perception of Sacred Space'. *Journal of Cultural Geography* 3, no. 2:94–107.

_____, and R.L. Layton. 1976. 'The Mormon Village: Analysis of a Settlement Type'. *The Professional Geographer* 28:136–41.

Jakle, J.A. 1985. *The Tourist: Travel in Twentieth Century North America*. Boston: Unwin Hyman.

_____. 1987. *The Visual Elements of Landscape*. Amherst: University of Massachusetts Press.

_____. 1994. *The Gas Station in America*. Baltimore: The Johns Hopkins University Press.

_____, S. Brunn, and C.C. Roseman. 1976. *Human Spatial Behavior*. North Scituate, MA: Duxbury.

James, P.E. 1964. *One World Divided: A Geographer Looks at the Modern World*, 2nd edn. Lexington, MA: Xerox College Publishing.

Jameson, F. 1991. *Postmodernism, or, The Cultural Logic of Late Capitalism*. Durham, NC: Duke University Press.

Jamieson, D. 1985. 'Against Zoos'. In *In Defense of Animals*, edited by P. Singer, 108–17. Cambridge, MA: Blackwell.

_____. 1997. 'Zoos Revisited'. In *The Philosophy of the Environment*, edited by T.D.J. Chappell, 180–92. Edinburgh: Edinburgh University Press.

Jarosz, L. 1993. 'Defining and Explaining Tropical Deforestation: Shifting Cultivation and Population Growth in Colonial Madagascar (1846–1940)'. *Economic Geography* 69:366–79.

Jeans, D.N. 1974. 'Changing Formulations of the Man–Environment Relationship in Anglo–American Geography'. *Journal of Geography* 73:36–40.

_____. 1981. 'Mapping the Regional Patterns of Australian Society: Some Preliminary Thoughts'. *Australian Historical Geography Bulletin* 2:1–6.

_____. 1988. 'The First World War Memorials in New South Wales: Centres of Meaning in the Landscape'. *Australian Geographer* 19:259–67.

Jensen, R., and H. Burgess. 1997. 'Mythmaking: How Introductory Psychology Texts Present B.F. Skinner's Analysis of Cognition'. *The Psychological Record* 47:221–32.

Johnson, N. 1995. 'Cast in Stone: Monuments, Geography, and Nationalism'. *Environment and Planning D: Society and Space* 13:51–65.

Johnson, R. 1996. 'Revising Culture Through 'Women and Nature': An Introduction to the Special Issue'. *Women's Studies* 25:v–xi.

Johnston, J.A. 1979. 'Image and Reality: Initial Assessments of Soil Fertility in New Zealand, 1839–55'. *Australian Geographer* 14:160–5.

_____. 1981. 'The New Zealand Bush: Early Assessments of Vegetation'. *New Zealand Geographer* 37:19–24.

Johnston, L. 1997. 'Queen(s') Street or Ponsonby Poofters? Embodied HERO Parade Sites'. *New Zealand Geographer* 53, no. 2:29–33.

Johnston, R.J. 1985. 'Introduction: Exploring the Future of Geography'. In *The Future of Geography*, edited by R.J. Johnston, 3–26. New York: Methuen.

_____. 1990. 'The Challenge for Regional Geography: Some Proposals for Research Frontiers'. In *Regional Geography: Current Developments and Future Prospects*, edited by R.J. Johnston, J. Hauer, and G.A. Hoekveld, 122–39. New York: Routledge.

_____. 1991. *A Question of Place: Exploring the Practice of Human Geography*. Cambridge, MA: Blackwell.

_____. 1997. *Geography and Geographers: Anglo-American Geography Since 1945*, 5th edn. New York: Wiley.

_____, D. Gregory, and D.M. Smith, eds. 1994. *The Dictionary of Human Geography*, 3rd edn. Cambridge, MA: Blackwell.

Jones, E.L. 1987. *The European Miracle: Environments, Economics and Geopolitics in the History of Europe and Asia*, 2nd edn. New York: Cambridge University Press.

Jordan, T.G. 1966. *German Seed in Texas Soil*. Austin: University of Texas Press.

_____. 1978. 'Perceptual Regions in Texas'. *Geographical Review* 68:293–307.

_____. 1982. *Texas Graveyards: A Cultural Legacy*. Austin: University of Texas Press.

_____. 1983. 'A Reappraisal of Fenno–Scandian Antecedents for Midland American Log Construction'. *Geographical Review* 73:58–94.

_____. 1985. *American Log Buildings: An Old World Heritage*. Chapel Hill: University of North Carolina Press.

_____. 1996. *The European Culture Area: A Systematic Geography*, 3rd edn. New York: Harper Collins.

_____, M. Domosh, and L. Rowntree. 1997. *The Human Mosaic: A Thematic Introduction to Cultural Geography*, 7th edn. New York: Harper and Row.

_____, and M. Kaups. 1989. *The American Backwoods Frontier: An Ethnic and Ecological Interpretation*. Baltimore: The Johns Hopkins University Press.

_____, J.T. Kilpinen, and C.F. Gritzner. 1997. *The Mountain West: Interpreting the Folk Landscape*. Baltimore: The Johns Hopkins University Press.

Judd, C.M., and A.J. Ray, eds. 1980. *Old Trails and New Directions: Papers of the Third North American Fur Trade Conference*. Toronto: University of Toronto Press.

Juergensmeyer, M. 1995. 'Religious Nationalism: A Global Threat?' *Current History* 95, no. 604:372–6.

Kaplan, R.D. 1994. 'The Coming Anarchy'. *Atlantic Monthly* 273, no. 2:44–76.

_____. 1996. *The Ends of the Earth: A Journey at the Dawn of the 21st Century*. New York: Random House.

Kasperson, J.X., R.E. Kasperson, and B.L. Turner II. 1995. *Regions at Risk: Comparisons of Threatened Environments*. New York: United Nations University Press.

Kates, R.W. 1971. 'Natural Hazard in Human Ecological Perspective: Hypotheses and Models'. *Economic Geography* 47:438–51.

Kaups, M. 1966. 'Finnish Places Names in Minnesota: A Study in Cultural Transfer'. *Geographical Review* 56:377–97.

Kearns, K.C. 1974. 'Resuscitation of the Irish Gaeltacht'. *Geographical Review* 64:82–110.

Keil, R., D. Bell, P. Penz, and L. Fawcett, eds. 1998. *Political Ecology: Global and Local*. New York: Routledge.

Keith, M., and S. Pile, eds. 1993. *Place and the Politics of Identity*. New York: Routledge.

Kelly, K. 1974. 'The Changing Attitudes of Farmers to Forest in Nineteenth Century Ontario'. *Ontario Geography* 8:67–77.

Kinsley, D. 1994. *Ecology and Religion*. Englewood Cliffs, NJ: Prentice Hall.

Kirk, W. 1951. 'Historical Geography and the Concept of the Behavioral Environment'. *Indian Geographical Journal* 25:152–60.

_____. 1963. 'Problems in Geography'. *Geography* 48:357–71.

Kitchin, R.M. 1996. 'Increasing the Integrity of Cognitive Mapping Research: Appraising Conceptual Schemata of Environment–Behavior Interaction'. *Progress in Human Geography* 20:56–84.

_____, M. Blades, and R.G. Golledge. 1997. 'Relations Between Geography and Psychology'. *Environment and Behavior* 29:554–73.

Kjekshus, H. 1977. *Ecology Control and Economic Development in East Africa: The Case of Tanganyika*. Berkeley: University of California Press.

Klare, M.T. 1996. 'Redefining Security: The New Global Schisms'. *Current History* 95:353–8.

Knapp, G.W. 1991. *Andean Ecology: Adaptive Dynamics in Ecuador*. Boulder, CO: Westview.

Kneafsey, M. 1998. 'Tourism and Place Identity: A Case-study in Rural Ireland'. *Irish Geography* 31:111–23.

Kniffen, F. 1932. 'Lower California Studies III: The Primitive Cultural Landscape of the Colorado Delta'. *University of California Publication in Geography* 5:43–66.

_____. 1936. 'Louisiana House Types'. *Annals of the Association of American Geographers* 26:179–93.

_____. 1949. 'The American Agricultural Fair: The Pattern'. *Annals of the Association of American Geographers* 39:264–82.

_____. 1951a. 'The American Agricultural Fair: Time and Place'. *Annals of the Association of American Geographers* 41:42–57.

_____. 1951b. 'The American Covered Bridge'. *Geographical Review* 41:114–23.

_____. 1965. 'Folk Housing: Key to Diffusion'. *Annals of the Association of American Geographers* 55:549–57.

_____. 1974. 'Material Culture in the Geographic Interpretation of the Landscape'. In *The Human Mirror*, edited by M. Richardson, 252–67. Baton Rouge: Louisiana State University Press.

_____, and H. Glassie. 1966. 'Building in Wood in the Eastern United States: A Time–Place Perspective'. *Geographical Review* 56:40–65.

Knight, D. 1982. 'Identity and Territory: Geographical Perspectives on Nationalism and Regionalism'. *Annals of the Association of American Geographers* 72:514–31.

Knopp, L. 1995. 'Sexuality and Urban Space: A Framework for Analysis'. In *Mapping Desire: Geographies of Sexualities*, edited by D. Bell and G. Valentine, 149–61. New York: Routledge.

Knox, P., and J. Agnew. 1994. *The Geography of the World Economy*, 2nd edn. New York: Arnold.

Kobayashi, A. 1989. 'A Critique of Dialectical Landscape'. In *Remaking Human Geography*, edited by A. Kobayashi and S. Mackenzie, 164–83. Boston: Unwin Hyman.

_____. 1993. 'Multiculturalism: Representing a Canadian Institution'. In *Place/Culture/Representation*, edited by J.S. Duncan and D. Ley, 205–31. New York: Routledge.

Koelsch, W.A. 1969. 'The Historical Geography of Harlan H. Barrows'. *Annals of the Association of American Geographers* 59:632–51.

Kofman, E., and G. Young, eds. 1996. *Globalization: Theory and Practice*. New York: Pinter.

Kollmorgen, W.M. 1941. 'A Reconnaissance of Some Cultural–Agricultural Islands in the South'. *Economic Geography* 17:409–30.

_____. 1943. 'Agricultural–Cultural Islands in the South, Part II'. *Economic Geography* 19:109–17.

Kong, L. 1993a. 'Negotiating Conceptions of Sacred Space: A Case Study of Religious Buildings in Singapore'. *Transactions of the Institute of British Geographers* NS 18:342–58.

_____. 1993b. 'Ideological Hegemony and the Political Symbolism of Religious Buildings in Singapore'. *Environment and Planning D: Society and Space* 11:23–45.

_____. 1995. 'Music and Cultural Politics: Ideology and Resistance in Singapore'. *Transactions of the Institute of British Geographers* NS 20:447–59.

_____. 1996. 'Popular Music in Singapore: Exploring Local Cultures, Global Resources, and Regional Identities'. *Environment and Planning D: Society and Space* 14:273–92.

_____. 1997. 'A "New" Cultural Geography: Debates About Invention and Reinvention'. *Scottish Geographical Magazine* 113:177–85.

_____. 1999. 'Cemeteries and Columbaria, Memorials and Mausoleums: Narrative and Interpretation in the Study of Deathscapes in Geography'. *Australian Geographical Studies* 37:1–10.

Kroeber, A.L. 1917. 'The Superorganic'. *American Anthropologist* 19:163–213.

_____. 1928. 'Native Cultures of the Southwest'. *University of California Publications in American Archaeology and Ethnology* 23, no. 9:375–98.

_____. 1939. *Cultural and Natural Areas of Native North America*. Berkeley: University of California Press, Publications in American Archaeology and Ethnology, 38.

_____, and C. Kluckhohn. 1952. *Culture: A Critical Review of Concepts and Definitions*. Cambridge, MA: Papers of the Peabody Museum of American Archaeology and Ethnology, Harvard University, 47, 1.

_____, and T. Parsons. 1958. 'The Concepts of Culture and of Social System'. *American Sociological Review* 23:582–3.

Kuper, A. 1999. *Culture: The Anthropologists' Account*. Cambridge, MA: Harvard University Press.

Kuznar, L.A. 1997. *Reclaiming a Scientific Anthropology*. Walnut Creek, CA: Altimira Press.

Lamal, P.A. 1991. 'Preface'. In *Behavioral Analysis of Societies and Cultural Practices*, edited by P.A. Lamal, xiii–xiv. New York: Hemisphere Publishing Corporation.

Lambert, A.M. 1985. *The Making of the Dutch Landscape: An Historical Geography of the Netherlands*, 2nd edn. New York: Academic Press.

Lamme, A.J., III, and R.K. Oldakowski. 1982. 'Vernacular Areas in Florida'. *Southeastern Geographer* 22:100–9.

Landes, D.S. 1998. *The Wealth and Poverty of Nations: Why Some Are So Rich and Some So Poor*. New York: W.W. Norton.

Langton, J. 1979. 'Darwinism and the Behavioral Theory of Sociocultural Evolution: An Analysis'. *American Journal of Sociology* 85:288–309.

Leaf, M.J. 1979. *Man, Mind and Science: A History of Anthropology*. New York: Columbia University Press.

Lee, R. 1990. 'Regional Geography: Between Scientific Geography, Ideology, and Practice, (or What Use is Regional Geography?)'. In *Regional Geography: Current Developments and Future Prospects*, edited by R.J. Johnston, J. Hauer, and G.A. Hoekveld, 103–21. New York: Routledge.

_____, and J. Wills, eds. 1997. *Geographies of Economies*. New York: Arnold.

Lefebvre, H. 1991. *The Production of Space*. Cambridge, MA: Blackwell.

Lehr, J.C. 1973. 'Ukrainian Houses in Alberta'. *Alberta Historical Review* 21:9–15.

Leighly, J. 1954. 'Innovation and Area'. *Geographical Review* 44:439–41.

_____, ed. 1963. *Land and Life: A Selection from the Writings of Carl Ortwin Sauer*. Berkeley: University of California Press.

_____. 1976. 'Carl Ortwin Sauer, 1889–1975'. *Annals of the Association of American Geographers* 66:337–48.

_____. 1978. 'Town Names of Colonial New England in the West'. *Annals of the Association of American Geographers* 68:233–48.

_____. 1987. 'Ecology as Metaphor: Carl Sauer and Human Ecology'. *Professional Geographer* 39:405–12.

Lemon, J.T. 1972. *The Best Poor Man's Country: A Geographical Study of Early Southeastern Pennsylvania*. Baltimore: The Johns Hopkins University Press.

Leopold, A. 1949. *A Sand County Almanac: And Sketches Here and There*. New York: Oxford University Press.

Lester, A. 1996. *From Colonization to Democracy: A New Historical Geography of South Africa*. New York: I.B. Tauris Publishers.

Levison, M., R.G. Ward, and J.W. Webb. 1973. *The Settlement of Polynesia. A Computer Simulation*. New York: Oxford University Press.

Lewin, K. 1944. 'Constructs in Psychology and Ecological Psychology'. In *Authority and Frustration*, edited by K. Lewin, C.E. Meyers, J. Kalhorn, M.L. Farber, and J.R.P. French, 17–23. Iowa City: University of Iowa Press.

_____. 1951. *Field Theory in Social Science: Selected Theoretical Papers*, edited by D. Cartwright. New York: Harper.

Lewis, M.W. 1991. 'Elusive Societies: A Regional–Cartographical Approach to the Study of Human Relatedness'. *Annals of the Association of American Geographers* 81:605–26.

_____, and K.E. Wigen. 1997. *The Myth of Continents: A Critique of Metageography*. Berkeley: University of California Press.

Lewis, P. 1975. 'Common Houses, Cultural Spoor'. *Landscape* 19, no. 2:1–22.

_____. 1979. 'Axioms for Reading the Landscape'. In *The Interpretation of Ordinary Landscapes: Geographical Essays*, edited by D.W. Meinig, 11–32. New York: Oxford University Press.

_____. 1983. 'Learning from Looking: Geographic and Other Writing About the American Cultural Landscape'. *American Quarterly* 35:242–61.

Ley, D. 1974. *The Black Inner City as Frontier Outpost: Images and Behavior of a Philadelphia Neighborhood*. Washington, DC: Association of American Geographers, Monograph No. 7.

_____. 1977. 'Social Geography and the Taken-for-Granted World'. *Transactions of the Institute of British Geographers* NS 2:498–512.

_____. 1981. 'Behavioral Geography and the Philosophies of Meaning'. In *Behavioral Problems in Geography Revisited*, edited by K.R. Cox and R.G. Golledge, 209–30. New York: Methuen.

_____. 1998. 'Classics in Human Geography Revisited: Author's Response'. *Progress in Human Geography* 22:78–80.

_____, and K. Olds. 1988. 'Landscape as "Spectacle": World's Fairs and the Culture of Heroic Consumption'. *Environment and Planning D: Society and Space* 6:191–212.

_____, and M.S. Samuels. 1978. 'Contexts of Modern Humanism in Geography'. In *Humanistic Geography*, edited by D. Ley and M.S. Samuels, 1–18. Chicago: Maaroufa Press.

Leyshon, A., D. Matless, and G. Revill. 1995. 'The Place Of Music'. *Transactions of the Institute of British Geographers* NS 20:423–33.

Liepens, R. 1996. 'Reading Agricultural Power'. *New Zealand Geographer* 52, no. 2:3–10.

Longhurst, R. 1994. 'Reflections on and a Vision for Feminist Geography'. *New Zealand Geographer* 50:14–19.

_____. 1997. '(Dis)embodied Geographies'. *Progress in Human Geography* 21:486–501.

Louder, D.R., C. Morissonneau, and E. Waddell. 1983. 'Picking up the Pieces of a Shattered Dream: Quebec and French America'. *Journal of Cultural Geography* 4, no. 1:44–56.

Lovell, W.G. 1985. *Conquest and Survival in Colonial Guatemala*. Kingston: McGill–Queen's University Press.

_____. 1992. '"Heavy Shadows and Black Night": Disease and Depopulation in Colonial Spanish America'. *Annals of the Association of American Geographers* 82:426–43.

Lovelock, J. 1982. *Gaia: A New Look at Life on Earth*. New York: Oxford University Press.

Low, N., and B. Gleeson. 1998. *Justice, Society and Nature: An Exploration of Political Ecology*. New York: Routledge.

Lowenthal, D. 1968. 'The American Scene'. *Geographical Review* 58:61–88.

_____. 1997. 'European Landscape Transformations: The Residue'. In *Understanding Ordinary Landscapes*, edited by P. Groth and T.W. Bressi, 180–99. New Haven: Yale University Press.

_____, and H.C. Prince. 1964. 'The English Landscape'. *Geographical Review* 54:309–46.

_____, and H.C. Prince. 1965. 'English Landscape Tastes'. *Geographical Review* 55:186–222.

Lowie, R.H. 1937. *History of Ethnological Theory*. New York: Farrar and Rinehart.

Luebke, F.C. 1984. 'Regionalism and the Great Plains: Problems of Concept and Method'. *Western Historical Quarterly* 15:19–38.

Lutz, C.A., and J.L. Collins. 1993. *Reading National Geographic*. Chicago: University of Chicago Press.

Lynch, K. 1960. *The Image of the City*. Cambridge, MA: MIT Press.

Lyod, B., and L. Rowntree. 1978, 'Radical Feminists and Gay Men in San Francisco: Social Space in Dispersed Communities'. In *An Invitation to Geography*, 2nd edn, edited by D. Lanegran and R. Palm, 78–88. New York: McGraw–Hill.

Mabogunje, A. 1997. *State of the Earth: Contemporary Geographic Perspectives*. Malden, MA: Blackwell.

McCann, E.J. 1997. 'Where Do You Draw the Line? Landscape, Texts and the Politics of Planning'. *Environment and Planning D: Society and Space* 15:641–61.

McDowell, L. 1994. 'The Transformation of Cultural Geography'. In *Human Geography: Society, Space and Social Science*, edited by D. Gregory, R. Martin, and G. Smith, 146–73. New York: MacMillan.

_____. 1996. 'Off the Road: Alternative Views of Rebellion, Resistance and "The Beats"'. *Transactions of the Institute of British Geographers* NS 21:412–19.

_____. 1997. 'Women/Gender/Feminisms: Doing Feminist Geography'. *Journal of Geography in Higher Education* 21:381–400.

McEwan, C. 1996. 'Review Article: Gender, Culture and Imperialism'. *Journal of Historical Geography* 22:489–94.

McIlwraith, T.F. 1997. *Looking for Old Ontario: Two Centuries of Landscape Change*. Toronto: University of Toronto Press.

McKay, I. 1994. *The Quest of the Folk: Antimodernism and Cultural Selection in Twentieth-Century Nova Scotia*. Kingston: McGill–Queen's University Press.

McLeay, C. 1995. 'Musical Words, Musical Worlds: Geographical Imagery in the Music of U2'. *New Zealand Geographer* 51, no. 2:1–6.

_____. 1997a. 'Inventing Australia: A Critique of Recent Cultural Policy Rhetoric'. *Australian Geographical Studies* 35:40–6.

_____. 1997b. 'Popular Music and Expressions of National Identity'. *New Zealand Journal of Geography* 103:12–17.

Macphail, C.L. 1997. 'Poetry and Pass Laws: Humanistic Geography in Urban South Africa'. *South African Geographical Journal* 79:35–42.

Macpherson, A. 1987. 'Preparing for the National Stage: Carl Sauer's First Ten Years at Berkeley'. In *Carl O. Sauer: A Tribute*, edited by M.S. Kenzer, 69–89. Corvallis: Oregon State University Press.

McQuillan, D.A. 1978. 'Territory and Ethnic Identity: Some New Measures of an Old Theme in the Cultural Geography of the United States'. In *European Settlement and North American Development: Essays on Geographical Change in Honour and Memory of Andrew Hill Clark*, edited by J.R. Gibson, 3–24. Toronto: University of Toronto Press.

_____. 1990. *Prevailing Over Time: Ethnic Adjustment on the Kansas Prairies, 1875–1925*. Urbana: University of Illinois Press.

_____. 1993. 'Historical Geography and Ethnic Communities in North America'. *Progress in Human Geography* 17:355–66.

Malin, J.C. 1947. *The Grasslands of North America: Prolegomena to its History*. Lawrence, KS: J.C. Malin.

Mannion, J. 1974. *Irish Settlements in Eastern Canada: A Study of Culture Transfer and Adoption*. Toronto:

University of Toronto, Department of Geography, Research Publication No. 12.

Marsh, G.P. [1864] 1965. *Man and Nature, or Physical Geography as Modified by Human Action*, edited by D. Lowenthal. Cambridge, MA: Harvard University Press.

Marshall, H.W. 1995. *Paradise Valley, Nevada: The People and Buildings of an American Place*. Tucson: University of Arizona Press.

Martin, C. 1978. *Keepers of the Game: Indian-Animal Relationships and the Fur Trade*. Berkeley: University of California Press.

Martin, D.G. 1991. *Psychology: Principles and Applications*. Englewood Cliffs, NJ: Prentice Hall.

Martin, G.J. 1987a. 'Foreword'. In *Carl O. Sauer: A Tribute*, edited by M.S. Kenzer, ix–xvi. Corvallis: Oregon State University Press.

_____. 1987b. 'The Ecologic Tradition in American Geography'. *Canadian Geographer* 31:74–7.

_____, and P.E. James. 1993. *All Possible Worlds: A History of Geographical Ideas*, 3rd edn. New York: Wiley.

Masini, E. 1994. 'The Futures of Cultures: An Overview'. In *The Futures of Cultures*, compiled by Division of Studies and Programming, UNESCO, 9–28. Paris: UNESCO Publishing.

Mason, O. 1895. 'Influence of Environment upon Human Industries and Arts'. *Annual Report of the Smithsonian Institution* 639–65.

Massey, D.B. 1985. 'New Directions in Space'. In *Social Relations and Spatial Structures*, edited by D. Gregory and J. Urry, 9–19. New York: MacMillan.

Mathewson, K. 1996. 'High/Low, Back/Center: Culture's Stages in Human Geography'. In *Concepts in Human Geography*, edited by C. Earle, K. Mathewson, and M.S. Kenzer, 97–125. Lanham, MD: Rowman and Littlefield.

_____. 1998. 'Cultural Landscapes and Ecology, 1995–96: Of Oecumenics and Nature(s)'. *Progress in Human Geography* 22:115–28.

_____. 1999. 'Cultural Landscape and Ecology II: Regions, Retrospects, Revivals'. *Progress in Human Geography* 23:267–81.

Matthews, H., and M. Limb. 1999. 'Defining an Agenda for the Geography of Children: Review and Prospect'. *Progress in Human Geography* 23:61–90.

May, J. 1996. 'A Little Taste of Something More Exotic'. *Geography* 81:57–64.

Mayer, J.D. 1996. 'The Political Ecology of Disease'. *Progress in Human Geography* 20:441–56.

Mazrui, A.A. 1997. 'Islamic and Western Values'. *Foreign Affairs* 76, no. 5:118–32.

Mead, W.R. 1981. *An Historical Geography of Scandinavia*. New York: Academic Press.

Meggers, B.J. 1954. 'Environmental Limitation on the Development of Culture'. *American Anthropologist* 56:801–24.

Meigs, P. 1935. 'The Dominican Mission Frontier of Lower California'. *University of California Publication in Geography* 7.

Meinig, D.W. 1965. 'The Mormon Culture Region: Strategies and Patterns in the Geography of the American West, 1847–1964'. *Annals of the Association of American Geographers* 55:191–220.

_____. 1969. *Imperial Texas: An Interpretive Essay in Cultural Geography*. Austin: University of Texas Press.

_____. 1971. *Southwest: Three Peoples in Geographical Change*. New York: Oxford University Press.

_____. 1972. 'American Wests: Preface to a Geographical Introduction'. *Annals of the Association of American Geographers* 62:159–84.

_____. 1976. 'Spatial Models of a Sequence of Transatlantic Interactions'. In *International Geography, 76, Vol. 9, Historical Geography*, 30–5. Toronto: Pergamon.

_____. 1978. 'The Continuous Shaping of America: A Prospectus for Geographers and Historians'. *American Historical Review* 83:1186–1217.

_____, ed. 1979a. *The Interpretation of Ordinary Landscapes: Geographical Essays*. New York: Oxford University Press.

_____. 1979b. 'Reading the Landscape: An Appreciation of W.G. Hoskins and J.B. Jackson'. In *The Interpretation of Ordinary Landscapes: Geographical Essays*, edited by D.W. Meinig, 195–244. New York: Oxford University Press.

_____. 1979c. 'Symbolic Landscapes: Some Idealizations of American Communities'. In *The Interpretation of Ordinary Landscapes: Geographical Essays*, edited by D.W. Meinig, 164–92. New York: Oxford University Press.

_____. 1986. *The Shaping of America: A Geographical Perspective on 500 Years of History. Volume 1: Atlantic America, 1492–1800*. New Haven: Yale University Press.

_____. 1993. *The Shaping of America: A Geographical Perspective on 500 Years of History. Volume 2: Continental America, 1800–1867*. New Haven: Yale University Press.

_____. 1998. *The Shaping of America: A Geographical Perspective on 500 Years of History. Volume 3: Transcontinental America, 1850–1915*. New Haven: Yale University Press.

Melko, M. 1969. *The Nature of Civilizations*. Boston: Porter Sargent.

Mellor, M. 1997. *Feminism and Ecology*. New York: New York University Press.

Merchant, C. 1980. *The Death of Nature: Women, Ecology, and the Scientific Revolution*. San Francisco: Harper and Row.

_____. 1989. *Ecological Revolutions: Nature, Gender, and Science in New England*. Chapel Hill, NC: University of North Carolina Press.

_____. 1990. 'Gender and Environmental History'. *Journal of American History* 76:1117–21.

_____. 1995. *Earthcare: Women and the Environment*. New York: Routledge.

Merrens, H.R. 1969. 'The Physical Environment of Early America: Images and Image Makers in Colonial South Carolina'. *Geographical Review* 59:530–56.

Mikesell, M.W. 1960. 'Comparative Studies in Frontier History'. *Annals of the Association of American Geographers* 50:62–74.

_____. 1961. 'Northern Morocco: A Cultural Geography'. *University of California Publications in Geography* 14:1–136.

_____. 1967. 'Geographic Perspectives in Anthropology'. *Annals of the Association of American Geographers* 57:617–34.

_____. 1968. 'Landscape'. In *International Encyclopedia of the Social Sciences*, vol. 8, 249–64. New York: MacMillan and Free Press.

_____. 1969. 'The Borderlands of Geography as a Social Science'. In *Interdisciplinary Relationships in the Social Sciences*, edited by M. Sherif and C.W. Sherif, 227–48. Chicago: Aldine.

_____. 1976. 'The Rise and Decline of Sequent Occupance: A Chapter in the History of American Geography'. In *Geographies of the Mind: Essays in Historical Geosophy in Honor of John Kirtland Wright*, edited by D. Lowenthal and M.J. Bowden, 149–69. New York: Oxford University Press.

_____. 1977. 'Cultural Geography'. *Progress in Human Geography* 1:460–4.

_____. 1978. 'Tradition and Innovation in Cultural Geography'. *Annals of the Association of American Geographers* 68:1–16.

Miller, E.J.W. 1968. 'The Ozark Culture Region as Revealed by Traditional Materials'. *Annals of the Association of American Geographers* 58:51–77.

Miller, V.P., Jr. 1971. 'Some Observations on the Science of Cultural Geography'. *Journal of Geography* 70:27–35.

Mills, S.F. 1997. *The American Landscape*. Edinburgh: Keele University Press.

Mitchell, D. 1995. 'There's No Such Thing as Culture: Towards a Reconceptualization of the Idea of Culture in Cultural Geography'. *Transactions of the Institute of British Geographers* NS 20:102–16.

_____. 1996. 'Sticks and Stones: The Work of Landscape (A Reply to Judy Walton's "How Real(ist) Can You Get?")'. *Professional Geographer* 48:94–6.

Mitchell, R.D. 1977. *Commercialism and Frontier: Perspectives on the Early Shenandoah Valley*. Charlottesville: University Press of Virginia.

_____. 1978. 'The Formation of Early American Cultural Regions'. In *European Settlement and Development in North America*, edited by J.R. Gibson, 66–90. Toronto: University of Toronto Press.

Mogey, J. 1971. 'Society, Man and Environment'. In *Man and His Habitat*, edited by R.H. Buchanan, E. Jones, and D. McCourt, 79–92. New York: Barnes and Noble.

Mohan, G. 1994. 'Deconstruction of the Con: Geography and the Commodification of Knowledge'. *Area* 26:387–90.

Monk, J. 1992. 'Gender in the Landscape: Expressions of Power and Meaning'. In *Inventing Places: Studies in Cultural Geography*, edited by K. Anderson and F. Gale, 123–38. Melbourne: Longman Cheshire.

_____. 1996. 'Challenging the Boundaries: Survival and Change in a Gendered World'. In *Companion Encyclopedia of Geography: The Environment and Humankind*, edited by I. Douglas, R. Huggett, and M. Robinson, 888–905. New York: Routledge.

Montagu, A. 1997. *Man's Most Dangerous Myth: The Fallacy of Race*, 6th edn. Walnut Creek, CA: Altamira Press.

Morehouse, B.J. 1996. *A Place Called Grand Canyon: Contested Geographies*. Tucson: University of Arizona Press.

Morgan, L.H. [1877] 1974. *Ancient Society, or Researches in the Lines of Human Progress from Savagery through Barbarism to Civilization*. Gloucester, MA: Peter Smith.

Morgan, W.B., and R.P. Moss. 1965. 'Geography and Ecology: The Concept of the Community and Its Relationship to Environment'. *Annals of the Association of American Geographers* 55:339–50.

Morrill, R.L. 1965. *Migration and the Spread and Growth of Urban Settlement*. Lund: Gleerup, Lund Studies in Geography, Series B, No. 26.

Mosely, C., and R.E. Asher. 1994. *Atlas of the World's Languages*. New York: Routledge.

Muir, R. 1999. *Approaches to Landscape*. Boulder, CO: Barnes and Noble.

Mungall, C., and D.J. McLaren. 1990. *Planet Under Stress: The Challenge of Global Change*. New York: Oxford University Press.

Murdoch, J. 1997. 'Inhuman/Nonhuman/Human: Actor-Network Theory and the Prospects for a Nondualistic and Symmetrical Perspective on Nature and Society'. *Environment and Planning D: Society and Space* 15:731–56.

Myrdal, G. 1944. *An American Dilemma: The Negro Problem and Modern Democracy*. New York: Harper.

Myslik, W.D. 1996. 'Renegotiating the Social/Sexual Identities of Places'. In *BodySpace: Destabilizing Geographies of Gender and Sexuality*, edited by N. Duncan, 156–69. New York: Routledge.

Naess, A. 1989. *Ecology, Community and Lifestyle*, translated and edited by D. Rothenberg. New York: Cambridge University Press.

Nash, C. 1998a. 'Editorial: Mapping Emotion'. *Environment and Planning D: Society and Space* 16:1–9.

_____. 1998b. 'Narratives and Names: Irish landscape Meanings'. In *A European Geography*, edited by T. Unwin, 73–6. New York: Addison Wesley Longman.

Neisser, U. 1967. *Cognitive Psychology*. New York: Appleton Century Crofts.

_____. 1976. *Cognition and Reality: Principles and Implications of Cognitive Psychology*. New York: Freeman.

Netting, R. McC. 1977. *Cultural Ecology*. Menlo Park, CA: Cummings.

Newcomb, R.M. 1969. 'Twelve Working Approaches to Historical Geography'. *Yearbook of the Association of Pacific Coast Geographers* 31:27–50.

Newson, L.A. 1976. 'Cultural Evolution: A Basic Concept for Human and Historical Geography'. *Journal of Historical Geography* 2:239–55.

_____. 1996. 'Review of *Culture, Form and Place: Essays in Cultural and Historical Geography* edited by K. Mathewson'. *Progress in Human Geography* 20:278–9.

Newton, M. 1974. 'Cultural Preadaptation and the Upland South'. In *Man and Cultural Heritage: Papers in Honor of Fred B. Kniffen*, edited by H.J. Walker and W.G. Haag. *Geoscience and Man* 5:143–54. Baton Rouge: Louisiana State University, Department of Geography and Anthropology, Geoscience Publications.

Nietschmann, B.Q. 1973. *Between Land and Water: The Subsistence Ecology of the Miskito Indians, Eastern Nicaragua*. New York: Seminar Press.

Noble, A.G. 1984. *Wood, Brick, and Stone: The North American Settlement Landscape*, 2 vols. Amherst, MA: The University of Massachusetts Press.

North, D.C. 1977. 'The New Economic History After Twenty Years'. *American Behavioral Scientist* 21:187–200.

Norton, W. 1984. *Historical Analysis in Geography*. London: Longman.

_____. 1987. 'Humans, Land and Landscape: A Proposal for Cultural Geography'. *Canadian Geographer* 31:21–30.

_____. 1988. 'Abstract Cultural Landscapes'. *Journal of Cultural Geography* 8, no. 2:67–80.

_____. 1989. *Explorations in the Understanding of Landscape: A Cultural Geography*. Westport, CT: Greenwood.

_____. 1995. 'State Boundaries and Agricultural Change in the South Eastern Australian Wheat Belt: Counterfactual Analyses, 1891–1911'. *Australian Geographical Studies* 33:228–41.

_____. 1997a. 'Behavior Analysis and Cultural Geography'. *Journal of Cultural Geography* 16, no. 2:1–19.

_____. 1997b. 'Human Geography and Behavior Analysis: An Application of Behavior Analysis to the Explanation of the Evolution of Human Landscapes'. *The Psychological Record* 47:439–60.

_____. 1998. *Human Geography*, 3rd edn. Toronto: Oxford University Press.

Norwine, J., and T.D. Anderson. 1980. *Geography as Human Ecology*. Lanham, MD: University Press of America.

Nostrand, R.L. 1992. *The Hispano Homeland*. Norman, OK: University of Oklahoma Press.

_____, and L.E. Estaville, Jr. 1993. 'Introduction: The Homeland Concept'. *Journal of Cultural Geography* 13, no. 2:1–4.

_____, and S.B. Hilliard, eds. 1988. *The American South*. *Geoscience and Man* 25. Baton Rouge: Louisiana State University, Department of Geography and Anthropology, Geoscience Publications.

O'Connor, J. 1998. *Natural Causes: Essays in Ecological Marxism*. New York: Guilford.

O'Dwyer, B. 1997. 'Pathways to Homelessness: A Comparison of Gender and Schizophrenia in Inner-Sydney'. *Australian Geographical Studies* 35:294–307.

Oelschlaeger, M. 1991. *The Idea of Wilderness*. New Haven: Yale University Press.

Ohmae, K. 1993. 'The Rise of the Region State'. *Foreign Affairs* 76, no. 2:78–87.

Olwig, K. 1980. 'Historical Geography and the Society/Nature 'Problematic': The Perspective of J.F. Schouw, G.P. Marsh and E. Reclus'. *Journal of Historical Geography* 6:29–45.

_____. 1996a. 'Nature—Mapping the Ghostly Traces of a Concept'. In *Concepts in Human Geography*, edited by

C. Earle, K. Mathewson, and M.S. Kenzer, 63–96. Lanham, MD: Rowman and Littlefield.

_____. 1996b. 'Recovering the Substantive Nature of Landscape'. *Annals of the Association of American Geographers* 86:630–53.

Ortner, S.B. 1972. 'Is Female to Male as Nature Is to Culture?' *Feminist Studies* 1:5–31.

_____, and H. Whitehead. 1981. 'Introduction: Accounting for Sexual Meaning'. In *Sexual Meanings: The Cultural Construction of Gender and Sexuality*, edited by S.B. Ortner and H. Whitehead, 1–27. New York: Cambridge University Press.

Osborne, B.S. 1988. 'The Iconography of Nationhood in Canadian Art'. In *The Iconography of Landscape*, edited by D. Cosgrove and S. Daniels, 162–78. New York: Cambridge University Press.

_____. 1992. 'Interpreting a Nation's Identity: Artists as Creators of National Consciousness'. In *Ideology and Landscape in Historical Perspective*, edited by A.R.H. Baker and G. Biger, 230–54. New York: Cambridge University Press.

Ostergren, R.C. 1988. *A Community Transplanted: The Trans-Atlantic Experience of a Swedish Immigrant Settlement in the Upper Midwest, 1835–1915*. Madison: University of Wisconsin Press.

Ó Tuathail, G. 1995. 'Political Geography I: Theorizing History, Gender and World Order Amidst Crises of Global Governance'. *Progress in Human Geography* 19:260–72.

_____. 1996. *Critical Geopolitics: The Politics of Writing Global Space*. Minneapolis: University of Minnesota Press.

Pahl, R.E. 1965. 'Trends in Social Geography'. In *Frontiers in Geographical Teaching*, edited by R.J. Chorley and P. Haggett, 81–100. London: Methuen.

Pain, R.H. 1997. 'Social Geographies of Women's Fear of Crime'. *Transactions of the Institute of British Geographers* NS 22:231–44.

_____. 1999. 'The Geography of Fear'. *Geography Review* 12, no. 5:22–5.

Park, C.C. 1994. *Sacred Worlds: An Introduction to Geography and Religion*. New York: Routledge.

Park, R.E. 1915. 'The City: Suggestions for the Investigation of Human Behavior in the City Environment'. *American Journal of Sociology* 20:577–612.

_____. 1952. *Human Communities: The City and Human Ecology. The Collected Papers of Robert Ezra Park, Volume II*. Glencoe, Ill.: Free Press.

_____, and E.W. Burgess. 1921. *Introduction to the Science of Sociology*. Chicago: Chicago University Press.

Parr, H., and C. Philo. 1995. 'Mapping "Mad" Identities'. In *Mapping the Subject: Geographies of Cultural Transformation*, edited by S. Pile and N. Thrift, 199–225. New York: Routledge.

Parsons, J.J. 1979. 'The Later Sauer Years'. *Annals of the Association of American Geographers* 69:9–15.

_____. 1996. 'Mr. Sauer and the Writers'. *Geographical Review* 86:22–41

Pawson, E. 1992. 'Two New Zealands: Maori and European'. In *Inventing Places: Studies in Cultural Geography*, edited by K. Anderson and F. Gale, 15–33. New York: Wiley.

_____, and G. Banks. 1993. 'Rape and Fear in a New Zealand City'. *Area* 25:55–63.

Peach, C. 1999. 'Social Geography'. *Progress in Human Geography* 23:282–8.

Peet, R. 1977a. *Radical Geography*. Chicago: Maaroufa Press.

_____. 1977b. 'The Development of Radical Geography in the United States'. *Progress in Human Geography* 1:240–63.

_____. 1985. 'The Social Origins of Environmental Determinism'. *Annals of the Association of American Geographers* 75:309–33.

_____. 1996a. 'Structural Themes in Geographical Discourse'. In *Companion Encyclopedia of Geography: The Environment and Humankind*, edited by I. Douglas, R. Huggett, and M. Robinson, 860–87. New York: Routledge.

_____. 1996b. 'Discursive Idealism in the "Landscape-as-Text" School'. *Professional Geographer* 48:96–8.

_____, and N. Thrift. 1989. 'Political Economy and Human Geography'. In *New Models in Geography: Volume 1*, edited by R. Peet and N. Thrift, 3–29. Boston: Unwin Hyman.

_____. and M. Watts. 1996. 'Liberation Ecology: Development, Sustainability and Environment in an Age of Market Triumphalism'. In *Liberation Ecologies: Environment, Development, Social Movements*, edited by R. Peet and M. Watts, 1–45. New York: Routledge.

Pelto, P.J. 1966. *The Nature of Anthropology*. Columbus, OH: Merrill.

Penrose, J. 1997. 'Construction, De(con)struction and Reconstruction. The Impact of Globalization and Fragmentation on the Canadian Nation-State'. *International Journal of Canadian Studies* 16:15–49.

Peters, B.C. 1972. 'Oak Openings or Barrens: Landscape Evaluation on the Michigan Frontier'. *Proceedings of the Association of American Geographers* 4:84–6.

Peterson, J., and J. Anfison. 1985. 'The Indian and the Fur Trade: A Review of Recent Literature'. *Manitoba History* 10 (Autumn):10–18.

Phares, E.J. 1991. *Introduction to Personality*, 3rd edn. New York: Harper Collins.

Philo, C. 1988. 'Conference Report: New Directions in Cultural Geography'. *Journal of Historical Geography* 14:178–81.

_____. 1989. Enough to Drive One Mad: The Organization of Space in 19th Century Lunatic Asylums'. In *The Power of Geography*, edited by J. Wolch and M. Dear, 258–90. Boston: Unwin Hyman.

_____, compiler. 1991. *New Words, New Worlds: Reconceptualising Social and Cultural Geography*. Aberystwyth, Wales: Cambrian Printers.

Pickles, J. 1985. *Phenomenology, Science, and Geography*. New York: Cambridge University Press.

Pile, S. 1993. 'Human Agency and Human Geography Revisited: A Critique of "New Models" of the Self'. *Transactions of the Institute of British Geographers* 18:122–39.

_____. 1996. *The Body and the City: Psychoanalysis, Space and Subjectivity*. New York: Routledge.

_____, and M. Keith, eds. 1997. *Geographies of Resistance*. New York: Routledge.

_____, and N. Thrift, eds. 1995. *Mapping the Subject: Geographies of Cultural Transformation*. New York: Routledge.

Pillsbury, R. 1970. 'The Urban Street Pattern as a Cultural Indicator'. *Annals of the Association of American Geographers* 60:428–46.

_____. 1974. 'Carolina Thunder: A Geography of Southern Stock Car Racing'. *Journal of Geography* 73:39–47.

_____. 1989. 'A Mythology at the Brink: Stock Car Racing in the American South'. *Sport Place: An International Journal of Sports Geography* 3:3–12.

Pinkerton, J.P. 1997. 'Enviromanticism: The Poetry of Nature as Political Force'. *Foreign Affairs* 76, no. 3:2–7.

Platt, R.S. 1962. 'The Rise of Cultural Geography in America'. In *Readings in Cultural Geography*, edited by P.L. Wagner and M.W. Mikesell, 35–43. Chicago: University of Chicago Press.

Plummer, V. 1993. *Feminism and the Mastery of Nature*. New York: Routledge.

Pocock, D.C.D., ed. 1981. *Humanistic Geography and Literature*. London: Croom Helm.

Porter, P.W. 1965. 'Environmental Potentials and Economic Opportunities: a Background for Cultural Adaptation'. *American Anthropologist* 67:409–20.

_____. 1978. 'Geography as Human Ecology'. *American Behavioral Scientist* 22:15–39.

_____. 1979. *Food and Development in the Semi-Arid Zone of East Africa*. Syracuse: Syracuse University, Maxwell School of Citizenship and Public Affairs.

_____. 1987. 'Ecology as Metaphor: Sauer and Human Ecology'. *Professional Geographer* 39:414.

Powell, J.M. 1977. *Mirrors of the New World: Images and Image Makers in the Settlement Process*. Folkestone, England: Dawson.

_____. 1980. 'Taylor, Stefansson and the Arid Centre: An Historic Encounter of "Environmentalism" and Possibilism'. *Journal of the Royal Australian Historical Society* 66:163–83.

Pred, A. 1984. 'Place as Historically Contingent Process: Structuration and the Time Geography of Becoming Places'. *Annals of the Association of American Geographers* 74:279–97.

_____. 1985. 'The Social Becomes the Spatial, The Spatial Becomes the Social: Enclosures, Social Change and the Becoming of Place in the Swedish Province of Skåne', In *Social Relations and Spatial Structures*, edited by D. Gregory and J. Urry, 336–75. London: Macmillan.

Price, M., and M. Lewis. 1993a. 'The Reinvention of Cultural Geography'. *Annals of the Association of American Geographers* 83:1–17.

_____. 1993b. 'Reply: On Reading Cultural Geography'. *Annals of the Association of American Geographers* 83:520–2.

Prince, H.C. 1971. 'Real, Imagined and Abstract Worlds of the Past'. In *Progress in Geography*, vol. 3, edited by C. Board et al., 1–86. New York: Arnold.

Proctor, J.D. 1998. 'The Social Construction of Nature: Relativist Accusations, Pragmatist and Critical Responses'. *Annals of the Association of American Geographers* 88:352–76.

Pryce, W.T.R. 1975. 'Migration and the Evolution of Culture Areas: Cultural and Linguistic Frontiers in North-East Wales, 1750 and 1851'. *Transactions of the Institute of British Geographers* 65:79–107.

Pudup, M.B. 1988. 'Arguments Within Regional Geography'. *Progress in Human Geography* 12:369–90.

Pulvirenti, M. 1997. 'Unwrapping the Parcel: An Examination of Culture Through Italian Australian Home Ownership'. *Australian Geographical Studies* 35:32–9.

Pyle, G.F. 1969. 'The Diffusion of Cholera in the United States in the Nineteenth Century'. *Geographical Analysis* 1:59–75.

Quani, M. 1982. *Geography and Marxism*. Cambridge, MA: Blackwell.

Quinn, J.A. 1950. *Human Ecology*. New York: Prentice Hall.

Raby, S. 1973. 'Indian Land Surrenders in Southern Saskatchewan'. *Canadian Geographer* 17:36–52.

Radcliffe, S.A. 1997. 'Frontiers and Popular Nationhood: Geographies of Identity in the 1995 Ecuador–Peru Border Dispute'. *Political Geography* 17:273–93.

Radding, C. 1997. *Wandering Peoples: Colonialism, Ethnic Spaces, and Ecological Frontiers in Northwestern Mexico, 1700–1850*. Durham: Duke University Press.

Raglon, R. 1996. 'Women and the Great Canadian Wilderness: Reconsidering the Wild'. *Women's Studies* 25:513–31.

Raitz, K.B. 1973a. 'Ethnicity and the Diffusion and Distribution of Cigar Tobacco Production in Wisconsin and Ohio'. *Tijdschrifte voor Economische en Sociale Geografie* 64:293–306.

_____. 1973b. 'Theology on the Landscape: A Comparison of Mormon and Amish-Mennonite Land Use'. *Utah Historical Quarterly* 41:23–34.

_____. 1979. 'Themes in the Cultural Geography of European Ethnic Groups in the United States'. *Geographical Review* 69:79–94.

_____, and R. Ulack. 1981a. 'Appalachian Vernacular Regions'. *Journal of Cultural Geography* 2:106–19.

_____, and R. Ulack. 1981b. 'Cognitive Maps of Appalachia'. *Geographical Review* 71:201–13.

Raivo, P.J. 1997. 'The Limits of Tolerance: The Orthodox Milieu as an Element in the Finnish Cultural Landscape'. *Journal of Historical Geography* 23:327–39.

Rapoport, A. 1969. *House Form and Culture*. Englewood Cliffs, NJ: Prentice Hall.

Rappaport, R.A. 1963. 'Aspects of Man's Influence on Island Ecosystems: Alteration and Control'. In *Man's Place in the Island Ecosystem*, edited by F.R. Fosberg, 155–74. Honolulu: Bishop Museum Press.

Ray, A.J. 1974. *Indians in the Fur Trade: Their Role as Hunters, Trappers and Middlemen in the Lands Southwest of Hudson Bay, 1660–1870*. Toronto: University of Toronto Press.

_____. 1996. *I Have Lived Here Since the World Began: An Illustrated History of Canada's Native People*. Toronto: Lester Publishing, Key Porter Books.

Rees, J. 1989. 'Natural Resources, Economy and Society'. In *Horizons in Human Geography*, edited by D. Gregory and R. Walford, 364–94. New York: MacMillan.

Rees, R. 1984. *Land of Earth and Sky: Landscape Painting of Western Canada*. Saskatoon: Western Producer Prairie Books.

_____. 1988. *New and Naked Land: Making the Prairies Home*. Saskatoon: Western Producer Prairie Books.

Relph, E. 1970. 'An Inquiry into the Relations Between Phenomenology and Geography'. *Canadian Geography* 14:193–201.

_____. 1976. *Place and Placelessness*. London: Pion.

_____. 1981. *Rational Landscapes and Humanistic Geography*. Totowa, NJ: Barnes and Noble.

_____. 1984. 'Seeing, Thinking, and Describing Landscapes'. In *Environmental Perception and Behavior: An Inventory and Prospect*, edited by T.F. Saarinen, D. Seamon, and J.L. Sell, 209–23. Chicago: University of Chicago, Department of Geography, Research Paper 209.

_____. 1985. 'Geographical experiences and Being-in-the-World: The Phenomenological Origins of Geography'. In *Dwelling, Place and Environment: Towards a Phenomenology of Person and World*, edited by D. Seamon and R. Mugeraur, 15–31. New York: Columbia University Press.

_____. 1997. 'Sense of Place'. In *Ten Geographic Ideas That Changed the World*, edited by S. Hanson, 205–26. New Brunswick, NJ: Rutgers University Press.

Renfrew, C. 1988. *Archaeology and Language: The Puzzle of Indo-European*. New York: Cambridge University Press.

Renner, M. 1996. *Fighting for Survival: Environmental Decline, Social Conflict, and the New Age of Insecurity*. New York: Norton.

Reynolds, H. 1982. *The Other Side of the Frontier: Aboriginal Resistance to the European Invasion of Australia*. Ringwood, Australia: Penguin.

Rice, J.G. 1977. 'The Role of Culture and Community in Frontier Prairie Farming'. *Journal of Historical Geography* 3:155–75.

Ringer, G., ed. 1998. *Destinations: Cultural Landscapes of Tourism*. New York: Routledge.

Roberts, N. 1989. *The Holocene: An Environmental History*. Cambridge, MA: Blackwell.

Robinson, G.M. 1998. *Methods and Techniques in Human Geography*. New York: Wiley.

Rogers, E.M. 1962. *Diffusion of Innovations*. New York: Free Press of Glencoe.

Rogerson, R.J., and A. Gloyer. 1995. 'Gaelic Cultural Revival or Language Decline'. *Scottish Geographical Magazine* 111:46–53.

Rolston, H. III. 1997. 'Nature for Real: Is Nature a Social Construct'. In *The Philosophy of the Environment*, edited by T.D.J. Chappell, 38–64. Edinburgh: Edinburgh University Press.

Rooney, J.R.Jr. 1974. *A Geography of American Sport: From Cabin Creek to Anaheim*. Reading, MA: Addison Wesley.

_____. 1993. 'The Golf Construction Boom'. *Sport Place: An International Journal of Sports Geography* 7:15–22.

_____, and P.L. Butt. 1978. 'Beer, Bourbon, and Boone's Farm: A Geographical Examination of Alcoholic Drink in the United States'. *Journal of Popular Culture* 11:832–55.

_____, and R. Pillsbury. 1992. *Atlas of American Sport.* New York: Macmillan.

_____, W. Zelinsky, and D.R. Loudon, eds. 1982. *This Remarkable Continent: An Atlas of United States and Canadian Society and Cultures.* College Station: Texas A&M University Press.

Rose, A.J. 1972. 'Australia as a Cultural Landscape'. In *Australia as Human Setting*, edited by A. Rapoport, 58–74. Sydney: Angus and Robertson.

Rose, C. 1981. 'William Dilthey's Philosophy of Historical Understanding: A Neglected Heritage of Contemporary Humanistic Geography'. In *Geography, Ideology and Social Concern*, edited by D.R. Stoddart, 99–133. Totowa, NJ: Barnes and Noble.

Rose, G. 1993. *Feminism and Geography: The Limits of Geographical Knowledge.* Minneapolis: University of Minnesota Press.

_____. 1997. 'Situating Knowledges: Positionality, Reflexivities and Other Tactics'. *Progress in Human Geography* 21:305–20.

_____, V. Kinnaird, M. Morris, and C. Nash. 1997. 'Feminist Geographies of Environment, Nature and Landscape'. In *Feminist Geographies: Explorations in Diversity and Difference*, edited by Women and Geography Study Group of the Royal Geographical Society with the Institute of British Geographers, 146–90. Harlow, England: Longman.

_____, et al. 1997. 'Introduction'. In *Feminist Geographies: Explorations in Diversity and Difference*, edited by Women and Geography Study Group of the Royal Geographical Society with the Institute of British Geographers, 1–12. Harlow, England: Longman.

Rosenau, P.M. 1992. *Post-Modernism and the Social Sciences: Insights, Inroads, and Intrusions.* Princeton: Princeton University Press.

Rossi, I., and E. O'Higgins. 1980. 'Unit 1: Theories of Culture and Anthropological Methods'. In *People in Culture: A Survey of Cultural Inquiry*, edited by I. Rossi, 31–78. New York: Praeger.

Rostlund, E. 1956. 'Twentieth-Century Magic'. *Landscape* 5:23–6.

Routledge, P. 1992. 'Putting Politics in its Place: Baliapal, India, as a Terrain of Resistance'. *Political Geography* 11:588–611.

_____. 1997. 'The Imagineering of Resistance: Pollok Free State and the Practice of Postmodern Politics'.

Transactions of the Institute of British Geographers NS 22:359–76.

Rowntree, L.B. 1996. 'The Cultural Landscape Concept in American Cultural Geography'. In *Concepts in Human Geography*, edited by C. Earle, K. Mathewson, and M.S. Kenzer, 127–59. Lanham, MD: Rowman and Littlefield.

_____, K. Foote, and M. Domosh. 1989. 'Cultural Geography'. In *Geography in America*, edited by G. Gaile and C. Wilmott, 209–17. New York: Merrill.

Roxby, P.M. 1930. 'The Scope and Aims of Human Geography'. *Scottish Geographical Magazine* 46:276–99.

Rupesinghe, K. 1996. 'Governance and Conflict Resolution in Multi-ethnic Societies'. In *Ethnicity and Power in the Contemporary World*, edited by K. Rupesinghe and V.A. Tishkov, 10–31. New York: United Nations University Press.

Rushton, G. 1979. 'Commentary: On "Behavioral and Perception Geography"'. *Annals of the Association of American Geographers* 69:463–4.

Russell, R.J., and F.B. Kniffen. 1951. *Culture Worlds.* New York: MacMillan.

Ryan, S. 1996. *The Cartographic Eye: How Explorers Saw Australia.* New York: Cambridge University Press.

Saarinen, T.F. 1974. 'Environmental Perception'. In *Perspectives on Environment*, edited by I.R. Manners and M.W. Mikesell, 252–89. Washington: Association of American Geographers, Commission on College Geography Publication No. 13.

_____. 1979. 'Commentary–Critique of the Bunting-Guelke Paper'. *Annals of the Association of American Geographers* 69:464–8.

Sahlins, M. 1974. *Stone Age Economics.* London: Tavistock.

Said, E. 1978. *Orientalism.* New York: Columbia University Press.

_____. 1993. *Culture and Imperialism.* London: Chatto and Windus.

Salter, C.L. 1971a. 'The Mobility of Man'. In *The Cultural Landscape*, edited by C.L. Salter, 1–4. Belmont, CA: Duxbury.

_____. 1971b. 'Introductory Comments to 'Some Curious Analogies in Explorer's Preconceptions of Virginia' by G. Dunbar'. In *The Cultural Landscape*, edited by C.L. Salter, 18. Belmont: Duxbury.

_____. 1972. 'A Speculative Cultural Geography: A Free-for-All Approach to Introductory Cultural Geography'. *Journal of Geography* 71:533–40.

_____. 1977. 'Learning Through Landscape'. *The California Geographer* 17:1–9.

Samuels, M.S. 1978. 'Existentialism and Human Geography'. In *Humanistic Geography: Prospects and*

Problems, edited by D. Ley and M.S. Samuels, 22–40. London: Croom Helm.

_____. 1979. 'The Biography of Landscape: Cause and Culpability'. In *The Interpretation of Ordinary Landscapes*, edited by D.W. Meinig, 51–88. New York: Oxford University Press.

_____. 1981. 'An Existential Geography'. In *Themes in Geographic Thought*, edited by M.E. Harvey and B.P. Holly, 115–32. London: Croom Helm.

Sanderson, S.K. 1990. *Social Evolutionism: A Critical History*. Cambridge, MA: Blackwell.

_____, ed. 1995. *Civilizations and World Systems: Studying World-Historical Change*. Walnut Creek, CA: Altimira Press.

_____, and T.D. Hall. 1995. 'World System Approaches to World–Historical Change'. In *Civilizations and World Systems: Studying World-Historical Change*, edited by S.K. Sanderson, 95–108. Walnut Creek, CA: Altimira Press.

Sauer, C.O. 1924. 'The Survey Method in Geography and its Objectives'. *Annals of the Association of American Geographers* 14:17–33.

_____. 1925. 'The Morphology Of Landscape'. *University of California Publications in Geography* 2:19–53.

_____. 1927. 'Recent Developments in Cultural Geography'. In *Recent Developments in the Social Sciences*, edited by E.C. Hayes, 154–212. Philadelphia: J.B. Lippincott.

_____. 1931. 'Cultural Geography'. In *Encyclopedia of the Social Sciences*, vol. 6, 621–4. New York: MacMillan.

_____. 1935. *Aboriginal Populations of North-western Mexico*. Ibero–Americana 10. Berkeley: University of California Press.

_____. 1940. *Culture Regions of The World: Outline of Lectures*. Berkeley: University of California at Berkeley, Department of Geography.

_____. 1941a. 'Foreword to Historical Geography'. *Annals of the Association of American Geographers* 31:1–24.

_____. 1941b. 'The Personality of Mexico'. *Geographical Review* 31:353–64.

_____. 1952. *Agricultural Origins and Dispersals*. New York: American Geographical Society.

_____. 1963. 'Historical Geography and the Western Frontier'. In *Land and Life: A Selection from the Writings of Carl Sauer*, edited by J. Leighly, 45–52. Berkeley: University of California Press.

_____. 1968. 'Human Ecology and Population'. In *Population and Economics*, edited by P. DePrez, 207–14. Winnipeg: University of Manitoba Press.

_____. 1969. *Seeds, Spades, Hearths, and Herds: The Domestication of Animals and Foodstuffs*. Cambridge, MA: MIT Press.

_____. 1970. 'Plants, Animals and Man'. In *Man and His Habitat*, edited by R.H. Buchanan, E. Jones, and D. McCourt, 34–61. London: Routledge and Kegan Paul.

_____. 1971. *Sixteenth Century North America*. Berkeley: University of California Press.

_____. 1974. 'The Fourth Dimension of Geography'. *Annals of the Association of American Geographers* 64:189–92.

_____. 1981. *Selected Essays, 1963–1975*. Berkeley, CA: Turtle Island Foundation.

_____. 1985. *North America: Notes on Lectures by Professor Carl O. Sauer at the University of California at Berkeley, 1936*. Northridge, CA.: California State University, Northridge, Department of Geography, Occasional Paper No. 1.

Sayer, A. 1979. 'Epistemology and Conceptions of People and Nature in Geography'. *Geoforum* 10:19–43.

_____. 1982. 'Explanation in Economic Geography'. *Progress in Human Geography* 6:68–88.

_____. 1992a. *Method in Social Science: A Realist Approach*, 2nd edn. New York: Routledge.

_____. 1992b. 'Radical Geography and Marxist Political Economy: Towards a Re–evaluation'. *Progress in Human Geography* 16:343–60.

_____, and M. Storper. 1997. 'Guest Editorial Essay— Ethics Unbound: For a Normative Turn in Social Theory'. *Environment and Planning D: Society and Space* 15:1–17.

Scarre, C., and B.M. Fagan. 1997. *Ancient Civilizations*. New York: Longman.

Schaefer, F. 1953. 'Exceptionalism in Geography: A Methodological Examination'. *Annals of the Association of American Geographers* 43:226–49.

Schein, R.H. 1997. 'The Place of Landscape: A Conceptual Framework for Interpreting an American Scene'. *Annals of the Association of American Geographers* 87:660–80.

Schmitt, R. 1987. *Introduction to Marx and Engels: A Critical Reconstruction*. Boulder, CO: Westview.

Schnaitter, R. 1986. 'A Coordination of Differences: Behaviorism, Mentalism, and the Foundations of Psychology'. In *Approaches to Cognition: Contrasts and Controversies*, edited by T.J. Knapp and L.C. Robertson, 291–315. Hillsdale, NJ: Lawrence Erlbaum.

Schnore, L.F. 1961. 'Geography and Human Ecology'. *Economic Geography* 37:207–17.

Schutz, A. 1967. *The Phenomenology of the Social World*, translated by G. Walsh and F. Lennert. Evanston, Ill.: Northwestern University Press.

Seamon, D. 1979. *A Geography of the Lifeworld*. London: Croom Helm.

Seig, L. 1963. 'The Spread of Tobacco: A Study in Cultural Diffusion'. *Professional Geographer* 15:17–21.

Semple, E. 1911. *Influences of Geographic Environment*. New York: Henry Holt.

Senn, P.R. 1971. *Social Science and its Methods*. Boston: Holbrook Press Inc.

Shiva, V. 1988. *Staying Alive: Women, Ecology and Development*. London: Zed Books.

Shortridge, B.G., and J.R. Shortridge. 1995. 'Cultural Geography of American Foodways: An Annotated Bibliography'. *Journal of Cultural Geography* 15, no. 2:79–108.

_____, and J.R. Shortridge, eds. 1998. *The Taste of American Place: A Reader on Regional and Ethnic Foods*. Lanham, MD: Rowman and Littlefield.

Shortridge, J.R. 1980. 'Vernacular Regions in Kansas'. *American Studies* 21:73–94.

_____. 1995. *Peopling the Plains: Who Settled Where in Frontier Kansas*. Lawrence: University Press of Kansas.

Shurmer-Smith, P., and K. Hannam. 1994. *Worlds of Desire, Realms of Power: A Cultural Geography*. New York: Arnold.

Sibley, D. 1992. 'Outsiders in Society and Space'. In *Inventing Places: Studies in Cultural Geography*, edited by K. Anderson and F. Gale, 107–22. Melbourne: Longman Cheshire.

_____. 1995. *Geographies of Exclusion*. New York: Routledge.

Simmons, I.G. 1988. 'The Earliest Cultural Landscapes of England'. *Environmental Review* 12:105–16.

_____. 1997. *Humanity and Environment: A Cultural Ecology*. London: Longman.

Simon, J., and H. Kahn, eds. 1984. *The Resourceful Earth*. Cambridge, MA: Blackwell.

Singer, P. 1993. *Practical Ethics*, 2nd edn. New York: Cambridge University Press.

Skelton, T., and G. Valentine, eds. 1998. *Cool Places: Geographies of Youth Cultures*. New York: Routledge.

Skinner, B.F. 1969. *Contingencies of Reinforcement: A Theoretical Analysis*. New York: Appleton Century Crofts.

Slater, T., and P. Jarvis, eds. 1982. *Field and Forest: An Historical Geography of Warwickshire and Worcestershire*. Norwich, England: Geobooks.

Slocum, T.A., and E.C. Butterfield 1994. 'Bridging the Schism Between Behavioral and Cognitive Analyses'. *The Behavior Analyst* 17:59–73.

Sluyter, A. 1997. 'On "Buried Epistemologies: The Politics of Nature in (Post)colonial British Columbia": On Excavating and Burying Epistemologies'. *Annals of the Association of American Geographers* 87:700–2.

Smil, V. 1987. *Energy, Food, Environment: Realities, Myths, Options*. New York: Oxford University Press.

_____. 1993. *Global Ecology: Environmental Changes and Social Flexibility*. New York: Routledge.

Smith, A.D. 1986. *The Ethnic Origins of Nations*. Cambridge, MA: Blackwell.

Smith, B.D. 1995. *The Emergence of Agriculture*. New York: Scientific American Library.

Smith, C.T. 1965. 'Historical Geography: Current Trends and Prospects'. In *Frontiers in Geographical Teaching*, edited by R.J. Chorley, 118–43. Methuen: London.

Smith, J.S. 1999. 'Anglo Intrusion on the Old Sangre de Cristo Land Grant'. *Professional Geographer* 51:170–83.

Smith, R. 1997. *The Norton History of the Human Sciences*. New York: W.W. Norton.

Smith, T.R., J.W. Pellegrino, and R.G. Golledge. 1982. 'Computational Process Modeling of Spatial Cognition and Behavior'. *Geographical Analysis* 14:305–25.

Soja, E.W. 1989. *Postmodern Geographies*. London: Verso.

Solot, M. 1986. 'Carl Sauer and Cultural Evolution'. *Annals of the Association of American Geographers* 76:508–20.

Sopher, D.E. 1967. *Geography of Religions*. Englewood Cliffs, NJ: Prentice Hall.

_____. 1972. 'Place and Location: Notes on the Spatial Patterning of Culture'. *Social Science Quarterly* 53:321–37.

_____. 1981. 'Geography and Religions'. *Progress in Human Geography* 5:510–24.

Southwick, C.H. 1996. *Global Ecology in Human Perspective*. New York: Oxford University Press.

Sowell, T. 1994. *Race and Culture: A World View*. New York: Basic Books.

_____. 1996. *Migrations and Cultures: A World View*. New York: Basic Books.

Spate, G.H.K. 1952. 'Toynbee and Huntington: A Study in Determinism'. *Geographical Journal* 118:406–28.

Spencer, C., and M. Blades. 1986. 'Pattern and Process: A Review Essay on the Relationship Between Behavioral Geography and Environmental Psychology'. *Progress in Human Geography* 10:230–48.

Spencer, J.E. 1960. 'The Cultural Factor in Underdevelopment'. In *Geography and Economic Development*, edited by N. Ginsburg, 35–48. Chicago: University of Chicago, Department of Geography, Research Paper.

_____. 1978. 'The Growth of Cultural Geography'. *American Behavioral Scientist* 22:79–92.

_____, and R.J. Horvath. 1963. 'How Does an Agricultural Region Originate?'. *Annals of the Association of American Geographers* 53:74–92.

_____, and W.L. Thomas, Jr. 1973. *Introducing Cultural Geography*. New York: Wiley.

Speth, W.W. 1967. 'Environment, Culture and the Mormon in Early Utah'. *Yearbook of the Association of Pacific Coast Geographers* 29:53–67.

_____. 1987. 'Historicism: The Disciplinary World View of Carl O. Sauer'. In *Carl O. Sauer: A Tribute*, edited by M.S. Kenzer, 11–39. Corvallis: Oregon State University Press.

Squire, S.J. 1994. 'Accounting for Cultural Meanings: The Interface Between Geography and Tourism Studies Re-Examined'. *Progress in Human Geography* 18:1–16.

Stanislawski, D. 1946. 'The Origin and Spread of the Grid Pattern Town'. *Geographical Review* 36:105–20.

Steiman, L.B. 1998. *Paths to Genocide: Antisemitism in Western History*. New York: St Martin's Press.

Stein, H.F., and G.L. Thompson. 1993. 'The Sense of Oklahomaness: Contributions of Psychogeography to American Culture'. *Journal of Cultural Geography* 13, no. 2:63–91.

Steward, J. 1936. 'The Economic and Social Basis of Primitive Bands'. In *Essays in Anthropology Presented to A.L. Kroeber*, edited by R.H. Lowie, 331–45. Berkeley: University of California Press.

_____. 1938. *Basin-Plateau Aboriginal Sociopolitical Groups*. Washington, DC: Smithsonian Institution.

_____. 1955. *Theory of Culture Change: The Methodology of Multilinear Evolution*. Urbana: University of Illinois Press.

Stilgoe, J.R. 1982. *Common Landscape of America: 1580–1845*. New Haven: Yale University Press.

Stoddart, D.R. 1965. 'Geography and the Ecological Approach'. *Geography* 50:242–51.

_____. 1966. Darwin's Impact on Geography'. *Annals of the Association of American Geographers* 56:683–98.

_____. 1967. 'Organism and Ecosystem as Geographical Models'. In *Models in Geography*, edited by R.J. Chorley and P. Haggett, 511–48. London: Methuen.

_____. 1986. *On Geography*. Cambridge, MA: Blackwell.

Straussfogel, D. 1997. 'World Systems Theory: Toward a Heuristic and Pedagogic Conceptual Tool'. *Economic Geography* 73:118–30.

Stump, R.W. 1986. 'Introduction'. *Journal of Cultural Geography* 7, no. 1:1–3.

Svobodová, H., ed. 1990. *Cultural Aspects of Landscape*. Wageningen, Netherlands: Pudoc.

Symanski, R. 1974. 'Prostitution in Nevada'. *Annals of the Association of American Geographers* 64:357–77.

_____. 1981. *The Immoral Landscape: Female Prostitution in Western Societies*. Toronto: Butterworths.

Tarrow, S. 1992. 'Mentalities, Political Cultures, and Collective Action Frames'. In *Frontiers in Social Movement Theory*, edited by A. Morris and C. Mueller, 174–202. New Haven: Yale University Press.

Tatham, G. 1951. 'Environmentalism and Possibilism'. In *Geography in the Twentieth Century*, edited by G. Taylor, 128–62. New York: Philosophical Library.

Taylor, P.J. 1991. 'The English and Their Englishness: "A Curiously Mysterious, Elusive and Little Understood People"'. *Scottish Geographical Magazine* 107:146–61.

_____. 1992. 'Understanding Global Inequalities: A World-Systems Approach'. *Geography* 77:10–21.

_____. 1994a. *Political Geography: World-economy, Nation-state, and Locality*, 2nd edn. New York: Wiley.

_____. 1994b. 'The State as Container: Territoriality in the Modern World-System'. *Progress in Human Geography* 18:151–62.

_____. 1996. *The Way the Modern World Works: World Hegemony to World Impasse*. New York: Wiley.

Taylor, T.G. 1928. *Australia in Its Physiographic and Economic Aspects*, 5th edn. Oxford: Clarendon.

_____. 1937. *Environment, Race, and Migration*. Toronto: University of Toronto Press.

_____. 1951. *Australia: A Study of Warm Environments and Their Effect on British Settlement*, 6th enlarged edn. New York: Dutton.

Teather, E.K. 1998. 'Themes from Complex Landscapes: Chinese Cemeteries and Columbaria in Urban Hong Kong'. *Australian Geographical Studies* 36:21–36.

Teich, M., R. Porter, and B. Gustafsson, eds. 1997. *Nature and Society in Historical Context*. New York: Cambridge University Press.

Tenbrunsel, A.E., K.A. Wade-Benzoni, D.M. Messick, and M.H. Bazerman. 1997. 'Introduction'. In *Environment, Ethics, and Behavior*, edited by M.H. Bazerman, D.M. Messick. A.E, Tenbrunsel, and K.A. Wade Benzoni, 1–9. San Francisco: The New Lexington Press.

Tetlock, P.E., and A. Belkin, eds. 1996. *Counterfactual Thought Experiments in World Politics: Logical, Methodological, and Psychological Perspectives*. Princeton: Princeton University Press.

Thomas, W.L., Jr. 1956. 'Introductory'. In *Man's Role in Changing the Face of the Earth*, edited by W.L. Thomas

Jr., C.O. Sauer, M. Bates, and L. Mumford, xxi–xxxviii. Chicago: University of Chicago Press.

_____, C.O. Sauer, M. Bates, and L. Mumford. 1956. *Man's Role in Changing the Face of the Earth*. Chicago: University of Chicago Press.

Thompson, E.P. 1968. *The Making of the English Working Class*. London: Gollancz.

Thompson, G. F. 1995a. 'A Message to the Reader'. In *Landscape in America*, edited by G.F. Thompson, xi–xiv. Austin: University of Texas Press.

_____, ed. 1995b. *Landscape in America*. Austin: University of Texas Press.

Thomson, G.M. 1975. *The North West Passage*. London: Martin Secker and Warburg.

Thornthwaite, C.W. 1940. 'The Relation of Geography to Human Ecology'. *Ecological Monograph* 10:343–8.

Thrift, N. 1994. 'Taking Aim at the Heart of the Region'. In *Human Geography: Society, Space and Social Science*, edited by D. Gregory, R. Martin, and G. Smith, 200–31. New York: MacMillan.

Tough, F. 1996. *As Their Natural Resources Fail: Native Peoples and the Economic History of Northern Manitoba, 1870–1930*. Vancouver: University of British Columbia Press.

Touraine, A. 1981. *The Voice and the Eye: An Analysis of Social Movements*. New York: Cambridge University Press.

Toynbee, A. 1934–61. *A Study of History*, 12 vols. New York: Oxford University Press.

Trépanier, C. 1991. 'The Cajunization of French Louisiana: Forging a Regional Identity'. *Geographical Journal* 157:161–71.

Trigger, B.G. 1982. 'Response of Native Peoples to European Contact'. In *Early European Settlement and Exploitation in Atlantic Canada*, edited by G.M. Story, 139–55. St John's: Memorial University of Newfoundland.

_____. 1985. *Natives and Newcomers: Canada's 'Heroic Age' Reconsidered*. Kingston: McGill–Queen's University Press.

_____.1998. *Sociocultural Evolution: Calculation and Contingency*. Malden, MA: Blackwell.

Tuan, Y.-F. 1971. 'Geography, Phenomenology and the Study of Human Nature'. *Canadian Geographer* 15:181–92.

_____. 1972. 'Structuralism, Existentialism and Environmental Perception'. *Environment and Behavior* 4:319–42.

_____. 1974. *Topophilia: A Study of Environmental Perception, Attitudes, and Values*. Englewood Cliffs, NJ: Prentice Hall.

_____. 1975. 'Images and Mental Maps'. *Annals of the Association of American Geographers* 65:205–13.

_____. 1977. *Space and Place: The Perspective of Experience*. Minneapolis: University of Minnesota Press.

_____. 1982. *Segmented Worlds and Self*. Minneapolis: University of Minnesota Press.

_____. 1989. *Morality and Imagination: Paradoxes of Progress*. Madison: University of Wisconsin Press.

_____. 1991a. 'A View Of Geography'. *Geographical Review* 81:99–107.

_____. 1991b. 'Language and the Making of Place: A Narrative–Descriptive Approach'. *Annals of the Association of American Geographers* 81:684–96.

_____. 1997. *Alexander von Humboldt and His Brother: Portrait of an Ideal Geographer in Our Time*. Los Angeles, University of California at Los Angeles, Department of Geography, Alexander von Humboldt Lecture.

Turner, B.L., II., et al. 1990. *The Earth as Transformed by Human Action: Global and Regional Changes in the Biosphere in the Past 300 Years*. New York: Cambridge University Press.

Turner, B.S. 1994. *Orientalism, Postmodernism and Globalism*. New York: Routledge.

Turner, F.J. 1961. *Frontier and Section*, edited by R.A. Billington. Englewood Cliffs, NJ: Prentice Hall.

Turner, T. 1993. `Anthropology and Multiculturalism: What Is Anthropology That Multiculturalists Should Be Mindful of It?'. *Cultural Anthropology* 8:411–29.

Tylor, E.B. 1916. *Anthropology: An Introduction to the Study of Man and Civilization*. New York: D. Appleton.

_____. 1924. *Primitive Culture: Researches into the Development of Mythology, Philosophy, Religion, Language, Art, and Custom*. New York: Brentano's.

Unstead, J.F. 1922. 'Geography and Historical Geography'. *Geographical Journal* 41:55–9.

Unwin, T. 1992. *The Place of Geography*. New York: Longman.

Valentine, G. 1989. 'The Geography of Women's Fear'. *Area* 21:385–90.

_____. 1993. '(Hetero)sexing Space: Lesbian Perceptions and Experiences of Everyday Spaces'. *Environment and Planning D: Society and Space* 11:395–413.

_____. 1996. '(Re)negotiating the Heterosexual Street'. In *BodySpace: Destabilizing Geographies of Gender and Sexuality*, edited by N. Duncan, 146–55. New York: Routledge.

Vayda, A.P., and R.A. Rappaport. 1968. 'Ecology, Cultural and Noncultural'. In *Introduction to Cultural Anthropology*, edited by J.A. Clifton, 477–97. Boston: Houghton Mifflin.

Veblen, T.T. 1977. 'Native Population Decline in Totoncapán, Guatemala'. *Annals of the Association of American Geographers* 67:484–99.

Vibert, E. 1997. *Trader's Tales: Narratives of Cultural Encounters in the Columbia Plateau, 1807–1846*. Norman: University of Oklahoma Press.

von Maltzahn, K.E. 1994. *Nature as Landscape: Dwelling and Understanding*. Kingston: McGill-Queen's University Press.

Wacker, P.O. 1968. *The Musconetcong Valley of New Jersey: A Historical Geography*. New Brunswick: Rutgers University Press.

_____. 1975. *Land and People: A Cultural Geography of Pre-Industrial New Jersey, Origins and Settlement Patterns*. New Brunswick: Rutgers University Press.

_____, and P.G.E. Clemens. 1995. *Land Use in Early New Jersey: A Historical Geography*. Newark: New Jersey Historical Society.

Wagner, P.L. 1958a. 'Nicoya: A Cultural Geography'. *University of California Publications in Geography* 12:195–250.

_____. 1958b. 'Remarks on the Geography of Language'. *Geographical Review* 48:86–97.

_____. 1960. *The Human Use of the Earth*. Glencoe, Ill.: Free Press.

_____. 1972. *Environments and Peoples*. Englewood Cliffs, NJ: Prentice Hall.

_____. 1974. 'Cultural Landscapes and Regions: Aspects of Communication'. In *Man and Cultural Heritage: Papers in Honor of Fred B. Kniffen*, edited by H.J. Walker and W.G. Haag. *Geoscience and Man* 5:133–42. Baton Rouge: Louisiana State University, Department of Geography and Anthropology, Geoscience Publications.

_____. 1975. 'The Themes of Cultural Geography Rethought'. *Yearbook of the Association of Pacific Coast Geographers* 37:7–14.

_____. 1988. 'Why Diffusion'. In *The Transfer and Transformation of Ideas and Material Culture*, edited by P.J. Hugill and D.B. Dickson, 179–93. College Station: Texas A&M University Press.

_____. 1990. 'Review of *Explorations in the Understanding of Landscape: A Cultural Geography* by W. Norton'. *Journal of Geography* 89:40–1.

_____. 1994. 'Foreword: Culture and Geography: Thirty Years of Advance'. In *Re-reading Cultural Geography*, edited by K.E. Foote, P.J. Hugill, K. Mathewson, and J.M. Smith, 3–8. Austin: University of Texas Press.

_____. 1996. *Showing Off: The Geltung Hypothesis*. Austin: University of Texas Press.

_____, and M.W. Mikesell. 1962. 'The Themes of Cultural Geography'. In *Readings in Cultural Geography*, edited by P.L. Wagner and M.W. Mikesell, 1–24. Chicago: University of Chicago Press.

Waitt, G. 1997. 'Selling Paradise and Adventure: Representations of Landscape in the Tourist Advertising of Australia'. *Australian Geographical Studies* 35:47–60.

Walker, H.J., and R.A. Detro, eds. 1990. *Cultural Diffusion and Landscapes: Selections by Fred B. Kniffen*. Geoscience and Man 27. Baton Rouge: Louisiana State University, Department of Geography and Anthropology, Geoscience Publications.

Walker, R. 1997. 'Unseen and Disbelieved: A Political Economist Among Cultural Geographers'. In *Understanding Ordinary Landscapes*, edited by P. Groth and T.W. Bressi, 162–73. New Haven: Yale University Press.

Wallerstein, I. 1974a. *The Modern World-System: Capitalist Agriculture and the Origins of the European World-Economy in the Sixteenth Century*. New York: Academic Press.

_____. 1974b. 'The Rise and Future Demise of the World-Capitalist System: Concepts for Comparative Analysis'. *Comparative Studies in Society and History* 16:387–415.

_____. 1980. *The Modern World-System II: Mercantilism and the Consolidation of the European World-Economy, 1600–1750*. New York: Academic Press.

_____. 1989. *The Modern World-System III: The Second Era of Great Expansion of the Capitalist World-Economy, 1730–1840s*. San Diego: Academic Press.

_____. 1996. 'The Global Picture'. In *The Age of Transition: Trajectory of the World System, 1945–2025*, coordinated by T.K. Hopkins and I. Wallerstein, 209–25. Atlantic Highlands, NJ: Zed Books.

Walmsley, D.J., and G.J. Lewis. 1984. *Human Geography: Behavioral Approaches*. London: Longman.

Walton, J. 1995. 'How Real(ist) Can You Get?'. *Professional Geographer* 47:61–5.

Warren, K.J. 1997. 'Taking Empirical Data Seriously: An Ecofeminist Philosophical Perspective'. In *Ecofeminism: Women, Culture, Nature*, edited by K.J. Warren, 3–20. Bloomington: Indiana University Press.

Waterman, S. 1998. 'Carnivals for Élites? The Cultural Politics of Arts Festivals'. *Transactions of the Institute of British Geographers* NS 22:54–74.

Watson, J.B. 1913. 'Psychology as the Behaviorist Views It'. *Psychological Review* 20:158–77.

Watson, J.W. 1951. 'The Sociological Aspects of Geography'. In *Geography in the Twentieth Century*, edited by G. Taylor, 463–99. New York: Philosophical Library.

_____. 1983. 'The Soul of Geography'. *Transactions of the Institute of British Geographers* NS 8:385–99.

Watts, M. 1983. *Silent Violence: Food, Famine, and Peasantry in Northern Nigeria*. Berkeley: University of California Press.

Watts, S.J., and S.J. Watts. 1978. 'On the Idealist Alternative in Geography and History'. *Professional Geographer* 30:123–7.

Webb, N.L. 1990/91. 'Deconstruction and Human Geography: Exploring Four Basic Themes'. *South African Geographer* 18:123–33.

Webb, W.P. 1931. *The Great Plains*. New York: Grosset and Dunlap.

_____. 1964. *The Great Frontier*. Austin: University of Texas Press.

Webber, M.J., and D.L. Rigby. 1996. *The Golden Age Illusion: Rethinking Postwar Capitalism*. New York: Guilford.

Werlen, B. 1993. *Society, Action and Space: An Alternative Human Geography*. New York: Routledge.

Wertz, F.J. 1998. 'The Role of the Humanistic Movement in the History of Psychology'. *Journal of Humanistic Psychology* 38:42–70.

West, R.C., and J.P. Augelli. 1966. *Middle America: Its Lands and Peoples*. Englewood Cliffs, NJ: Prentice Hall.

West Edmonton Mall. n.d. 'Map and Directory: The Wonder of It All'.

Western, J. 1981. *Outcast Cape Town*. Minneapolis: University of Minnesota Press.

Whatmore, S., and L. Thorne. 1998. 'Wild(er)ness: Reconfiguring the Geographies of Wildlife'. *Transactions of the Institute of British Geographers* NS 23:435–54.

White, C.L. 1948. *Human Geography: An Ecological Study of Society*. New York: Appleton Century Crofts.

_____, and G.T. Renner. 1936. *Geography: An Introduction to Human Ecology*. New York: Appleton Century.

White, L., Jr. 1967. 'The Historical Roots of our Ecological Crisis'. *Science* 155:1203–7.

White, L.A. 1949. *The Science of Culture: A Study of Man and Civilization*. New York: Grove Press.

White, R. 1990. 'Environmental History, Ecology, and Meaning'. *Journal of American History* 76:1111–16.

Whittlesey, D. 1929. 'Sequent Occupance'. *Annals of the Association of American Geographers* 19:162–5.

Widdis, R. 1993. 'Saskatchewan: The Present Cultural Landscape'. In *Three Hundred Prairie Years*, edited by H. Epp, 142–57. Regina: Canadian Plains Research Centre.

Wilkinson, D. 1987. 'Central Civilization'. *Comparative Civilizations Review* 17:31–59.

_____. 1994. 'Civilizations *Are* World Systems!'. *Comparative Civilizations Review* 30:59–71.

Willems-Braun, B. 1997a. 'Buried Epistemologies: The Politics of Nature in (Post)colonial British Columbia'. *Annals of the Association of American Geographers* 87:3–31.

_____. 1997b. 'Reply: On Cultural Politics, Sauer, and the Politics of Citation'. *Annals of the Association of American Geographers* 87:703–8.

Williams, M. 1974. *The Making of the South Australian Landscape*. New York: Academic Press.

_____. 1983. 'The Apple of My Eye: Carl Sauer and Historical Geography'. *Journal of Historical Geography* 9:1–28.

_____. 1987. 'Historical Geography and the Concept of Landscape'. *Journal of Historical Geography* 15:92–104.

Williams, R. 1958. *Culture and Society, 1780–1950*. New York: Columbia University Press.

_____. 1976. *Keywords: A Vocabulary of Culture and Society*. New York: Oxford University Press.

_____. 1977. *Marxism and Literature*. New York: Oxford University Press.

Wilmer, F. 1997. 'Identity, Culture, and Historicity'. *World Affairs* 160, 1:3–16.

Wilson, A. 1992. *The Culture of Nature*. Cambridge, MA: Blackwell.

Wilson, A.G. 1981. *Geography and the Environment: Systems Analytical Methods*. New York: Wiley.

Wilson, C.R. 1998. 'Introduction'. In *The New Regionalism*, edited by C.R. Wilson, ix–xxiii. Jackson: University of Mississippi Press.

Wilson, W.J. 1987. *The Truly Disadvantaged. The Inner City, the Underclass, and Public Policy*. Chicago: University of Chicago Press.

Winchester, H. 1992. 'The Construction and Deconstruction of Women's Roles in the Urban Landscape'. In *Inventing Places: Studies in Cultural Geography*, edited by K. Anderson and F. Gale, 139–56. Melbourne: Longman Cheshire.

Wishart, D., A. Warren, and R.H. Stoddard. 1969. 'An Attempted Definition of a Frontier Using a Wave Analogy'. *Rocky Mountain Social Science Journal* 6:73–81.

Wisner, B. 1978. 'Does Radical Geography Lack an Approach to Environmental Relations?' *Antipode* 10:84–95.

Wissler, C. 1917. *The American Indian: An Introduction to the Anthropology of the New World*. New York: McMurtrie.

_____. 1923. *Man and Culture*. New York: Crowell.

Withers, C.W.J. 1988. *Gaelic Scotland: The Transformation of a Culture Region.* London: Routledge.

_____. 1995. 'How Scotland Came to Know Itself: Geography, National Identity and the Making of a Nation'. *Journal of Historical Geography* 21:371–97.

Wolch, J.R., and J. Emel. 1995. 'Bringing the Animals Back In'. *Environment and Planning D: Society and Space* 13:632–6.

Wolf, E. 1982. *Europe and the People Without History.* Berkeley: University of California Press.

Wolpert, J. 1964. 'The Decision Process in Spatial Context'. *Annals of the Association of American Geographers* 54:537–58.

Women and Geography Study Group of the Royal Geographical Society with the Institute of British Geographers. 1997. *Feminist Geographies: Explorations in Diversity and Difference.* Harlow, England: Longman.

Worster, D. 1984. 'History as Natural History: An Essay on Theory and Method'. *Pacific Historical Review* 53:1–19.

_____. 1988. 'Doing Environmental History'. In *The Ends of the Earth: Perspectives on Modern Environmental History,* edited by D. Worster, 289–307. New York: Cambridge University Press.

_____. 1990. Transformations of the Earth: Toward an Agroecological Perspective in History'. *Journal of American History* 76:1087–1106.

_____. 1993. 'The Ecology of Order and Chaos'. *Environmental History Review* 14:1–18.

Wright, E.O. 1983. 'Gidden's Critique of Marxism'. *New Left Review* 138:11–35.

Wright, J.K. 1947. 'Terra Incognitae: The Place of Imagination in Geography'. *Annals of the Association of American Geographers* 37:1–15.

Wyckoff, W. 1999. *Creating Colorado: The Making of a Western American Landscape, 1860–1940.* New Haven: Yale University Press.

Yaeger, P. 1996. 'Introduction: Narrating Space'. In *The Geography of Identity,* edited by P. Yaeger, 1–38. Ann Arbor: University of Michigan Press.

Yapa, L.S. 1977. 'The Green Revolution: A Diffusion Model'. *Annals of the Association of American Geographers* 67:350–9.

_____. 1996. 'Innovation Diffusion and Paradigms of Development'. In *Concepts in Human Geography,* edited by C. Earle, K. Mathewson, and M.S. Kenzer, 231–70. Lanham, MD: Rowman and Littlefield.

Yeoh, B.S.A., and P. Teo. 1996. 'From Tiger Balm Gardens to Dragon World: Philanthropy and Profit in the Making of Singapore's First Cultural Theme Park'. *Geografiska Annaler* 76B:27–42.

Yeung, H.W.-C. 1997 'Critical Realism and Realist Research in Human Geography: A Method or a Philosophy in Search of a Method'. *Progress in Human Geography* 21:51–74.

Young, G.L. 1974. 'Human Ecology as an Interdisciplinary Concept: A Critical Inquiry'. *Advances in Ecological Research* 8:1–105.

_____. 1983. 'Introduction'. In *Origins of Human Ecology,* edited by G.L. Young, 1–9. Stroudsburg, PA: Hutchinson Press.

Young, I.M. 1993. 'Together in Difference: Transforming the Logic of Group Potential Conflict'. In *Principled Positions,* edited by J. Squires, 121–50. London: Lawrence and Wishart.

Zdorkowski, R.T., and G.O. Carney. 1985. 'This Land Is My Land: Oklahoma's Changing Vernacular Regions'. *Journal of Cultural Geography* 5, no. 2:97–106.

Zelinsky, W. 1961. 'An Approach to the Religious Geography of the United States: Patterns of Church Membership in 1952'. *Annals of the Association of American Geographers* 51:139–93.

_____. 1967. 'Classical Town Names in the United States: The Historical Geography of an American Idea'. *Geographical Review* 57:463–95.

_____. 1973. *The Cultural Geography of the United States.* Englewood Cliffs, NJ: Prentice Hall.

_____. 1980. 'North America's Vernacular Regions'. *Annals of the Association of American Geographers* 70:1–16.

_____. 1983. 'Nationalism in the American Place-Name Cover'. *Names* 30:1–28.

_____. 1988. *Nation into State: The Shifting Symbolic Foundations of American Nationalism.* Chapel Hill: University of North Carolina Press.

_____. 1992. *The Cultural Geography of the United States,* revised edition. Englewood Cliffs, NJ: Prentice Hall.

_____. 1994. *Exploring the Beloved Country: Geographic Forays Into American Society and Culture.* Iowa City: University of Iowa Press.

_____. 1996. 'Review of *Re-Reading Cultural Geography* edited by K.E. Foote, P.J. Hugill, K. Mathewson, and J.M. Smith'. *Annals of the Association of American Geographers* 86:750–3.

_____. 1997. 'Seeing Beyond the Dominant Culture'. In *Understanding Ordinary Landscapes,* edited by P. Groth and T.W. Bressi, 157–61. New Haven: Yale University Press.

Zimmerer, K.S. 1994. 'Human Ecology and the "New Ecology": The Prospect and Promise of Integration'. *Annals of the Association of American Geographers* 84:108–25.

_____. 1996. 'Ecology as Cornerstone and Chimera in Human Geography'. In *Concepts in Human Geography*, edited by C. Earle, K. Mathewson, and M.S. Kenzer, 161–88. Lanham, MD: Rowman and Littlefield.

_____, and K.R. Young, eds. 1998. *Nature's Geography: New Lessons for Conservation in Developing Countries*. Madison: University of Wisconsin Press.

Subject Index

ideology. *See* religion
images differing from reality, 210,
215, 217–19; examples, 219,
220–4, 300
imperialism, 85, 278
Industrial Revolution, 36–7; cultural
impacts, 36–7, 264–5; environmental impacts, 36; explanations for, 261, 263
inequalities, global, 261, 263–4, 318
information, in postmodern world,
38
initial occupance proposal, 116
innovation, diffusion. *See* diffusion of
culture
intertextuality, 243, 272
Ireland, historical cultural geography
in, 104
Irish settlement in eastern Canada,
97–8
isolation, and developing a cultural
hearth, 115

knowledge: associated with power,
38; social construction of, 5, 238

Lamarckianism, 43, 62
Landscape magazine, 99, 100–2, 289,
290, 303
landscape reading, 99–104, 295–301;
different groups read differently, 298–9; *Landscape* magazine
tradition, 99, 100–2; local landscape histories, 99, 100; making
the American landscape, 102–3;
making the Irish landscape,
104; making the Ontario landscape, 103–4; personality of
place, 302–3; reading the urban
landscape, 298–300
landscapes, 2, 3, 10, 110; assumptions of dominant societies,
277–8; behavior and, 23, 24,
211–24, 279–81; ethnic group
identity, 135, 136; evolution, 22,
23, 91, 100, 101, 105; of exclusion, 284; expression of culture,
290–5, 311; folk culture landscapes, 303–4; gender in,
279–81, 300; Hispano homeland
region, 120, 132–5, 148; human
adjustments to environment,
155; and image distortion,

217–24; interpreting. *See* landscape reading; Mormon cultural region, 130, 131–2; ordinary.
See ordinary landscapes; popular culture. *See* popular culture
landscape; and regions, 22, 23,
110; of sexuality, 281–3; six
tenets of contemporary landscape studies, 289; symbolic, 24,
65, 102, 187, 288–301; use of
term, 110, 148
landscape school, 9–10, 12–13, 22, 24,
25, 53–4, 138; ecological content, 155–6; and historical geography, 75, 91–104, 105; revisions
of the approach, 74–5; Sauer's
key ideas, 71–2; three related
themes, 74; view of human and
nature relationship, 28. *See also*
Sauer, Carl
land surveys, 136–7
language, 18; classifying into families, 143; and ethnic groups,
257–8; Foucault's ideas, 240;
importance for cultural identity,
291; naming places, 290, 291,
292, 293, 294
lebensraum, 247
limits to growth thesis, 165
linguistic turn, 238–9, 240, 297
literature of resistance, 285, 286

magazines, geographic: *Landscape*
magazine. *See Landscape* magazine; popular geographic magazines, 302
Malayan rubber landscape, 123
male dominance, 20, 31, 41–2, 279,
281, 282, 283
Maoris, 85, 86
Marxism, 10, 233–6; basic concerns,
235; in geography, 235–6; social
theory, 233–4
materialism, 41
mechanistic view, 6–7
memorials, 301, 302
Mennonite populations, 136
mental map. *See* images differing
from reality
mercantilism, 35
Mesolithic cultures, 31
Mexico, Sauer on cultural regions,
98–9

migration, and diffusion of culture,
78–9
model of humans: behavioral perspective, 204–5, 209–10, 275;
humanistic perspective, 202,
205–9, 210, 275, 276
modernism, 7, 15, 37, 180, 242
modernity, 242, 266
Monte Carlo simulation models,
79–80
monuments, 280, 294, 300–1
Mormon cultural region, 114–15,
120; behavior analytic interpretation, 212–13; homeland,
129–32; landscapes, 130, 131–2;
sacred sites, 294, 297, 298
multicivilizational view, 145–6
multiculturalism, 263, 287, 288
multilinear evolution, 63
music, identity, place, 304–5, 306–7

naming places, 290, 291, 292, 293,
294; creating landscape and
claiming ownership, 291, 294
narratives in historical geography, 93
national and local identities, 259–60,
300, 301
nationalism: and ethnic groups, 260;
and state. *See* nation states
nationality, 19
National Socialist Party in Germany,
247–8
nation states, 37, 259–60, 264; demise
of, 265–6
naturalism, 6–7; evolutionary, 60–1
natural selection, 44
nature: culture of, 58; definition, 5;
European attitude towards, 84;
invention and reinvention of,
59; social construction of, 28,
182
neighborhood effect of diffusion, 81
Neolithic cultures, 32
new cultural geography: challenges
Sauerian tradition, 13–14, 267,
272, 317; concepts of culture, 14,
24–5
new ecology, 169–71
niche, ecological, 160
non-equilibrium view of nature,
169–71
North West Passage, 219

Names Index